Impact Studies

Series Editor: Christian Koeberl

Editorial Board

Springer

Berlin
Heidelberg
New York
Hong Kong
London
Milan
Paris
Tokyo

C. Wylie Poag
Christian Koeberl
Wolf Uwe Reimold

The Chesapeake Bay Crater

Geology and Geophysics of a Late Eocene Submarine Impact Structure

With 207 Figures, 42 Tables and a CD-ROM

ESF IMPACT

Springer

Dr. C. Wylie Poag
U.S. Geological Survey
384 Woods Hole Road
Woods Hole, MA 02543-1598
USA
Email: *wpoag@usgs.gov*

Dr. Wolf Uwe Reimold
School of Geosciences
University of the Witwatersrand
P.O. Wits 2050
Johannesburg, South Africa
Email: *reimoldw@geosciences.wits.ac.za*

Dr. Christian Koeberl
Department of Geological Sciences
University of Vienna
Althanstrasse 14
1090 Vienna
Austria
Email: *christian.koeberl@univie.ac.at*

ISBN 3-540-40441-4 Springer-Verlag Berlin Heidelberg New York

Cataloging-in-Publication Data applied for

Bibliographic information published by die Deutsche Bibliothek
Die Deutsche Bibliothek lists this publication in the Deutsche Nationalbibliographie; detailed bibliographic data is available in the Internet at <http://dnb.ddb.de>.

Springer-Verlag Berlin Heidelberg New York
a member of BertelsmannSpringer Science+Business Media GmbH

http://www.springer.de

Printed in Germany

Cover Design: *Kirchner*, Heidelberg
Typesetting: Camera-ready by the authors

Printed on acid free paper 32/2132 AO – 5 4 3 2 1 0

We dedicate this book to David J. Roddy (1932–2002), one of the pioneers of impact cratering studies. For 40 years, Dave was a driving force in the analysis of natural hypervelocity impact structures and the mechanics of nuclear explosion cratering. The wealth of data and observations he contributed remains fundamental to our understanding of the physics of impact cratering and the shock-wave deformation of the Earth's crustal materials.

Preface

"... bangs have replaced whimpers and the geological record has become much more exciting than it was thought to be." Derek Ager (1993) The New Catastrophism. Cambridge University Press, Cambridge, p xix

Scientific and public interest in asteroids, comets, and meteorite impacts has never been more intense than right now. Much of this interest stems from the fervent debates surrounding the causes of the Cretaceous-Tertiary mass extinctions and their possible relationships to a giant bolide impact in Mexico's Yucatán Peninsula. Recent spectacular impacts on Jupiter, and several near misses of our own planet by Near-Earth Objects have intensified professional and popular discussion of society's imperative need to understand the process and effects of bolide impacts. In the United States, the scientific community and the public, as well, were startled to learn, in 1994, that the largest impact structure in this country had been detected beneath Virginia's portion of the Chesapeake Bay. Seismic surveys and deep coring revealed a huge crater, 85 kilometers in diameter and more than a kilometer deep, stretching from Yorktown, Virginia, to 15 kilometers out onto the shallow continental shelf. Several of Virginia's major population centers, including Norfolk, Hampton, and Newport News, are located on the western rim of the crater, and still experience residual effects of the original collision, 36 million years after the impact took place.

Exploration and documentation of the Chesapeake Bay impact structure has proceeded in three phases. Phase one was characterized by mainly serendipitous discoveries. Initial clues to its presence came from deep-sea cores collected by scientists aboard the drillship *Glomar Challenger*, during a coring cruise off the coast of Atlantic City, New Jersey, in 1983. Diagnostic evidence of an impact, in the form of microtektites and impact-shocked minerals, showed up in a few centimeters of late Eocene chalk, dated at ~35 million years old. The thickness and coarse-grained nature of the impact debris indicated that the impact site must have been relatively close to the core site. Three years later (1986), the first of four stratigraphic coreholes in southeastern Virginia recovered additional impact-generated debris, containing diagnostic shock-metamorphosed minerals. The geologic age of the debris was identical to the microtektite-bearing debris cored off New Jersey. In 1994, acquisition of multichannel seismic reflection data from commercial oil companies revealed that two of the Virginia coreholes had penetrated the massive body of impact breccia that fills an enormous impact crater buried beneath Chesapeake Bay. Each of these three milestone events was the result of chance – surprise discoveries made during geologic investigations of unrelated phenomena.

Phase two of the crater documentation was marked by the acquisition of additional seismic reflection data to clarify the detailed structure and morphology of the impact structure. When added to the previous data set, the new surveys yielded a database of >2000 kilometers of seismic reflection profiles. These seismic data clearly revealed that the Chesapeake Bay structure is a complex, peak-ring/central-peak structure, with many features similar to those of other large terrestrial and planetary impact structures, but which displays several unique aspects, as well.

Firm knowledge of the crater's structure and morphology allowed the third phase of exploration to begin in 2000. Phase three emphasized the careful selection of new core sites to answer specific questions regarding impact processes and resultant impact-generated deposits.

Now, 20 years after the New Jersey core discoveries, researchers have established the principal structural, morphological, depositional, and paleoenvironmental features of the Chesapeake Bay impact and its resultant structure. This volume contains the first synthesis of our current knowledge of this fascinating cosmic event and its aftermath. It is our hope that the broad spectrum of data and interpretations we offer herein will enhance the understanding and appreciation of bolide impacts as crucial events in the geologic and biologic evolution of our planet.

C. Wylie Poag
US Geological Survey
Woods Hole,
Massachussetts, USA

Christian Koeberl
University of Vienna
Vienna, Austria

W. Uwe Reimold
University of the
Witwatersrand
Johannesburg, South
Africa

Acknowledgments

We are indebted to a host of colleagues who contributed data, analyses, interpretations, and advice, during our roughly 12-year study of the Chesapeake Bay impact crater. The list is headed by Debbie Hutchinson, Steve Colman, Tommy O'Brien, Barry Irwin, Dave Nichols, Jeremy Loss, John Evans, and Nancy Soderberg, who constituted the shipboard science party that collected seismic reflection data aboard the *R/V Seaward Explorer* (1996). Texaco, Inc. (particularly Parish Erwin) contributed the multichannel seismic reflection profiles that originally defined the major features of the crater. Rusty Tirey, John Grow, and Pete Popenoe collected the early USGS seismic reflection data before we knew the crater was there. Dave Foster and John Diebold helped to acquire and process the seismic data collected by the *R/V Maurice Ewing* (1998). Dave Powars, Bob Mixon, Scott Bruce, and Don Queen carried out the initial coring programs that provided ground truth for the seismic reflection analyses. Marie-Pierre Aubry provided critical analyses of calcareous nannofossils. Gene Shoemaker, Jens Ormö, Filippos Tsikalas, Henning Dypvik, Richard Grieve, Peter Schultz, Kevin Pope, Bill Glass, Ron Stanton, Jeff Williams, Dave Folger, Glen Izett, and Michael Rampino provided expert advice and much needed encouragement during this project. Tom Aldrich, Joe Newell, and the crews of the *R/V Seaward Explorer* and *R/V Maurice Ewing* facilitated collection of seismic data in Chesapeake Bay. John Costain, Carl Bowin, Larry Poppe, Warren Agena, Myung Lee, Dann Blackwood, Dick Norris, Ed Mankinen, Judy Commeau, Louie Kerr, and Jeff Plescia provided critical data, data analysis or processing, scientific advice, and technical assistance. Chuck Pillmore provided the Manson seismic profile; Lubomir Jansa provided the Montagnais seismic line. The Deep Sea Drilling Project (DSDP) and Ocean Drilling Program (ODP) provided offshore cores. The National Geographic Research Committee provided funding to the senior author for the single channel survey of the crater. Steve Curtin provided downhole logs. Core drilling in the Chesapeake Bay impact crater has been a cooperative effort among the Hampton Roads Planning District Commission, the NASA Langley Research Center, the Virginia Department of Environmental Quality, the Geology Department of the College of William and Mary, and the USGS. The Chesapeake Coring Team (Greg Gohn, Dave Powars, Scott Bruce, Laurel Bybell, Tom Cronin, Lucy Edwards, Norm Frederiksen, Wright Horton, Glen Izett, Gerry Johnson, Joel Levine, Randy McFarland, Jim Quick, Steve Schindler, Jean Self-Trail, Matt Smith, Bob Stamm, Rob Weems, Art Clark, and Don Queen) acquired and described the NASA Langley, North, and Bayside cores. Becca Drury, Michael Taylor, Andy McIntire, Laura Hayes, Philip Molzer, Emily Denham, Daniel Boamah and Kassa Amare provided computer skills and laboratory assistance. Philip Molzer also collected a new set of gravity data on the Delmarva Peninsula and carried out the gravity modeling of the crater. VeeAnn Cross constructed the 3-D structural model of the crater.

We are especially indebted to the reviewers, David Crawford, Henning Dypvik, David Foster, Jens Ormö, Larry Poppe, Scott Snyder, Filippos Tsikalas, and Buck Ward, for significant improvements to the manuscript.

The USGS Coastal and Marine Geology Program and Earth Surface Processes Program supported Poag's crater research. Koeberl's geochemical and petrographic studies were supported by the Austrian Science Foundation (FWF) project Y58-GEO. Reimold's research was supported by the National Research Foundation of South Africa and a grant from the University of the Witwatersrand to the Impact Cratering Research Group. This is ICRG Contribution No. 45.

Contents

CD-ROM

1 Introduction

The list of impact craters documented on Earth is short. Only about 165 genuine impact structures have been identified to date (Table 1.1). Even so, the number is steadily increasing at the rate of ~3–5 per year (Grieve et al. 1995; Earth Impact Database at http://www.unb.ca/passc/Impact/Database/). In stark contrast, most other rocky planets and satellites of our solar system are pockmarked by thousands to hundreds of thousands of impact features (Beatty et al. 1999). Nevertheless, impact specialists acknowledge that Earth, too, has undergone billions of years of bolide bombardment (Melosh 1989; Schoenberg et al. 2002). The most intense bombardment, however, took place during Earth's earliest history (~3.8–4 Ga; Ryder 1990; Cohen et al. 2000; Ryder et al. 2000). Traces of most terrestrial impacts have been completely erased or strongly altered by the dynamic processes of a thick atmosphere, deep ocean, and mobile crust, a combination unique to our planet. Planetary geologists now recognize that processes associated with bolide impacts are fundamental to planetary accretion and surface modification (Melosh 1989; Peucker-Ehrenbrink and Schmitz 2001). Incoming meteorites may have been primary sources for Earth's water, and, perhaps, even organic life as we know it (Thomas et al. 1997; Kring 2000). There is little doubt that impacts played a major role in the evolution of Earth's biota (Ryder et al. 1996; Hart 1996).

Only eight impact structures with diameters of 80 km or larger have been discovered on Earth (Fig. 1.1; Table 1.1). Three of these, Vredefort, Sudbury, and Chicxulub, are classified as multiring basins, a structural/morphological category encompassing craters generally 200 km or more in diameter and characterized by three or more structural rings (Spudis 1993; Melosh 1989; Morgan et al. 1997). That leaves only the Chesapeake Bay crater and five other structures (Acraman, Manicouagan, Morokweng, Popigai, Puchezh Katunki) as representatives of the 80–100-km-diameter class of complex impact craters (having two structural rings). The Chesapeake Bay crater (along with the Popigai crater, virtually a twin with regard to size and age; Masaitis et al. 1999; Poag et al. 1999; Whitehead et al. 2002) is the largest known complex impact structure created since the end of the Jurassic Period (145 Ma). Moreover, the Chesapeake Bay crater is the largest of 30 craters documented in the United States (Koeberl and Anderson 1996b), surpassing the second largest (Iowa's Manson crater) by more than a factor of two.

Among Earth's complex craters, only a handful are relatively undeformed and accessible. Examples prior to this study include the Manson, Mjølnir, Lockne, Popigai, and Ries craters. If one focuses only on submarine impact structures, the list is even shorter. Though several structures currently exposed on continental

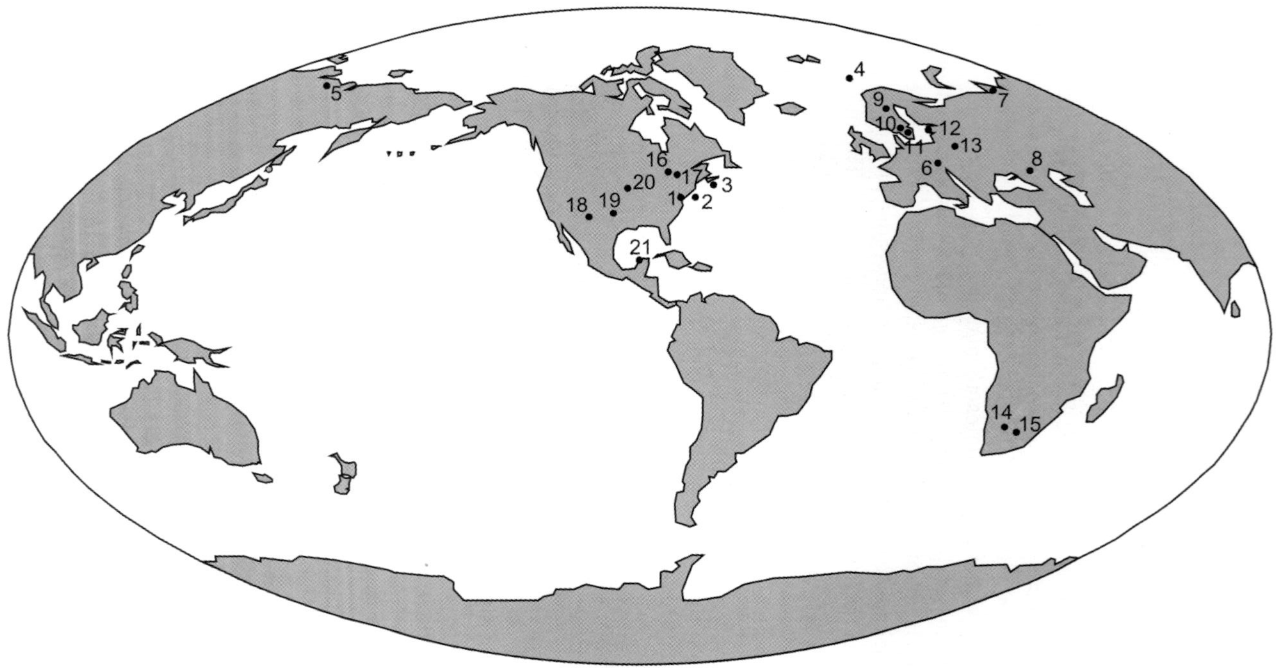

1 Chesapeake Bay
2 Toms Canyon
3 Montagnais
4 Mjølnir
5 Popigai
6 Ries
7 Kara/Ust Kara
8 Kamensk
9 Lockne
10 Granby
11 Tvären
12 Kärdla
13 Kaluga
14 Morokweng
15 Vredefort
16 Sudbury
17 Brent
18 Barringer (Meteor)
19 Ames
20 Manson
21 Chicxulub

Fig. 1.1. Geographic distribution of 21 terrestrial impact structures (craters and basins) discussed in this volume. See Table 1.1 for additional information about individual structures.

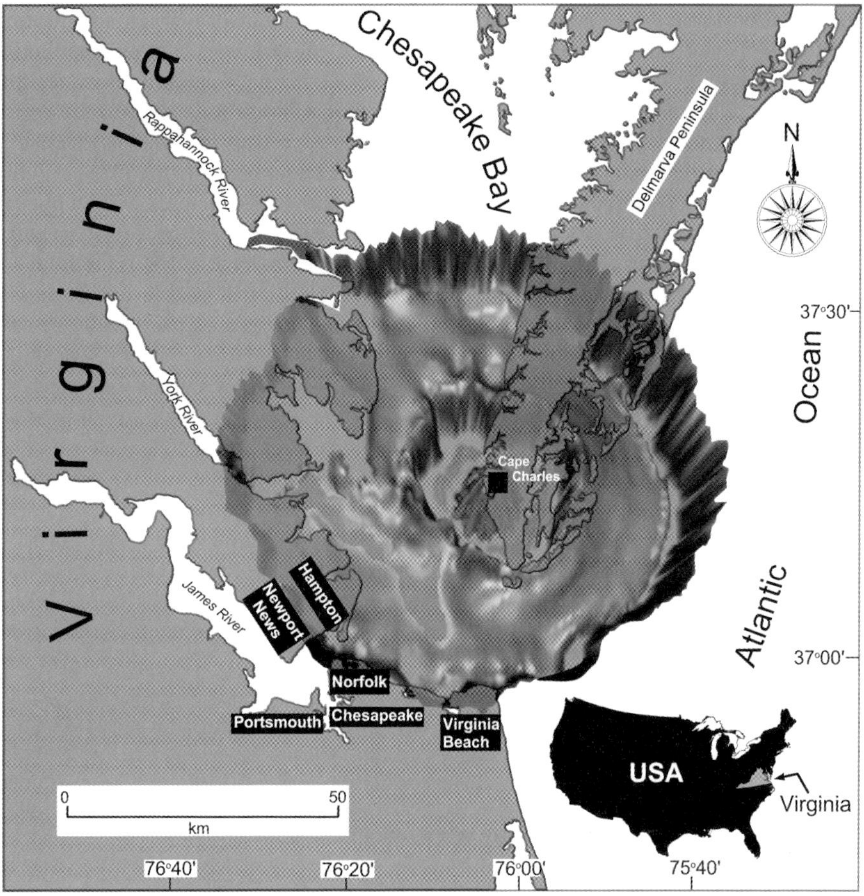

Fig. 1.2. Computer-generated 3-D perspective of Chesapeake Bay impact crater, showing location beneath lower part of Chesapeake Bay, its surrounding peninsulas, and inner part of adjacent Atlantic Continental Shelf. Six principal cities shown on southwest margin of crater encompass densest human population in Virginia.

platforms are thought to have been submarine in origin, only four impact craters are still wholly or partly covered by oceanic waters. In order of decreasing diameter, these four are Chesapeake Bay (85-km diameter), Montagnais (45-km diameter), Mjølnir (40-km diameter), and Ust Kara (25-km diameter). A fifth submarine structure, Toms Canyon crater (22-km diameter), is considered by us to be also of impact origin, though that conclusion requires additional confirmation (Poag and Poppe 1998; Fig. 1.1; Table 1.1).

The Chesapeake Bay structure (Fig. 1.2) is unique among both subaerial and submarine impact craters on Earth by virtue of the following combination of features: (1) its location on a passive continental margin has preempted the kinds of tectonic or orogenic disruption or distortion typical of many large terrestrial craters; (2) its original location on a relatively deep continental shelf allowed marine deposition to resume immediately following the impact, which buried it rapidly and completely, thereby preventing subsequent erosion of any principal feature except the distal margins of the surrounding apron of impact debris; (3) the upper part of the breccia body inside the crater was derived from the washback of impact-generated tsunami waves; (4) that same breccia body encompasses a large volume of impact-generated brine; (5) numerous smaller structures, which appear to be secondary craters, are present within a few tens of kilometers of the primary crater, a phenomenon that, in itself, sets the Chesapeake Bay crater apart from all other known impact structures on our planet; and (6) the crater underlies a densely populated urban corridor, whose two million citizens are still affected by crater-related phenomena, 36 million years after the impact.

In several earlier reports, Poag and his collaborators have established the general aspects of the Chesapeake Bay crater's structure and morphology, as well as the large-scale characteristics of the crater-filling impact breccia (Poag et al. 1992, 1994b, 1999; Koeberl et al. 1996; Poag 1996, 1997a; Poag and Aubry 1995; Poag and Foster 2000; Poag et al. 2001). The Chesapeake Bay structure is a complex, peak-ring/central-peak crater, 85 km in average diameter, and ~1.3–2.0 km deep at maximum estimated depth. The crater interior features a low-relief peak ring (~300 m maximum height) and a rugged central peak (~1,000 m maximum height). At twice the area of the State of Rhode Island and as deep as the Grand Canyon, the Chesapeake Bay crater (along with Popigai) is the sixth largest impact crater currently known on the globe.

In hindsight, it is clear that telltale sedimentary and structural evidence of a buried, giant impact structure in southeastern Virginia first came to light in the 1940s through geohydrological studies (Poag 1996, 1997a, 1999c). These studies, mainly sediment and ground-water analyses from shallow boreholes, were carried out by the US Geological Survey (USGS) (Cederstrom 1945a,b,c, 1957). At that time, however, the extraterrestrial implications of the evidence were not appreciated. More than 50 years passed before this early evidence could be unequivocally linked to a late Eocene bolide impact (Poag et al. 1992, 1994b; Koeberl et al. 1996; Poag 1999c). In the interim, however, several authors (USGS scientists in particular) published a large database of subsurface stratigraphic analyses derived mainly from >200 uncored boreholes (Cederstrom 1945a,b,c; Richards 1945, 1967; Cushman and Cederstrom 1949; Maher 1965, 1971; Brown et al. 1972; Teifke 1973; Gibson 1983; Gohn 1988; Poag and Ward 1993; Fig. 1.3; CD-ROM.1; Table 1.2). These subsurface studies, along with extensive studies of outcrop stratigraphy (Ward and Krafft 1984; Owens and Gohn 1985; Ward and Strickland 1985; Mixon et al. 1989), firmly established a regional structural and stratigraphic framework for the sedimentary rocks of southeastern Virginia outside the crater rim. Regional surveys of gravity and magnetics, coupled with sparse deep well data and a few onshore seismic reflection surveys, provided a complementary geological framework of crystalline basement rocks beneath the Virginia Coastal Plain (Ewing et al. 1937; Woollard et al. 1957; LeVan and Pharr 1963; Taylor et al. 1968; Sabet 1973; Johnson 1977; Hawarth et al. 1980; Costain and

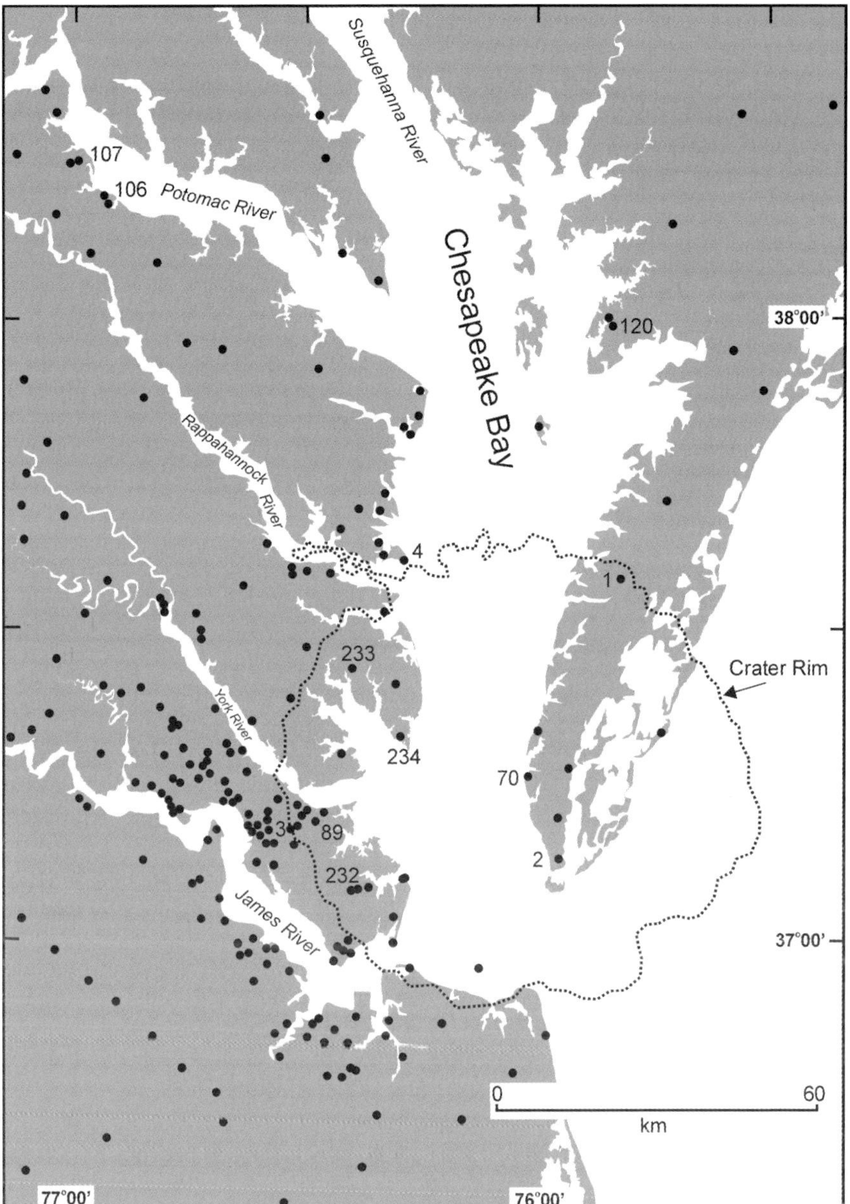

Fig. 1.3. Geographic distribution of 234 boreholes used in our study of Chesapeake Bay impact crater. Numbered boreholes are discussed in text. See Table 1.2 and CD-ROM.1 for name, number, latitude, and longitude of each borehole.

Glover 1976–1982; Lyons and O'Hara 1982; Dysart et al. 1983; Thomas et al. 1989).

While documenting the Chesapeake Bay structure and establishing its impact origin, Poag and his colleagues published data from five continuously cored boreholes and ~300 km of multichannel seismic reflection profiles (Poag et al. 1992, 1994b, 1999; Poag 1996, 1997a; Powars and Bruce 1999; Powars et al. 2001; Gohn in press). Definition and understanding of the structure and morphology of the crater and its associated features have improved progressively with the acquisition of each new data set (Figs. 1.4, 1.5).

This book affords an opportunity to synthesize the large body of geological data (including sedimentological, paleontological, geochemical, paleomagnetic, and petrographic analyses) and geophysical data (including seismic reflection surveys, gravity surveys, and downhole logging) amassed over the past 16 years (1986–2002). In doing so, we analyze in greater detail much of the old data, refine and(or) reinterpret previously published inferences, and present new interpretations based on abundant new (unpublished) data. Among the latter, in particular, we have analyzed approximately 1,700 km of new seismic reflection profiles, 63 new gravity stations on the Delmarva Peninsula, a 90-km-long continuous marine gravity survey over the crater center (1,587 measurements), and >1,780 m (>5,840 ft) of core from three new deep, continuously cored boreholes. We have obtained petrographic analyses of >100 new samples from the cored sections of crater-fill breccia, have analyzed several hundred micropaleontological samples from the thick marine clay bed (Chickahominy Formation) that caps the breccia, and we provide new descriptions and illustrations of whole and split core sections. We document a fallout layer inside the crater, and recognize a <1–3-kyr-long dead zone immediately above the fallout layer. We also offer new interpretations of stable isotopic and paleomagnetic data, and initial interpretations of the impact's short-term and long-term effects on postimpact paleoenvironments at the crater and at other sites around the globe. In addition, we present a comprehensive numerical and field-based conceptual model for cratering processes and the resultant impact-generated depositional regimes at Chesapeake Bay.

As a supplement to this volume, we include a CD-ROM, which incorporates color versions of 29 selected figures from the main text and items too large or complex for page-sized illustrations. These items include: color maps for borehole locations, seismic survey tracklines, crater structure, and basement structure; a 3-D computer model of crater structure; scaled depth-sections along selected seismic profiles; a depth-scaled stratigraphic section showing core lithologies and downhole geophysical logs; and 18 selected seismic reflection profiles.

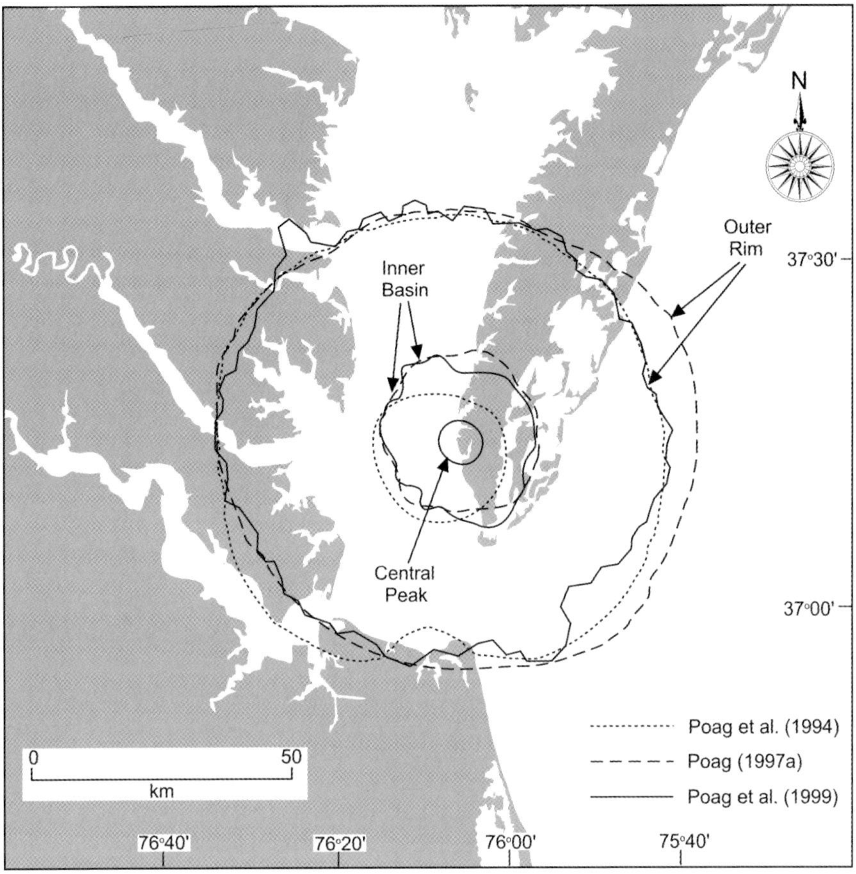

Fig. 1.4. Locations of outer rim of Chesapeake Bay impact crater published in previous articles by Poag et al. (1994b), Poag (1997a), and Poag et al. (1999). Note increased detail in peripheral geometry as additional data were acquired. See Chapter 4 for discussion of current interpretation of crater outline.

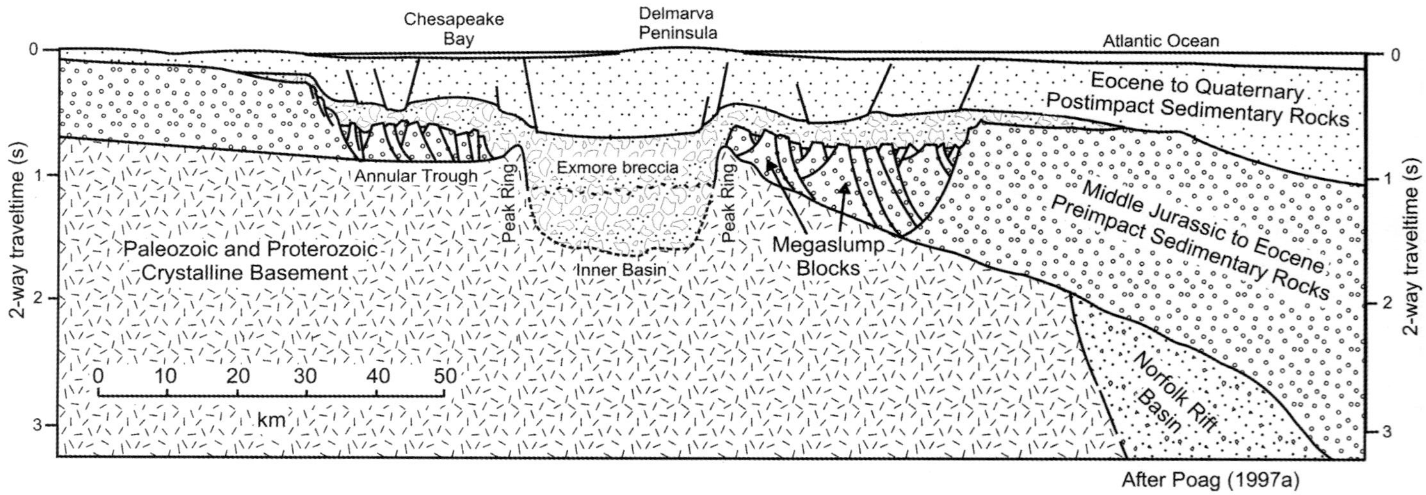

Fig. 1.5A. Previously published interpretation of cross sectional structure and morphology of Chesapeake Bay impact crater, based on time-scaled seismic reflection profiles and boreholes. Modified from Poag (1997a); no central peak recognized.

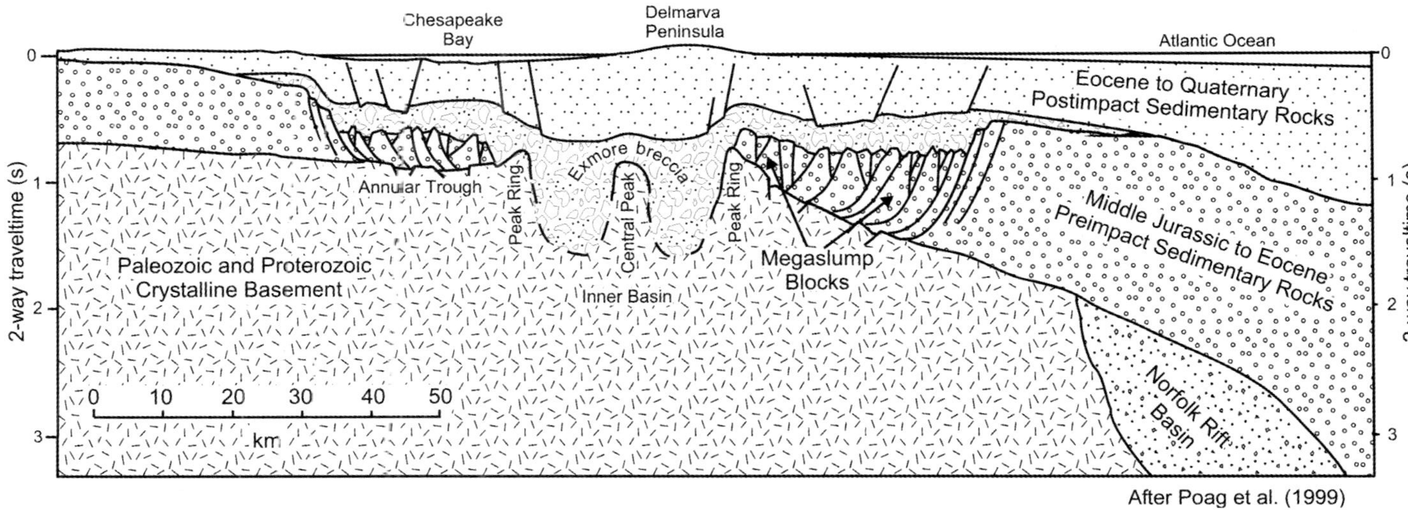

Fig. 1.5B. Previously published interpretation of cross sectional structure and morphology of Chesapeake Bay impact crater, based on time-scaled seismic reflection profiles and boreholes. Modified from Poag et al. (1999); central peak recognized on nearshore end of seismic profile SEAX-7 near mouth of Cape Charles harbor.

Table 1.1. Impact structures (simple and complex craters; multiring basins) documented on Earth. List modified from Grieve et al. (1995).

No.	Name	Location	Latitude	Longitude	Age [Ma]	Diam.[km]	Type
1	Acraman	South Australia	32°01' S	135°27' E	586	90	complex
2	Ames	Oklahoma, USA	36°15' N	98°12' W	470 ± 30	16	buried, complex
3	Amguid	Algeria	26°05' N	04°23' E	<0.1	0.45	simple
4	Aorounga	Chad	19°15" N	19°15' E	<345	12.6	complex
5	Aouelloul	Mauritania	20°15' N	12°41' W	3.1 ± 0.3	0.39	simple
6	Araguainha Dome	Brazil	16°47' S	52°59' W	247 ± 5.5	40	complex
7	Avak	Alaska, USA	71°15' N	156°38' W	>95	12	buried, complex
8	Azuara*	Spain	41°10' N	00°55' W	~40	30	complex
9	B.P. Structure	Libya	25°19' N	24°20' E	<120	2	complex
10	Barringer (Meteor)	Arizona, USA	35°02' N	111°01' W	0.049 ± 0.003	1.186	simple
11	Beaverhead	Montana, USA	44°36' N	113°00' W	~600	60?	complex
12	Beenchime-Salaaty	Russia	71°00' N	121°40' E	40 ± 20	8	complex
13	Bigach	Kazakhstan	48°34' N	82°01" E	6 ± 3	5 ± 3	complex
14	Boltysh	Ukraine	48°45' N	32°10' E	65.17 ± 0.64	24	buried, complex
15	Bosumtwi	Ghana	06°30' N	01°25' W	1.03 ± 0.02	10.5	complex
16	Boxhole	Northern Territory, Australia	22°37' S	135°12' E	0.03 ± 0.0005	0.17	simple
17	Brent	Ontario, Canada	46°05' N	78°29' W	450 ± 30	3.8	simple
18	Calvin*	Michigan USA	41°50 N	85°57' W	450 ± 10	8.5	buried, complex
19	Campo del Cielo	Argentina	27°38' S	61°42" W	<0.004	0.05	simple
20	Carswell	Saskatchewan, Canada	58°27' N	109°30' W	115 ± 10	39	complex
21	Charlevoix	Quebec, Canada	47°32 N	70°18' W	357 ± 15	54	complex
22	Chesapeake Bay	Virginia, USA	37°17' N	76°01' W	~35.6	85	buried, complex
23	Chicxulub	Yucatán, Mexico	21°20' N	89°30' W	64.98 ± 0.05	195	buried, multiring
24	Chiyli	Kazakhstan	49°10' N	57°51' E	46 ± 7	5.5	complex
25	Chukcha	Russia	75°42' N	97°48' E	<70	6	complex

*Impact origin doubtful

Table 1.1. (cont.)

No.	Name	Location	Latitude	Longitude	Age [Ma]	Diam.[km]	Type
26	Clearwater East	Quebec, Canada	56°05' N	74°07' W	290 ± 20	26	complex
27	Clearwater West	Quebec, Canada	56°13' N	74°30' W	290 ± 20	36	complex
28	Cloud Creek	Wyoming, USA	43° 7' N	106° 45' W	190 ± 30	7	buried, complex
29	Connolly Basin	Western Australia	23°32' S	124°45' E	<60	9	complex
30	Couture	Quebec, Canada	60°08' N	75°20' W	430 ± 25	8	?
31	Crooked Creek	Missouri, USA	37°50' N	91°23' W	320 ± 80	7	complex
32	Dalgaranga	Western Australia	27°38' S	117°17' E	0.027	0.021	simple
33	Decaturville	Missouri, USA	37°54' N	92°43' W	<300	6	complex
34	Deep Bay	Saskatchewan, Canada	56°24' N	102°59' W	100 ± 50	13	complex
35	Dellen	Sweden	61°48' N	16°48' E	89.0 ± 2.7	19	complex
36	Des Plaines	Illinois, USA	42°03' N	87°52' W	<280	8	buried, complex
37	Dobele	Latvia	56°35' N	23°15' E	290 ± 35	4.5	buried, complex
38	Eagle Butte	Alberta, Canada	49°42' N	110°30' W	<65	10	complex
39	Elbow	Saskatchewan, Canada	50° 59' N	106°43' W	395 ± 25	8	complex
40	El'gygytgyn	Russia	67°30' N	172°05' E	3.5 ± 0.5	18	complex
41	Flynn Creek	Tennessee, USA	36°17' N	85°40' W	360 ± 20	3.55	complex
42	Gardnos	Norway	60°39' N	09°00' E	500 ± 10	5	complex
43	Glasford	Illinois, USA	40°36' N	89°47' W	<430	4	buried, complex
44	Glover Bluff	Wisconsin, USA	43°58' N	89°32' W	<500	8	complex
45	Goat Paddock	Western Australia	18°20' S	126°40' E	<50	5.1	complex
46	Gosses Bluff	Northern Territory, Australia	23°49' S	132°19' E	142.5 ± 0.8	22	complex
47	Gow	Saskatchewan, Canada	56°27' N	104°29' W	<250	5	complex
48	Goyder	Northern Territory, Australia	13°29' S	135°02' E	>136	3	simple
49	Granby	Sweden	58°25' N	14°56' E	470	3	simple
50	Gusev	Russia	48°26' N	40°32' E	49.15 ± 0.18	3	twin, buried, simple

Table 1.1. (cont.)

No.	Name	Location	Latitude	Longitude	Age [Ma]	Diam.[km]	Type
51	Gweni-Fada	Chad	17°25' N	21°45' E	<345	14	complex
52	Haughton	Northwest Territories, Canada	75°22' N	89°41' W	23 ± 1	24	complex
53	Haviland	Kansas, USA	37°35' N	99°10' N	<0.001	0.015	simple
54	Henbury	Northern Territory, Australia	24°34' S	133°08' E	<0.005	0.157	cluster, simple
55	Highbury	Zimbabwe	17°04' S	30°07' E	1034?	20	complex
56	Holleford	Ontario, Canada	44°28' N	76°38' W	550 ± 100	2.35	simple?
57	Ile Rouleau	Quebec, Canada	50°41' N	73°53' W	<300	15	complex
58	Ilumetsa	Estonia	57°58' N	27°25' E	>0.002	0.08	cluster, simple
59	Ilyinets	Ukraine	49°07' N	29°06' E	400	8	buried, complex
60	Iso-Naakkima	Finland	62°11' N	27°09' E	>1000	3	simple
61	Jänisjärvi	Russia	61°58' N	30°55' E	700 ± 5	14	complex
62	Kaalijärvi	Estonia	58°24' N	22°40' E	0.004 ± 0.001	0.11	cluster, simple
63	Kalkkop	South Africa	32°43' S	24°34' E	0.25	0.64	simple
64	Kaluga	Russia	54°30' N	36°12' E	380	15	buried, complex
65	Kamensk	Russia	48°21' N	40°30' E	49.15 ± 0.18	25	buried, complex
66	Kara	Russia	69°06' N	64°09' E	70.3 ± 2.2	65	buried, complex
67	Kara-Kul	Tajikistan	39°01' N	73°27' E	<5	52	complex
68	Kärdla	Estonia	59°01' N	22°46' E	455 ± 1	7	buried, complex
69	Karikkoselkä	Finland	63°13' N	25°15' E	<1.88	1.5	simple
70	Karla	Russia	54°55' N	48°02' E	5 ± 1	10	buried, complex
71	Kelly West	Northern Territory, Australia	19°56' S	133°57' E	>550	10	complex
72	Kentland	Indiana, USA	40°45' N	87°24' W	<97	13	complex
73	Kgagodi	Botswana	22°29' S	27°34' E	<60?	3.4	simple
74	Kursk	Russia	51°42' N	36°00' E	250 ± 80	6	buried, complex
75	La Moinerie	Quebec, Canada	57°26' N	66°37' W	400 ± 50	8	complex

Table 1.1. (cont.)

No.	Name	Location	Latitude	Longitude	Age [Ma]	Diam.[km]	Type
76	Lappajärvi	Finland	63°12' N	23°42' E	77.3 ± 0.4	23	complex
77	Lawn Hill	Queensland, Australia	18°40' S	138°39' E	>515	18	complex
78	Liverpool	Northern Territory, Australia	12°24'S	134°03' E	150 ± 70	1.6	simple
79	Lockne	Sweden	63°00' N	14°48' E	>455	7.5	complex
80	Logancha	Russia	65°31' N	95°56' E	40 ± 20	20	complex
81	Logoisk	Belarus	54°12' N	27°48' E	30 ± 5	15	buried, complex
82	Lonar	India	19°58' N	76°31' E	0.052 ± 0.006	1.83	simple
83	Lumparn	Finland	60°09' N	20°06' E	~1000	9	complex
84	Macha	Russia	60°06' N	117°35' E	<0.007	0.3	simple
85	Manicouagan	Quebec, Canada	51°23' N	68°42' W	214 ± 1	100	complex
86	Manson	Iowa, USA	42°35' N	94°35' W	74.1 ± 0.1	37	buried, complex
87	Maple Creek	Saskatchewan, Canada	49°48' N	109°6' W	<75	6	complex
88	Marquez	Texas, USA	31°17' N	96°18' W	58.3 ± 2	13	complex
89	Middlesboro	Kentucky, USA	36°37' N	83°44' W	<300	6	complex
90	Mien	Sweden	56°25' N	14°52' E	121.0 ± 2.3	9	simple
91	Mishina Gora	Russia	58°43' N	28°03' E	300 ± 50	2.5	simple
92	Mistastin	Newfound/Labrador, Canada	55°53' N	63°18' W	38 ± 4	28	complex
93	Mizarai	Lithuania	54°01' N	23°54' E	500 ± 20	5	buried, simple
94	Mjølnir	Barents Sea, Norway	73°48' N	29°40' E	143.1 ± 0.8	40	buried, complex
95	Montagnais	Scotian Shelf, Canada	42°53' N	64°13' W	50.50 ± 0.76	45	buried, complex
96	Monturaqui	Chile	23°56' S	68°17' W	<1	0.46	simple
97	Morasko	Poland	52°29' N	16°54' E	0.01	0.1	cluster, simple
98	Morokweng	South Africa	26°28' S	23°32' E	145.0 ± 0.8	70-80	buried, complex
99	Mount Toondina	South Australia	27°57' S	135°22' E	<110	4	simple?
100	Neugrund	Estonia	59°20' N	23°40' E	?	6-8	complex?

Table 1.1. (cont.)

No.	Name	Location	Latitude	Longitude	Age [Ma]	Diam.[km]	Type
101	New Quebec	Quebec, Canada	61°17' N	73°40' W	1.4 ± 0.1	3.44	simple
102	Newporte	North Dakota, USA	48°58' N	101°58' W	<500	3.2	buried, simple
103	Nicholson	Northwest Territories, Canada	62°40' N	102°41' W	<400	12.5	complex
104	Oasis	Libya	24°35' N	24°24' E	<120	≤18	complex
105	Obolon'	Ukraine	49°35' N	32°55' E	169 ± 7	20	buried, complex
106	Odessa	Texas, USA	31°45' N	102°29' W	<0.05	0.168	simple
107	Ouarkziz	Algeria	29°00' N	07°33' W	<70	3.5	complex
108	Picaninny	Western Australia	17°32' S	128°25' E	<360	7	simple
109	Pilot	Northwest Territories, Canada	60°17' N	111°01" W	445 ± 2	6	simple?
110	Popigai	Russia	71°38' N	111°11' E	35.7 ± 0.2	85	complex
111	Presqu'île	Quebec, Canada	49°43' N	74°48' W	<500	24	complex
112	Puchezh-Katunki	Russia	56°58' N	43°43' E	167 ± 3	80	buried, complex
113	Ragozinka	Russia	58°44' N	61°48' E	46 ± 3	9	buried, complex
114	Red Wing	North Dakota, USA	47°36' N	103°33' W	200 ± 25	9	buried, complex
115	Riachao Ring	Brazil	07°43' S	46°39' W	<200	4.5	complex
116	Ries	Germany	48°53' N	10°37' E	15 ± 1	24	buried, complex
117	Rio Cuarto	Argentina	32°52' S	64°14' W	<0.1	4.5	cluster, simple
118	Rochechouart	France	45°50' N	00°56' E	214 ± 8	23	highly eroded
119	Rock Elm	Wisconsin, USA	44°43' N	92°14' W	<505	6	simple
120	Roter Kamm	Namibia	27°46' S	16°18' E	3.7 ± 0.3	2.5	simple
121	Rotmistrovka	Ukraine	49°11' N	32°45' E	120 ± 10	2.7	buried, simple
122	Sääksjärvi	Finland	61°24' N	22°24' E	~560	6	simple
123	Saint Martin	Manitoba, Canada	51°7' N	98°32' W	220 ± 32	40	complex
124	Serpent Mound	Ohio, USA	39°02' N	83°24' W	<320	8	complex
125	Serra da Cangalha	Brazil	08°05' S	46°52' W	<300	12	complex

Table 1.1. (cont.)

No.	Name	Location	Latitude	Longitude	Age [Ma]	Diam.[km]	Type
126	Shunak	Kazakhstan	47°12' N	72°42' E	45 ± 10	2.8	simple
127	Sierra Madera	Texas, USA	30°36' N	102°55' W	<100	13	complex
128	Sikhote Alin	Russia	46°07' N	134°40' E	fell in 1947	0.027	cluster, simple
129	Siljan	Sweden	61°02' N	14°52' E	368 ± 1.1	52	complex
130	Slate Islands	Ontario, Canada	48°40' N	87°00' W	~450	30	complex
131	Sobolev	Russia	46°12' N	138°54' E	<0.001	0.053	simple
132	Söderfjärden	Finland	63°02' N	21°35' E	~600	5.5	buried, complex
133	Spider	Western Australia	16°44' S	125°05' E	>570	13	complex
134	Steen River	Alberta, Canada	59°30' N	117°38' W	95 ± 7	25	buried, complex
135	Steinheim	Germany	48°41' N	10°04' E	15 ± 1	3.8	complex
136	Strangways	Northern Territory, Australia	15°12' S	133°35' E	<470	25	complex
137	Suavjärvi	Russia	63°07' N	33°23' E	2400	16	complex
138	Sudbury	Ontario, Canada	46°36' N	81°11' W	1850 ± 3	250	multiring
139	Suvasvesi N	Finland	62°42' N	28°00' E	<1000	4	complex?
140	Tabun-Khara-Obo	Mongolia	44°06' N	109°36' E	150 ± 20	1.3	simple
141	Talemzane	Algeria	33°19' N	04°02' E	<3	1.75	simple
142	Teague	Western Australia	25°02' S	120°53' E	1630 ± 5	30	complex
143	Tenoumer	Mauritania	22°55' N	10°24' W	2.5 ± 0.5	1.9	simple
144	Terny	Ukraine	48°15' N	33°30' E	280 ± 10	11	complex
145	Tin Bider	Algeria	27°36' N	05°07' E	<70	6	complex
146	Tookoonooka	Queensland, Australia	27°07' S	142°50' E	128 ± 5	55	complex
147	Tswaing (Saltpan)	South Africa	25°24' S	28°05' E	0.220 ± 0.052	1.13	simple
148	Tvären	Sweden	58°46' N	17°25' E	>455	2	simple
149	Upheaval Dome	Utah, USA	38°26' N	109°54' W	<65	10	complex

Table 1.1. (cont.)

No.	Name	Location	Latitude	Longitude	Age [Ma]	Diam.[km]	Type
150	Ust-Kara*	Russia	69°15' N	65°23' E	70.3 ± 2.2	25	complex
151	Vargeao Dome	Brazil	26°50' S	52°07' W	<70	12	complex
152	Veevers	Western Australia	22°58' S	125°22' E	<1	0.08	simple
153	Vepriai	Lithuania	55°05' N	24°34' E	>160 ± 5	7.5	buried, complex
154	Viewfield	Saskatchewan, Canada	49°35' N	103°4' W	190 ± 20	2.5	simple
155	Vredefort	South Africa	27°00' S	27°30' E	2023 ± 4	300	multiring
156	Wabar	Saudi Arabia	21°30' N	50°28' E	0.006 ± 0.002	0.116	simple
157	Wanapitei	Ontario, Canada	46°45' N	80°45' W	37 ± 2	7.5	complex
158	Wells Creek	Tennessee, USA	36°23' N	87°40' W	200 ± 100	12	buried, complex
159	West Hawk	Manitoba, Canada	49°46' N	95°11' W	100 ± 50	2.44	simple
160	Wetumpka	Alabama, USA	32°31' N	86°10' W	81.0 ± 1.5	6.5	complex
161	Wolfe Creek	Western Australia	19°10' S	127°48' E	<0.3	0.875	simple
162	Woodleigh	Australia	26°03' S	114°39' E	<300	40	complex
163	Zapadnaya	Ukraine	49°44' N	29°00' E	165 ± 5	3.2	buried, complex
164	Zeleny Gai	Ukraine	48°04' N	32°54' E	80 ± 20	3.5	buried, complex
165	Zhamanshin	Kazakhstan	48°24' N	60°58' E	1.0 ± 0.1	13	complex

*Uncertain

See also, Earth Impact Database at http://www.unb.ca/passc/Impact/Database/

Table 1.2. Borehole data used to interpret the stratigraphic framework of the Chesapeake Bay impact crater and surrounding coastal-plain sections of Virginia and southern Maryland, and, in particular, to calibrate the network of seismic reflection profiles. Depths in feet (') and meters (m).

#	Name/ Location	Year Drilled	Latitude/ Longitude	Top Chickahominy	Top Exmore/ Mattaponi	Total Depth	Well Head Elevation	Basement Elevation/ Lithology	Cored or Drilled	Relative Location	Reference
1	USGS Exmore Corehole	1986	37° 35' 08" 75° 49' 09"	-1082' -329.8 m	-1180.6' -359.9 m	-1366' -416.4 m	+30' +9.1 m	Not reached	C	Annular through	1, 2, 3
2	Va. Dept. Env. Qual. Kiptopeke Corehole	1989	37°08' 07" 75°57' 08"	-1073' -327.1 m	-1283' -391.1 m	-1993' -607.5 m	+7' +2.1 m	Not reached	C to -1323' -403.3 m	Inner basin	2, 3
3	Va. Dept. Env. Qual. Newport News Park II Corehole	1990	37°12' 08" 76°34' 11"	-240' -73.2 m	-339.3' -103.4 m	-570' -173.7 m	+52' +15.9 m	Not reached	C	Washback channel?	2, 3
4	USGS Windmill Point Corehole	1992	37°36' 50" 76°16' 55"	-439' -133.8 m	-530.5' -161.7 m	-743.8' -226.7 m	+4' +1.2 m	Not reached	C	Breccia apron	2, 3
5	USGS Corehole, Solomons Island, MD	1986	38°19' 49" 76°27' 52"	Absent	Absent	-621' -189.3 m	+15' +4.6 m	Not reached	C	Outside crater	4
6	USGS Oak Grove Corehole	1976	38° 10' 54" 76° 09' 50"	Absent	Absent	-1200' -365.7 m	+180' +54.9 m	Not reached	C	Outside crater	5
7	St.M. - Df 84 Lexington Park, MD	1932	38°15' 48" 76°27' 21"	Absent	Absent	-1569.6' -783.3 m	+108.39' +33 m	-2514.6' -766.5 m diabase	D & C	Outside crater	6
8	Ohio Oil, L.G. Hammond, Salisbury, MD	1944	38°20' 48" 75°21' 54"	Absent	Absent	-5498' -1675.8 m	+54' +16.5m	-5429.8' -1655 m gneiss	D & C	Distal breccia	7
9	J&J Enterprises, E.G. Taylor, Atlantic, VA	1971	37°53' 18" 75°30' 54"	Absent	Absent	-6230' -1911.7 m	+42', +12.8 m	-6144' -1872.7 m gneiss	D	Outside crater	8

Table 1.2. (cont.)

#	Name/ Location	Year Drilled	Latitiude/ Longitude	Top Chickahominy	Top Exmore/ Mattaponi	Total Depth	Well Head Elevation	Basement Elevation/ Lithology	Cored or Drilled	Relative Location	Reference
10	Som - Ce 95, near Princess Anne, MD	1989	38°09' 21" 75°42' 32"	Absent	Absent	-1502' -457.8 m	+13' +4 m	Not reached	C -78 to -1104 m	Outside crater	9
11	VA - VB - P - 3	pre 1972	36°51' 20" 75°58' 35"	Absent	Absent	-1583' -482.5 m	+5' +1.5 m	Not reached	D	Breccia apron	7
12	VA - VB - T - 4	pre 1972	36°57' 40" 76°07' 20"	-830' -253 m	-870' -265.2 m	-1500' -457.2 m	+35' +10.7 m	Not reached	D	Annular trough	7
13	VA - MID - T - 7	pre 1972	37°32' 20" 76°19' 50"	-439' -133.8 m	-479' -146 m	-785' -239.3 m	+5' +1.5 m	Not reached	D	Breccia apron	7
14	VA - MID - P - 9	pre 1972	37°35' 45" 76°26' 20"	-364' -110.9 m	-404' -123.1 m	-539' -164.3 m	+15' +4.6 m	Not reached	D	Breccia apron	7
15	VA - MID - P - 10	pre 1972	37°36' 00" 76°29' 25"	-306' -93.3 m	-332' -101.2 m	-498' -151.8 m	+30' +9.1 m	Not reached	D	Breccia apron	7
16	VA - MID - P - 3	pre 1972	37°36' 20" 76°31' 30"	-300' -91.4 m	-330' -100.6 m	-710' -216.4 m	+40' +12.2 m	Not reached	D	Breccia apron	7
17	VA - MID - P - 8	pre 1972	37°38' 30" 76°34' 30"	Absent	Absent	-535' -163.1 m	+7' +2.1 m	Not reached	D	Outside crater	7
18	VA - RI - P - 6	pre 1972	35°57' 50" 76°45' 00"	Absent	Absent	-570' -173.7 m	+130' +39.6 m	Not reached	D	Outside crater	7
19	VA - WES - T - 8	pre 1972	38°05' 30" 76°49' 15"	Absent	Absent	-521' -158.8 m	+120' +36.6 m	Not reached	D	Outside crater	7
20	VA - WES - T - 7	pre 1972	38°12' 00" 76°55' 40"	Absent	Absent	-638' -194.5 m	+15' +4.6 m	Not reached	D	Outside crater	7
21	VA - KG - P - 15	pre 1972	38°20' 00" 77°02' 15"	Possibly present	Possibly present	-990' -301.8 m	+20' +6.1 m	Not reached	D	Secondary crater?	7
22	VA - KG - P - 16	pre 1972	38°02' 10" 71°03' 45"	Absent	Absent	-565' -172.2 m	+110' +33.5 m	Not reached	D	Outside crater	7

Table 1.2. (cont.)

#	Name/ Location	Year Drilled	Latitude/ Longitude	Top Chickahominy	Top Exmore/ Mattaponi	Total Depth	Well Head Elevation	Basement Elevation/ Lithology	Cored or Drilled	Relative Location	Reference
23	VA - NOR - T - 12	pre 1972	36°52' 20" 76°12' 00"	-610' -185.9 m	-667' -203.3 m	-2567' -782.4 m	+15 +4.6 m	-2567' -782.4 m granite?	D	Breccia apron	7
24	VA - NAN - P - 8	pre 1972	36°50' 30" 76°27' 00"	Absent	-340' -103.6 m	-634' -193.2 m	+20 +6.1 m	Not reached	D	Breccia apron	7
25	VA - NAN - P - 14	pre 1972	36°42' 45" 76°39' 35"	Absent	-185' -56.4 m	-652' -198.7 m	+65' +19.8 m	Not reached	D	Breccia apron	7
26	VA - IW - P - 8	pre 1972	36°41' 15" 76°04' 30"	Absent	Absent	-880' -268.2 m	+40' +12.2 m	-828' -252.4 m ?	D	Breccia apron	7
27	VA - SO - T - 6	pre 1972	36°38' 05" 77°05' 00"	Absent	Absent	-290' -88.4 m	+85' +25.9 m	Not reached	D	Outside crater	7
28	VA - SO - P - 3	pre 1972	36°35' 05" 77°02' 00"	Absent	Absent	-428' -130.5 m	+40' +12.2 m	-340' -103.6 m ?	D	Outside crater	7
29	NC - NOR - T - 8	pre 1972	36°32' 30" 77°19' 00"	Absent	Absent	-255' -77.7 m	+50' +15.2 m	-235' -71.6 m ?	D	Outside crater	7
30	NC - NOR - T - 9	pre 1972	36°31' 20" 77°22' 30"	Absent	Absent	-185' -56.4 m	+100' +30.5 m	-150' -45.7 m ?	D	Outside crater	7
31	NC - NOR - T - 10	pre 1972	36°30' 05" 77°24' 55"	Absent	Absent	-144' -43.9 m	+100 +30.5 m	-123' -37.5 m ?	D	Outside crater	7
32	VA - CHE - P - 5	pre 1972	36°43' 30" 76°20' 15"	Absent	Absent	-806' -245.7 m	+14 +4.3 m	Not reached	D	Outside crater	7
33	VA - POR - P - 10	pre 1972	36°48' 00" 76°23' 30"	Absent	Absent	-638' -194.5 m	+15' + 4.6 m	Not reached	D	Breccia apron	7

Table 1.2. (cont.)

#	Name/ Location	Year Drilled	Latitiude/ Longitude	Top Chickahominy	Top Exmore/ Mattaponi	Total Depth	Well Head Elevation	Basement Elevation/ Lithology	Cored or Drilled	Relative Location	Reference
34	VA - CHE - P - 11	pre 1972	36°50' 30" 76°24' 00"	Absent	Absent	-597' -182 m	+3' +0.9 m	Not reached	D	Breccia apron	7
35	VA - NAN - P -13	pre 1972	36°02' 30" 76°28' 25"	Absent	Absent	-633' -192.9 m	+22' +6.7 m	Not reached	D	Breccia apron	7
36	VA - IW - P - 13	pre 1972	37°00' 05" 76°07' 50'	Absent	Absent	-414' -126.2 m	+40' +12.2 m	Not reached	D	Breccia apron	7
37	VA - SUR - P - 1	pre 1972	37°10' 00" 76°41' 55"	-201' -61.3 m	-212' -64.6 m	-379' -115.5 m	+15' +4.6 m	Not reached	D	Washback channel?	7
38	VA - JC - T - 10	pre 1972	37°13' 00" 76°45' 30"	Absent	-166' -50.6 m	-355' -108.2 m	+10' +3.1 m	Not reached	D	Washback channel?	7
39	VA - CC - T - 4	pre 1972	37°22' 05" 77°02' 35"	Absent	0.0'? 0.0 m?	-90' -27.4 m	+70' +21.3 m	Not reached	D	Washback channel?	7
40	VA - CC - P - 3	pre 1972	37°28' 00" 77°11' 00"	Absent	Absent	-244' -74.4 m	+120' +36.6 m	Not reached	D	Outside crater	7
41	VA - NK - P - 6	pre 1972	37°34' 30" 77°09' 00"	Absent	Absent	-350' -106.7 m	+150' + 45.7 m	Not reached	D	Outside crater	7
42	VA - HEN - P - 12	pre 1972	37°31' 30" 77°18' 00"	Absent	Absent	-130' -39.6 m	+160' +48.8 m	Not reached	D	Outside crater	7
43	VA - HAN - P - 12	pre 1972	37°37' 30" 77°21' 15"	Absent	Absent	-157' -47.9 m	+190' +57.9 m	Not reached	D	Outside crater	7
44	VA - PG - P - 1	pre 1972	37°16' 15" 77°11' 15"	Absent	Absent	-127' -38.7 m	+125' +38.1 m	Not reached	D	Outside crater	7
45	VA - PG - P - 3	pre 1972	37°13' 20" 77°16' 45"	Absent	Absent	-135' -41.1 m	+120' +36.6 m	Not reached	D	Outside crater	7
46	VA - PG - T - 8	pre 1972	37°11' 00" 77°19' 00"	Absent	Absent	+40' +12.2 m	+140' +42.7 m	Not reached	D	Outside crater	7

Table 1.2. (cont.)

#	Name/ Location	Year Drilled	Latitude/ Longitude	Top Chickahominy	Top Exmore/ Mattaponi	Total Depth	Well Head Elevation	Basement Elevation/ Lithology	Cored or Drilled	Relative Location	Reference
47	MD - SOM - T - 10	pre 1972	38°00' 20" 75°50' 45"	Absent	Absent	-1505' -458.7 m	+5' +1.5 m	Not Reached	D	Outside crater	7
48	VA - NOD - T - 8	pre 1972	37°49' 50" 76°17' 15"	Absent	Absent	-650' -198.1 m	+5' +1.5 m	Not reached	D	Outside crater	7
49	VA - NOD - T - 7	pre 1972	37°43' 30" 76°19' 30"	Absent	Absent	-424' -129.2 m	+2' +0.6 m	Not reached	D	Outside crater	7
50	VA - LAN - P - 1	pre 1972	37°37' 45" 76°20' 30"	-440' -134 m	-500' -152.4 m	-734' -223.7 m	+3' +0.9 m	Not reached	D	Breccia apron	7
51	VA - GLO - P - 1	pre 1972	37°21' 10" 76°36' 35"	-269' -82 m	-305' -93 m	-417' -127.1 m	+7' +2.1 m	Not reached	D	Washback channel?	7
52	VA - YK - T - 9	pre 1972	37°19' 15" 76°39' 30"	-197' -60 m	-213' -64.9 m	-395' -120.1 m	+75' +22.9 m	Not reached	D	Washback channel?	7
53	VA - SUS - T - 2	pre 1972	36°09' 20" 77°01' 30"	Absent	Absent	-90' -27.4 m	+100' +30.5 m	Not reached	D	Washback channel?	7
54	VA - YK - T - 6	pre 1972	37°14' 00" 76°33' 00"	-264' -80.5 m	-310' -94.5 m	-458' -139.6 m	+50' +15.2 m	Not reached	D	Washback channel?	7
55	VA - KW - T - 4	pre 1972	37°32' 40" 76°48' 20"	Absent	-126' -38.4 m	-1659' -505.7 m	+30' +9.1 m	Top Trias. -1255' -382.5 m	D	Washback channel?	7
56	VA - KQ - P - 2	pre 1972	37°48' 10" 77°03' 10"	Absent	Absent	-278' -84.7 m	+192' +58.5 m	Not reached	D	Outside crater	7
57	VA - KQ - T - 5	pre 1972	37°54' 10" 77°06' 20"	Absent	Absent	-11' -3.4 m	+175' +53.3 m	Not reached	D	Outside crater	7
58	VA - CHE - P - 6	pre 1972	36°38' 30" 76°22' 00"	Absent	Absent	-701' -213.7 m	+15' +4.6 m	Not reached	D	Washback channel?	7
59	VA - NON - T - 3	pre 1972	37°12' 10" 75°57' 15"	Absent	Absent	-441' -134.4 m	+20' +6.1 m	Not reached	D	Inner basin	7

Table 1.2. (cont.)

#	Name/ Location	Year Drilled	Latitiude/ Longitude	Top Chickahominy	Top Exmore/ Mattaponi	Total Depth	Well Head Elevation	Basement Elevation/ Lithology	Cored or Drilled	Relative Location	Reference
60	VA - NAN - C - 21	pre 1972	36°45' 40" 76°40' 30"	Absent	Absent	-228' -69.5 m	+60' +18.3 m	Not reached	D	Outside crater	7
61	VA - IW - P - 7	pre 1972	36°48' 00" 76°45' 00"	Absent	Absent	-371' -113.1 m	+82' +25 m	Not reached	D	Outside crater	7
62	VA - IW - C - 11	pre 1972	36°51' 00" 76°48' 45"	Absent	Absent	-190' -57.9 m	+50 +15.2 m	Not reached	D	Outside crater	7
63	VA - SO - C - 8	pre 1972	36°54' 20" 76°53' 30"	Absent	Absent	-148' -45.1 m	+80 +24.4 m	Not reached	D	Outside crater	7
64	VA - SO - T - 9	pre 1972	36°56' 20" 76°57' 00"	Absent	Absent	-115' -35.1 m	+75' +22.9 m	Not reached	D	Outside crater	7
65	VA - SUS - T - 3	pre 1972	37°02' 15" 77°05' 50"	Absent	Absent	-50' -15.2 m	+100' +30.5 m	Not reached	D	Outside crater	7
66	VA - SUS - T - 4	pre 1972	37°05' 40" 77°11' 00"	Absent	Absent	-35' -10.7 m	+105' +32 m	Not reached	D	Outside crater	7
67	VA - PG - T - 7	pre 1972	37°08' 15" 77°14' 45"	Absent	Absent	-10' -3.1 m	+100' +30.5 m	Not reached	D	Outside crater	7
68	VA - PG - T - 9	pre 1972	37°17' 30" 77°18' 00"	Absent	Absent	-25' -7.6 m	+90' +27.4 m	Not reached	D	Outside crater	7
69	USGS Corehole Haynesville No. 2	1985	37°57' 13" 76°40' 26"	Absent	Absent	-556' -169.5 m	+87' +26.5 m	Not reached	C	Outside crater	7
70	#82 Cape Charles NY, Phil., Norfolk RR	1910	37°16' 76°01'	-335' -102.1m	-355' -108.2 m	-1790' -545.6 m	+20' +6.1 m	Not reached	D	Inner basin	10
71	#183 Tangier Island	1925 ?	37°50' 76°00	Absent	Absent	-858' -261.5 m	+2' ? +0.6 m	Not reached	D	Outside crater	11
72	#180 Cobbs Island	?	37°20' 75°44	Not reached	Not reached	-1001' -305.1 m	+9' +2.7 m	Not reached	D	Annular trough	11

Table 1.2. (cont.)

#	Name/ Location	Year Drilled	Latitude/ Longitude	Top Chickahominy	Top Exmore/ Mattaponi	Total Depth	Well Head Elevation	Basement Elevation/ Lithology	Cored or Drilled	Relative Location	Reference
73	# 17 Ocran Dymer Fish Co.	pre-1945	36°39' 76°20'	-243' ? -74.1 m	-375' ? -114.2 m	-497.5' -151.6 m	+10' ? +3.1 m	Not reached	D	Breccia apron	10
74	# 18 Byrdton, Estate East Richlands	pre-1945	37°42' 76°20'	-210' ? -64.0 m	-335' ? -102.1 m	-800' -243.8 m	+10' +3.1 m	Not reached	D	Breccia apron	10
75	# 20 Reedville Blunden & Hinton	pre-1945	37°51' 76°15'	Absent?	-370' ? -112.8 m	>-785' >-239.3 m	0.0' 0.0 m	Not reached	D	Breccia apron	10
76	#21 Reedville McNeal & Dodson	pre-1945	37°49' 76°16'	-275' ? -83.8 m	-345' ? -105.2 m	-865' -263.7 m	+5' ? +1.5 m	Not reached	D	Breccia apron	10
77	# 42 West Point Chesapeake Camp Co.	1944	37°32' 76°48'	-183' ? -55.8 m	-395' ? -120.4 m	-657' -200.3 m	+10' +3.1 m	Not reached	D	Washback channel?	10
78	# 43 West Point Chesapeake Camp Co.	1944	37°32' 76°48'	-178' ? -54.3 m	-378' ? -115.2 m	-741' -225.9 m	+22' +6.7 m	Not reached	D	Washback channel?	10
79	# 44 Gloucester	pre-1945	37°29' 76°30'	-347' ? -105.8 m	-528' ? -160.9 m	-745' -227.1m	+70' +21.3 m	Not reached	D	Washback channel?	10
80	# 45 Severn J.N. Shakelford	pre 1945	37°19' 76°25'	-432' ? -131.7 m	-567 ? -172.8 m	-602' -183.5 m	+8' +2.4 m	Not reached	D	Washback channel?	12
81	# 46 Mathews Elkins Oil & Gas Co.	1929	37°25' 76°18'	-588' ? -179.2 m	-803 ? -244.8 m	-2318' -706.5 m	+7' +2.1 m	-2318' -706.5 m granite	D	Annular trough	10
82	# 58 (56) Williamsburg Dunbar Farm	1937	37°25' 76°45'	-230' ? -70.1 m	-280 ? -85.3 m	-610' -185.9 m	+60' +18.3 m	Not reached	D	Washback channel?	10, 14

Table 1.2. (cont.)

#	Name/ Location	Year Drilled	Latitiude/ Longitude	Top Chickahominy	Top Exmore/ Mattaponi	Total Depth	Well Head Elevation	Basement Elevation/ Lithology	Cored or Drilled	Relative Location	Reference
83	# 59 (49) Williamsburg J. Levinson	1944	37°15' 76°49'	-180 ? -54.9 m	-290' ? -88.4 m	-345' -105.2 m	+90' ? +27.4 m	Not reached	D	Washback channel?	10, 14
84	# 60 (23) Williamsburg R.W. Mahone	1940	37°16' 76°40'	-177' ? -53.9 m	-300' ? -91.4 m	-340' -103.6 m	+25' +7.6 m	Not reached	D	Washback channel?	10, 14
85	# 61 (22) Williamsburg Waller Pond	1943	37°16' 76°43'	-210' ? -64.0 m	-345' ? -105.2 m	-457.5' -139.5 m	+10' ? +3.1 m	Not reached	D	Washback channel?	10, 14
86	# 62 (13) Camp Peary	1943	37°18' 76°39'	-256' ? -78.0 m	-330' ? -100.6 m	-359' -109.4 m	+84' +25.6 m	Not reached	D	Washback channel?	10, 14
87	# 63 (7) Camp Peary	1942	37°19' 76°39'	-286' ? -87.2 m	-360' ? -109.7 m	-390' -118.9 m	+74' +22.6 m	Not reached	D	Washback channel?	10, 14
88	# 64 (24) Penniman Fuel Depot	1942	37°17' 76°37'	-242' ? -73.8 m	-362' ? -110.3 m	-430' -131.1 m	+83' +25.3 m	Not reached	D	Washback channel?	10, 14
89	# 65 (35) Yorktown Navy Mine Depot	1942	37°12' 76°30'	-300' ? -91.4 m	-380' ? -115.8 m	-540' -164.6 m	+80' +24.4 m	Not reached	D	Annular trough	10, 14
90	# 66 (3) Lœ Hall, Newport News Water Co.	1916	37°12' 76°34'	-281.5' ? -85.8 m	-457' ? -139.3 m	-516.5' -157.4 m	+10'? +3.1 m	Not reached	D	Washback channel?	10, 14
91	# 67 (8) Lœ Hall, Newport News Waterworks	1941	37°10' 76°31'	-307' ? -93.6 m	-429' ? -130.8 m	-455' -138.7 m	+15' +4.6 m	Not reached	D	Washback channel?	10, 14
92	# 68 (21) Ft. Eustis 2	1941	37°10' 76°34'	-262' ? -79.9 m	-443' ? -135.0 m	-658' -200.6 m	+34' +10.4 m	Not reached	D	Washback channel?	10, 14
93	# 69 (22) Ft. Eustis 3	1941	37°08' 76°33'	-300 '? -91.4 m	-409' ? -124.7 m	-519' -158.2 m	+31' +9.5 m	Not reached	D	Washback channel?	10, 14

Table 1.2. (cont.)

#	Name/ Location	Year Drilled	Latitiude/ Longitude	Top Chickahominy	Top Exmore/ Mattaponi	Total Depth	Well Head Elevation	Basement Elevation/ Lithology	Cored or Drilled	Relative Location	Reference
94	# 70 (20) Ft. Eustis 4	1941	37°08' 76°33	-269' ? -82.0 m	-431' ? -131.4	-512' -156.1 m	+31' +9.5 m	Not reached	D	Washback channel?	10, 14
95	# 71 (17) Ft. Eustis Dozier 1	1941	37°08' 76°33'	?	-430' -131.1 m	-440' -134.1 m	+7' +2.1 m	Not reached	D	Washback channel?	10, 14
96	# 72 (23) Mulberry Island	1942	37°08' 76°36'	-314' ? -95.7 m	-396' ? -120.7 m	-455'+ -138.7 m	+10' +3.1 m	Not reached	D	Washback channel?	10, 14
97	# 73 (30) Camp Patrick Henry	1942	37°00' 76°25'	-400' ? -121.9 m	-524' ? -159.7 m	-524' -159.7 m	+30' +9.1 m	Not reached	D	Washback channel?	10, 14
98	# 74 (29) Camp Patrick Henry	1942	37°00' 76°24'	-415' ? -126.5 m	-457' ? -139.3 m	-457' -139.3 m	+35' +10.7 m	Not reached	D	Washback channel?	10
99	# 78 (46) Newport News Levinson Meat Packing	1940	36°58' 76°26'	-405' ? -123.4 m	-485' ? -147.8 m	-890' -271.3 m	+10' +3.1 m	Not reached	D	Washback channel?	10, 14
100	# 79 (44) Newport News Va. Public Serv. Co.	1944	36°59' 76°04'	-413' ? -125.9 m	-592' ? -180.4m	-1070' -326.1 m	+12' +3.7 m	Not reached	D	Washback channel?	10, 14
101	# 80 (13) Newport News Buxton Hospital	pre 1945	36°59' 76°24'	-250' ? -76.2 m	-440' ? -134.1 m	-810' -246.9 m	+10' +3.1 m	Not reached	D	Washback channel?	10, 14
102	# 81 (8c) Fort Monroe	1902	37°00' 76°18'	-630' ? -192.0 m	-830' ? -253.0 m	-2244' -684.0 m	+10' +3.1 m	-2244' -684 m crystalline rock	D	Annular trough	10, 14

Table 1.2. (cont.)

#	Name/ Location	Year Drilled	Latitiude/ Longitude	Top Chickahominy	Top Exmore/ Mattaponi	Total Depth	Well Head Elevation	Basement Elevation/ Lithology	Cored or Drilled	Relative Location	Reference
103	W5521	?	36°51' 76°29'	?	?	-1990' -606.6 m	+15' +4.6 m	-1800' -548.6 m ?	D	Breccia apron	13
104	W6839	?	36°56' 77°10'	Absent	Absent	-400' -121.9 m	+110' +33.5 m	-400' -121.9 m gneiss?	D	Outside crater	13
105	59E-5 NASA Research Center, Langley AFB	1964	37°05' 38" 76°22' 43"	-585' -178.3 m	-735' -224.0 m	-2084' -635.2 m	+9' +2.7 m	-2080' -634 m gneiss?	D	Annular trough	3, 13
106	Washington's Birthplace	1946	38°11' 76°56'	-210' -64 m	-736' -224.33 m	-565' -172.2 m	+20' +6.1 m	Not reached	D	Secondary crater?	14
107	Colonial Beach	1946	38°15' 76°59'	-177 -54.0 m	-205' -62.5 m	-634' -193.2 m	+20' +6.1 m	Not reached	D	Secondary crater?	14
108	C28 Va. Tech Geothermal Oyster	1976- 1982	37°17' 75°56'	-1011.5'? -308.3 m?	Not reached	-1011.5' -308.3 m	+12' +3.7 m	Not reached	D	Inner basin	15
109	C56 Va. Tech Geothermal Eastville	1976- 1982	37°21' 76°00'	-1011.5' -308.3 m	Not reached	-1011' -308.3 m	+12' +3.7 m	Not reached	D	Inner basin	15
110	C24 Va. Tech Geothermal Willoughby Bay	1976- 1982	36°58' 76°16'	-897.3' -273.5 m	-962.9' -293.5 m	-1024.9' -312.4 m	+5' +1.5 m	Not reached	D	Annular trough	15
111	C25 Va. Tech Geothermal Portsmouth	1976- 1982	36°53' 76°28'	-716.3' -218.3 m	-916.3' -279.3 m	-1982.6' -604.3 m	+22' +6.7 m	-1805.4' -550.3 m metavolcanics	D & C	Breccia apron	15

Table 1.2. (cont.)

#	Name/ Location	Year Drilled	Latitude/ Longitude	Top Chickahominy	Top Exmore/ Mattaponi	Total Depth	Well Head Elevation	Basement Elevation/ Lithology	Cored or Drilled	Relative Location	Reference
112	C26 Va. Tech Geothermal Isle of Wight	1976-1982	36°56' 76°36'	-941.9' -287.1 m?	-1112.5' -339.1 m?	-1473.4' -449.1 m	+75' +22.9 m	-1319.2' -402.1 m granite	D & C	Breccia apron	15
113	C27 Va Tech Geothermal Langley AFB	1976-1982	37°06' 76°22'	-795.6' -242.5 m	-995.7' -303.5 m	-1473.4' -449.1 m	+5' +1.5 m	Not reached	D	Annular trough	15
114	C60 Va. Tech Geothermal Bunny's Bar	1975-1982	37°03' 76°18'	-790.4' -240.9 m	-885.5' -269.9 m?	-951.1' -289.9 m	+10' +3.1 m	Not reached	D	Annular trough	15
115	C55 Va. Tech Geothermal Tasley	1976-1982	37°43' 75°43'	Absent	Absent	-753.9' -229.8 m	+40' +12.2 m	Not reached	D	Outside crater	15
116	C59 Va. Tech Geothermal Smith Point	1976-1982	37°53' 76°15'	Absent	Absent	-934.7' -284.9 m	+10' +3.1 m	Not reached	D	Outside crater	15
117	Fentress Corehole	1980s	36°43' 76°09'	Absent	Absent	-2005' -611.1 m	+15' +4.6 m	Not reached	C to -1100' -335.3 m	Outside crater	3
118	Dismal Swamp Corehole	1980s	36°37' 76°44'	Absent	Absent	-1857' -566.0 m	+33' +10.1 m	-1850' -563.9 m arkose	C to -1080' -329.2 m	Outside crater	3
119	Jenkins Bridge Corehole	1980s	37°57' 75°35'	Absent	Absent	-1314' -400.5 m	+6' +1.8 m	Not reached	C except -484 to -1194' -146 to -364m	Outside crater	3

Table 1.2. (cont.)

#	Name/ Location	Year Drilled	Latitiude/ Longitude	Top Chickahominy	Top Exmore/ Mattaponi	Total Depth	Well Head Elevation	Basement Elevation/ Lithology	Cored or Drilled	Relative Location	Reference
120	DGT-1 Va. Tech Geothermal, near airport Crisfield, MD	1976-1982	38°00' 58" 75°49' 34"	Absent	Absent	-5573.5' -1698.8 m	+4' +1.2 m	-4080.7' -1243.8 m metavolcanics	D & C	Outside crater	15
121	Socony Vacuum Bethards Berlin, MD	1940s	38°18' 15" 75°16' 30"	Absent	Absent	-7116' -2169 m	+45' +13.7 m	-7070' -2154.9 m gabbro	D	Outside crater	7
122	CH-Ce 37 Charles Co, MD	pre 1978	38°30' 76°59'	Absent	Absent	~-1765' ~-538.0 m	+193' +58.8 m	-1765' -538.0 m arkose	D	Outside crater	16
123	St. M-T-27 Point Lookout, MD	pre 1978	38°14' 30" 76°29' 15"	Absent	Absent	-1409' -429.5 m	+ 100' +30.1 m	Not reached	D	Outside crater	16
124	St. M-P-22 Kitts Point, MD	pre 1978	38°03' 00" 76°19' 30"	Absent	Absent	-676' -206.0 m	+ 5' +1.5 m	Not reached	D	Outside crater	16
125	St.M.- Ff 36 Kitts Point, MD	pre 1978	38°06' 76°25'	Absent	Absent	-933' -284.4 m	+ 7' +2.1 m	Not reached	D	Outside crater	6
126	St.M.- Gg 14 Point Lookout, MD	pre 1978	38°02' 76°19'	Absent	Absent	-721' -219.8 m	+ 10' +3.1 m	Not reached	D	Outside crater	6
127	JSC Drilling Thompson No. 1	1968	38°16' 77°07'	Absent	Absent	?	~+180' ~+54.9 m	~-1312' ~-400 m arkose	D	Outside crater	17
128	Texaco Wilkins et ux. No. 1	1989	38°15' 77°01'	Absent	Absent	-10115' -3083.1 m in granite	~+20' ~+6.1 m	~-1457' ~-444 m arkose	D	Outside crater	17

Table 1.2. (cont.)

#	Name/ Location	Year Drilled	Latitiude/ Longitude	Top Chickahominy	Top Exmore/ Mattaponi	Total Depth	Well Head Elevation	Basement Elevation/ Lithology	Cored or Drilled	Relative Location	Reference
129	Texaco P.H. Gouldman No. 1	1991	38°07' 76°55'	Absent	Absent	-8015' -2443.0 m in schist	~+10' ~+3.1 m	~-1630' ~-497 m arkose	D	Outside crater	17
130	C-22 Va.Tech Geothermal Creeds Field	1980s	36°06' 23" 76°00' 26"	Absent	Absent	-969.5' -295.5 m	+5' +1.5 m	Not reached	D	Outside crater	15
131	C-23 Va. Tech Geothermal, Oceana	1980s	36°48' 09" 76°02' 30"	Absent	Absent	-974.1' -296.9 m	+10' +3.1 m	Not reached	D	Outside crater	15
132	Md # 1 near Marbury, MD	pre 1978	38°35' 77°09'	Absent	Absent	~-560' ~-170.7 m	~+90' ~+27.4 m	-560' -170.7 m schist	D	Outside crater	16
133	Md # 2 Prince Georges Co., MD	pre 1978	38°38' 76°42'	Absent	Absent	~-2439' ~-743.4 m	~+100' ~+30.1 m	-2439' -743.4 m ?	D	Outside crater	16
134	# 8 Chuckatuck-Cedarbrook Farm	pre 1945	36°51' 16" 76°33' 26"	?	?	-535' -163.1 m	+15' +4.6 m	Not reached	D	Breccia apron	12
135	# 37 Drivers-Monogram Farm	1943	36°49' 04" 76°32' 50"	-285' -86.9 m	-295' -89.5 m	-520' -158.5 m	+20' +6.1 m	Not reached	D	Breccia apron	12
136	# 5 Virginia Division of Forestry	?	35°48' 08" 75°23' 15"	?	?	-633' -192.9 m	+20' +6.1 m	Not reached	D	Breccia apron	3
137	# 6 City of Chesapeake Bowers Hill - Production	?	36°47' 02" 76°24' 55"	?	?	-979' -298.4 m	+21' +6.4 m	Not reached	D	Breccia apron	3

Table 1.2. (cont.)

#	Name/ Location	Year Drilled	Latitiude/ Longitude	Top Chickahominy	Top Exmore/ Mattaponi	Total Depth	Well Head Elevation	Basement Elevation/ Lithology	Cored or Drilled	Relative Location	Reference
138	# 10 MW4 Corehole	1980s	36°47' 10" 76°26' 52"	Absent	Absent	-983' -299.6 m	+17' +5.2 m	Not reached	C	Breccia apron	1
139	# 11 Lone Star Cement Corp.	?	36°48' 53" 76°17' 09"	?	?	-790' -240.8 m	+5' +1.5 m	Not reached	D	Breccia apron	3
140	# 12 City of Portsmouth	?	36°51' 15" 76°19' 17"	?	?	-1144' -348.7 m	+10' +3.1 m	Not reached	D	Breccia apron	3
141	# 81 Smithfield Ice Plant	1924	36°59' 05" 76°37' 21"	?	?	-311' -94.8 m	+10' +3.1 m	Not reached	D	Breccia apron	12
142	# 19 Rescue Water Co.	?	36°59' 39" 76°33' 30"	?	?	-528' -160.9 m	+22' +6.7 m	Not reached	D	Breccia apron	3
143	# 20 Town of Smithfield Red Point Heights	?	36°59' 12" 76°36' 50"	?	?	-477' -145.4 m	+35' +10.7 m	Not reached	D	Breccia apron	
144	# 21 Nimmo Well Chuckatuck	?	36°52' 18" 76°31' 30"	?	?	-543' -165.5 m	+35' +10.7 m	Not reached	D	Breccia apron	3
145	# 22 Tidewater Virginia Properties-Graymore Estate	?	36°57' 27" 76°31' 39"	?	?	-541' -164.9 m	+15' +4.6 m	Not reached	D	Breccia apron	3
146	# 54 Battery Park Water Co.	1908	36°59' 32" 76°29' 44"	-247' -75.3 m	-272' -82.9 m	-333' -101.5 m	+13' +4.0 m	Not reached	D	Breccia apron	12
147	# 108 Carrolton	1915	36°58' 02" 76°34' 48"	?	?	-382' -116.4 m	+8' +2.4 m	Not reached	D	Breccia apron	12
148	# 25 Tidewater Water Co.	?	36°52' 55" 76°23' 11"	?	?	-573' -174.7 m	+15' +4.6 m	Not reached	D	Breccia apron	3

Table 1.2. (cont.)

#	Name/ Location	Year Drilled	Latitiude/ Longitude	Top Chickahominy	Top Exmore/ Mattaponi	Total Depth	Well Head Elevation	Basement Elevation/ Lithology	Cored or Drilled	Relative Location	Reference
149	# 26 Newport News- City Hall Complex	?	36°58' 40" 76°25' 50"	-410' -125 m	-574' -175.0 m	-870' -265.2 m	+30' +9.1 m	Not reached	D	Breccia apron	3
150	# 9 Lamberts Point-Norfolk & Western RR	pre 1945	36°52' 26" 76°18' 56"	-387' -118.0 m	-397' -121.1 m	-606' -184.7 m	+10' +3.1 m	Not reached	D	Breccia apron	12
151	# 36 Va. Department of Environmental Quality	?	37°02' 36" 76°42' 59"	Absent	Absent	-615' -187.5 m	+85' +25.9 m	Not reached	D	Outside crater	3
152	# 3a Rushmere	1939	37°04' 34" 76°40' 05"	?	?	-381' -116.1 m	+5' +1.5 m	Not reached	D	Breccia apron	12
153	# 7 Burwells Bay	1908	37°03' 23" 76°40' 13"	?	?	-306' -93.3 m	+15' +4.6 m	Not reached	D	Breccia apron	12
154	# 25 Lone Star Cement Co. near Mogarts Beach	1917	37°00' 29" 76°36' 24"	?	?	-324' -98.8 m	+12' +3.7 m	Not reached	D	Breccia apron	12
155	# 42 Bacons Castle Test Well	1942	37°06' 10" 76°44' 13"	Absent	Absent	-985' -300.2 m	+70' +21.3 m	Not reached	D	Outside crater	12
156	# 24 North End Point	1896	37°06' 30" 75°17' 25"	-737' -224.6 m	-917' -279.5 m	-1169' -356.3 m	+3' +0.9 m	Not reached	D	Annular trough	14
157	# 2 Town of Claremont	1917	37°13' 21" 75°57' 06"	Absent	Absent	-313' -95.4 m	+90' +27.4 m	Not reached	D	Outside crater	3
158	# 4 Claremont O.E. Belding	1917	37°14' 20" 76°58' 32"	Absent	Absent	-270' -82.3 m	+17' +5.2 m	Not reached	D	Outside crater	12
159	# 50 First Colony	?	37°14' 34" 76°48' 15"	?	?	-464' -141.4 m	+30' +9.1 m	Not reached	D	Washback channel?	3

Table 1.2. (cont.)

#	Name/ Location	Year Drilled	Latitiude/ Longitude	Top Chickahominy	Top Exmore/ Mattaponi	Total Depth	Well Head Elevation	Basement Elevation/ Lithology	Cored or Drilled	Relative Location	Reference
160	# 51 Surry Court House # 2	?	37°08' 32" 76°50' 27"	?	?	-375' -114.3 m	+103' +31.4 m	Not reached	D	Washback channel?	3
161	# 26 Jamestown 4-H Club	1946	37°13' 41" 76°47' 28"	-135' -41.2 m	-180' -54.9 m	-265' -80.7 m	+10' +3.1 m	Not reached	D	Washback channel?	14
162	# 27a Jamestown	1906	37°13' 57" 76°47' 32"	-130' -39.6 m	-218' -66.5 m	-287' -87.4 m	+33' +10.1 m	Not reached	D	Washback channel?	14
163	# 51 Williamsburg Carolyn Tourist Court	1940	37°17' 13" 76°43' 22"	-207' -63.1 m	Not reached	-258' -78.7 m	+90' +27.4 m	Not reached	D	Washback channel?	14
164	# 55 Jamestown Corehole	?	37°13' 05" 76°46' 37"	?	?	-272' -82.9 m	+1' +0.3 m	Not reached	C	Washback channel?	3
165	# 56 York Public Utilities	?	37°14' 21" 76°38' 28"	?	?	-586' -178.6 m	+80' +24.4 m	Not reached	D	Washback channel?	3
166	# 57 Hog Island Nuclear Plant	?	37°09' 50" 76°41' 52"	?	?	-386' -117.7 m	+34' +10.4 m	Not reached	D	Washback channel?	3
167	# 58 Busch Gardens	?	37°13' 43" 76°40' 08"	?	?	-457' -139.3 m	+53' +16.2 m	Not reached	D	Washback channel?	3
168	# 59 Busch Gardens	?	37°14' 06" 76°38' 43"	?	?	-435' -132.6 m	+85' +25 9 m	Not reached	D	Washback channel?	3
169	# 60 Hog Island	?	37°11' 33" 76°40' 53"	?	?	-1235' -376.4 m	+5' +1.5 m	Not reached	D	Washback channel?	3
170	# 61 Virgi nia Electric Power Company	?	37°09' 51" 76°41' 57"	?	?	-385' -117.3 m	+35' +10.7 m	Not reached	D	Washback channel?	3

Table 1.2. (cont.)

#	Name/ Location	Year Drilled	Latitiude/ Longitude	Top Chickahominy	Top Exmore/ Mattaponi	Total Depth	Well Head Elevation	Basement Elevation/ Lithology	Cored or Drilled	Relative Location	Reference
171	# 62 Dow Badische	?	37°11' 20" 76°36' 54"	-291' -86.7 m	-344' -104.8 m	-1540' -469.4 m	+20' +6.1 m	Not reached	D	Washback channel?	3
172	# 64 Grove	?	37°12' 50" 76°36' 52"	-279' -85.0 m	-320' -97.5 m	-445' -135.7 m	+40' +12.2 m	Not reached	D	Washback channel?	3
173	# 67 City of Newport News Golf Course	?	37°11' 14" 76°31' 21"	-404' -123.1 m	-460' -140.2 m	-487' -148.4 m	+20' +6.1 m	Not reached	D	Washback channel?	3
174	# 68 Lee Hall Treatment Plant	?	37°10' 01" 76°33' 16"	-303' -92.4 m	-395' -120.4 m	-1315' -400.8 m	+35' +10.7 m	Not reached	D	Washback channel?	3
175	# 69 Upper Potomac Monitor Well	?	37°11' 29" 76°30' 38"	-408' -124.4 m	-588' -179.2 m	-1244' -379.2 m	+56' +17.1 m	Not reached	D	Washback channel?	3
176	# 70 Upper Potomac Production Well	?	37°10' 41" 76°35' 17"	-312' -95.1 m	-381' -116.1 m	-1120' -341.4 m	+30' +9.1 m	Not reached	D	Washback channel?	3
177	# 71 Middle Potomac Production Well	?	37°11' 12" 76°34' 13"	-308' -93.9 m	-369' -112.5 m	-1113' -339.2 m	+35' +10.7 m	Not reached	D	Washback channel?	3
178	# 72 Upper Potomac Monitor Well	?	37°10' 41" 76°35' 17"	?	?	-560' -170.7 m	+40' +12.2 m	Not reached	D	Washback channel?	3
179	# 73 Va.. Peninsula Econ. Development Council	?	37°11' 49" 76°35' 34"	-293' -89.3 m	-332' -101.2 m	-530' -161.5 m	+50' +15.2 m	Not reached	D	Washback channel?	3
180	# 39 Yorktown Colonial National Monument	1931	37°13' 36" 76°30' 33"	-375' -114.3 m	-524' -159.7 m	-722' -220.1 m	+50' +15.2 m	Not reached	D	Annular trough	14

Table 1.2. (cont.)

#	Name/ Location	Year Drilled	Latitiude/ Longitude	Top Chickahominy	Top Exmore/ Mattaponi	Total Depth	Well Head Elevation	Basement Elevation/ Lithology	Cored or Drilled	Relative Location	Reference
181	# 83 U.S. Naval Supply Center	?	37°13' 04" 76°29' 19"	Not reached?	Not reached?	-396' -120.7 m	+50' +15.2 m	Not reached	D	Annular trough	3
182	# 84 U.S. Navy Tank Farm	?	37°12' 51" 76°27' 08"	-427' -130.1 m	Not reached?	-440' -134.1 m	+10' +3.1 m	Not reached	D	Annular trough	3
183	# 41 York County	1943	37°11' 58" 76°28' 13"	-410' -125.0 m	Not reached?	-431' -131.4 m	+51' +15.5 m	Not reached	D	Annular trough	14
184	# 90 Charles City County	?	37°18' 45" 76°56' 13"	?	?	-303' -92.4 m	+35' +10.7 m	Not reached	D	Washback channel?	3
185	# 91 James City Water Service Authority	?	37°16' 10" 76°45' 43"	?	?	-307' -93.6 m	+90' +27.4 m	Not reached	D	Washback channel?	3
186	# 92 James City Water Service Authority	?	37°16' 25" 76°46' 20"	?	?	-220' -67.1 m	+90' +27.4 m	Not reached	D	Washback channel?	3
187	# 93 Powhatan Village Corp., E. of Chick. R.	?	37°16' 04" 76°52' 24"	?	?	-414' -126.2 m	+32' +97.5 m	Not reached	D	Washback channel?	3
188	# 94 Powhatan Village Corporation	?	37°21' 45" 76°49' 32"	?	?	-726' -221.3 m	+84' +25.6 m	Not reached	D	Washback channel?	3
189	# 95 James City Water Service Authority	?	37°21' 48" 76°46' 10"	?	?	-736' -224.3 m	+100' +30.1 m	Not reached	D	Washback channel?	3
190	# 96 James City Water Service Authority	?	37°18' 37" 76°47' 41"	?	?	-742' -226.2 m	+109' +33.2 m	Not reached	D	Washback channel?	3

Table 1.2. (cont.)

#	Name/ Location	Year Drilled	Latitiude/ Longitude	Top Chickahominy	Top Exmore/ Mattaponi	Total Depth	Well Head Elevation	Basement Elevation/ Lithology	Cored or Drilled	Relative Location	Reference
191	# 97 James City Water Service Authority	?	37°22' 01" 76°46' 17"	?	?	-188' -57.3 m	+112.5' +34.3 m	Not reached	D	Washback channel?	3
192	# 98 James City Water Service Authority	?	37°21' 48" 76°46' 10"	?	?	-200' -61.0 m	+100' +30.1 m	Not reached	D	Washback channel?	3
193	# 99 Eastern State Hospital	?	37°17' 49" 76°44' 18"	?	?	-501' -152.7 m	+90' +27.4 m	Not reached	D	Washback channel?	3
194	# 100 Carven Gardens	?	37°15' 12" 76°39' 24"	?	?	-501' -152.7 m	+90' +27.4 m	Not reached	D	Washback channel?	3
195	# 101 James River Estates	?	37°15' 38" 76°40' 06"	?	?	-422' -128.6 m	+80' +24.4 m	Not reached	D	Washback channel?	3
196	# 102 Ewell	?	36°19' 34" 76°44' 14"	?	?	-330' -100.6 m	+100' +30.1 m	Not reached	D	Washback channel?	3
197	# 103 Williamsburg Lodge	?	37°16' 05" 75°42' 03"	?	?	-430' -131.1 m	+70' +21.3 m	Not reached	D	Washback channel?	
198	# 104 Williamsburg Motor House	?	37°15' 56" 76°41' 51"	?	?	-445' -135.6 m	+55' +16.8 m	Not reached	D	Washback channel?	3
199	# 20 Camp Peary	1942	37°19' 25" 75°39' 13"	-264' -80.5 m	-305' -93.0 m	-393 -119.8 m	+41' +124.5 m	Not reached	D	Washback channel?	14
200	# 26 Pennimen Fuel Depot U.S. Navy	1918	37°16' 58" 76°36' 33"	?	-410' -125.0 m	-515' -157.0 m	+20' +6.1 m	Not reached	D	Washback channel?	14
201	# 114 City of Newport News	?	37°24' 28" 76°56' 15"	?	?	-768' -234.1 m	+10' +3.1 m	Not reached	D	Washback channel?	3

Table 1.2. (cont.)

#	Name/ Location	Year Drilled	Latitiude/ Longitude	Top Chickahominy	Top Exmore/ Mattaponi	Total Depth	Well Head Elevation	Basement Elevation/ Lithology	Cored or Drilled	Relative Location	Reference
202	# 115 Southern Properties	?	37°23' 59" 76°54' 04"	?	?	-183' -55.8 m	+95' +29.0 m	Not reached	D	Washback channel?	3
203	# 116 James City County Research Station	?	37°24' 51" 76°51' 33"	?	?	-905' -275.8 m	+90' +27.4 m	Not reached	D	Washback channel?	3
204	# 117 James City Water Service Authority	?	37°23' 12" 76°48' 06"	?	?	-701' -231.7 m	+106' +32.3 m	Not reached	D	Washback channel?	3
205	# 118 Yorkview Plantation	?	37°23' 10" 76°41' 14"	?	?	-503' -153.3 m	+50' +15.2 m	Not reached	D	Washback channel?	3
206	# 119 West End Station	?	37°26' 21" 76°04' 42"	?	?	-914' -278.6 m	+10' +3.1 m	Not reached	D	Washback channel?	3
207	# 120 Gloucester	?	37°23' 31" 76°31' 26"	?	?	-1775' -541.0 m	+75' +22.9 m	Not reached	D	Breccia appron	3
208	# 122 West Point-Chesapeake Corporation	?	37°32' 46" 76°48' 30"	?	?	-1252' -381.6 m	+27' +8.2 m	Not reached	D	Washback channel?	3
209	# 123 West Point-Chesapeake Corporation	?	37°31' 26" 76°45' 41"	Absent	Absent	-1255' -382.5 m	+15' +4.6 m	Not reached	D	Outside crater	3
210	# 124 Chesapeake Corporation	?	37°30' 08" 76°42' 56"	?	?	-100' -30.1 m	+51' +15.5 m	Not reached	D	Washback channel?	3
211	# 125 Barnhardt Farms	?	37°36' 30" 76°31' 26"	?	?	-702' -214.0 m	+40' +12.2 m	Not reached	D	Breccia apron	3

Table 1.2. (cont.)

#	Name/ Location	Year Drilled	Latitiude/ Longitude	Top Chickahominy	Top Exmore/ Mattaponi	Total Depth	Well Head Elevation	Basement Elevation/ Lithology	Cored or Drilled	Relative Location	Reference
212	# 126 Rappahannock Community College	?	37°33' 52" 76°37' 28"	Absent	Absent	-590' -197.8 m	+110' +33.5 m	Not reached	D	Outside crater	3
213	# 129 West Irvington Well # 2	?	37°39' 41" 76°25' 48"	Absent	Absent	-655' -199.6 m	+15' +4.6 m	Not reached	D	Outside crater	3
214	# 130 Town of Kilmarnock Well # 3	?	37°42' 12" 76°23' 09"	Absent	Absent	-707' -215.5 m	+65' +19.8 m	Not reached	D	Outside crater	3
215	# 131 USGS Essex Mill Pond Corehole	?	37°52' 30" 76°51' 04"	Absent	Absent	-214' -65.2 m	+11' +3.4 m	Not reached	D	Outside crater	3
216	# 134 USGS Clarks Mill Pond Corehole	?	37°55' 52" 75°28' 05"	Absent	Absent	-299' -91.1 m	+46' +14.0 m	Not reached	D	Outside crater	3
217	# 137 Waller Mill Park	?	37°18' 59" 76°42' 04"	?	?	-435' -132.6 m	+70' +21.3 m	Not reached	D	Washback channel?	3
218	# 40 Bacons Castle Estate	1917	37°06' 33" 75°43' 22"	?	-296' -90.2 m	-348' -106.1 m	+19' +5.8 m	Not reached	D	Washback channel?	12
219	# 8 King William County Aylett	pre 1957	~37°37' ~77°06'	-40' -12.2 m	-136' -41.5 m	-350' -106.7 m	+35' +10.7 m	Not reached	D	Washback channel?	14
220	# 9 King William County Aylett Mill	pre 1957	~37°46' ~77°06'	-125' -38.1 m	-193' -58.8 m	-278' -84.7 m	+22 +6.7 m	Not reached	D	Washback channel?	14

Table 1.2. (cont.)

#	Name/ Location	Year Drilled	Latitiude/ Longitude	Top Chickahominy	Top Exmore/ Mattaponi	Total Depth	Well Head Elevation	Basement Elevation/ Lithology	Cored or Drilled	Relative Location	Reference
221	# 14 King William County Court House	pre 1957	~37°41' ~77°01'	-50' -15.2 m	-161' -49.1 m	-289' -88.1 m	+140' +42.7 m	Not reached	D	Washback channel?	14
222	# 34 King William County, Cohoke	1946	~37°35' ~76°57'	-114.5' -34.9 m	-212' -64.6 m	-569' -173.4 m	+6' +1.8 m	Not reached	D	Washback channel?	14
223	# 48 King William County, Grimes Landing	1950	~37°39' ~77°06'	-105' -32.0 m	-124' -37.8 m	-190' -57.9 m	+40' +12.2 m	Not reached	D	Washback channel?	14
224	# 49 King William County, Rumford	1950	~37°44' ~77°05'	-72' -22.0 m	-127' -38.7 m	-199' -60.6 m	+128' +39.0 m	Not reached	D	Washback channel?	14
225	# 59 King William County, Manquin	pre 1957	~37°43' ~77°09'	+10' +3.1 m	-130' -39.6 m	-378' -115.2 m	+130' +39.6 m	Not reached	D	Washback channel?	14
226	# 66 King William County, Aylett	1951	~37°47' ~77°06'	?	-140' -42.7 m	-350' -106.7 m	+20' +6.1 m	Not reached	D	Washback channel?	14
227	# 1 King William County Walkerton	pre 1957	~37°44' ~77°01'	-110' -33.5 m	-182' -55.5 m	-365' -111.3 m	+10' +3.1 m	Not reached	D	Washback channel?	14
228	# 12 New Kent County Cumberland Landing	1946	~37°32' ~76°58'	-60' -18.3 m	-160' -48.8 m	-290' -88.4 m	+10' +3.1 m	Not reached	D	Washback channel?	14
229	# 40 New Kent County Providence Forge	1943	~37°27' ~77°03'	-42' -12.8 m	-172' -52.4 m	-253' -77.1 m	+28' +8.5 m	Not reached	D	Washback channel?	14

Table 1.2. (cont.)

#	Name/ Location	Year Drilled	Latitiude/ Longitude	Top Chickahominy	Top Exmore/ Mattaponi	Total Depth	Well Head Elevation	Basement Elevation/ Lithology	Cored or Drilled	Relative Location	Reference
230	# 34 Charles City County Charles City School	1948	~37°21' ~77°04'	-20' -6.1 m	-150' -45.7 m	-175' -53.3 m	+45' +13.7 m	Not reached	D	Washback channel?	14
231	# 75 Charles City County Charles City	1933	~37°21' ~77°04'	-20' -6.1 m	-90' -27.4 m	-175' -53.3 m	+60' +18.3 m	Not reached	D	Washback channel?	14
232	USGS-NASA Langley Corehole	2000	37°05'44" 76°23'09"	-593.4' -180.87 m	-736.4' -224.46 m	-2075.9' -632.73 m	+7.89' +2.41 m	-2046.8' -623.87 m granite	C	Annular trough	18
233	USGS North Corehole	2001	37°26'41" 76°24'02"	-495.3' -150.97 m	-724.3' -220.77 m	-1412.5 -430.53 m	+15' +4.57 m	Not reached	C	Annular trough	19
234	USGS Bayside Corehole	2001	37°19'34" 76°17'33"	-698' -212.75 m	-913' -278.28 m	-2389' -728.17 m	+2' +0.61 m	-2323.7' -708.26 m granite	C	Annular trough	19

All boreholes are in Virginia, unless designated MD (Maryland)
In depth columns 5-9, a number such as 20', indicates 20 feet
Elevations are measured above (+) or below (-) mean sea level

1 Powars et al. (1992)
2 Poag (1997a)
3 Powars and Bruce (1999)
4 Gibson and Bybell (1994)
5 Reinhardt et al. (1980)
6 Hansen and Wilson (1984)
7 Brown et al. (1972)
8 Robbins et al. (1975)
9 Hansen and Wilson (1990)
10 Cederstrom (1945a)
11 Sinnott and Tibbits (1968)
12 Cederstrom (1945b)
13 Mixon et al. (1989)
14 Cederstrom (1957)
15 Virginia Tech Geothermal Web Site
16 Hansen (1978)
17 Milici et al. (1995)
18 Powars et al. (2001)
19 Gohn (in press)

2 Geological Framework of Impact Site

2.1 Crystalline Basement Rocks

2.1.1 Regional Tectonostratigraphy

Crystalline basement rocks beneath the Virginia Coastal Plain include a variety of plutonic, volcanic, and metamorphic rocks that constitute distal parts of the Appalachian orogen (Thomas et al. 1989; Fig. 2.1). The inner edge of the coastal plain sedimentary wedge laps westward onto greenschist-facies rocks of the Eastern Slate Belt, which, in outcrop, form the Virginia segment of the Appalachian Piedmont Province. Composition of slate-belt rocks beneath the inner Virginia Coastal Plain ranges chiefly from muscovite-biotite-quartz-albite schist to fine-grained phyllitic quartzite and phyllitic biotite-muscovite-quartz rock. These metamorphic bodies are intruded by plutons of fine-grained granodiorite, diorite, gabbro, tonalite, and monzogranite (Gleason 1982). Fine-grained felsic volcanic rocks, along with thin-bedded, calcareous, distal turbidites and massive-to-laminated, very fine-grained tuff, are part of the subsurface slate-belt complex in easternmost Virginia (Robbins et al. 1975; Gleason 1979).

Rocks of the Goochland terrane (Fig. 2.1), an internal basement massif of Proterozoic age, project under the Virginia Coastal Plain, and have been generally called granite by drillers. Most of the Goochland rocks, however, are metamorphic bodies, which include the State Farm Gneiss (granodioritic to tonalitic orthogneiss; Glover et al. 1978), the Sabot Amphibolite, the Maidens Gneiss (garnet-biotite-quartz-plagioclase), and the Montpelier Metanorthosite (Clement and Bice 1982; Farrar 1984).

Principal granitic plutons drilled beneath the Virginia Coastal Plain are the Petersburg (biotite-quartz-microcline-plagioclase granite) and Portsmouth (postmetamorphic quartz-microcline-plagioclase-biotite granite) Granites. Petersburg Granite has a zircon U-Pb age of ~330 Ma (Wright et al. 1975), and Portsmouth Granite has a whole-rock Rb-Sr isochron age of ~263 Ma (Russell and Russell 1980). Both of these intrusive units are marked by conspicuous negative gravity anomalies, and several other similar negative anomalies are interpreted to represent additional granitic plutons (Thomas et al. 1989).

Horton et al. (1991) analyzed the tectonostratigraphic terranes accreted during the Paleozoic, which form the east-central sector of the Appalachian orogen (Fig. 2.1). The chief terrane in southeastern Virginia is the Chesapeake Block, a broad region represented by only sparse borehole data from beneath coastal plain sedimentary beds. Drilling in the Chesapeake Block has yielded mainly metamorphic

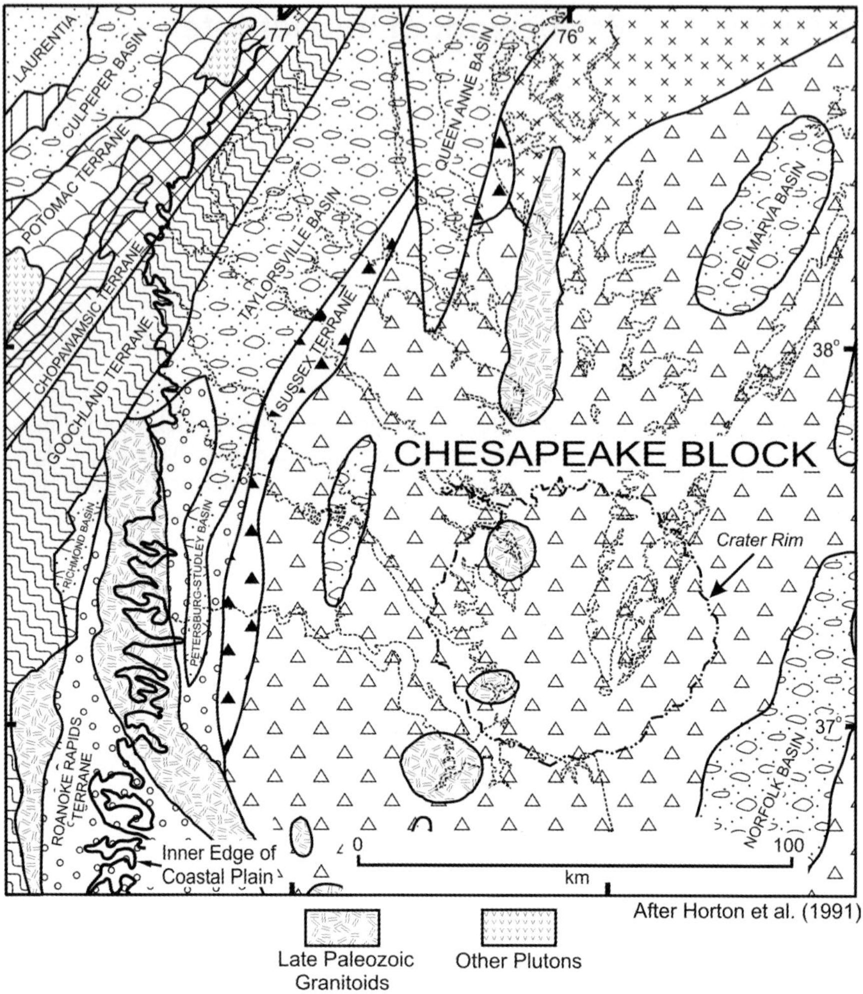

Fig. 2.1. Tectonostratigraphic terranes recognized within crystalline basement rocks underlying the Coastal Plain and Piedmont Provinces of southeastern Virginia (modified from Horton et al. 1991). Note that Chesapeake Bay impact crater occupies broad Chesapeake Block and western sector of annular trough is underlain by two prominent granitoid intrusions. Culpeper, Taylorsville, Richmond, Petersburg-Studley, Delmarva, and Norfolk basins are sediment-filled Triassic-Jurassic rift grabens (or half grabens) formed by extensional stress in Proterozoic-Paleozoic igneous and metasedimentary basement rocks.

rocks of greenschist facies (argillite, chloritic schist, phyllitic metavolcanic rock, and serpentinized gabbro). It is the Chesapeake Block into which the Chesapeake Bay impact crater was excavated. The Chesapeake Block has been intruded by several post-accretion granitoid plutons, two of which underlie the western sector of the crater (Fig. 2.1), and have been sampled by drilling (boreholes 81, 105, 232, 234; Figs. 1.3, CD-ROM.1; Table 2.1).

The assemblage of Appalachian crystalline terranes was disrupted during the Late Triassic and Early Jurassic by continental rifting, which preceded opening of the Atlantic Ocean basin (Manspeizer 1988). Four half-graben rift basins, filled with coarse siliciclastic sequences assigned to the Newark Supergroup, can be identified on seismic reflection profiles used in this study: Taylorsville basin; Queen Anne basin; Norfolk basin; and an unnamed (also not yet mapped) rift basin seen on line 11A of Milici et al. (1995), on the Northern Neck Peninsula (see Chapter 3).

Table 2.1. Nineteen most significant boreholes in southeastern Virginia and southern Maryland that penetrated crystalline basement rocks. See Fig. 2.2, Table 1.2, CD-ROM.1.

No.	Name	Elevation	Lithology	Inside Crater
7	St.M. -Df 84	-766.5 m	Diabase	No
8	Ohio Oil-Larry G. Hammond	-1655 m	Gneiss	No
9	J&J Enterprises-E.G. Taylor	-1872.7 m	Gneiss	No
24	VA-NOR-T-12	-782.4 m	Granite?	No
26	VA-IW-P-8	-252.4 m	?	No
28	VA-SO-P-3	-103.6 m	?	No
81	#46 Mathews, Elkins Oil & Gas	-706.5 m	Granite?	Yes
102	#81 Ft. Monroe	-684 m	Granite?	Yes
103	W5521	-548.6 m	?	No
104	W6839	-121.9 m	?	No
105	59-E-5	-634 m	Gneiss?	Yes
111	C-25	-550.3 m	Metavolcanics	No
112	C-26	-402.1m	Granite	No
120	DGT-1, Crisfield Airport	-1243.8 m	Metavolcanics	No
121	Socony Vacuum-Bethards	-2154.9 m	Gabbro	No
127	Md. No. 1 near Marbury	-170.7 m	Schist	No
128	Md. No. 2 Prince Georges Co.	-743.4 m	?	No
232	USGS-NASA Langley	-634 m	Granite	Yes
234	USGS Bayside	-728.17 m	Granite	Yes

2.1.2 Crystalline Basement Rocks in Boreholes

A total of 25 boreholes have sampled the crystalline basement rocks in or near the study area, five of which are located inside the crater rim (Fig. 2.2; Table 2.1). Granitoid rocks are present in all five of these basement sections. Lithologic descriptions of crystalline basement from boreholes 81, 102, and 105 came from very old driller's logs, however, and their accuracy is questionable. The only reliable samples of *in situ* basement rocks derived from inside the crater come from the NASA Langley corehole (borehole 232; Fig. 2.2; Table 1.2) and the Bayside corehole (borehole 234; Fig. 2.2; Table 2.1). At the NASA Langley site, basement rocks from the outer edge of the crater's annular trough are composed of Proterozoic metagranite (~0.6 Ga; Horton et al. 2001). Crystalline basement from near

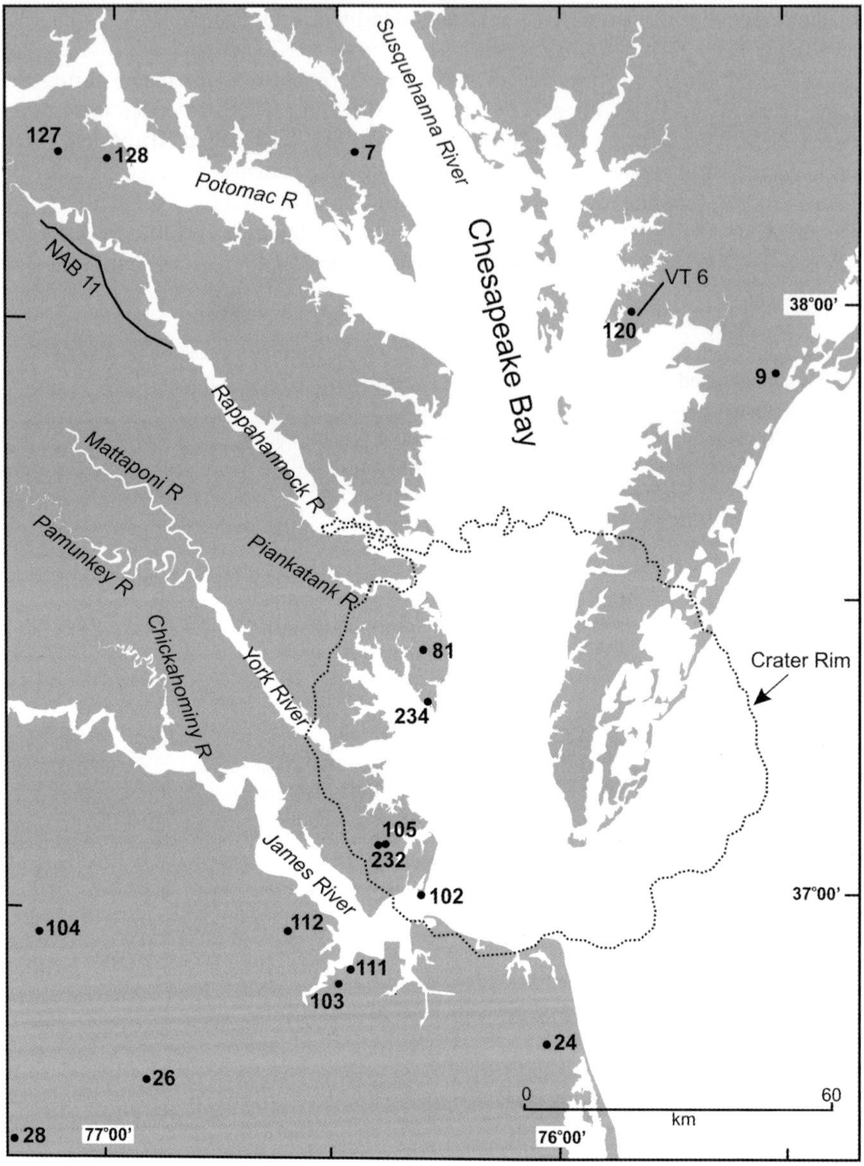

Fig. 2.2. Geographic distribution of 17 boreholes in which crystalline basement has been penetrated in southeastern Virginia and southern Maryland, and location of onshore seismic reflection profiles NAB-11 and VT-6. Boreholes 81, 105, 232, and 234 were drilled on or close to the granitoid intrusions shown in Fig. 2.1. See Table 2.1 for more information about these boreholes.

the inner edge of the annular trough at Bayside is of similar granitic composition and age as that in the NASA Langley core (Horton et al. 2002).

2.1.3 Regional Configuration of Crystalline Basement Surface

The oldest published structure map of the crystalline basement's upper surface (Brown et al. 1972; Fig. 2.3A) was contoured on the basis of only ten wells widely scattered through the study area (plus several others outside the study area). No seismic reflection profiles were available in 1972. The most prominent feature of the 1972 map is a west-trending embayment north of the Potomac River, defined by the 1000–7000-ft contours. The embayment is bounded to the south by a southeast-trending structural nose, for which there is little control. Farther south, the basement surface is essentially monoclinal. A minor structural nose defined by the 2000–2500-ft contours south of Norfolk (based on sparse well data outside the study area) was designated the Ft. Monroe high by Richards and Straley (1953). Gibson (1967) called this structure the Norfolk arch. In the northeast corner of the study area, the basement gradient steepens into the deepest part of the Salisbury embayment, but little was known at the time about the offshore geology.

By 1978, 11 more basement wells had been drilled in the study area (total of 21; Fig. 2.3B), and several additional ones were drilled outside the area. The additional wells, supplemented by eight composite seismic reflection profiles and one seismic refraction station, allowed Hansen (1978) to revise the structure map (Fig. 2.3B). The revision steepened the gradient of the basement slightly. For example, the northern segment of the 3000-ft contour was shifted westward, so that it crosses the western bayshore north of the Potomac River, but the major features are virtually unchanged from Brown et al.'s (1972) version. The northern embayment is nearly identical, but the adjacent nose is less prominent, and its axis is shifted southward by 10–15 km. The "Norfolk arch" is hardly detectable, indicated only by a small eastward deflection of the 2000-ft contour. The steepened gradient in the northeast corner of the study area is indicated by much straighter contours than on the Brown et al. (1972) map, though based on the same well data.

Twenty years after Brown et al.'s (1972) map was published, Powars et al. (1992) used a fourfold increase in well control (total of 43 wells in the study area and several more outside it) to upgrade the basement map (Fig. 2.3C). Powars et al. flattened the gradient in the north, as seen in the eastward shift in the 3000-ft contour. The axis of the northern embayment is moved somewhat farther north, and the adjacent nose is shifted another ~5 km to the south from Hansen's (1978) placement. The steep-gradient contours in the northeast corner have been almost completely straightened and are extended southwestward onto the continental shelf. Powars et al. (1992) used the increased well control south of Norfolk to accentuate the "Norfolk arch" into a distinct east-trending structural nose defined by the 2000–3000-ft contours. Accentuation of the "arch" produced a small, adjacent, parallel embayment to the north.

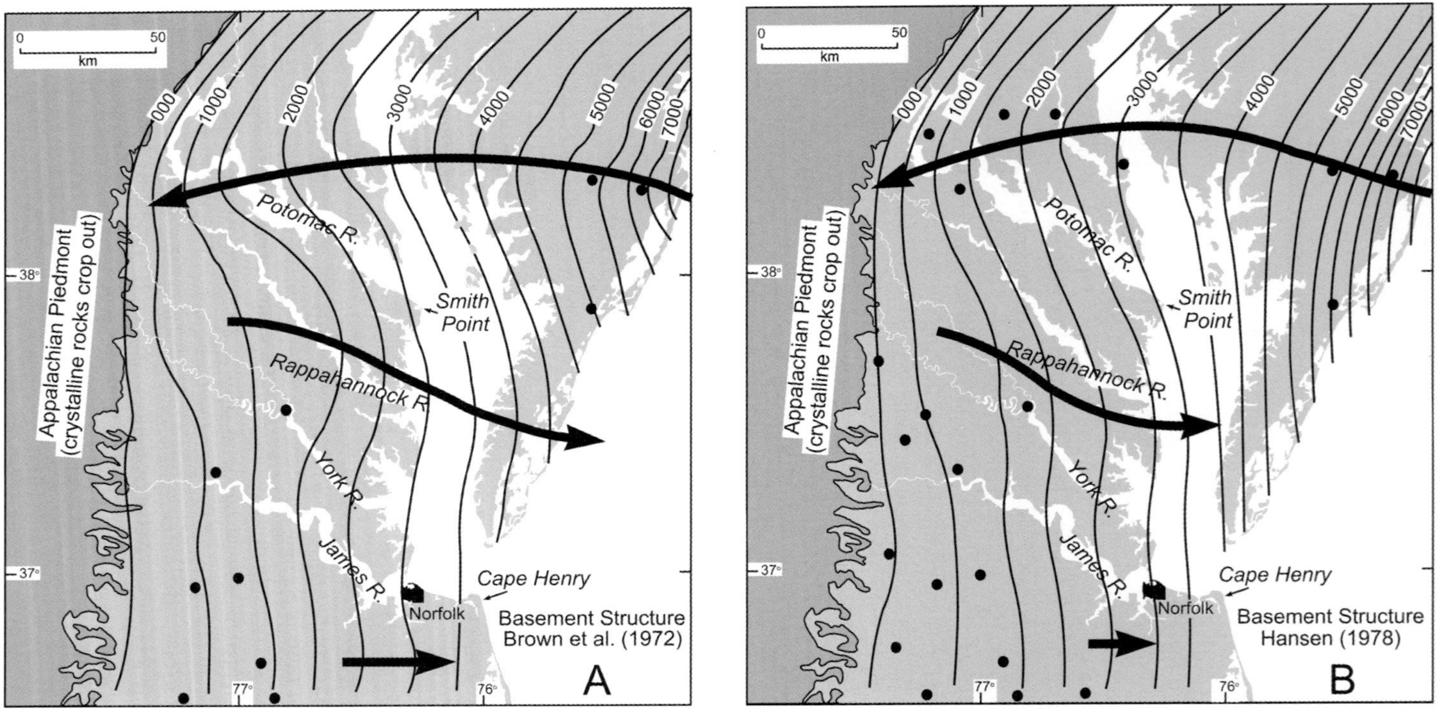

Fig. 2.3A,B. Two earlier published versions of structure of crystalline basement for southeastern Virginia and southern Maryland, showing how interpretation of basement geometry changed as data proliferated through six-year period (1972–1978). Contour interval 500 ft; solid circles designate boreholes that penetrated crystalline basement rocks. Arrows indicate main structural axes. See text for discussion of principal features.

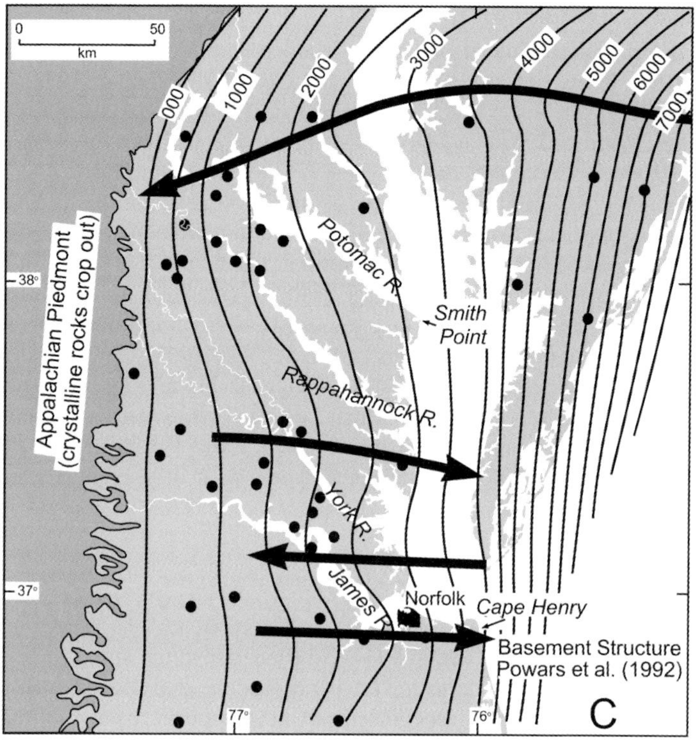

Fig. 2.3C. A 1992 published version of structure of crystalline basement for southeastern Virginia and southern Maryland, showing how interpretation of basement geometry changed as data proliferated through fourteen-year period (1978–1992). Contour interval 500 ft; solid circles designate boreholes that penetrated crystalline basement rocks. Arrows indicate main structural axes. See text for discussion of principal features.

2.2 Coastal Plain Sedimentary Rocks

2.2.1 General Stratigraphic Framework

Sedimentary deposits of the Atlantic Coastal Plain constitute a seaward-thickening wedge of dominantly unconsolidated to poorly consolidated siliciclastic sands, silts, and clays of both marine and nonmarine origin. The deposits range in age from Early Cretaceous to Holocene, and generally comprise the updip lithofacies of an ~18-km-thick column of sediments that fills the southern end of an elongate offshore basin known as the Baltimore Canyon trough (Maher 1965; Poag 1985; Grow et al. 1988; Poag and Ward 1993). In Virginia, a hierarchy of 27 formal formations has been developed for these deposits over a period of more than 100 years (Darton 1891; Clark 1896; Cederstrom 1945a,b,c, 1957; Richards 1945, 1967; Cushman and Cederstrom 1949; Bennett and Collins 1952; Otton 1955; Murray 1961; Brown et al. 1972; Oaks and Coch 1973; Tiefke 1973; Ward and Blackwelder 1980; Gibson 1983; Ward and Krafft 1984; Mixon 1985; Owens and Gohn 1985; Ward and Strickland 1985; Meng and Harsh 1988; Powars et al. 1992; Poag and Ward 1993; Powers and Bruce 1999; Powars 2000; Figs. 2.4–2.7).

Modern emphasis on unconformity-bounded depositional units (Vail et al. 1977; North American Commission on Stratigraphic Nomenclature 1983) has resulted in additional formalization of alloformations, which can be traced across the entire US Atlantic margin north of Cape Hatteras (onshore and offshore), including the outcrops and subsurface beds of southeastern Virginia (Poag and Ward 1993; Poag and Commeau 1995; Figs. 2.4, 2.5). Almost no sequence-stratigraphic analyses exist, however, for the Virginia Coastal Plain. The single published study identified 15 depositional units that correlate with third-order sequences of the Exxon model (Poag and Commeau 1995; Fig. 2.6). Virginia systems tracts are dominated by transgressive and highstand types, with little or no record of lowstand deposits.

2.2.2 Preimpact Deposits

Preimpact sediments of the Virginia Coastal Plain constitute an ~1 to 1.5-km-thick interval of Lower Cretaceous to lowest upper Eocene deposits. Seven preimpact coastal plain formations are recognized on the composite basis of lithology and biozonation (Fig. 2.4).

2.2.2.1 Potomac Formation

Oldest outcropping sedimentary strata in the Virginia Coastal Plain are mainly nonmarine, quartz sand- and silt-dominated lithofacies of the Potomac Formation (Ward and Kraft 1984; Mixon et al. 1989; Powars and Bruce 1999; Powars 2000; Fig. 2.4). Potomac beds rest unconformably on metasedimentary and metaigneous rocks of the crystalline basement, and can be examined in outcrop along their updip contact with the Appalachian Piedmont. The Potomac Formation extends into the subsurface, and underlies the entire study area, except where excavated by the Chesapeake Bay bolide impact. Potomac strata constitute the oldest sediments above crystalline basement rocks, except for a possibly Triassic or Jurassic subsurface unit (Crisfield unit) inferred by some investigators (Hansen 1978; Dysart et al. 1983; Hansen and Wilson 1984) from seismic profiles. The Potomac Formation is by far the thickest sedimentary unit recognized on the Virginia Coastal Plain, and reaches as much as 1.3 km in the J&J Enterprises, Taylor # 1 well on the Delmarva Peninsula (borehole 9; Figs. 1.3, CD-ROM.1; Table 1.2). From a thickness of ~600 m at the western margin of the crater, the Potomac thickens to >1 km at the eastern (downdip) margin (Grow et al. 1988). No allostratigraphic units have yet been proposed for US East Coast Lower Cretaceous units.

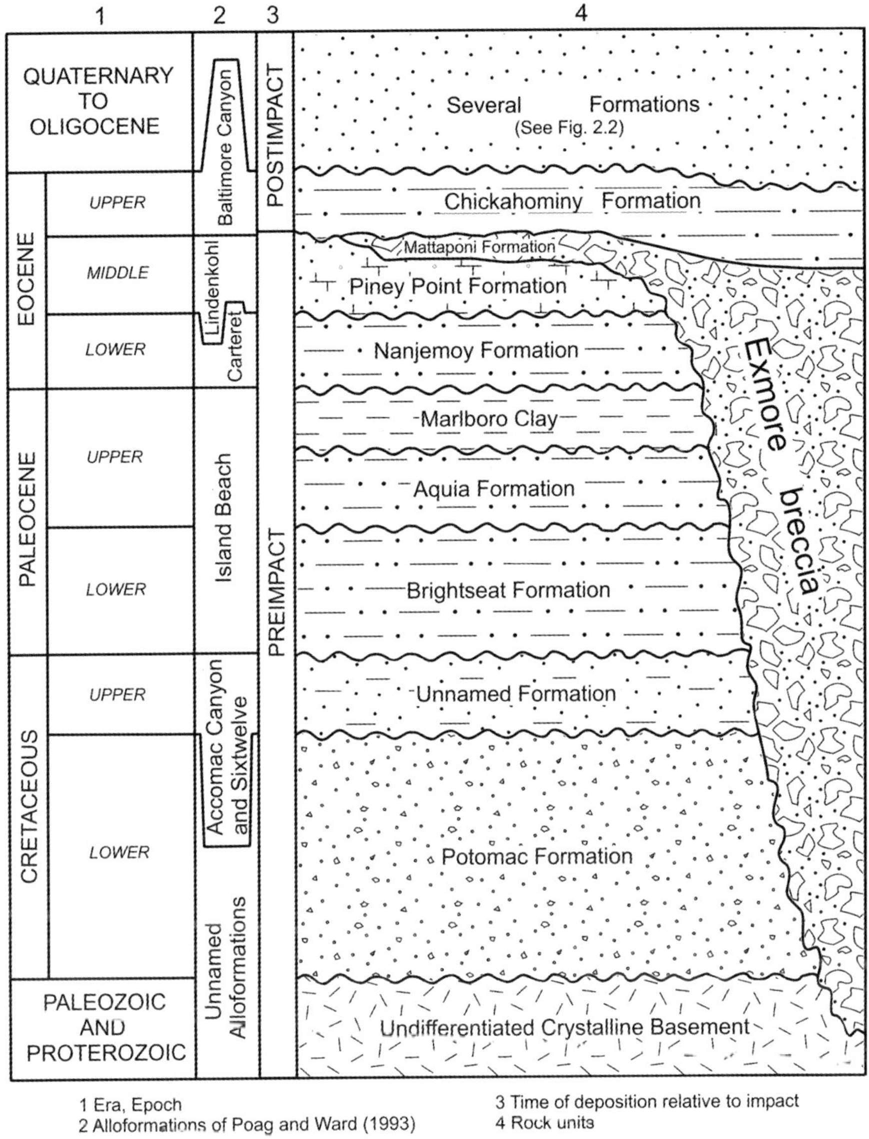

Fig. 2.4. General stratigraphic succession of crystalline and mainly preimpact sedimentary formations in southeastern Virginia, showing their relation to each other and to impact deposits. Undulating horizontal lines indicate unconformities.

2.2.2.2 Unnamed Upper Cretaceous Beds

Upper Cretaceous beds have not been found cropping out in the Virginia Coastal Plain. A few downdip wells have encountered Late Cretaceous (Cenomanian and Campanian) megafossils and microfossils, however, in nonmarine (red beds), deltaic (micaceous, lignitic, glauconitic, quartz sand), and marine beds (shelly, glauconitic silt, clay, and quartz sand) ranging from 40 to 110 m in thickness (Clark and Miller 1912; Powars et al. 1992; Powars and Bruce 1999; Powars 2000). Poag and Ward (1993) included some beds of this interval in the Sixtwelve Alloformation (Fig. 2.4).

2.2.2.3 Brightseat Formation

Oldest Cenozoic deposits of the Virginia Coastal Plain belong to the lower Paleocene Brightseat Formation (Bennett and Collins 1952), a dominantly subsurface unit consisting of mainly clayey, sparsely glauconitic, quartz sand (Fig. 2.4). In outcrop, Brightseat beds are confined to the northeastern part of the state, and are known only along the Potomac and Rappahannock Rivers (Ward 1984), but equivalent strata have been reported in the subsurface from the Oak Grove corehole (borehole 6; Figs. 1.3, CD-ROM.1; Table 1.2; Reinhardt et al. 1980) and the Dismal Swamp corehole (borehole 118, Figs. 1.3, CD-ROM.1; Table 1.2; Powars et al. 1992; Powars and Bruce 1999; Powars 2000). These two coreholes are separated by a distance of 180 km. The Brightseat is included in the Island Beach Alloformation of Poag and Ward (1993) (Fig. 2.4).

2.2.2.4 Aquia Formation

Clayey, silty, shell-rich, glauconitic, quartz sands of the upper Paleocene Aquia Formation (Clark and Martin 1901; Figs. 2.4, 2.6) crop out in river banks in a continuous arc from north of Baltimore on upper Chesapeake Bay to around Hopewell, Virginia, on the James River (Ward 1984). Equivalent beds are present throughout the subsurface (Powars et al. 1992; Powars and Bruce 1999; Powars 2000), and are included in the Island Beach Alloformation (Poag and Ward 1993; Fig. 2.4).

2.2.2.5 Marlboro Clay

A second upper Paleocene unit is the Marlboro Clay (Clark and Martin 1901; Glaser 1971; Fig. 2.4). The Marlboro Clay is present at scattered outcrops on the western side of Chesapeake Bay from southern Maryland to the James River in Virginia (Ward 1984), and is widespread in the subsurface (Powars et al. 1992; Powars and Bruce 1999; Powars 2000). The thin, silver-gray to pale red plastic clays, interbedded with yellowish-gray to reddish silts of the Marlboro, are part of the Island Beach Alloformation of Poag and Ward (1993) (Fig. 2.4).

2.2.2.6 Nanjemoy Formation

Oldest Eocene strata in Virginia belong to the lower Eocene Nanjemoy Formation (Clark and Martin 1901; Figs. 2.4, 2.6). Glauconitic sands with variable amounts of clay and silt characterize the Nanjemoy in outcrop (Ward 1984) and subsurface occurrences (Powars et al. 1992; Powars and Bruce 1999; Powars 2000), which extend throughout the Virginia Coastal Plain. The Nanjemoy Formation was included by Poag and Ward (1993) in the Carteret Alloformation (Fig. 2.4).

2.2.2.7 Piney Point Formation

The olive gray, clayey, poorly sorted, glauconitic, fossil-rich sand of the Piney Point Formation represents middle Eocene deposition. This formation was originally described from rotary cuttings derived from a well on Piney Point, Maryland, on the Potomac River (Otton 1955). The Piney Point characteristically contains thick beds and lenses dominated by rich accumulations of oyster shells [*Cubitostrea sellaeformis* (Conrad 1832)], which are cemented into concrete-hard layers of glauconitic, bioclastic limestone. The Piney Point is best exposed on the Pamunkey River, a tributary of the York River (Virginia), but also occurs along the James River and in the subsurface (Ward 1984; Powars et al. 1992; Powars and Bruce 1999; Powars 2000). The Piney Point Formation is included in the Lindenkohl Alloformation of Poag and Ward (1993) (Fig. 2.4).

2.2.2.8 Unnamed Upper Eocene Deposits

The bulk of late Eocene deposition in the crater is represented by the Exmore breccia and the postimpact Chickahominy Formation (Fig. 2.4). The presence of late Eocene foraminifers and calcareous nannofossils within the Exmore breccia, however, indicates that some late Eocene deposits of unknown lithology were already present in the target area prior to impact. We speculate that these early late Eocene deposits were marine clays similar to those of the basal Chickahominy Formation.

2.2.3 Postimpact Deposits

Marine sedimentation resumed at the impact site immediately following deposition of the impact-generated Exmore breccia and its capping layer of fallout debris (see Chapter 6 for detailed discussion of Exmore breccia and fallout layer). The crater is now covered by 200–550 m of postimpact sediments, principally siliciclastic silts and sands of marine origin. The postimpact sedimentary column is thickest over the crater, because of increased accommodation space produced by compaction and subsidence of the water-saturated breccia (Poag 1997a). Twenty postimpact formations are formally recognized in southeastern Virginia (Figs. 2.4–2.6).

2.2.3.1 Chickahominy Formation

The upper Eocene Chickahominy Formation (Figs. 2.4–7) lies above the Exmore breccia at all the sites cored within the crater. The Chickahominy is an entirely subsurface unit, which extends only a few kilometers outside the crater. Using cable-tool cuttings from the type well, in York County, Virginia (borehole 89; Figs. 1.3, CD-ROM.1; Table 1.2), Cushman and Cederstrom (1949) originally described the Chickahominy as dominantly blue, brown, and dull gray clays with abundant glauconite. In the seven continuous coreholes drilled in and near the crater, however, the Chickahominy Formation is composed mainly of hard, massive to laminated, silty to sandy, highly fossiliferous, greenish-gray marine clay, containing variable amounts of finely comminuted glauconite and mica. Silt-filled, sand-filled, and pyrite-filled burrows are common in the upper and lower few meters of the formation. Cores and seismic reflection profiles indicate that the thickness of the Chickahominy varies considerably (20–220 m) over the crater, compared to the more uniform thickness of most other postimpact units, and represents a deep-water (outer neritic to upper bathyal) basin-fill deposit. The presence of relatively deep-water microfaunas in its updip occurrences indicates that the Chickahominy Formation may have been originally much more widespread, and that its landward margin has been partly eroded during repeated Cenozoic lowstands of sea level (see Chapter 7 for more details regarding the Chickahominy Formation). Poag and Ward (1993) included the Chickahominy Formation as part of the Baltimore Canyon Alloformation (Fig. 2.4).

2.2.3.2 Delmarva Beds

The name Delmarva beds is an informal designation for lower Oligocene, micaceous, clayey, silty, glauconitic sands, which have been cored inside the Chesapeake Bay crater (boreholes 1, 2, 232, 233, 234; Figs. 1.3, CD-ROM.1; Table 1.2) and in the Fentress corehole, south of the crater (borehole 117; Figs. 1.3, CD-ROM.1; Table 1.2; Powars et al. 1992; Powars and Bruce 1999; Powars 2000; Powars et al. 2001; Gohn in press). The Delmarva unit is the only lower Oligocene deposit known in Virginia, and is known only in the subsurface, in and near the crater. The Delmarva beds represent the first downlapping, coarse-grained siliciclastic deposits that began to fill bathymetric lows created by the crater depression and its subsiding breccia fill. Because of this infilling, the Delmarva beds vary widely in thickness. Maximum cored thickness is 7.28 m (23.9 ft), but the unit reaches an estimated maximum of ~20 m over the inner basin of the crater. The Delmarva beds are part of the Baltimore Canyon Alloformation (Poag and Ward 1993; Figs. 2.5, 2.6).

2.2.3.3 Old Church Formation

Upper Oligocene glauconitic sands of the Old Church Formation crop out in Virginia along the Pamunkey River (its type section), a tributary to the York River, and at two other quarry locations (Ward 1984, 1985; Fig. 2.2; CD-ROM.1). The

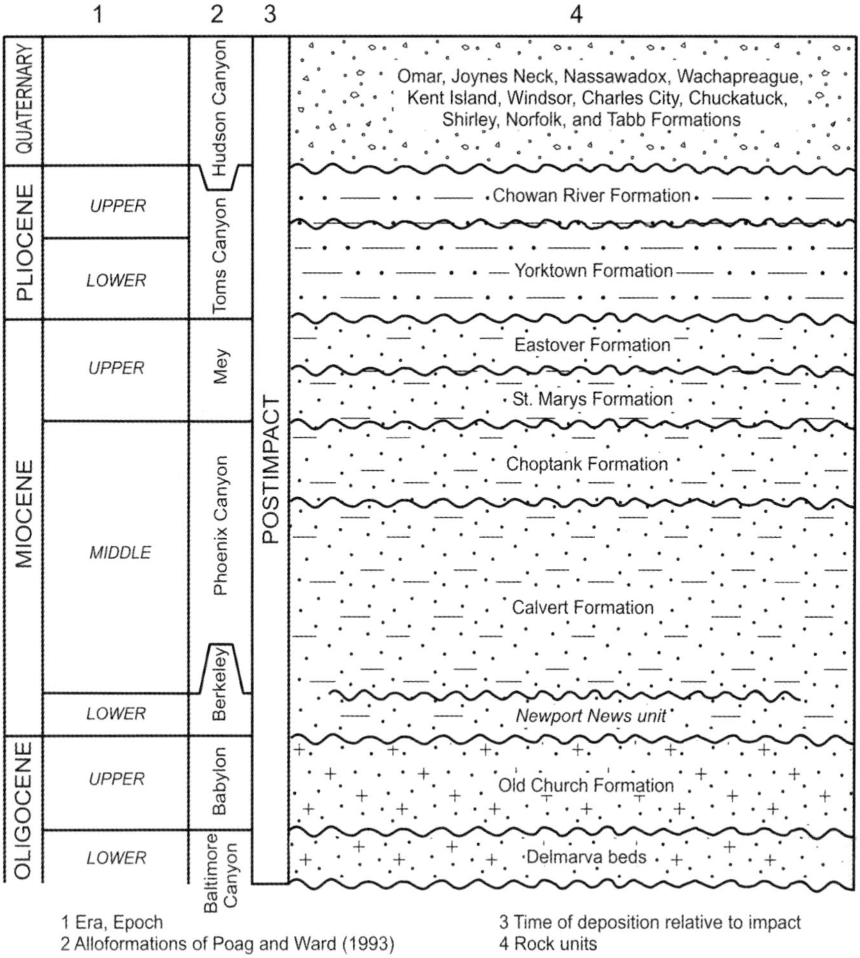

Fig. 2.5. General stratigraphic succession of postimpact deposits younger than the Chickahominy Formation in southeastern Virginia. Undulating horizontal lines indicate unconformities.

Old Church also has been sampled from the subsurface in the Exmore, Kiptopeke, NASA Langley, North, and Bayside coreholes (boreholes 1, 2, 232, 233, 234; Figs. 1.3, CD-ROM.1; Table 1.2) and at a few other sites (Powars et al. 1992; Powars and Bruce 1999; Powars 2000; Powars et al. 2001; Gohn in press). Poag and Ward (1993) included the Old Church as part of the Babylon Alloformation (Figs. 2.5, 2.6).

2.2.3.4 Calvert Formation

Oldest Miocene strata in Virginia are included in the thick, richly fossiliferous, silty, fine sands of the Calvert Formation (Shattuck 1902, 1904; Clark and Miller 1912; Ward and Blackwelder 1976; Ward 1984). Calvert strata crop out widely in Virginia riverbanks, and are well known from many subsurface locations, as well (Ward 1992; Powars and Bruce 1999; Powars 2000). At least three different members are recognized from outcrops, and can be distinguished as distinct depositional units on seismic reflection profiles. The lower part of the Calvert is of early Miocene age, and has been informally designated the Newport News unit in its subsurface expression (Powars and Bruce 1999; Powars et al. 2001; Gohn in press). The Newport News unit is included in the Berkeley Alloformation of Poag and Ward (1993; Fig. 2.5). The upper part of the Calvert Formation, on the other hand, is of middle Miocene age, and has been assigned to the Phoenix Canyon Alloformation (Poag and Ward 1993; Fig. 2.5).

2.2.3.5 Choptank Formation

Middle Miocene strata also are represented by the sandy, shell-rich Choptank Formation (Shattuck 1902, 1904; Ward 1984). The Choptank crops out in a more restricted region than the Calvert Formation, mainly from the Rappahannock River northward, and is poorly known in the subsurface (Powars et al. 1992; Powars and Bruce 1999; Powars et al. 2001). Poag and Ward (1993) included the Choptank Formation in the Phoenix Canyon Alloformation (Fig. 2.5).

2.2.3.6 St. Marys Formation

Upper Miocene strata referable to the St. Marys Formation crop out in Virginia from the Mattaponi River northward (Ward 1984, 1992). The St. Marys, represented by dominantly silty clays, silty shelly clays, and shelly sands, also is widespread in the subsurface (Powars et al. 1992; Powars and Bruce 1999; Powars 2000; Powars et al. 2001). Poag and Ward (1993) included the St. Marys Formation in the Mey Alloformation (Fig. 2.5).

2.2.3.7 Eastover Formation

Additional upper Miocene strata are included in the sandy Eastover Formation (Ward and Blackwelder 1980), which crops out widely over the Virginia Coastal Plain (Ward 1984, 1992), and also is well known from the subsurface (Powars et al. 1992; Powars and Bruce 1999; Powars 2000). The Eastover Formation was included in the Mey Alloformation by Poag and Ward (1993) (Fig. 2.5).

2.2.3.8 Yorktown Formation

Early and early late Pliocene deposition in Virginia is represented by shelly, clayey, phosphatic sands and silty, clayey, very fine sands assigned to the Yorktown Formation (Clark and Miller 1906, 1912; Mansfield 1928; Johnson and Goodwin 1969; Ward and Blackwelder 1980; Ward 1984), which crops out widely south of the Rappahannock River, and is widely distributed in the subsurface (Powars et al. 1992; Powars and Bruce 1999; Powars 2000; Powars et al. 2001). Nonmarine equivalents of the Yorktown are present as far north as the Potomac River. The Yorktown Formation is included in the lower part of the Toms Canyon Alloformation of Poag and Ward (1993; Fig. 2.5).

2.2.3.9 Chowan River Formation

Shelly, silty sands and crossbedded sands and silts of late Pliocene age are exposed in borrow pits in Newport News, Norfolk, and Chesapeake in southern Virginia, and also are known from the subsurface in that region (Powars et al. 1992; Powars and Bruce 1999; Powars 2000). These strata are assigned to the Chowan River Formation (Blackwelder 1981). Poag and Ward (1993) included the Chowan River Formation in the upper part of the Toms Canyon Alloformation (Fig. 2.5).

2.2.3.10 Quaternary Formations

A variety of alluvial, estuarine, and back-barrier deposits of Quaternary age constitute the surficial and shallow subsurface strata of the Virginia Coastal Plain (Coch 1968; Bick and Coch 1969; Oaks and Coch 1973; Johnson 1976; Mixon 1985; Mixon et al. 1989; Powars et al. 1992; Powars and Bruce 1999; Powars 2000). Crossbedded sands, gravels, cobbles, silty sands, shelly sands, and sands rich in organic matter are widespread around the bay margin, but good exposures are limited mainly to borrow pits. Eleven formations of Quaternary age in Virginia (Omar Formation, Joynes Neck Sand, Nassawadox Formation, Wachapreague Formation, Kent Island Formation, Windsor Formation, Charles City Formation, Chuckatuck Formation, Shirley Formation, Norfolk Formation, Tabb Formation) are included in the Hudson Canyon Alloformation of Poag and Ward (1993) (Fig. 2.5).

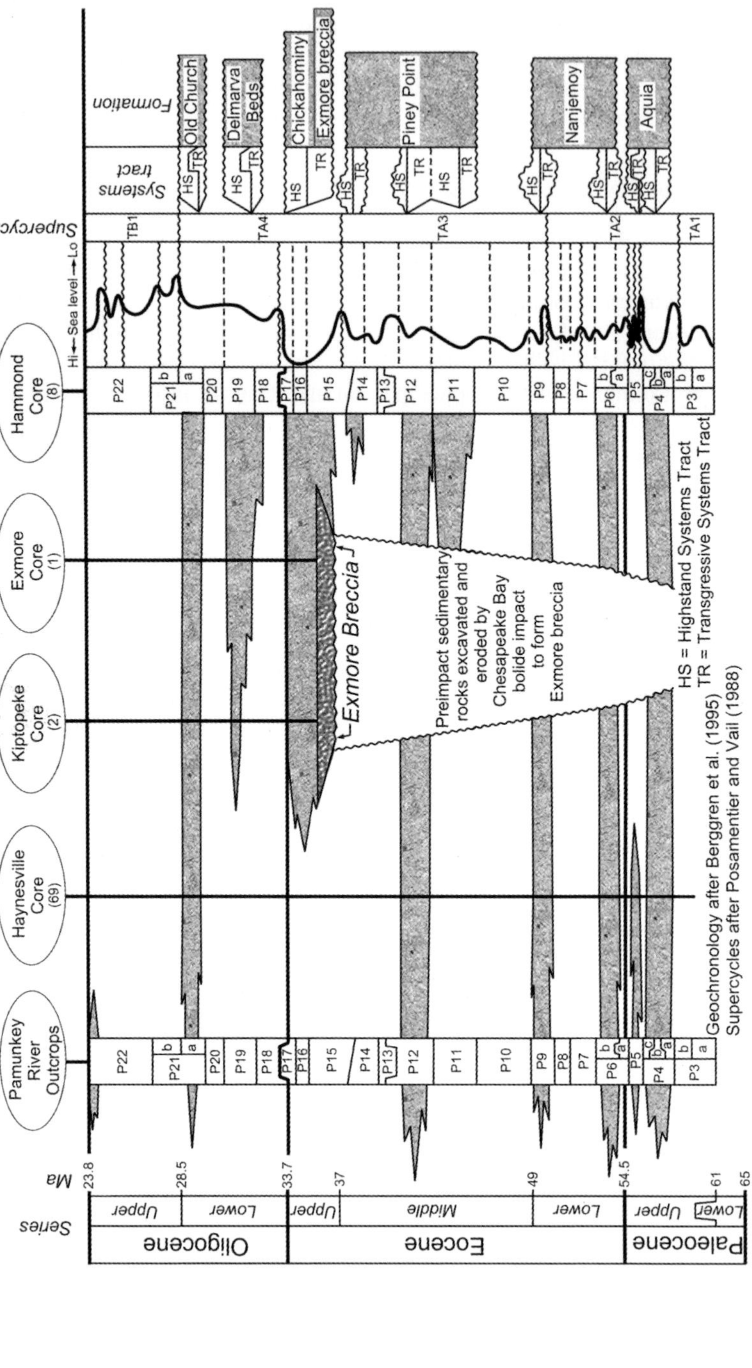

Fig. 2.6. Interpreted sequence stratigraphy for Paleogene formations of southeastern Virginia and southern Maryland (modified from Poag and Commeau 1995). Numbers in parentheses under core labels refer to boreholes shown in Fig. 1.3 and Table 1.2. See Fig. 2.7B for location of section.

2.3 Sequence Stratigraphy

Poag and Commeau (1995) showed that each normal Cenozoic depositional sequence in the Virginia Coastal Plain is bounded by an erosional unconformity (see also Ward and Strickland 1985; Figs. 2.4–6). During each hiatus, erosion removed all (or most) alluvial and fluvial sediments that may have accumulated during sea-level lowstands (Ward and Krafft 1984). In combination with subsequent shoreface erosion and ravinement during the following sea-level rise, each sequence boundary has become a composite surface, which separates marine facies below from marine facies above (Darby 1984; Kidwell 1984). In other words, only highstand and transgressive systems tracts are represented in the preserved record, and lowstand systems tracts are missing. Sea level underwent a series of short-lived eustatic fluctuations during the late Eocene, but on the U.S. Atlantic margin, late Eocene sea levels were higher than at any other time in the Cenozoic (Poag and Ward 1993; Poag and Sevon 1989; Fig. 2.6). When the Chesapeake Bay bolide struck, during the middle part of supercycle TA3 of the Exxon sequence stratigraphic model (Posamentier and Vail 1988; Fig. 2.6), a marine transgression was underway, and relatively deep water (~300 m) covered the impact site (see Chapter 13 for further documentation of target-site paleodepths).

Though the Exmore breccia is a major depositional unit, bounded below by a remarkable unconformity (maximum hiatus of >500 myr), that unconformity is a local feature, which does not constitute a sequence boundary in the sense of the Exxon model. In a broad sense, the Exmore breccia is merely an unusual lithofacies deposited in the initial stages of an early Eocene transgressive systems tract. Following emplacement of impact-related deposits and accumulation of a thin (19 cm) dead zone (see Chapters 7, 13), normal marine deposition resumed at the impact site, and the silty clays of the Chickahominy Formation began to accumulate in deep water as the marine transgression continued. The upper part of the Chickahominy Formation represents the subsequent highstand systems tract. The lower boundary of the Chickahominy Formation appears to be a conformable surface in some parts of the crater, but the upper boundary is an unconformable sequence boundary. The unconformity formed during an early Oligocene regression, the subsequent lowstand, and the following marine transgression.

2.4 Paleogeography of Impact Site

The general paleogeographic-paleoceanographic setting of ground zero at the time of impact can be estimated from the fossil record (mainly microfossils) contained in the Exmore breccia and in the overlying Chickahominy Formation. The oldest planktonic and benthic foraminifera and calcareous nannofossils in the Chickahominy Formation, which began to collect on the seafloor <1–3 kyr after impact, indicate marine paleodepths of ~300 m (see Chapter 13 for discussion). Subsequent Chickahominy microfaunas and nannofloras show that similar outer neritic to upper bathyal biotopes persisted at the impact site until the end of the Eocene (Poag and Aubry 1995).

If the water depth today were ~300 m deep over the crater center, marine waters would flood the entire coastal plain and the adjoining Piedmont, pushing the shoreline against the flanks of the Appalachian Mountains, >100 km west of the crater (Figs. 2.7, 2.8). Poag and Sevon (1989) and Poag (1993) showed that late Eocene deposition rates along the Atlantic margin were among the lowest of the Cenozoic (Fig. 2.9). These unusually low deposition rates were chiefly responsible for minimal late Eocene flexural subsidence rates in southeastern Virginia (see discussion under subheading 2.5). The slow subsidence of this continental margin would have created a late Eocene coastal plain and continental shelf with a gentler slope than at any time following the impact (including the present). Therefore, the late Eocene shoreline would have extended farther westward onto the Piedmont than the current position of the 300-m elevation. The late Eocene shelf break, as indicated on seismic profiles, would have been approximately at the present shelf-break location (Poag and Ward 1993). This puts the impact site on the inner third of the continental shelf, even though the high late Eocene sea level combined with minimal deposition and subsidence rates to create outer neritic to upper bathyal depths at the site (Figs. 2.7, 2.8).

In addition, late Eocene paleoclimate was much warmer and wetter than modern climate at the impact site (Wolfe 1978, 1992; Prothero 1994; Bestland et al. 1996; Poag 1997b). Tropical rainforests covered the slopes of the Appalachians, and carbonate ramp deposits (bioclastic limestones of the Piney Point Formation and chalks and micritic limestones of the Lindenkohl and Baltimore Canyon Alloformations) had accumulated across the broad continental shelf since the middle Eocene (Poag and Ward 1993; Fig. 2.5).

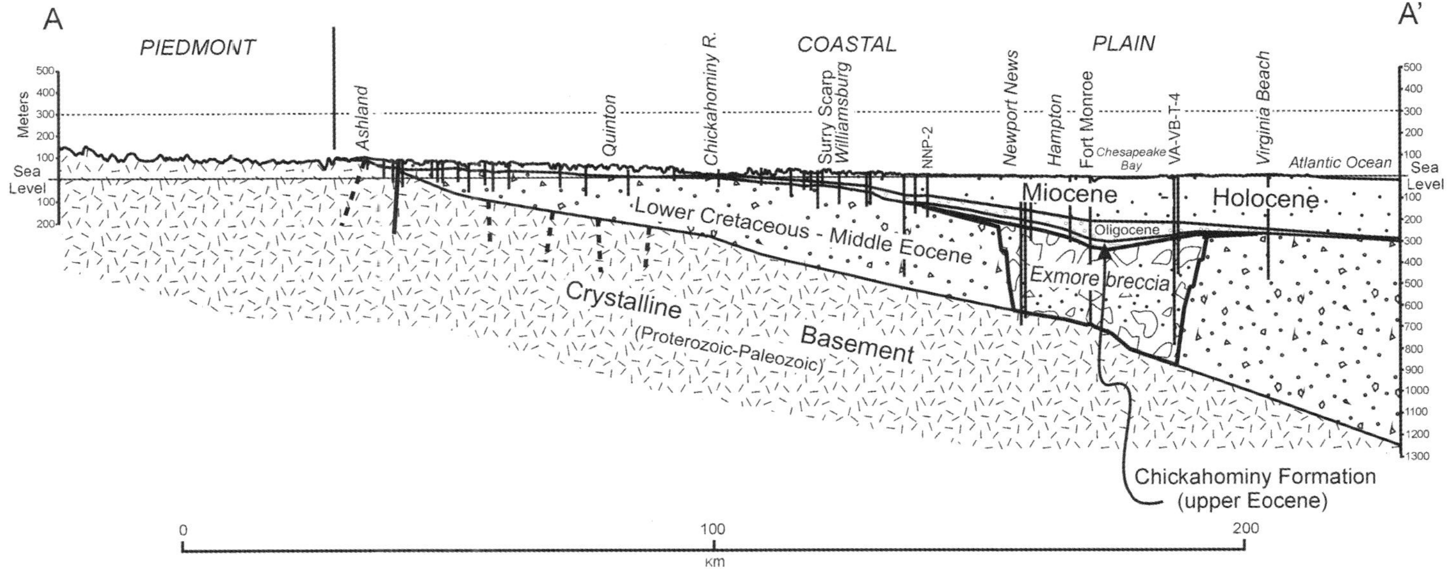

Fig. 2.7A. Topographic and general geologic cross section *A-A'* through southeastern Virginia showing location of Chesapeake Bay impact crater. Note that a 300-m rise in sea level (estimated paleodepth at time of Chesapeake Bay impact) would flood entire coastal plain and most of Appalachian Piedmont. See Table 1.2 for additional data about labeled boreholes.

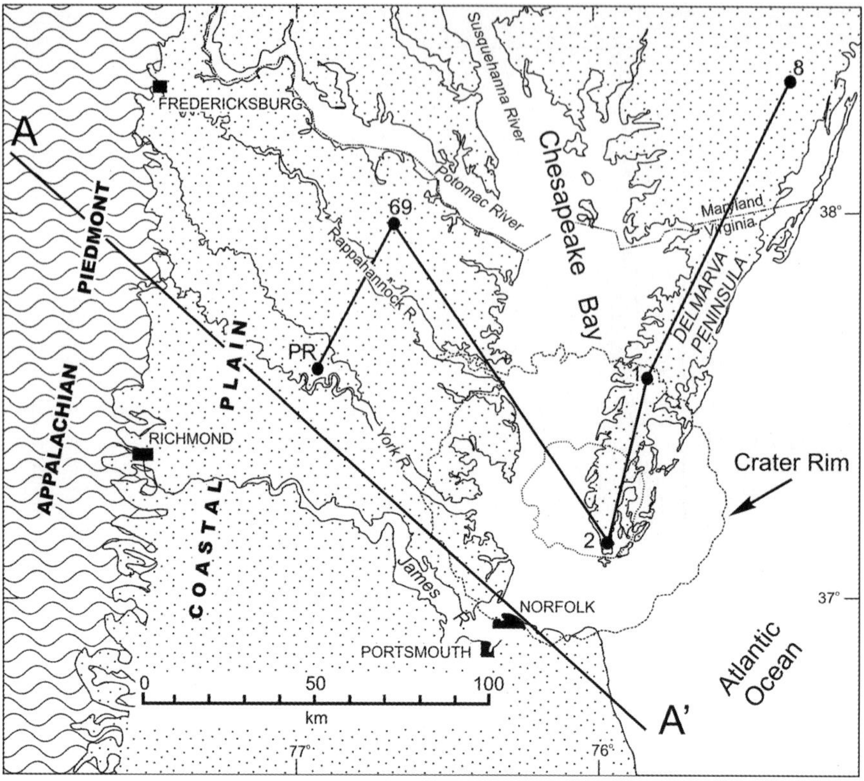

Fig. 2.7B. Map of locations for line of section *A-A'* for Fig. 2.7A and also for Fig. 2.6 (*PR* on map refers to Pamunkey River outcrops).

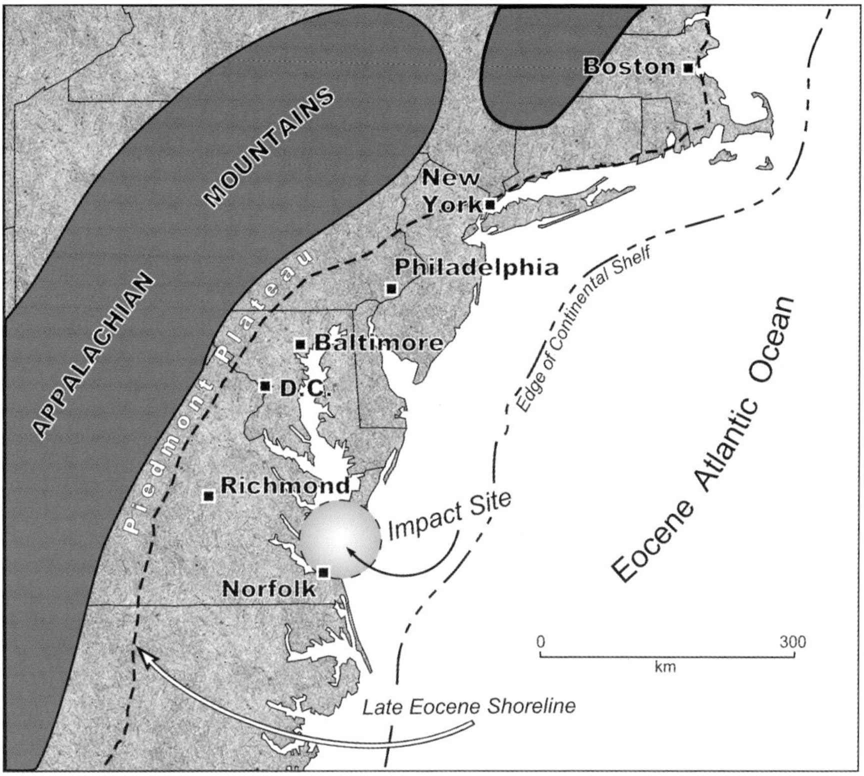

Fig. 2.8. Generalized late Eocene paleogeography for northern segment of US East Coast at time of Chesapeake Bay impact, showing location of impact site relative to shoreline and edge of continental shelf.

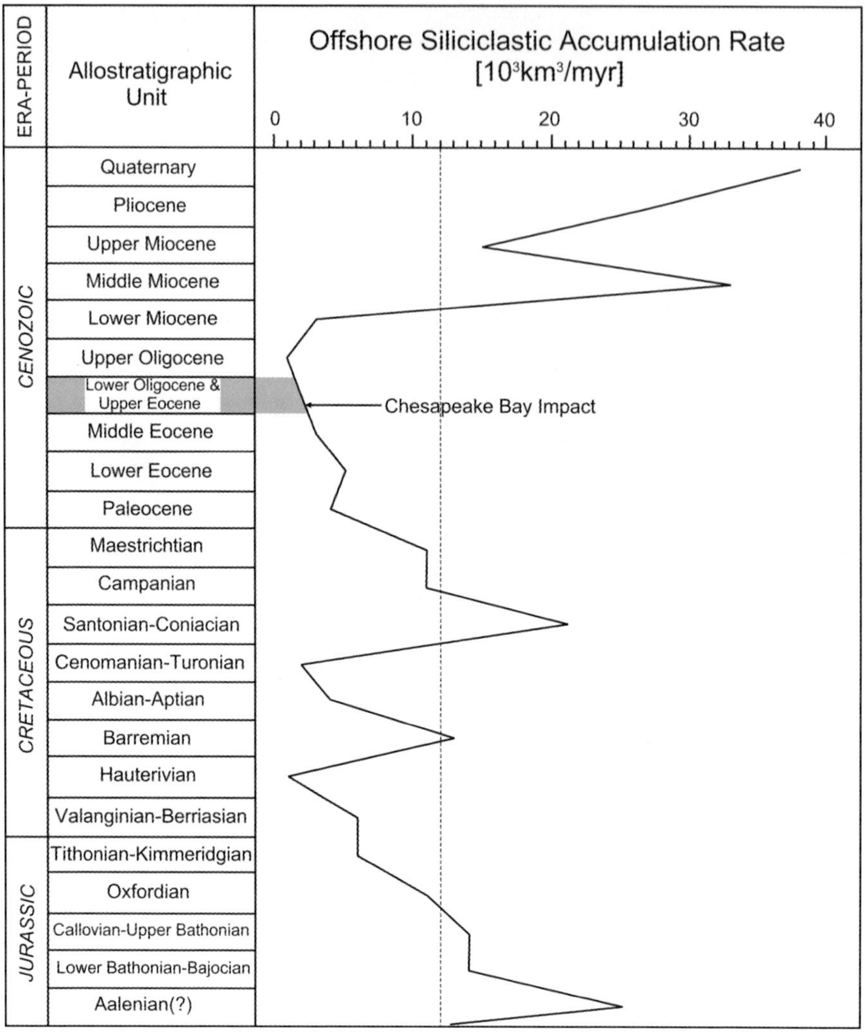

Fig. 2.9. Siliciclastic sediment accumulation rates for the US mid-Atlantic offshore region, showing unusually low rates for late Eocene at time of Chesapeake Bay impact (from Poag and Sevon 1989; Poag 1993). Accumulation rate measured as relative volume of sediment (thousands of km^3 per myr).

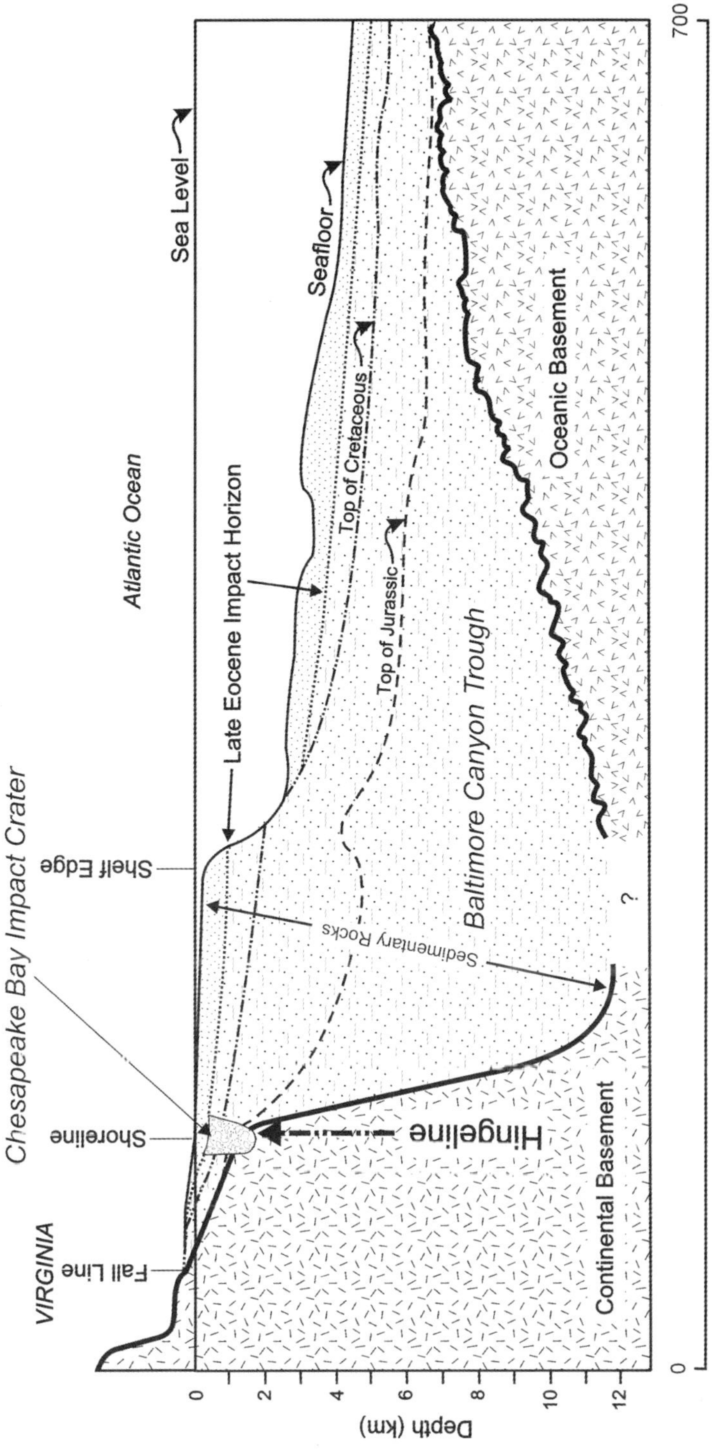

Fig. 2.10. Stylized cross section of Virginia continental margin showing location of Chesapeake Bay impact crater relative to principal structural, stratigraphic, and morphologic features. Hingeline marks axis along which rate of basin subsidence increases significantly, thereby producing main sedimentary depocenter of offshore Baltimore Canyon trough. See text for further explanation.

2.5 Subsidence of Virginia Continental Margin

Variation in postrift (Mesozoic-Cenozoic) sediment thickness across the US Atlantic margin has allowed recognition of three major structural-subsidence regimes (Watts and Steckler 1979; Steckler et al. 1988). On the landward side is a seaward-thickening wedge of Upper Jurassic to Holocene sediments. In Virginia, this wedge consists mainly of sand and silt deposits, which feather out near the Piedmont Fall Line and reach a maximum total thickness of ~1–1.5 km near the present shoreline (Fig. 2.10). This landward sedimentary wedge rests unconformably on continental crust, whose subsidence is attributed almost wholly to flexural downwarping caused by sediment loading (Watts and Steckler 1979). Beneath the inner continental shelf, the sediment thickness increases rapidly just seaward of the shoreline, across a hinge zone comprised of a series of faults and flexures. The hinge zone marks a major transition between undeformed continental crust landward, and rifted crust (thinned, heated), whose relatively rapid subsidence is attributed to thermotectonic processes that formed the offshore Baltimore Canyon trough. Maximum postrift sediment thickness in this trough (11–12 km) is reached between this hinge zone and the present shelf edge, which approximates the western edge of true oceanic crust. The Chesapeake Bay bolide struck approximately on the subsidence hinge zone in Virginia. This location gives a decided eastward tilt to the crater, due to 36 myr of postimpact flexural subsidence, which was driven almost entirely by postimpact sediment loading (Watts and Steckler 1979).

2.6 Initial Evidence of East Coast Impact

Initial evidence of a bolide impact in the western North Atlantic region came from distal ejecta that constitute part of the North American tektite strewn field. This ejecta field includes tektites, microtektites, coesite, stishovite, and shocked grains of several minerals, which display well-documented planar deformation features (PDFs; Glass 1989; Koeberl 1989; Glass and Wu 1993; see Chapter 6 for further discussion of shock deformation features). The strewn field, as known at that time, was documented from deep-sea cores and surface outcrops, and was extrapolated to cover a ~9-million-km^2 area of the western North Atlantic, Caribbean Sea, Gulf of Mexico, Barbados, Cuba, and the coastal plains of Texas (bediasites) and Georgia (georgiaites) (Fig. 2.11A). As an alternative, Koeberl (1989) suggested that the North American tektites might be distributed in irregular rays, extending radially outward from the impact site, rather than forming a continuous thin blanket (Fig. 2.11B).

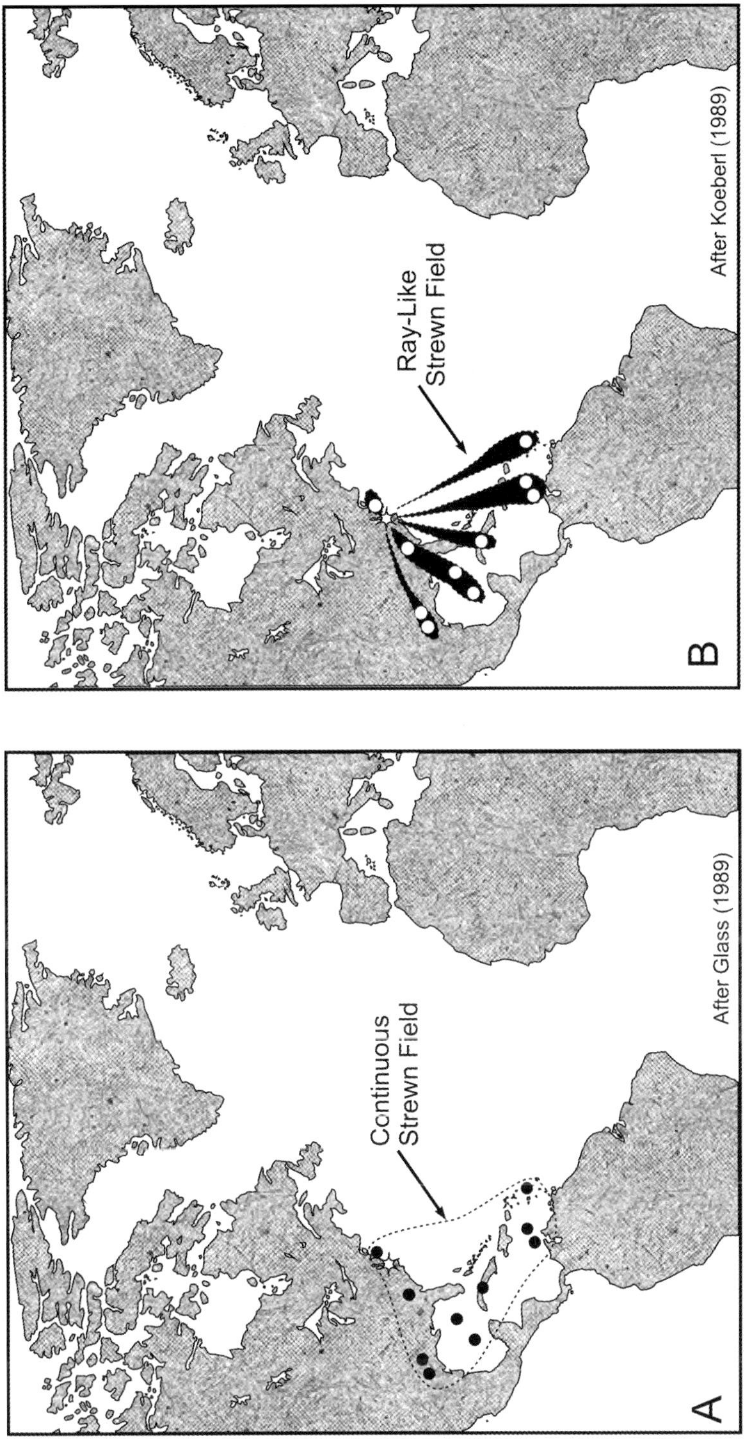

Fig. 2.11. Geographic distribution of North American tektite strewn field in western North Atlantic, Caribbean Sea, and Gulf of Mexico regions. *A* Continuous ejecta distribution inferred by Glass (1989); *B* Distribution in radially oriented ejecta rays inferred by Koeberl (1989). See Chapter 9 for further discussion of North American tektite strewn field.

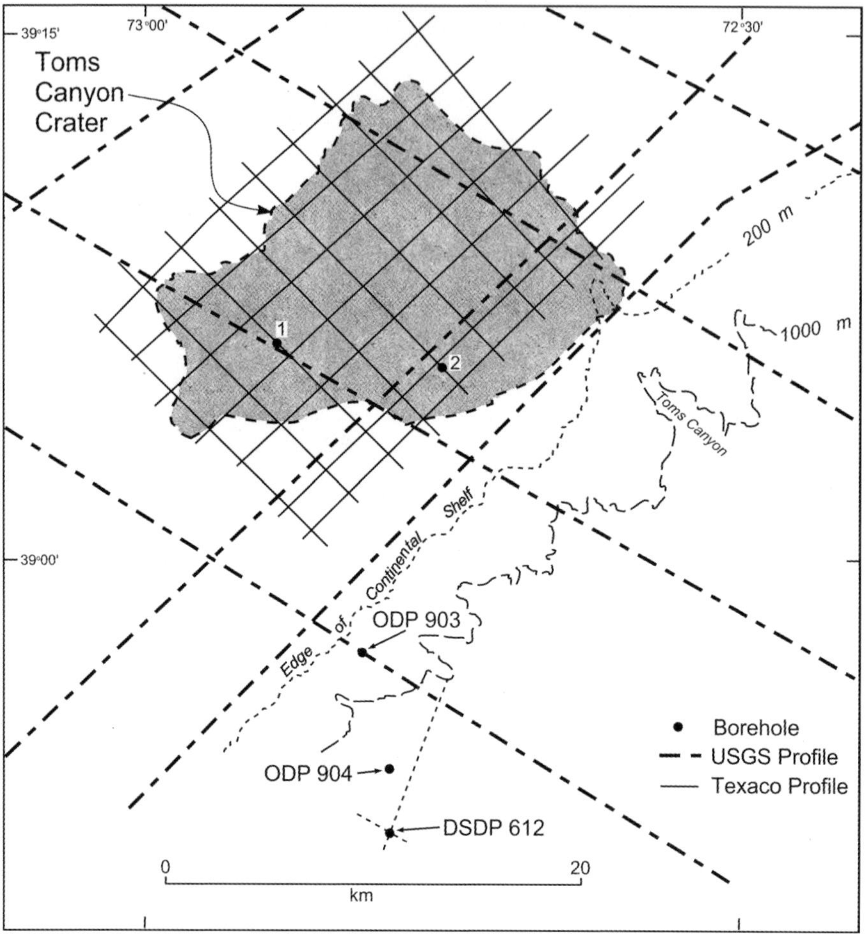

Fig. 2.12. Location of Toms Canyon impact crater (irregular shaded area), two exploratory wells drilled into crater (*1* = Tenneco; *2* = Houston Oil and Minerals), and three nearby coreholes (DSDP Site 612; ODP Sites 903, 904) from which North American tektite debris has been documented, and distribution of seismic reflection profiles crossing and surrounding crater (modified from Poag and Poppe 1998). *ODP* = Ocean Drilling Program; *DSDP* = Deep Sea Drilling Project.

First attempts to determine the location of a source crater for the North American ejecta relied on the physical properties of the ejecta layer at Deep Sea Drilling Project (DSDP) Site 612 (Figs. 2.12, 2.13). Site 612 was drilled on the continental slope, 120 km east of Atlantic City, New Jersey by scientists of DSDP Leg 95 (Poag, Watts, et al. 1985; Thein 1985). Because the Site 612 ejecta were coarser and occurred in a thicker layer than at other sites in the North American strewn field, Glass (1989) and Koeberl (1989) concluded that the responsible impact must have taken place within a few hundred kilometers of Site 612. The size and location of the Chesapeake Bay crater, discovered five years later, made it the prime candidate for the source of the North American tektite strewn field (Poag et al. 1994). A preliminary geochemical analysis of clasts from the Exmore breccia led Koeberl et al. (1996) to confirm this likelihood. An additional possible source for the New Jersey ejecta was suggested by Poag and Aubry (1995) and Poag and Poppe (1998). These authors proposed that the Toms Canyon structure, a small, elliptical (22 km maximum diameter), possible impact crater, may have supplied all or part of the ejecta cored at DSDP Site 612 and at nearby ODP Sites 903 and 904, cored in 1992 (McHugh et al. 1996; Glass et al. 1998; Figs. 2.12, 2.13). The Toms Canyon structure is only 35 km north of the three deep-sea core sites (Fig. 2.12). The Toms Canyon connection has been disputed by McHugh et al. (1996) and Glass et al. (1998), who considered the Chesapeake Bay crater as the most likely source of all the ejecta known from offshore New Jersey.

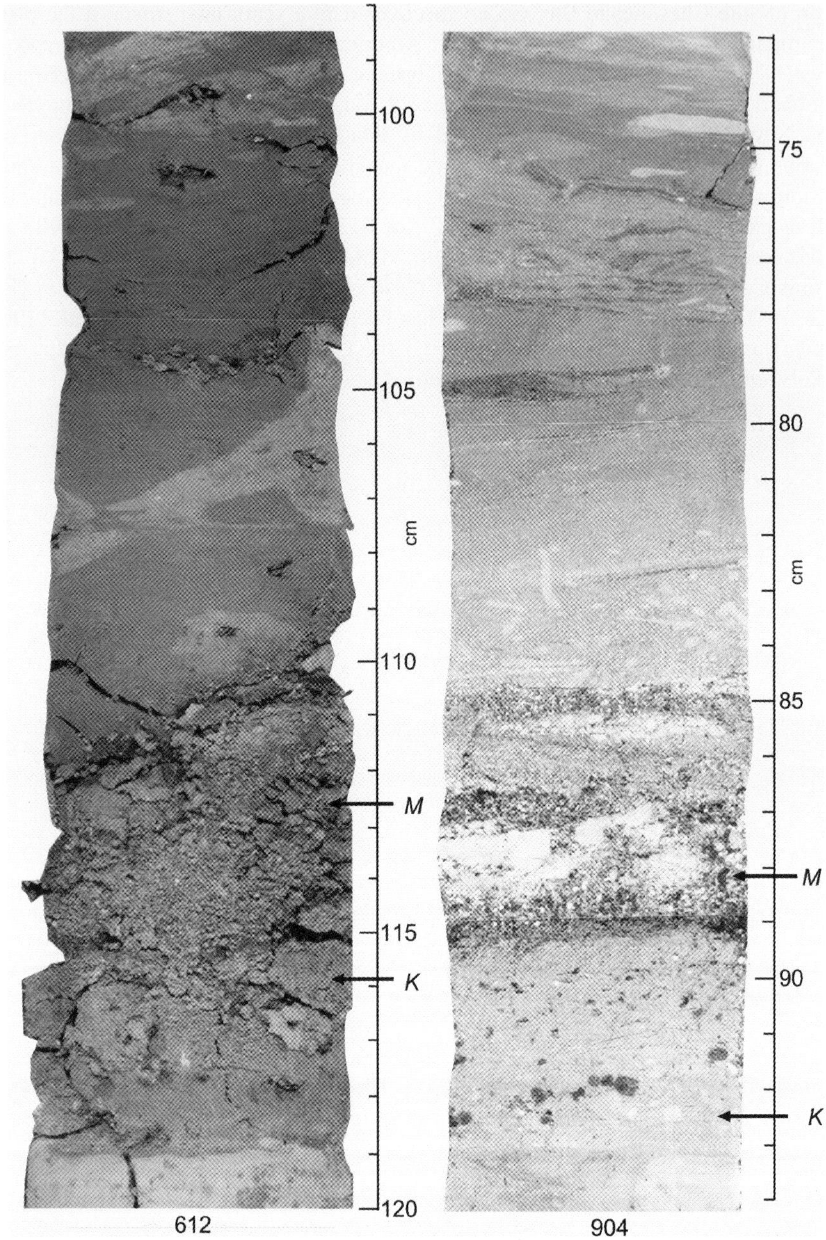

Fig. 2.13. Photographs of archive half of cores from DSDP Site 612 and ODP Site 904, showing stratigraphic position and visual expression of two ejecta layers (*M* = microtektite layer derived from Chesapeake Bay impact; *K* = microkrystite layer derived from Popigai impact). Depth scale in cm below top of core section. See CD-ROM for color version of this figure.

2.7 Onshore Borehole Evidence

2.7.1 Noncored Boreholes

Ejecta and impact breccia from the Chesapeake Bay impact were recovered from boreholes in southeastern Virginia more than fifty years ago, though their impact origin was not suspected until 1992 (Poag et al. 1992). Early publications by Cederstrom (1945a,b, 1957) described a persistent subsurface interval of highly diverse, brightly colored, variegated rock types containing a mixture of Cretaceous, Paleocene, and Eocene foraminifera. To this interval, which he identified in more than fifty wells, Cederstrom assigned the name Mattaponi Formation (Figs. 2.4, 2.14; Table 1.2). Stratigraphic and geographic inconsistencies and ambiguities in its definition have caused most subsequent authors to ignore or misuse the term Mattaponi (Brown et al. 1972; Tiefke 1973; Ward 1984; Hansen 1978; Meng and Harsh 1988), and it is now seldom used (Poag 1997a; Powars and Bruce 1999). By correlating the Mattaponi with more recently drilled sections, we can now extrapolate the presence of the Exmore breccia in many of Cederstrom's well logs (e.g., Poag 1997a; Powars and Bruce 1999).

A particularly widely cited and comprehensive borehole study was published by Brown et al. (1972). These authors constructed subsurface cross sections, which extrapolated the onshore stratigraphy to the undrilled strata underneath Chesapeake Bay prior to discovery of the impact crater or its breccia fill. Figure 2.15 compares a Brown et al. (1972) cross section with our reinterpretation, which acknowledges the presence of the impact crater (see also Powars and Bruce 1999).

2.7.2 Cored Boreholes

Modern sampling of the Exmore breccia was initiated inadvertently through continuous coring carried out collaboratively by the USGS and the Virginia State Water Control Board (now the Virginia Department of Environmental Quality; Powars et al. 1990, 1991, 1992; Powars and Bruce 1999; Powars 2000; Fig. 1.3; Table 1.2). Though originally interpreted as a debriite or channel-fill deposit (Powars et al. 1992), biostratigraphic and petrographic analyses later confirmed the late Eocene age and impact origin of the breccia (Poag et al. 1992; Koeberl et al. 1996). It was not until the acquisition of multichannel seismic reflection data from Texaco, Inc., however, that investigators fully realized that two of the coreholes had been drilled into a giant impact crater (Poag et al. 1994; Poag 1999c).

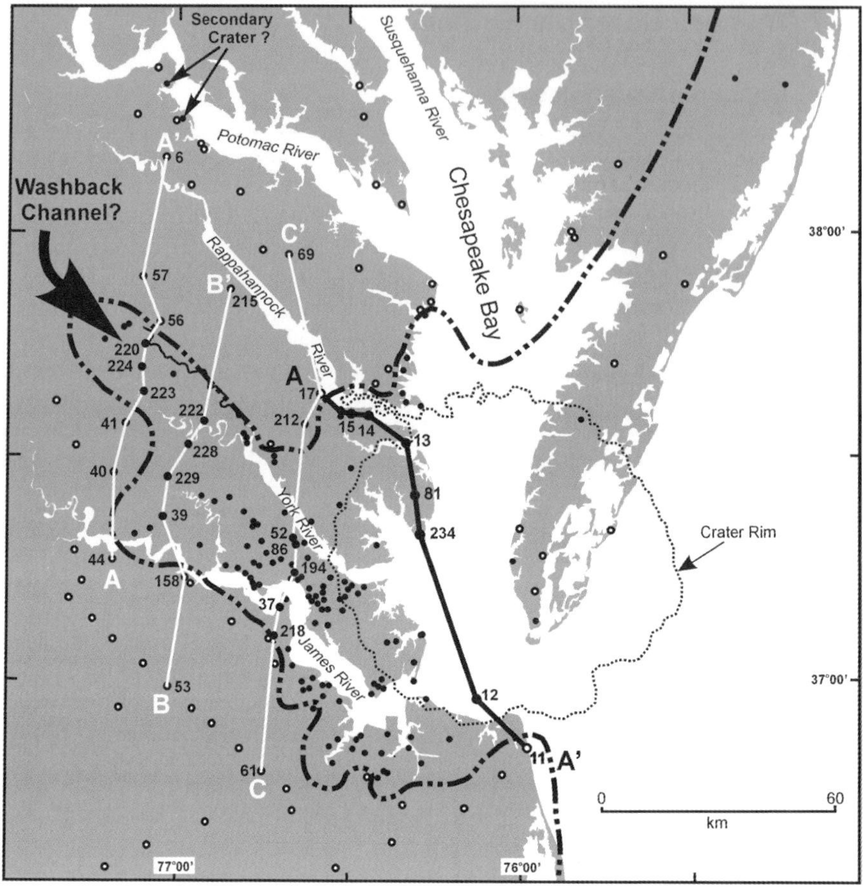

Fig. 2.14. Geographic distribution of boreholes where Mattaponi Formation and Exmore breccia have been documented (solid black dots) and where these units are missing or were not reached (open circles). Also shown are locations of stratigraphic cross sections and numbered boreholes shown in Fig. 2.15 (heavy black line) and Fig. 5.5 (white lines). Heavy dash-dot line represents approximate radial extent of Exmore breccia away from crater. See CD-ROM.1 for more accurate distribution of boreholes.

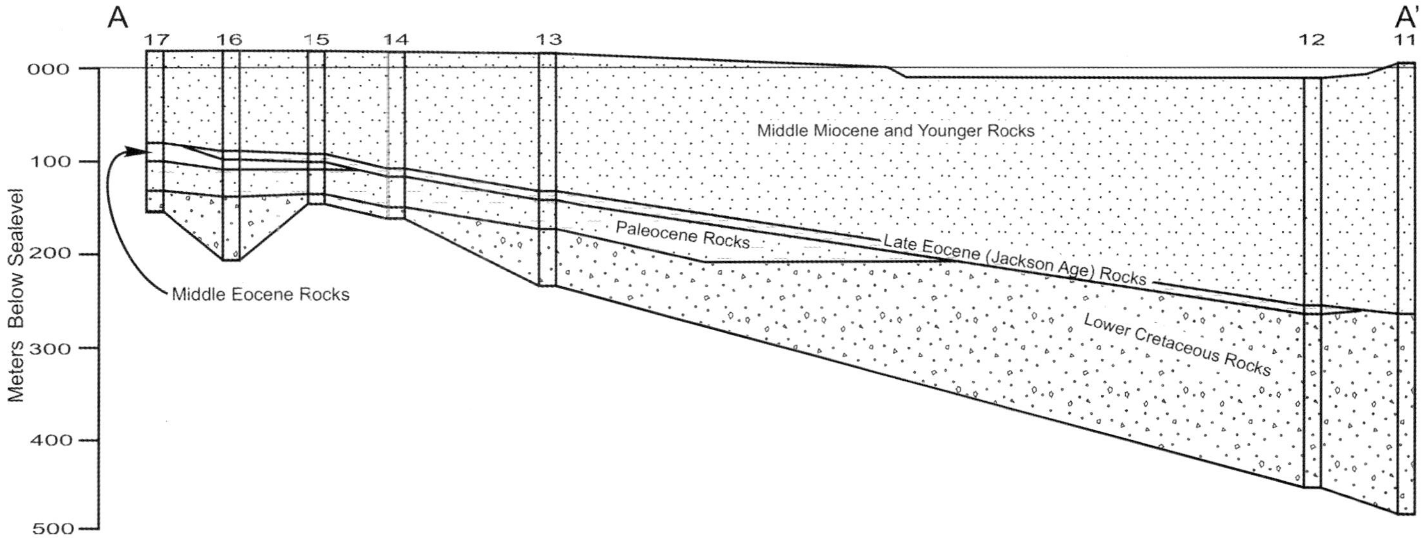

Fig. 2.15A. Interpretation of stratigraphic cross section across southeastern Virginia published by Brown et al. (1972) prior to crater discovery. See Fig. 2.14 for profile location.

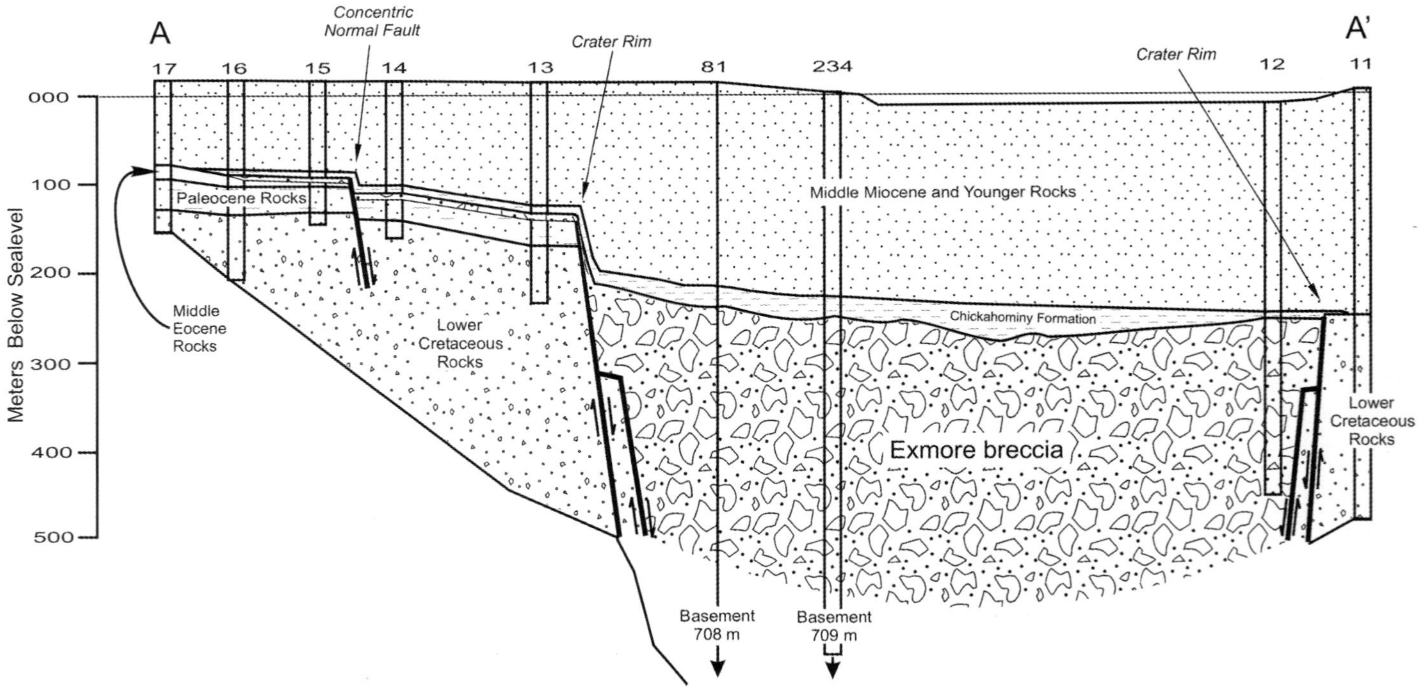

Fig. 2.15B. Brown et al.'s (1972) stratigraphic cross section across southeastern Virginia revised by us to show presence of crater. Compare also with Powars and Bruce (1999). See Fig. 2.14 for profile location.

3 Geophysical Framework of Impact Site

3.1 Seismic Investigations of Virginia Coastal Plain

Surprisingly few seismic reflection profiles have been collected during the >100-year history of geological and geophysical studies of southeastern Virginia. A few scattered onshore surveys have been published (Dysart et al. 1983; Bayer and Milici 1987; Milici et al. 1995), but otherwise, the seismic architecture of this coastal province has virtually been ignored. It was not until Poag et al. (1994) began to publish marine seismic profiles from Chesapeake Bay, that the seismostratigraphic and seismostructural framework of the impact crater could be constructed. In this book, we draw liberally from the database of marine seismic reflection profiles, collected mainly by the US oil industry and the USGS, to document the structure and morphology of this submarine impact structure.

3.2 Seismic Signature of Crystalline Basement Rocks

The seismic expression of the crystalline basement surface in most of the study area is the upper member of a couplet of high-amplitude, nearly continuous reflections, clearly manifest on most seismic profiles, and nearly evenly spaced apart over the entire region. We refer to this reflection as AB (acoustic basement; Fig. 3.1A). Hansen (1978) first identified this couplet signature on profiles collected for a subsurface study of southern Maryland. The profiles were collected by the Vibroseis method, in which a truck-mounted vibrator provides the acoustic source. Hansen used a natural-amplitude synthetic seismogram from a well 45 km northwest (updip) of his Vibroseis trackline to construct a best-fit stratigraphic model. From the model he inferred that the upper member of the reflection couplet represented the top surface of a high-velocity, indurated, sedimentary layer (interval velocity of 2.5 km/s). Hansen inferred that the subjacent lower-velocity interval (interval velocity of 1.9 km/s) represented a saprolite bed. The lower reflection of the high-amplitude couplet, Hansen interpreted to represent the upper surface of fresh crystalline basement rocks (interval velocity of 4.5 km/s). Hansen (1978) calculated individual thicknesses of 15–30 m for both the indurated sedimentary layer and the saprolite bed. Hansen noted that the indurated layer was not omnipresent, but patchily distributed in southern Maryland. He attributed the patchiness to erosion.

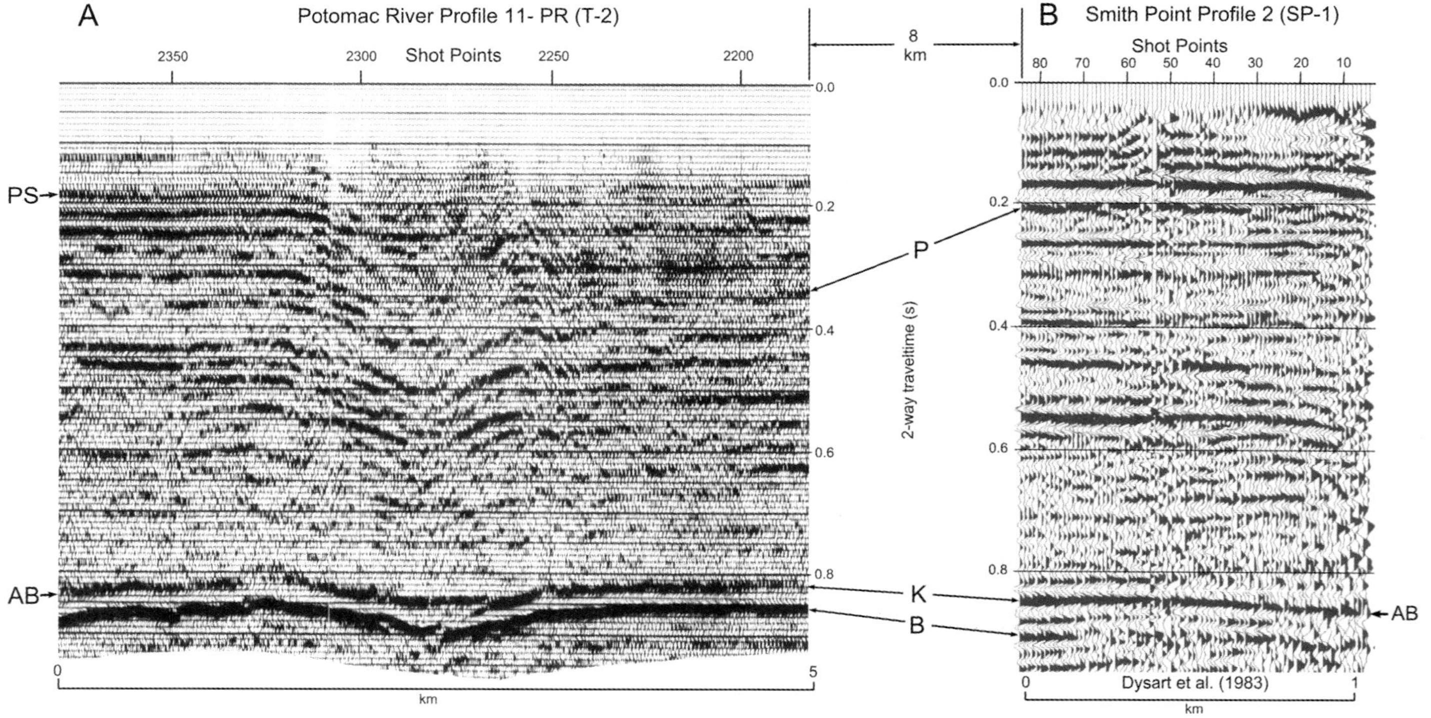

Fig. 3.1. Comparison of seismic reflection signatures between Potomac River dip profile 11-PR (**A**; labeled T-2 on Fig. 3.3) and published strike profile of Dysart et al. (1983) from Smith Point, Virginia (**B**; labeled SP-1 on Fig. 3.3). Profiles are ~8 km apart, as measured along structural and depositional strike. Note much better definition of basement reflection couplet (*AB*) on profile 11-PR. See text for explanation of labeled primary reflections *PS*, *AB*, *P*, *K*, *B*.

Five years later, Dysart et al. (1983) identified a similar couplet of reflections on two Vibroseis profiles collected in southeastern Virginia (Fig. 3.1B). Dysart et al. (1983), however, interpreted the two reflections to be the boundaries of two, rather than three, different rock units. They termed the upper reflection K, and ascribed it to the top of an indurated sedimentary unit of unknown (possibly Triassic) age, whose estimated thickness was 72–111 m. The indurated unit was not cored in the vicinity of their Vibroseis study, but Dysart et al. (1983) correlated it with a 75-m indurated layer drilled 125 km to the south, at Portsmouth, Virginia, and with the top of the Wastegate Formation drilled 40 km east at Crisfield, Maryland (Hansen 1978; Costain et al. 1981).

Dysart et al. (1983) called the lower reflection of the basement couplet, B. They ascribed it to the top of crystalline basement, and correlated it with metavolcanic basement rocks drilled in the Crisfield, Maryland borehole (Hansen 1978; Costain 1979). Dysart et al. (1983) calculated interval velocities of 4.1–4.7 km/s for the indurated layer at Smith Point, similar to the velocities Hansen (1978) estimated for the crystalline basement in southern Maryland. Dysart et al. (1983) calculated an interval velocity of 6.4 km/s for the upper part of the Smith Point crystalline basement, about 50% higher than Hansen's (1978) estimate for southern Maryland.

In 1984, Hansen and Wilson published more Vibroseis data from southern Maryland, and correlated their results with drilling data from a deep basement-penetrating borehole St.M. Df 84 (borehole 7 herein; Fig. 2.2; Table 1.2). In their study, Hansen and Wilson (1984) modeled the basement surface as a single reflection, although the familiar couplet is clearly present on the published Vibroseis profile (Hansen 1978; Fig. 3.2A herein), which is only 4 km east of the DF 84 well. They correlated the lower basement reflection with a diabase section, 16 m of which was drilled in the St.M. Df 84 well. The upper reflection of the couplet they correlated with the top of a sedimentary unit correlative with the Wastegate Formation, of possible Triassic or Jurassic age.

In a fourth Vibroseis study, Milici et al. (1995) published additional profiles from southeastern Virginia and southern Maryland (Fig. 3.2), and correlated them with the profiles of Hansen (1978) and Dysart et al. (1983). Milici et al. (1995) also recognized a basement reflection couplet, or more precisely, a variety of different but correlative reflection couplet(s), which, in their view, represented different "basement" events and lithologies. On their profile NAB-11A, for example (Fig. 3.2B), which crosses the coarse, siliciclastic rift deposits of the Taylorsville basin, a reflection couplet marks the boundary between rift deposits and postrift deposits. Both of these latter deposits, however, are composed of sedimentary strata, as sampled by numerous deep wells in the area. Milici et al. (1995) assigned the interval encompassed by the basement couplet to an indurated zone of Lower Cretaceous sedimentary rocks, and projected the interval 45 km downdip (eastward) to correlate with Dysart et al.'s (1983) unit K. However, Milici et al. (1995) believed the indurated zone on NAB-11A to be considerably younger than the Wastegate Formation in the Crisfield, Maryland, borehole, with which Dysart et al. (1983) had originally correlated their reflection K.

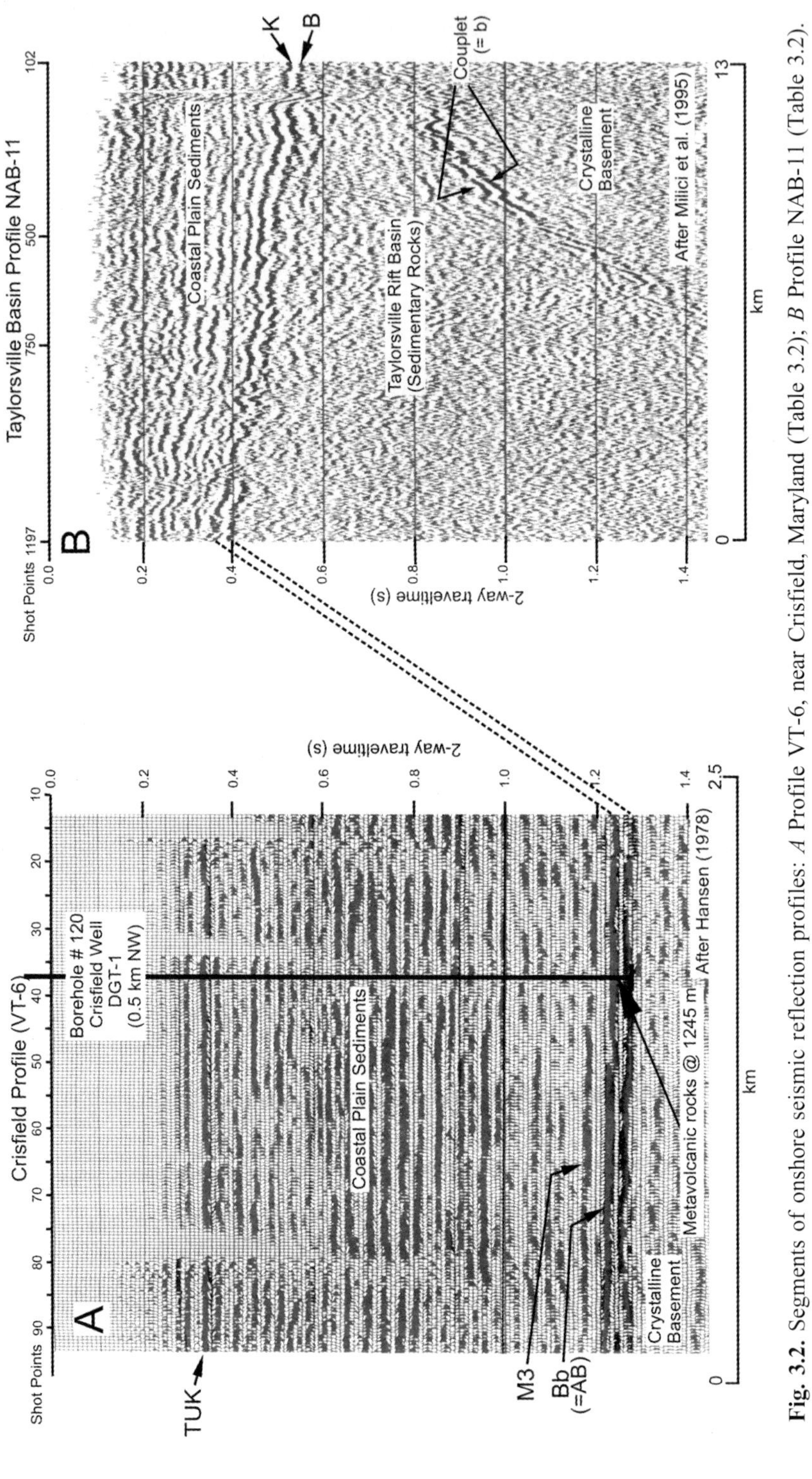

Fig. 3.2. Segments of onshore seismic reflection profiles: *A* Profile VT-6, near Crisfield, Maryland (Table 3.2); *B* Profile NAB-11 (Table 3.2). Figure shows seismic signatures and correlation of principal reflections from east (*A*) and west (*B*) sides of Chesapeake Bay (both profiles outside impact crater). See Fig. 2.2 for profile locations and text for further explanation.

On the Delmarva Peninsula, Milici et al. (1995) modeled crystalline basement as a single high-amplitude reflection on the basis of a synthetic seismogram from the Crisfield borehole DGT-1 (Fig. 3.2A). They correlated this reflection with the top of metavolcanic rocks in the bottom of the borehole, and designated the reflection as Bb (Fig. 3.2A), inferring that B and b are coincident in this area, though they are separate reflections elsewhere. The stratigraphically next highest high-amplitude reflection in Milici et al.'s (1995) analysis marks the top of the inferred Jurassic section, and was referred to by those authors as reflection Mark 3 (M3, Fig. 3.2A). Mark 3 can be distinguished on the Delmarva Peninsula Vibroseis profiles, but the unit whose top it represents pinches out updip; thus Mark 3 is not equivalent to K as Dysart et al. (1983) had inferred. Clearly, the near-basement reflection couplets in different parts of the study area cannot be assumed to represent coeval units or equivalent lithologic units from profile to profile. In areas underlain by rift basins, for example, the crystalline basement surface (b) can be kilometers deeper than our AB horizon (Fig. 3.2B).

Our interpretation of the Maryland Vibroseis profiles calibrated with seismic and lithologic data from the Crisfield borehole DGT-1 (designated borehole 120, herein; Fig. 1.3) indicates that reflection Bb of Milici et al. (1995) is equivalent to our AB, and represents the top of the metavolcanic section drilled in the DGT-1 well at 1245 m depth (Fig. 3.2A). Thus, except where rift basins are present, we interpret reflection AB to represent the top of crystalline basement.

3.3 Chesapeake Bay Seismic Reflection Profiles

For this study, we used 2,018 km of onshore and offshore seismic reflection profiles, derived from 11 different sources, and collected over a 25-year period (1975–2000; Fig.3.3; CD-ROM.2; Tables 3.1, 3.2). For seismostratigraphic analysis, we used standard interpretation techniques described by Vail et al. (1977) and Wilgus et al. (1988).

Four principal sets of seismic reflection data document the structure and morphology of Chesapeake Bay impact crater (Fig. 3.3; CD-ROM.2; Tables 3.1, 3.2). The key data set was collected in 1986, by Teledyne Exploration Company, for a partnership whose principal members were Texaco, Inc. and Exxon Exploration Co. Texaco donated the data to the USGS, and we refer to these data as the Texaco profiles (designated by T on Fig. 3.3). The Texaco profiles constitute 310 km of migrated, 48-fold, multichannel data collected within the southern part of Chesapeake Bay and the lower parts of the James, York, Rappahannock, and Potomac Rivers. The Texaco profiles give an unambiguous record of the crystalline basement configuration (faults, peak ring, central peak), the displaced sedimentary megablocks that overlie crystalline basement, the Exmore breccia that fills the crater, preimpact sedimentary strata surrounding the crater, and all overlying postimpact sediments except the uppermost 80 m (Poag 1996, 1997a; Poag et al. 1999; Powars and Bruce 1999). All the principal structural, morphological, and gross depositional features of a peak-ring/central-peak impact crater are displayed on the

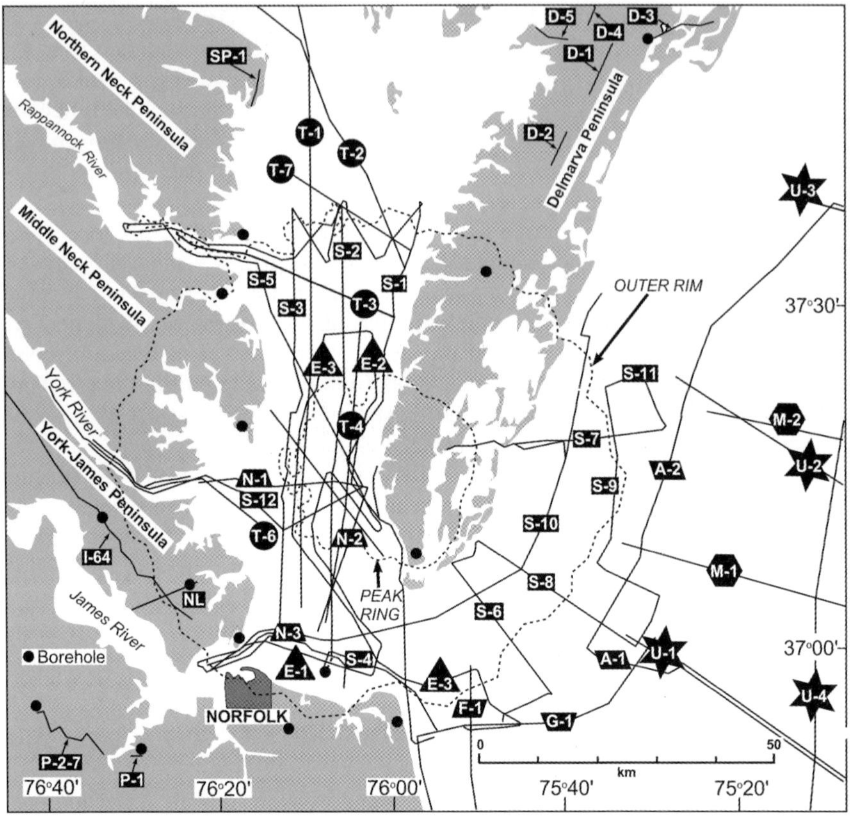

Fig. 3.3. Geographic distribution of onshore and offshore seismic reflection tracklines showing positions at which seismic profiles cross outer rim and inner edge of peak ring of Chesapeake Bay crater. Solid dots indicate key boreholes for calibrating seismic stratigraphy. See Table 3.2, CD-ROM.2, and CD-ROM.8–16 for more detail.

Texaco profiles.

A second set of multichannel profiles was collected in 1982 by the USGS. They are called the *Neecho* profiles, after the research vessel used to collect them, and are designated N (Fig. 3.3; CD-ROM.2; Tables 3.1, 3.2; Poag 1996). Three *Neecho* profiles (68 line km) in the lower bay and the mouth of the James River intersect the Texaco profiles and provide further documentation of all the impact features except the displaced megablocks. These Neecho profiles also record reflections from the uppermost 80 meters of postimpact sediments, which the Texaco profiles lack.

A third set of multichannel profiles (220 line km) was collected in 1998, by the USGS and Lamont-Doherty Earth Observatory, using the *R/V Maurice Ewing*. These profiles are herein designated E (Figs. 3.3, 3.4; CD-ROM.2; Tables 3.1, 3.2). This set of profiles images the postimpact sediments, preimpact sediments,

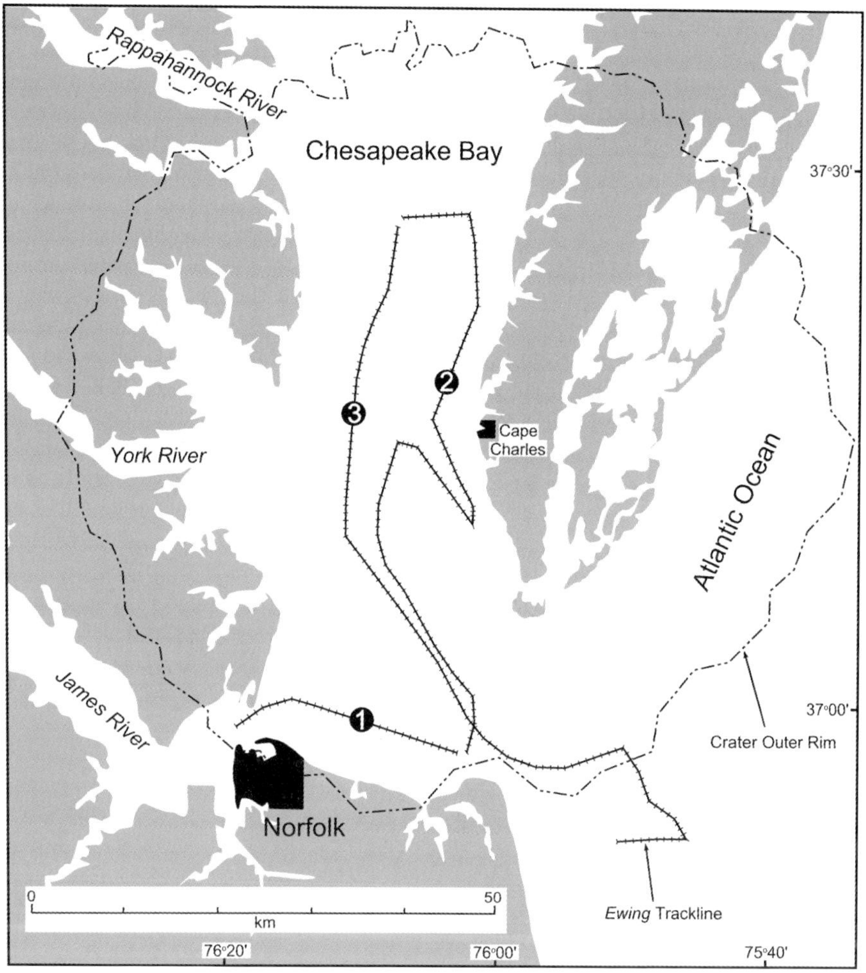

Fig. 3.4. Trackline map for seismic reflection survey of Chesapeake Bay impact crater carried out by USGS and Lamont-Doherty Earth Observatory in 1998 using *R/V Maurice Ewing* (*E*-1, *E*-2, and *E*-3 of Fig. 3.3; see Table 3.2 and CD-ROM.2, 15, and 16).

and the upper surface of crystalline basement, including the peak ring and a clear expression of the central peak.

A fourth set of reflection data comprises 875 km of single-channel profiles collected in 1996 by the USGS in collaboration with the National Geographic Society (NGS). These profiles are designated S (Fig. 3.3; Tables 3.1, 3.2), after the *Seaward Explorer*, the vessel used for the survey (Poag et al. 1999). The USGS-NGS profiles emphasize details of the postimpact sedimentary beds, such as sequence boundaries and compaction faults, but give little information about the displaced megablocks. They also clearly delineate the configuration of the crystalline basement inside the bay, but failed to image the basement surface beneath the thicker sediments of the inner continental shelf east of the Delmarva Peninsula.

Five additional sets of offshore profiles do not cross the crater, but are near enough to help constrain its boundaries and(or) to document the depth to basement (Fig. 3.3; CD-ROM.2; Tables 3.1, 3.2). Offshore, the USGS collected 140 line km of multichannel data just seaward of the crater rim, which reveal the depth to crystalline basement and the stratigraphy of overlying sediments (Grow et al. 1988). These profiles are designated U. The US Minerals Management Service (MMS) also contracted for several multichannel surveys east of the crater rim (65 line km), and these profiles are designated M. Three earlier sets of single-channel data collected by the USGS (1975, 1976) also constrain the eastern rim of the crater, and are designated A (after the ship *Atlantis II*; 140 line km), G (after the ship *Gyre*; 60 line km), and F (after the ship *Fay*; 25 line km; Fig. 3.3; CD-ROM.2; Tables 3.1, 3.2). These early single-channel profiles provide reflections from the sedimentary section only; they do not image the crystalline basement.

Table 3.1. Trackline lengths for seismic reflection profiles collected over and near Chesapeake Bay impact crater.

Source [Ship* or Organization]	Designated Prefix	Total Trackline Length [km]
Texaco	T	310
USGS-OCS	U	140
*Atlantis II**	A	140
*Fay**	F	25
MMS-OCS	M	65
*Gyre**	G	60
*Neecho**	N	68
*Seaward Explorer**	S	875
*Maurice Ewing**	E	220
Vibroseis (on land)	D, P, SP	100
USGS (on land)	NL	~15
TOTAL		2018

Table 3.2. Seismic reflection profiles that either cross Chesapeake Bay impact crater or help constrain location of outer rim and(or) depth to crystalline basement.

Number	Original Designation	Data Type	Contents	Reference
T-1*	Texaco 1-CB	Multichannel 48-fold CDP	Crosses crater	Poag (1996, 1997)
T-2*	Texaco 11-PR	Multichannel 48-fold CDP	Crosses crater	This volume
T-3	Texaco 10-RR	Multichannel 48-fold CDP	Crosses crater	Poag (1996, 1997)
T-4*	Texaco 8-S-CB-E	Multichannel 48-fold CDP	Crosses crater	Poag (1996, 1997)
T-5	Texaco 7-CB-H	Multichannel 48-fold CDP	Crosses crater	Poag (1996, 1997)
T-6	Texaco 13-YR	Multichannel 48-fold CDP	Crosses crater	Poag (1996, 1997)
T-7*	Texaco 9-CB-F	Multichannel 48-fold CDP	Crosses crater	Poag (1996, 1997)
N-1	*Neecho* 1	Multichannel 6-fold CDP	Crosses crater	This volume
N-2	*Neecho* 2	Multichannel 6-fold CDP	Crosses crater	This volume
N-3	*Neecho* 3	Multichannel 6-fold CDP	Crosses crater	This volume
S-1*	SEAX 8-7-6	Single channel	Crosses crater	This volume
S-2*	SEAX 9-10	Single channel	Crosses crater	This volume
S-3*	SEAX 16-4a-4	Single channel	Crosses crater	This volume
S-4	SEAX 17-9	Single channel	Crosses crater	This volume
S-5	SEAX 13-14-15-4a-16	Single channel	Crosses crater	This volume

Table 3.2. (cont.)

Number	Original Designation	Data Type	Contents	Reference
S-6	SEAX 13-14-6-27	Single channel	Crosses crater	This volume
S-7	SEAX 3-15-(N-1)-7	Single channel	Crosses crater	This volume
S-8	SEAX 25-26	Single channel	Crosses crater	This volume
S-9	SEAX 22-23-24	Single channel	Crosses crater	This volume
S-10	SEAX 1-2-18	Single channel	Crosses crater	This volume
S-11	SEAX 20-21	Single channel	Crosses crater	This volume
S-12	SEAX 5	Single channel	Crosses crater	This volume
S-13	SEAX 11-12	Single channel	Crosses crater	This volume
E-1	*Ewing* 1	Multichannel 3-fold CDP	Crosses inner basin	This volume
E-2*	*Ewing* 2	Multichannel 3-fold CDP	Crosses inner basin	This volume
E-3*	*Ewing* 3	Multichannel 3-fold CDP	Crosses inner basin & outer rim	This volume
F-1	*Fay* 19	Single-channel	Crosses crater	Poag (1996)
U-1	USGS 11	Multichannel 48-fold CDP	Constrains crater	Klitgord et al. (1994)
U-2	USGS 28	Multichannel 48-fold CDP	Constrains crater	Grow and Klitgord (1988)
U-3	USGS 3	Multichannel 48-fold CDP	Constrains crater	Klitgord et al. (1994)
U-4	USGS 12	Multichannel 48-fold CDP	Constrains crater	Klitgord et al. (1994)

Table 3.2. (cont.)

Number	Original Designation	Data Type	Contents	Reference
M-1	MMS	Multichannel 48-fold CDP	Constrains crater	This volume
M-2	MMS	Multichannel 48-fold CDP	Constrains crater	This volume
A-1	*Atlantis* II	Single-channel	Constrains crater	Poag (1996)
A-2	*Atlantis* II	Single-channel	Constrains crater	Poag (1996)
G-1	*Gyre* 81	Single-channel	Constrains crater	Poag (1996)
SP-1	Smith Point 1	VibroSeis	Documents basement	Dysart et al. (1989)
P-1	Portsmouth 1	VibroSeis	Documents basement	Costain and Glover (1976–1982)
P2-7	Portsmouth 2, 3, 4, 5, 6, 7	VibroSeis	Documents basement	Costain and Glover (1976–1982)
I-64	Interstate Highway I-64	VibroSeis?	Documents basement & outer rim	Bayer and Milici (1987)
D-1	Delmarva 1	VibroSeis	Documents basement	Hansen (1978)
D-2	Delmarva 2	VibroSeis	Documents basement	Hansen (1978)
D-3	Delmarva 3	VibroSeis	Documents basement	Hansen (1978)
D-4	Delmarva 4	VibroSeis	Documents basement	Hansen (1978)
D-5	Delmarva 5	VibroSeis	Documents basement	Hansen (1978)

Table 3.2. (cont.)

Number	Original Designation	Data Type	Contents	Reference
VT-6	Virginia Tech 6	VibroSeis	Documents basement	Milici et al. (1995)
VT-7	Virginia Tech 7	VibroSeis	Documents basement	Milici et al. (1995)
VT-13	Virginia Tech 13	VibroSeis	Documents basement	Milici et al. (1995)
NAB 11A	Virginia Tech NAB 11A	VibroSeis	Documents basement	Milici et al. (1995)
NL	USGS-5	High-resolution Multichannel	Documents basement	Gohn (in press)

* Indicates profile included on accompanying CD-ROM

Onshore, four sets of Vibroseis profiles have been collected outside the crater rim (total of 100 line km), and provide depths to crystalline basement and reflections from the overlying sedimentary section. The oldest set of onshore profiles was published by Hansen (1978), and is designated D (Fig. 3.3; Tables 3.1, 3.2). The second set was published by Dysart et al. (1983), and is designated SP (Fig. 3.3; Tables 3.1, 3.2). The third set of onshore profiles was published by Milici et al. (1995), and is designated VT or NAB (Fig. 3.3; Tables 3.1, 3.2). The fourth set of profiles onshore was collected by John Costain (Virginia Tech) for a regional geothermal study (see http://rglsun1.geol.vt.edu/geothermal.html). These profiles are designated P (Fig. 3.3; Tables 3.1, 3.2).

An additional USGS high-resolution, multichannel, seismic reflection profile (designated NL) was constructed from data collected along an ~15-km transect from the NASA Langley corehole northwestward to a point ~10 km outside the outer rim of the crater (Gohn in press). This profile clearly images the basement reflection couplet (AB), the preimpact stratal reflections outside the crater, and postimpact stratal reflections inside and outside the crater. The profile also resolved the Exmore breccia and the displaced megablocks in the annular trough, which were cored at the NASA Langley site.

3.4 Depth Conversion of Seismic Two-way Traveltimes

We used three types of data to help convert the two-way traveltimes to depth: (1) stacking velocities [root-mean-square (RMS) values] from Vibroseis and marine seismic reflection surveys; (2) velocity profiles derived from interval transit-time logs run in the NASA Langley borehole; and (3) especially the subsurface elevations of key stratigraphic boundaries determined from boreholes drilled near the seismic tracklines. We focused on two key horizons for the conversion: (1) the top of acoustic basement, AB (Fig. 3.1A,B) (acoustic basement is composed of varying crystalline lithologies, as noted above); and (2) the base of the postimpact sediments, PS (Fig. 3.1A,B); inside and near the crater, this horizon corresponds to the upper surface of the Exmore breccia.

Dysart et al. (1983) derived RMS velocities in the sedimentary section from the ground surface to reflection B at Smith Point, ~30 km north of the crater rim on the west side of Chesapeake Bay; there, B is approximately 0.9 s deep (2-way traveltime; Fig. 3.1B). The RMS values range from 2175 to 2350 m/s, with an average of approximately 2350 m/s. This would yield a depth conversion factor of 0.1 s = 113 m. Dysart et al. (1983) did not publish RMS velocities related to reflection K, however. Our correlation of borehole stratigraphy with the seismic profiles indicates average velocities of ~2000 m/s to K. Therefore, we calculated the depth to equivalent horizon AB using the relationship 0.1 s = 100 m.

Klitgord and Schneider (1994) provided an unusually large database of offshore seismic velocities calculated from marine reflection profiles collected east of the crater rim (USGS-OCS, MMS-OCS; Fig. 3.3; Tables 3.1, 3.2). They derived velocity values from normal moveout analysis of these offshore profiles, combined with sonic logs and velocity checkshot studies in numerous industry boreholes, and wide-angle data from two-ship seismic experiments. On the shallow ends of offshore profiles nearest the crater (profiles U-1, U-2, U-3; Fig. 3.3), where horizon AB is ~2.0 s below sea level, RMS values average ~5000 m/s, which yields a depth conversion of 0.1 s = 125 m.

The RMS velocities derived by Dysart et al. (1983) for the interval from the ground surface to horizon PS (Fig. 3.1B) range from 1550 to 1625 m/s, averaging 1592 m/s. This yields a depth conversion of 0.1 s = 796 m. Stratigraphic correlation of horizon PS between the boreholes and the seismic profiles indicates a nearly identical depth conversion of 0.1 s = 800 m. For convenience, we used the latter value.

We have interpolated RMS values (assuming linear variation with depth) above and below horizons AB and PS, and between their onshore and offshore end-member values, to produce the depth sections and structure maps illustrated herein.

3.5 Gravity Evidence

3.5.1 Database

To analyze gravity anomaly data in the vicinity of the impact crater, we combined four sources of data (Fig. 3.5): (1) land and marine gravity anomalies compiled by Carl Bowin (Woods Hole Oceanographic Institution; 3,941 stations; personal communication, 1998); (2) land and marine gravity anomalies compiled by John Costain (Virginia Tech; 14,240 stations; personal communication, 1998); (3) marine gravity collected by *R/V Maurice Ewing* in Chesapeake Bay (1,587 stations; USGS cruise EW9809, Oct 15–16, 1998; Fig. 3.4); and (4) land gravity data collected by Phillip Molzer (USGS) on the Delmarva Peninsula and its southeastern islands in September, 1998 (63 stations). Data from Bowin and Costain were provided as Bouguer anomalies, compiled from a variety of sources, and apparently generated with the same crustal density of 2.67g/cm^3. The *R/V Ewing* data were measured with a Bell gravimeter at one sample/second, post-processed with a 6-minute and 8-minute gaussian filter, and averaged to one-minute intervals. Navigation positioning was measured with three GPS transceivers. Data collected on the Delmarva Peninsula were measured with a Lacoste and Romberg gravity meter and tied to gravity points from the Bowin and Costain data sets.

The position of each gravity station on the Delmarva survey was measured with an Ashtech GPS receiver. Positions were corrected in post-processing with reference to the National Geodetic Survey's fixed GPS station CHR1 located at Cape Henry, Virginia (Fig. 3.5).

We merged data sets from Bowin and Costain, and removed duplicate points. *R/V Ewing* data are referenced to a gravity station at the dock in Portsmouth, Virginia, which is in the Bowin-Costain data set, thus aligning the *Ewing* measurements with the older data. Data collected on and about the Delmarva Peninsula in

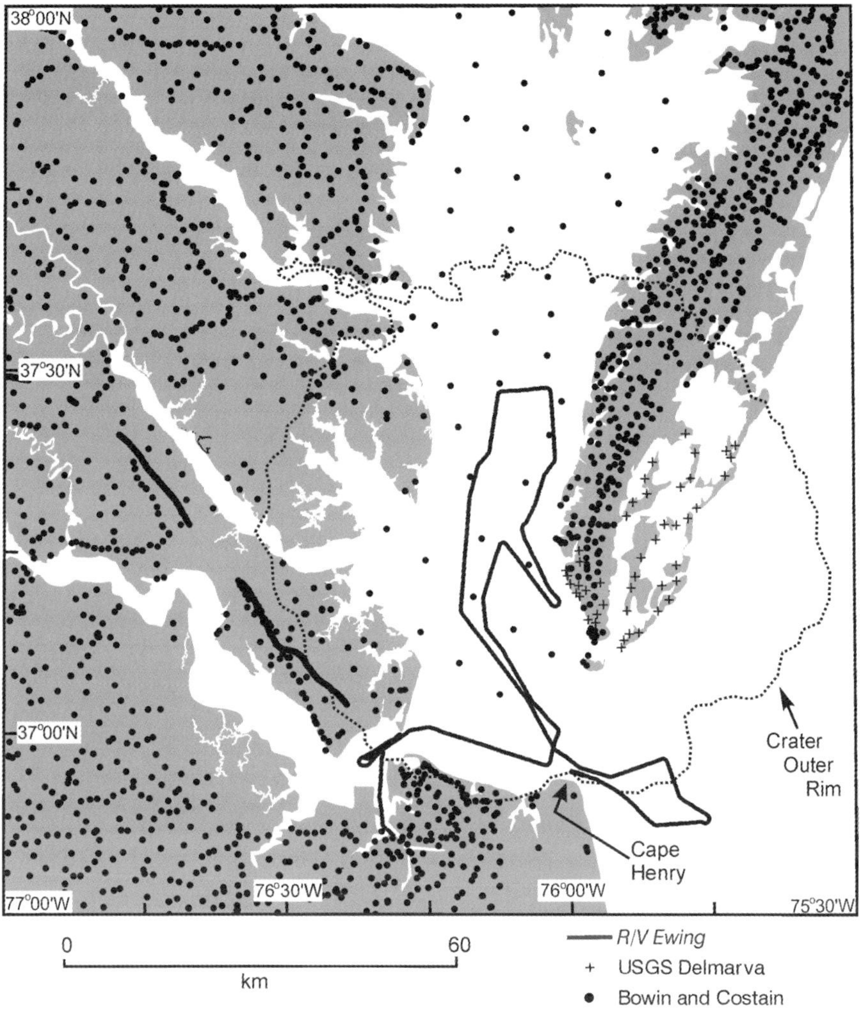

Fig. 3.5. Geographic distribution of gravity stations collected previous to and expressly for this study.

1998 also were tied to the Bowin-Costain data set at the time of the survey.

3.5.2 Interpretation

Visual examination of a gridded image of these data reveals a general eastward decline in gravity anomalies, probably associated with the subsidence of the basement surface beneath the thick sedimentary column on the continental margin (Fig. 3.6A). To remove this regional trend, we fitted a planar surface to the data

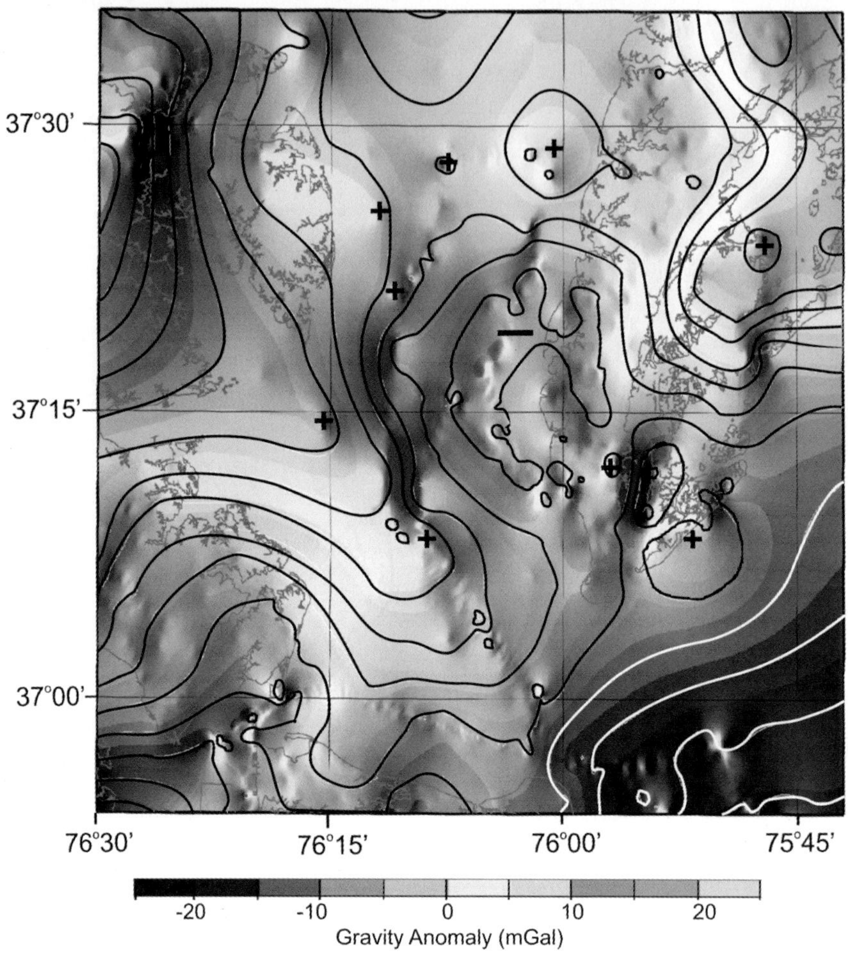

Fig. 3.6A. Simple Bouguer gravity anomaly map of the study area. Distribution of relative gravity highs (+) reflects presence of subcircular crystalline peak ring encircling gravity low (-) of the inner basin.

by least squares, with iterative data reweighting (using trend2d software of Generic Mapping Tool; Wessel and Smith 1991). Subtracting the fitted surface from the input data produced a set of residual gravity anomalies (Fig. 3.6B). We gridded the residual data set at an increment of 0.001 degrees for further analysis.

The spatial distribution of residual gravity anomalies supports the structural and morphological interpretations derived from our seismostratigraphic analyses. The principal features identified from the gravity surveys are (Fig. 3.6B): (1) a subcircular bull's-eye negative anomaly correspondent with the seismically defined inner

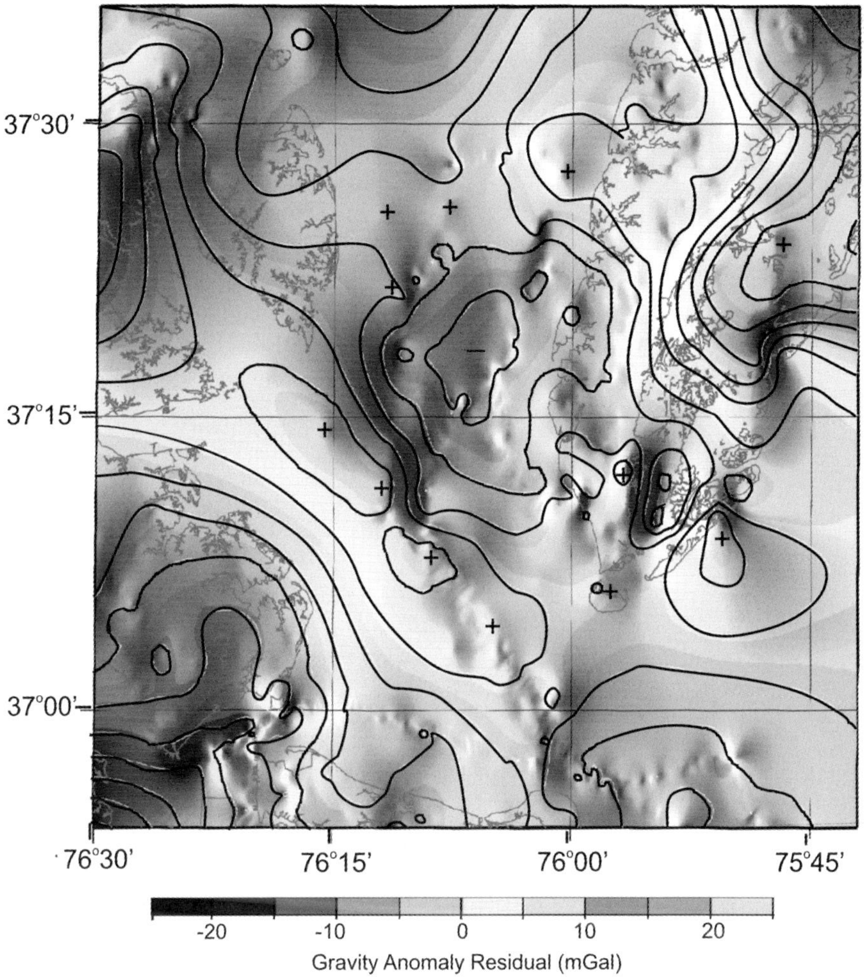

Fig. 3.6B. Residual gravity anomaly map of the study area. Distribution of relative gravity highs (+) reflects presence of subcircular crystalline peak ring encircling gravity low (-) of the inner basin.

basin; (2) a ring of positive anomalies correlative with the peak ring inferred from the seismic profiles (see Chapter 4 for further discussion and illustration of gravity data).

4 The Primary Crater

4.1 Crater Structure and Morphology

4.1.1 Seismic Interpretation

Our extensive network of seismic reflection profiles clearly documents that the structural-morphological features of the outer rim, annular trough, and displaced megablocks are expressed principally by preimpact sedimentary rocks, and to a much lesser degree, by crystalline basement rocks (Figs. 1.5, 4.1, 4.2; CD-ROM.3–6). In contrast, the peak ring, inner basin, and central peak of the Chesapeake Bay crater are strongly developed within rocks of the crystalline basement (Figs. 1.5, 4.1, 4.2; CD-ROM.3–6).

4.1.1.1 Outer Rim

The outer rim of the Chesapeake Bay crater is a steep, roughly circular fault scarp constructed almost entirely of sedimentary rocks. On all seismic profiles, the outer rim is manifest as an abrupt loss of coherent, continuous to subcontinuous, moderate- to high-amplitude horizontal reflections, which characterize the preimpact sedimentary section outside the rim (Figs. 4.3–4.19). On typical profiles, the coherency loss marks the steep normal fault scarp, formed by massive failure, slumping, and sliding of the sedimentary section near the maximum lateral limit of strongest ground-shock effects and of surgeback effects from subsequent water-column collapse. The rim scarp appears to extend all the way to the crystalline basement surface on most profiles, at which level the failed sediments become detached along a surface or zone of décollement. At or near the base of the outer rim scarp are huge megaslump and megaslide blocks, kilometers long, some of which have been horizontally displaced for short distances toward the crater center. Some of these displaced megablocks have rotated several hundred meters from their original near-horizontal positions (Figs. 4.3, 4.7B). Other megablocks appear to have simply dropped vertically downward, as their basal strata were disrupted by the impact; these blocks display little or no evidence of horizontal displacement (Fig. 4.3B).

The crater outer rim is crossed by seismic profiles at 61 locations (Figs. 3.3, 4.3–4.19; Table 4.1), providing good structural and morphologic control around the full 360° of the crater circumference, although there are wide gaps between some profiles. General features are similar on each profile, but in detail, morphologic variability is marked. The most extensively imaged part of the outer rim is

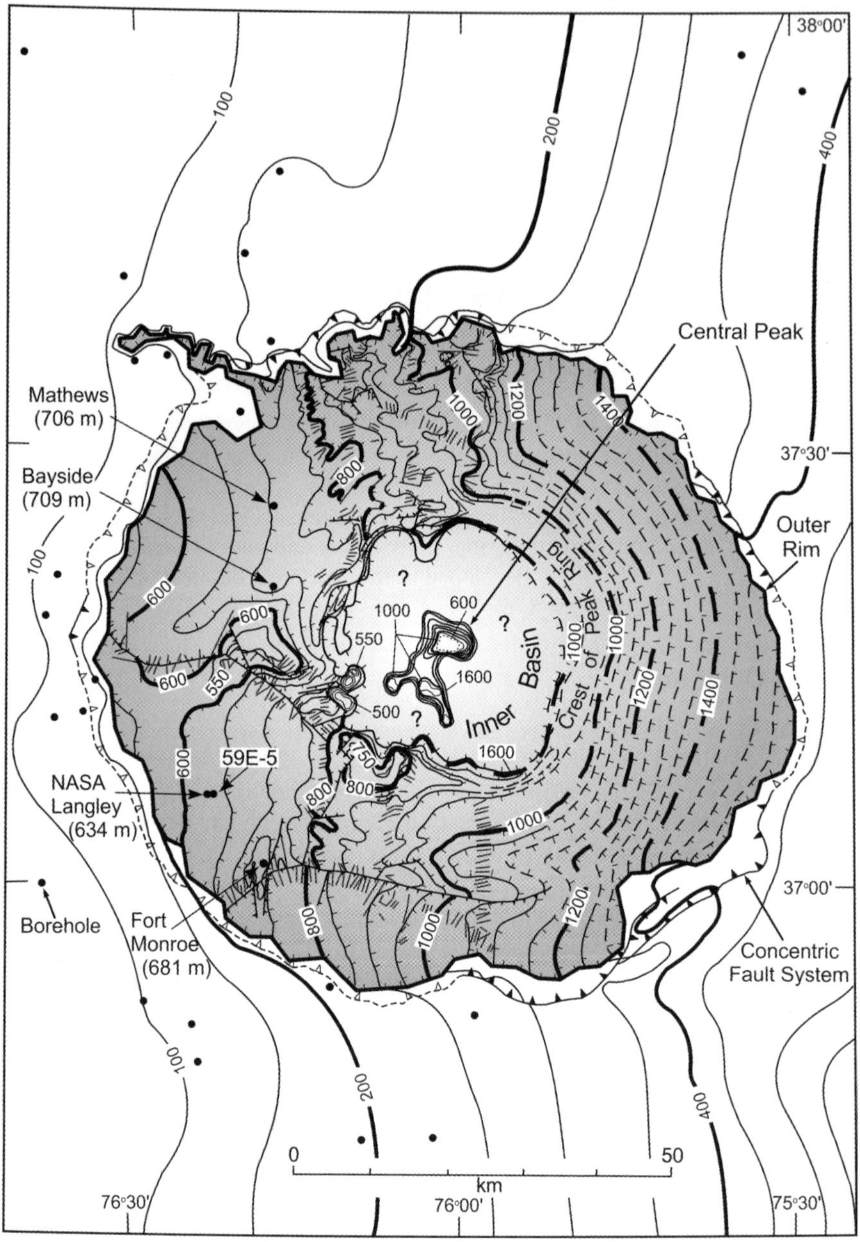

Fig. 4.1. Structural map of Chesapeake Bay impact crater constructed from seismic reflection and borehole data. Shaded area represents top of crystalline basement. Boreholes shown encountered either crystalline basement (inside crater; drill depth shown), or Exmore breccia (inside or outside crater), or unconformable surface correlative to Exmore breccia (outside crater). Contour interval 50 m. See CD-ROM.3 for sheet-sized color version.

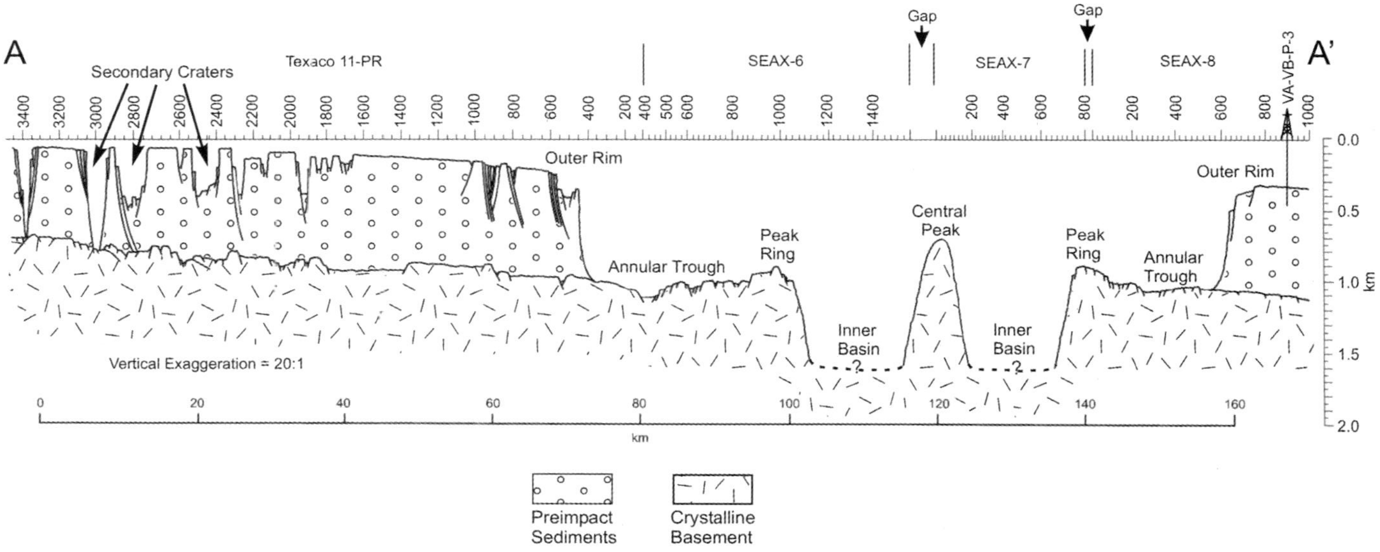

Fig. 4.2A. Depth-scaled cross section (derived from composited seismic reflection profiles) through lower part of Potomac River and Chesapeake Bay (*A-A'*), showing location and principal features of Chesapeake Bay primary and secondary impact craters. Exmore breccia, displaced megablocks, and postimpact deposits have been removed. See also CD-ROM.6.

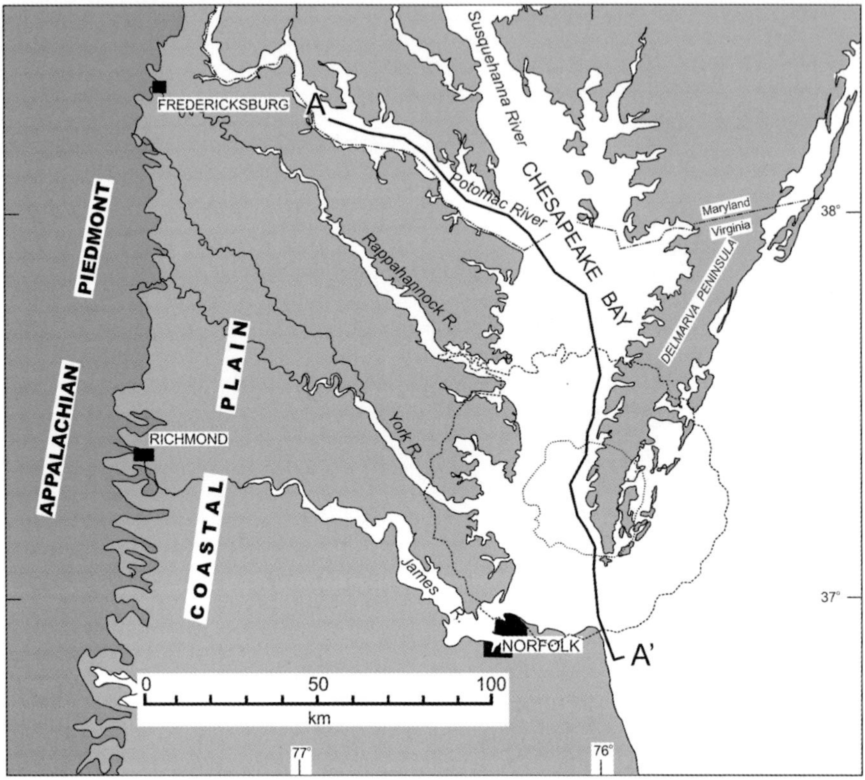

Fig. 4.2B. Location map for cross section *A-A'* shown in Fig. 4.2A.

the section within and directly east of the Rappahannock River. In this area are 21 rim crossings, most of which are diagonal to the rim as seen in map view (Fig. 4.4). These data clearly indicate that the outer rim is not a smooth, circular escarpment, but is broken up into blocky segments (Figs. 4.1, 4.4, 4.5; CD-ROM.3, 4). These segments are roughly 2–4 km on a side, though their precise geometries cannot be determined from the spacing of our tracklines. Two elongate narrow blocks form slender promontories that jut 2–3 km inward toward the center of the crater (Figs. 4.1, 4.4; CD-ROM.3, 4).

Of particular interest on the northwest segment of the crater rim is a 15-km-long, 1- to 3-km-wide, canyon-like feature, which extends up the Rappahannock River (Figs. 4.1, 4.6, 4.7; CD-ROM.3). Here, three subparallel and partly overlapping seismic profiles (SEAX-12, SEAX-13, 10-RR; Fig. 4.6) show the geometric complexity of the canyon, as the seismic tracklines cross in and out of the canyon in very short lateral distances (0.25–6 km; average 1.9 km; Fig. 4.7). We infer that the canyon is a collapse zone related to a radial basement fault that appears to extend beneath the Rappahannock River (see Chapter 9 for further discussion of the radial faults).

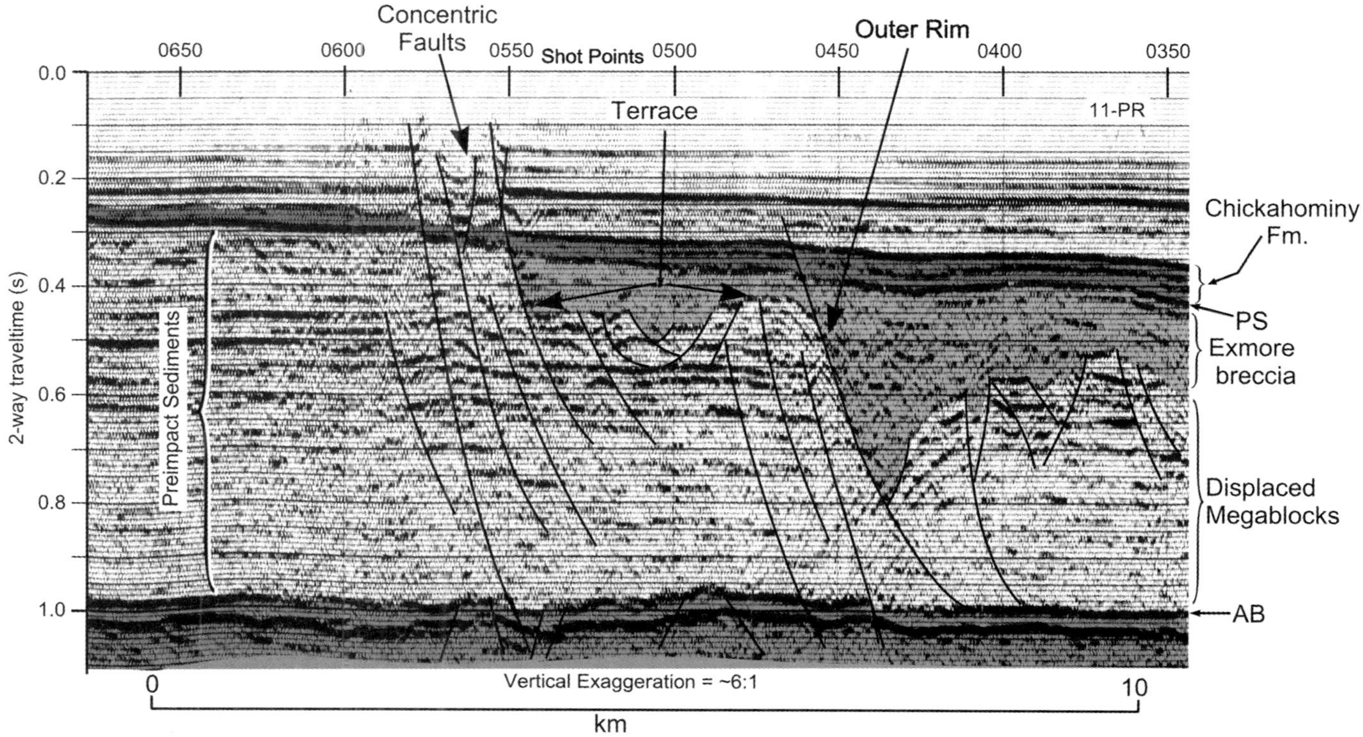

Fig. 4.3A. Interpreted segment of seismic reflection profile 11-PR crossing northern part of outer rim of Chesapeake Bay impact crater; see Fig. 4.4 for location, CD-ROM.14 for full-scale profile, and Table 3.2 for further information about seismic profile. *PS* designates base of postimpact sedimentary section (approximate boundary between Exmore breccia and Chickahominy Formation); *AB* (acoustic basement) designates upper surface of crystalline basement.

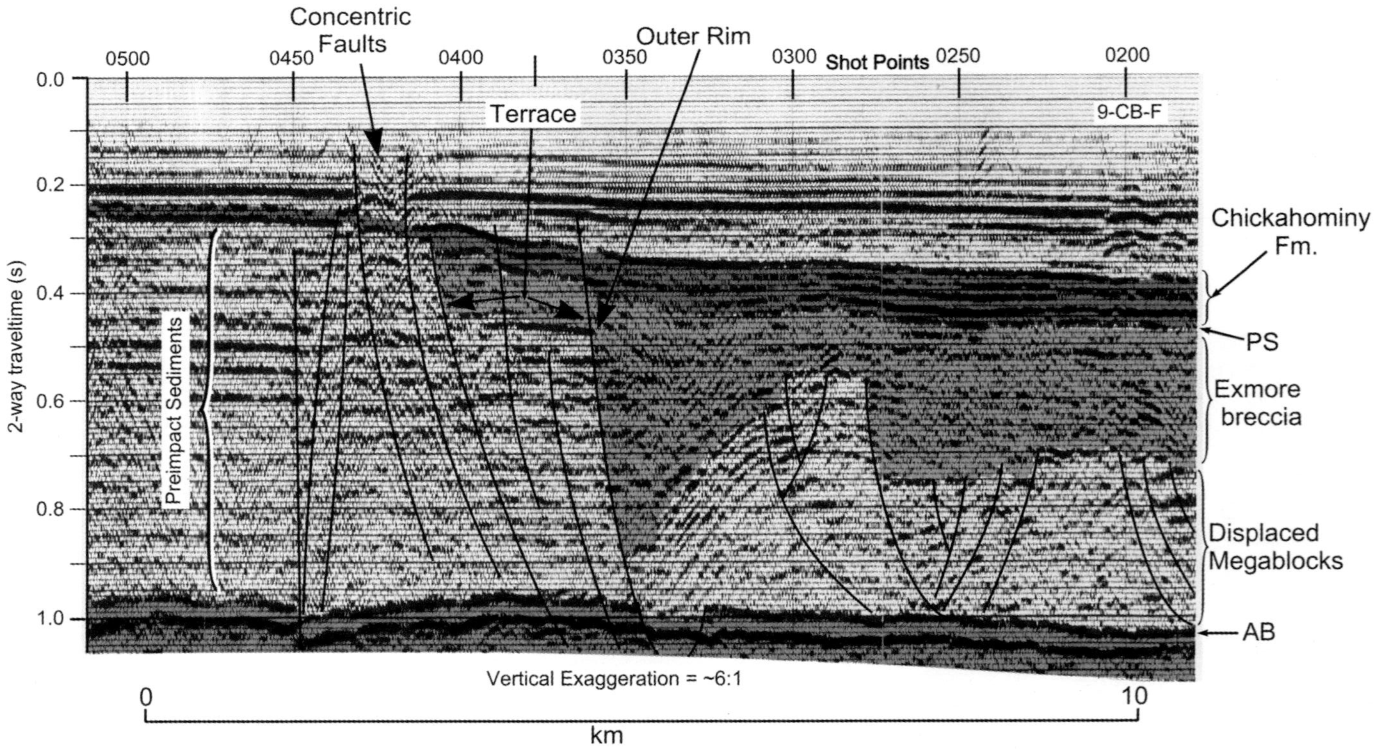

Fig. 4.3B. Interpreted segment of seismic reflection profile T-9-CB-F crossing northern part of outer rim of Chesapeake Bay impact crater; see Fig. 4.4 for location, CD-ROM.13 for full-scale profile, and Table 3.2 for further information. *PS* designates base of postimpact sediments (approximate boundary between Exmore breccia and Chickahominy Formation); *AB* (acoustic basement) designates upper surface of crystalline basement.

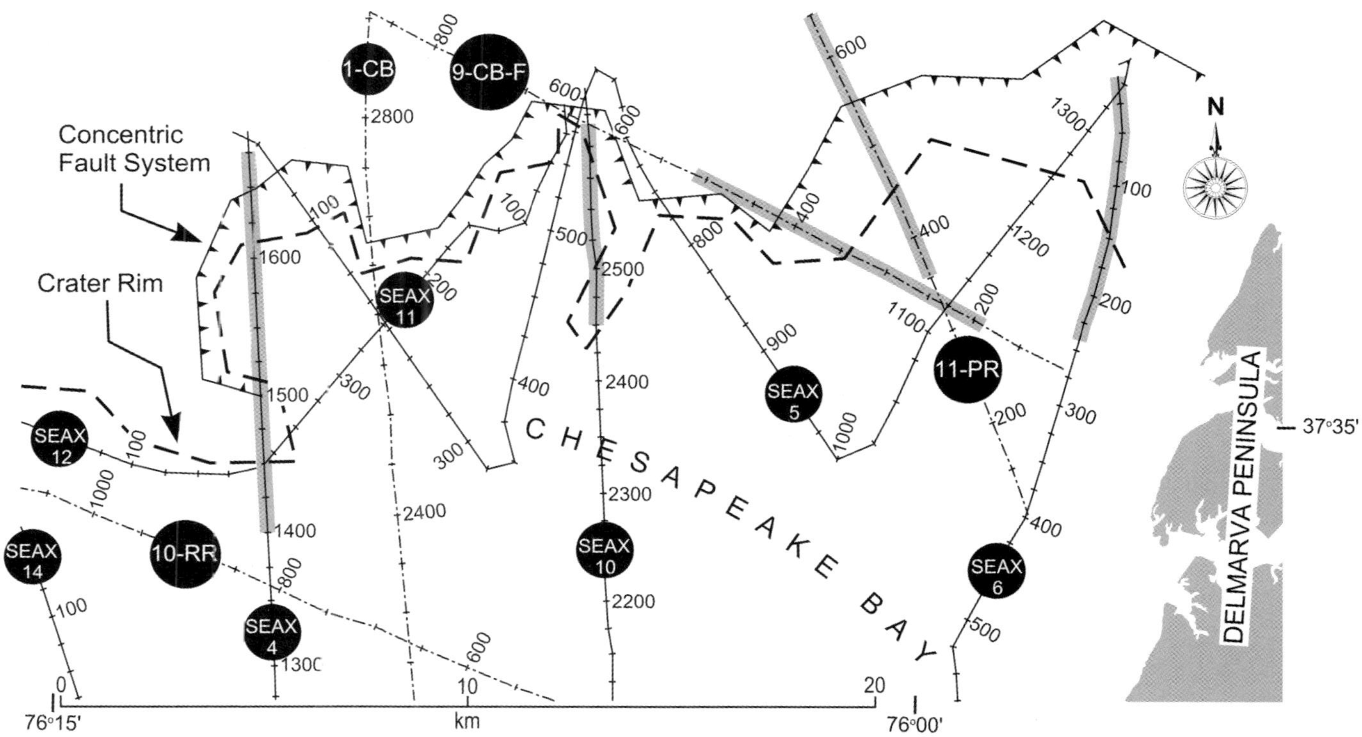

Fig. 4.4. Detailed trackline map, showing shot points for profiles along northern rim of Chesapeake Bay impact crater between mouth of Rappahannock River and western shore of Delmarva Peninsula. Shaded lines indicate segments of profiles 11-PR and T-9-CB-F (shown in Fig. 4.3A,B), profiles S-10 (SEAX-10) and S-4 (SEAX-4) (shown in Fig. 4.5A,B), and profile S-6 (SEAX-6; shown in Fig. 4.14).

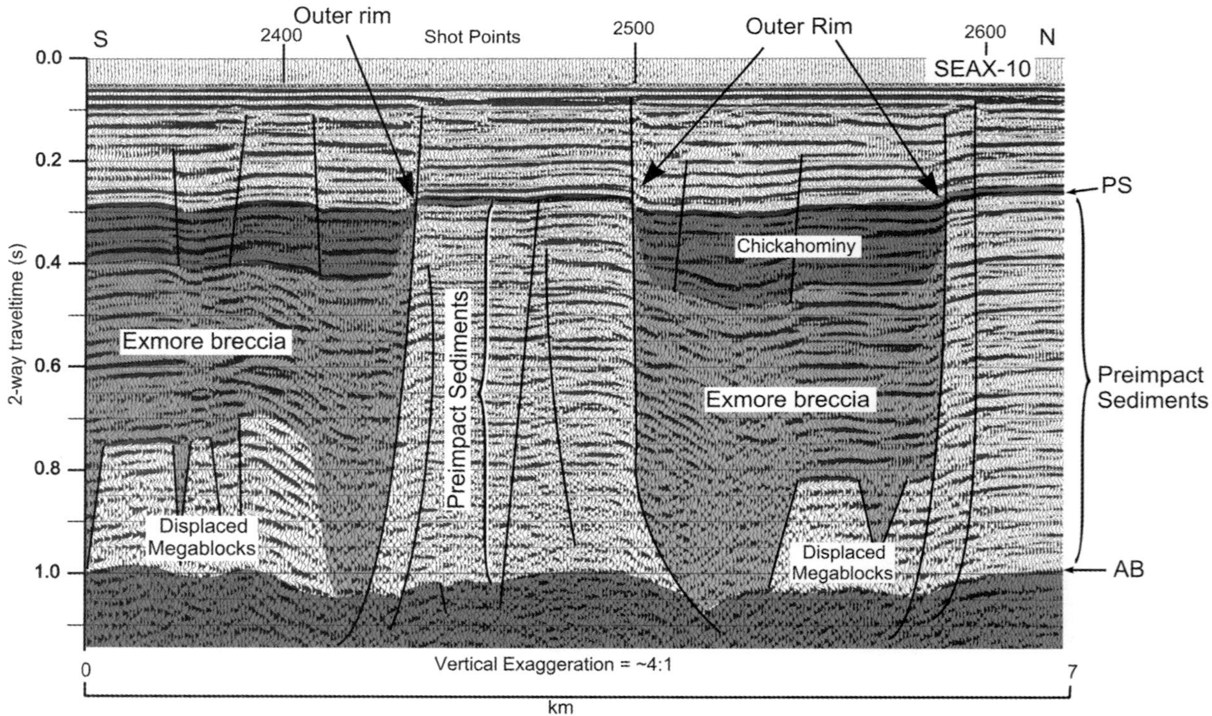

Fig. 4.5A. Interpreted segment of seismic reflection profile SEAX-10 crossing northern part of outer rim of Chesapeake Bay impact crater. Profile crosses prominent blocks of preimpact sediment, which project out into the crater as part of irregularly faulted and slumped outer rim. See Fig. 4.4 for precise location and CD-ROM.10 for full-scale profile; reflection abbreviations as in Fig. 4.3.

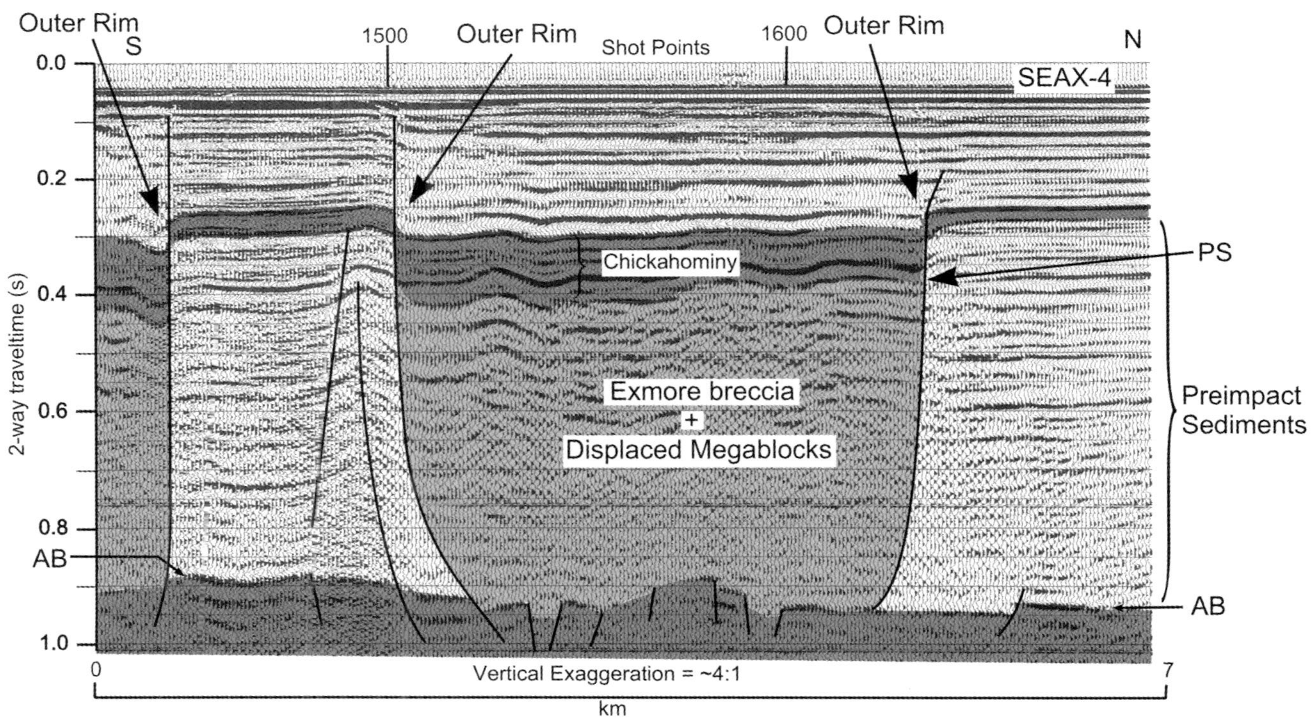

Fig. 4.5B. Interpreted segment of seismic reflection profile SEAX-4 crossing northern part of outer rim of Chesapeake Bay impact crater. Profile crosses prominent blocks of preimpact sediments, which project out into the crater as part of irregularly faulted and slumped outer rim. See Fig. 4.4 for precise location and CD-ROM.8 for full-scale profile; reflection abbreviations as in Fig. 4.3.

Table 4.1. Vertical structural [1]relief at outer rim of Chesapeake Bay impact crater.

Seismic Trackline	Shot Point	Relief [m]	Seismic Trackline	Shot Point	Relief [m]
S-1	460	>1200	S-12	597	570
S-2	502	420	S-12	545	530
S-3	137	420	S-12	515	580
S-4	1625	610	S-12	435	470
S-4	1505	540	S-12	220	470
S-4	1450	470	S-13	65	480
S-5	110	700	S-13	240	430
S-5	580	700	S-13	390	480
S-5	768	730	S-13	460	440
S-5	1255	710	S-13	520	450
S-6	155	760	S-16	435	450
S-8	665	660	S-17	050	530
S-9	370	620	S-19	165	>1200
S-10	2598	740	S-22	560	>1200
S-10	2501	700	S-22	940	>1200
S-10	2442	700	S-25	455	>1200
S-11	385	570	S-27	455	>1200
S-11	360	600	N-3	[2]1940 hrs	420
S-11	180	570	1-CB	340	630
S-11	148	580	11-PR	440	700
S-11	060	630	9-CB-F	340	700
S-11	030	650	10-RR	1150	550
S-11	005	720	10-RR	1400	520
S-12	870	460	10-RR	1430	450
S-12	830	440	10-RR	1480	430
S-12	755	480	10-RR	1560	470
S-12	750	470	13-YR	2140	420
S-12	710	500	I-64	4075	300
S-12	635	520	E-3	2430	1250

[1]Relief measured from sedimentary lip to crystalline floor on seismic reflection profiles
[2]*hrs* indicates location on seismic trackline is measured in clock time rather than in shot points

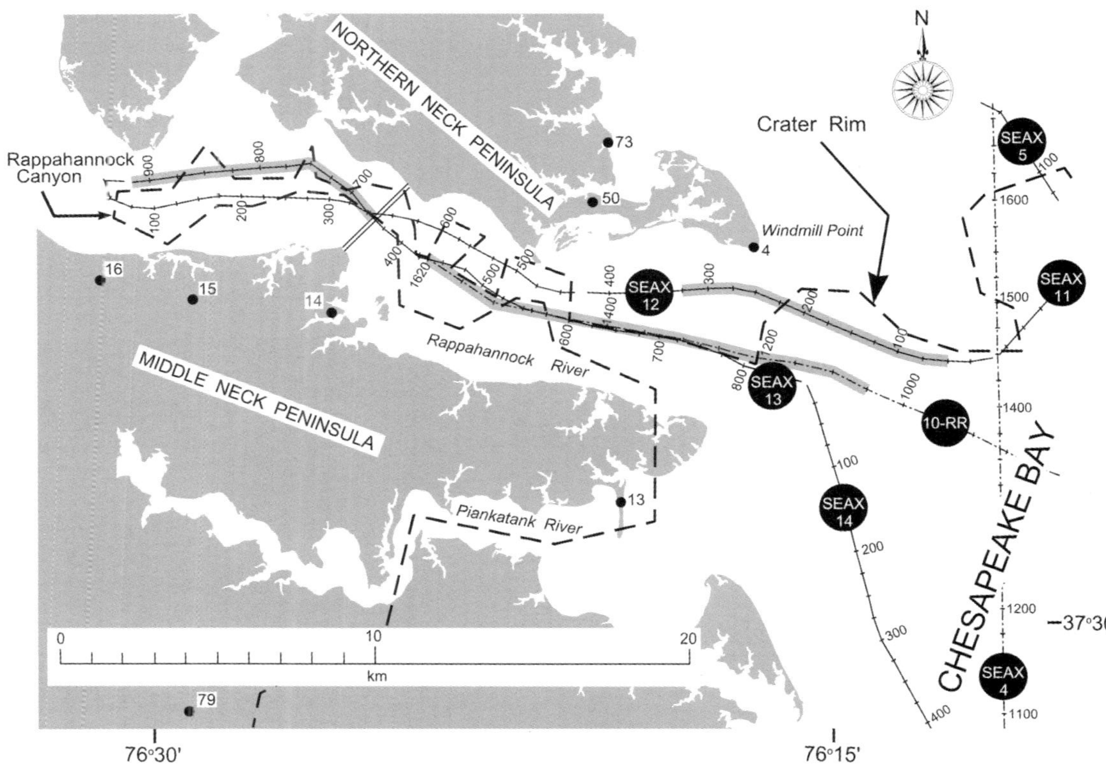

Fig. 4.6. Detailed trackline map, showing shot points for profiles within lower reach of Rappahannock River, which image Rappahannock Canyon. Shaded lines indicate segments of profiles SEAX-12 (S-12) and T-10-RR (10-RR) shown in Fig. 4.7A,B.

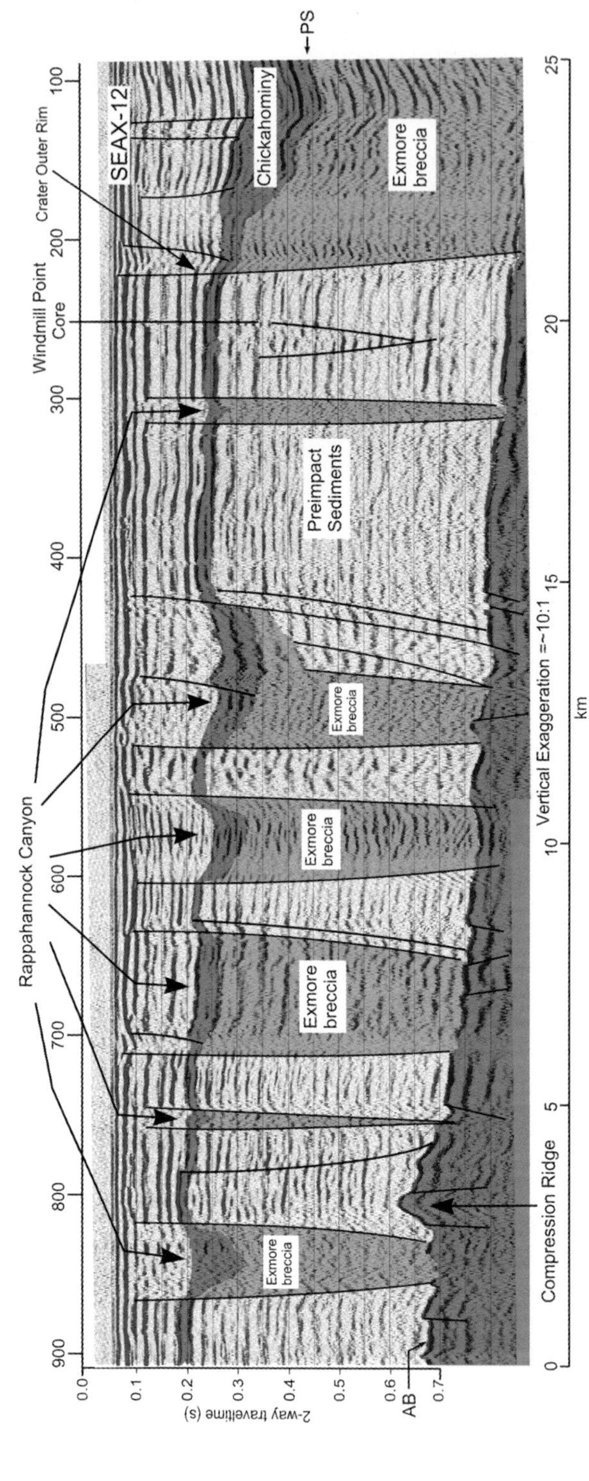

Fig. 4.7A. Interpreted segment of seismic reflection profile SEAX-12, which crosses in and out of Rappahannock Canyon. See Fig. 4.6 for precise location; reflection abbreviations as in Fig. 4.3.

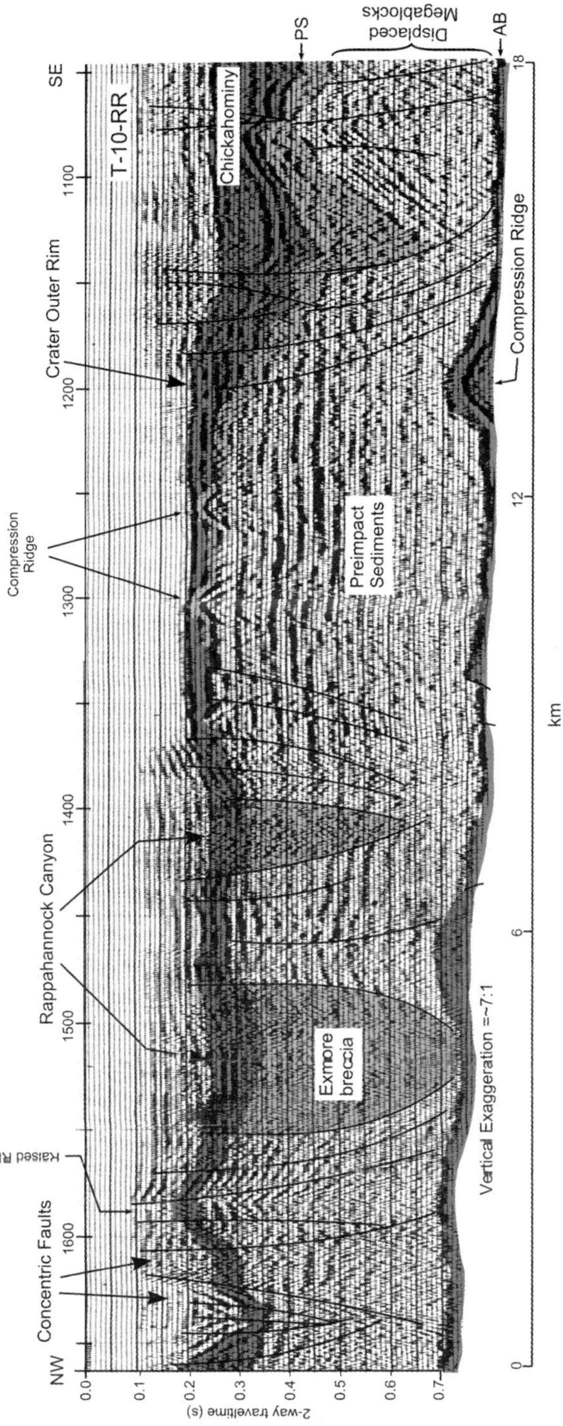

Fig. 4.7B. Interpreted segment of seismic reflection profile T-10-RR, which crosses in and out of Rappahannock Canyon. See Fig. 4.6 for precise location; reflection abbreviations as in Fig. 4.3.

The Rappahannock Canyon is a unique feature in our data set; nothing similar has been observed on profiles in other parts of the crater. In contrast, for example, three subparallel profiles in the York River (SEAX-2, SEAX-3, 13-YR; 0.25 km apart; Figs. 4.8, 4.9A,B) cross the outer rim of the crater at approximately the same relative position, evidence that this segment of the outer rim is nearly perpendicular to the river channel. In the James River, three profiles (SEAX-16, SEAX-17, Neecho-2; approximately 0.2–1.3 km apart; Figs. 4.10, 4.11) also cross the crater outer rim at the same relative positions. We infer that this segment of the outer rim also is perpendicular to the river channel, and extends southeastward beneath the Norfolk Naval Base and northwestward under the city of Hampton.

Poag (1997a) reported that the southeast segment of the crater's outer rim (diagonally southeast from the Rappahannock Canyon) appeared to have massively collapsed (based on single-channel profile F-1; Fig. 3.3; Table 3.2), which carved out a broad embayment in this segment of the crater rim. One of the new *Ewing* multichannel profiles (E-3; Figs. 4.12, 4.13) substantiates Poag's original interpretation. On a few seismic profiles, particularly to the north, west, and southeast, mass failure of the outer rim is limited to the upper few tens of meters of the preimpact sedimentary section. This shallow mass failure has formed narrow (0.5–4 km) terraces concentric to the outer rim in these areas (Figs. 4.3A,B, 4.4). The outer edges of the terraces are marked by auxiliary normal faults approximately concentric to the primary rim fault.

Auxiliary concentric normal faults also are present in some locations where the preimpact sediments display only incipient disruption. The most notable example of this type of fault parallels the rim in the crater's southeast quadrant (Fig. 4.1; CD-ROM.3). These auxiliary concentric faults indicate that the impact shock and subsequent water-column collapse were strong enough to initiate additional sedimentary failure outside the crater rim, but that failure was incomplete, resulting in only minor vertical offsets. Evidence of auxiliary concentric faults at nearly every rim crossing suggests that these features probably encircle the entire crater rim.

Structural relief at the crater's outer rim, as measured from the sedimentary lip of the crater to the crater floor (crystalline basement surface) varies from a minimum ~300–420 m on the updip (west) side (profiles along Interstate Highway 64 and in the Rappahannock, York, and James Rivers) to a measured maximum of ~760 m on the northern rim (profile S-6; Table 4.1). Extrapolated maximum relief of ~1,200 m occurs on the downdip (east) side where the basement is too deep to have been imaged by our single-channel seismic data.

An important and diagnostic feature of the outer rim is the marked thickening and structural sagging that takes place within the postimpact sedimentary units, as they cross into the crater (Fig. 4.14; Table 4.2). This cross-rim thickening varies from 80 m beneath the York River (profile S-3; Fig. 4.9A) to 190 m near the mouth of the Rappahannock River (profile S-13; Fig. 4.6). Average increase in postimpact sediment thickness at the rim is 119 m. This represents an into-the-crater thickness increase of 24–130 percent; average increase is 59 percent (Table 4.2). This change takes place within a lateral distance of 0.7–10 km, averaging 3.0 km.

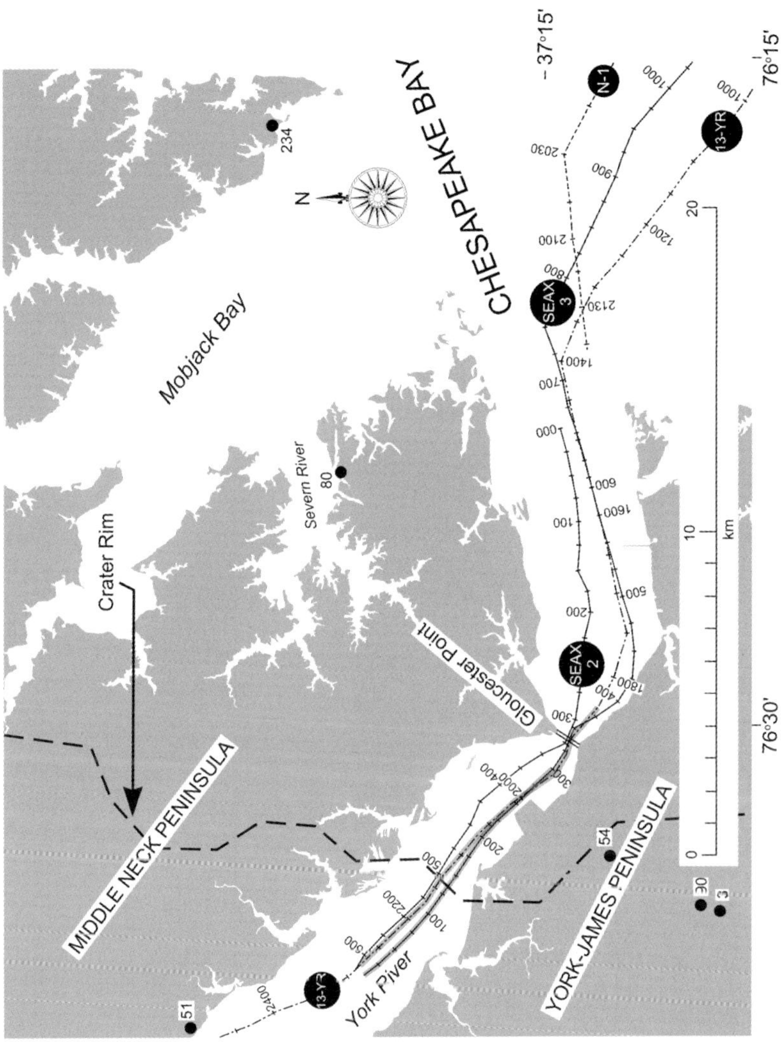

Fig. 4.8. Detailed trackline map, showing shot points for profiles within lower reach of York River. Shaded lines designate segments of profiles SEAX-3 (S-3) and 13-YR shown in Fig. 4.9A,B. Reflection abbreviations as in Fig. 4.3.

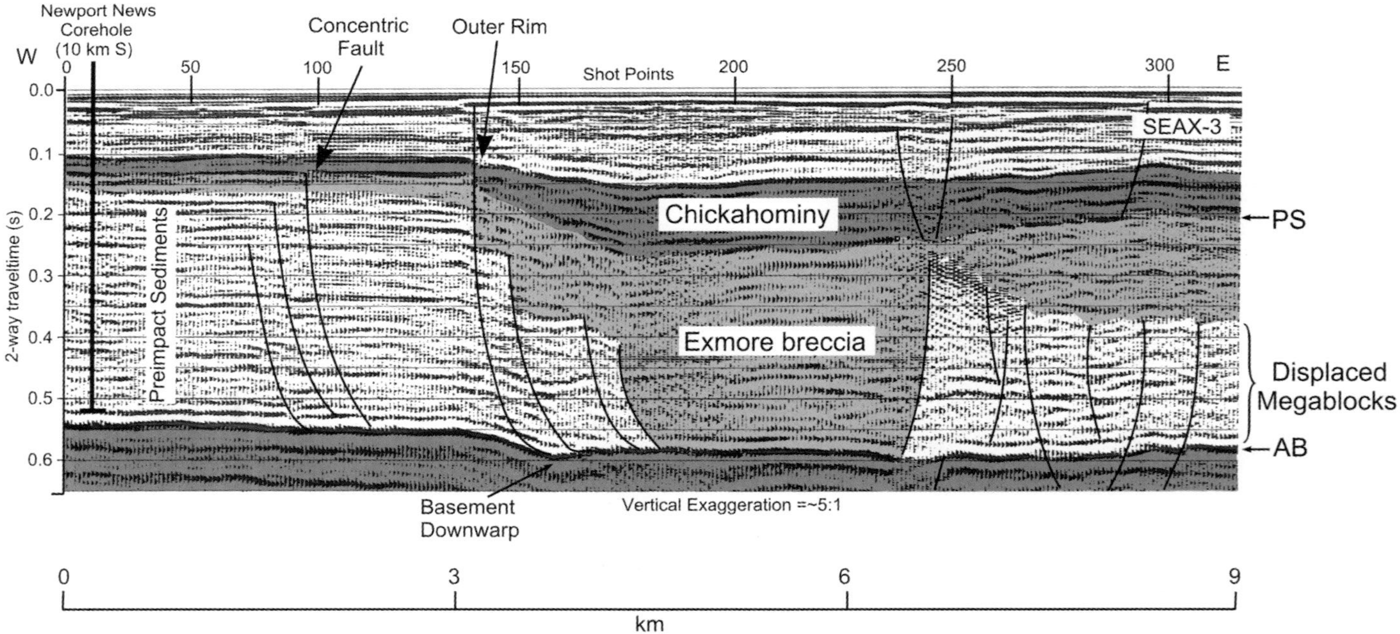

Fig. 4.9A. Interpreted segment of seismic reflection profile SEAX-3 (S-3) crossing outer rim of Chesapeake Bay impact crater within the lower reach of York River. See Fig. 4.8 for precise location. Reflection abbreviations as in Fig. 4.3.

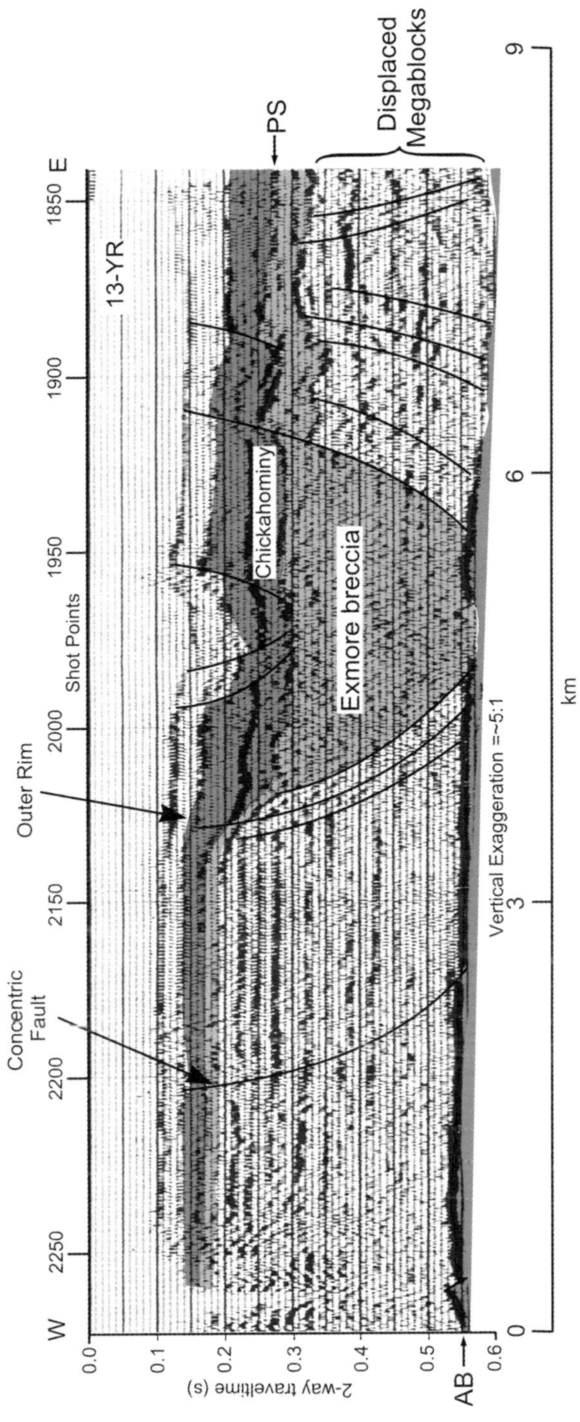

Fig. 4.9B. Interpreted segment of seismic reflection profile 13-YR crossing outer rim of Chesapeake Bay impact crater within the lower reach of York River. See Fig. 4.8 for precise location. Reflection abbreviations as in Fig. 4.3.

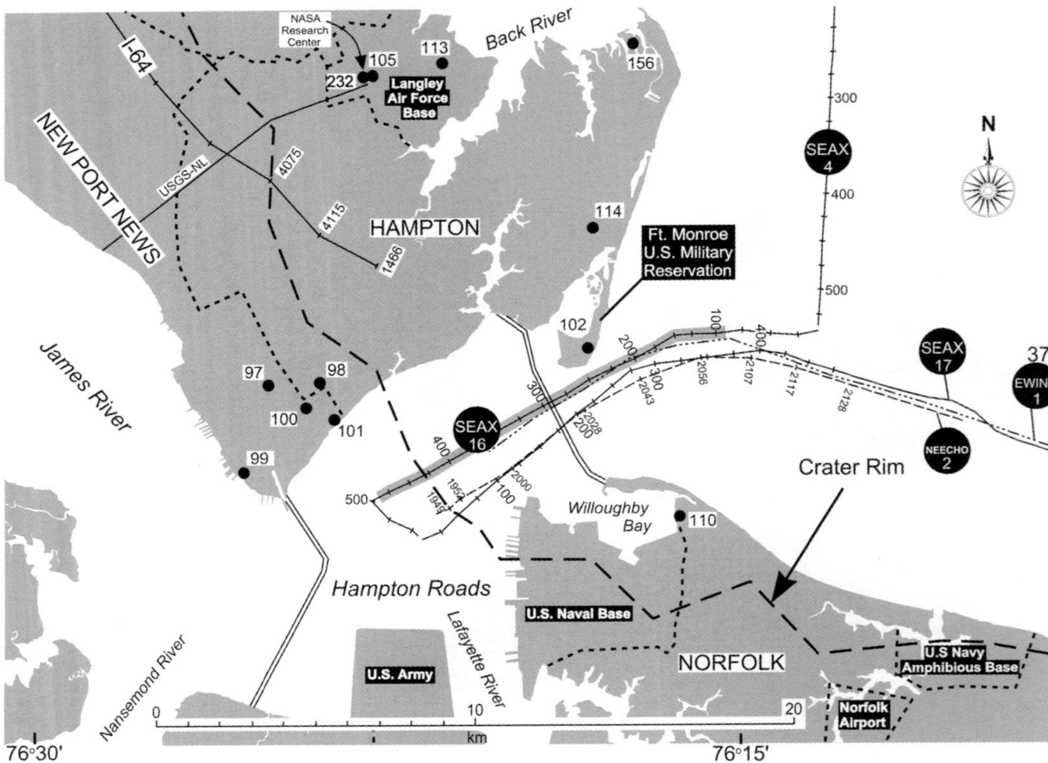

Fig. 4.10. Detailed trackline map, showing shot points for profiles within lower reach of James River (Hampton Roads) and along Interstate Highway 64 (I-64), and trackline for USGS-NL from NASA Langley corehole to James River. Shaded line indicates segment of profile SEAX-16 (S-16) shown in Fig. 4.11. Numbered solid black dots represent boreholes (see Table 1.2 and CD-ROM.1).

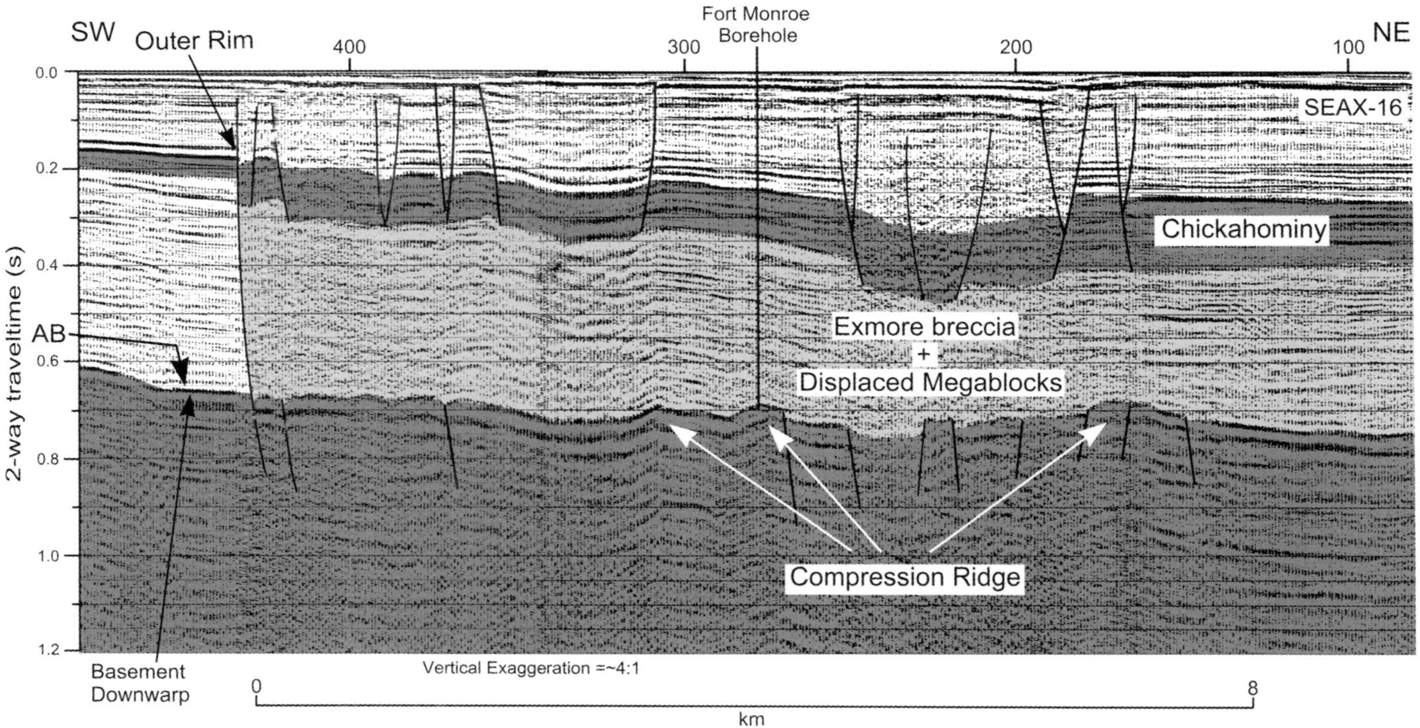

Fig. 4.11. Interpreted segment of seismic reflection profile SEAX-16 (S-16) crossing outer rim of Chesapeake Bay impact crater in lower reach of James River (Hampton Roads). See Fig. 4.10 for precise location and CD-ROM.8 for full-scale profile. Reflection abbreviations as in Fig. 4.3.

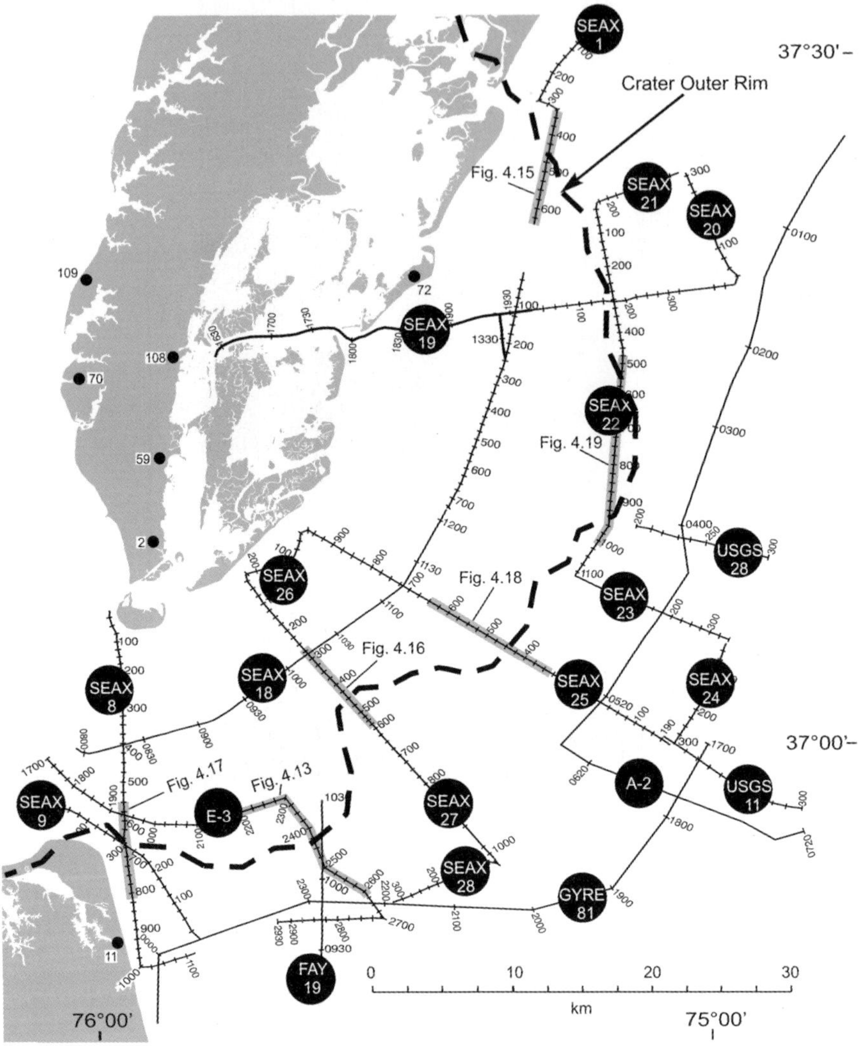

Fig. 4.12. Detailed trackline map, showing shot points for seismic reflection profiles crossing outer rim of Chesapeake Bay crater on inner continental shelf and in mouth of Chesapeake Bay. Shaded lines indicate segments of profiles shown in Fig. 4.13 and Figs. 4.15–4.19. Solid black dots represent boreholes (see Table 1.2 and CD-ROM.1).

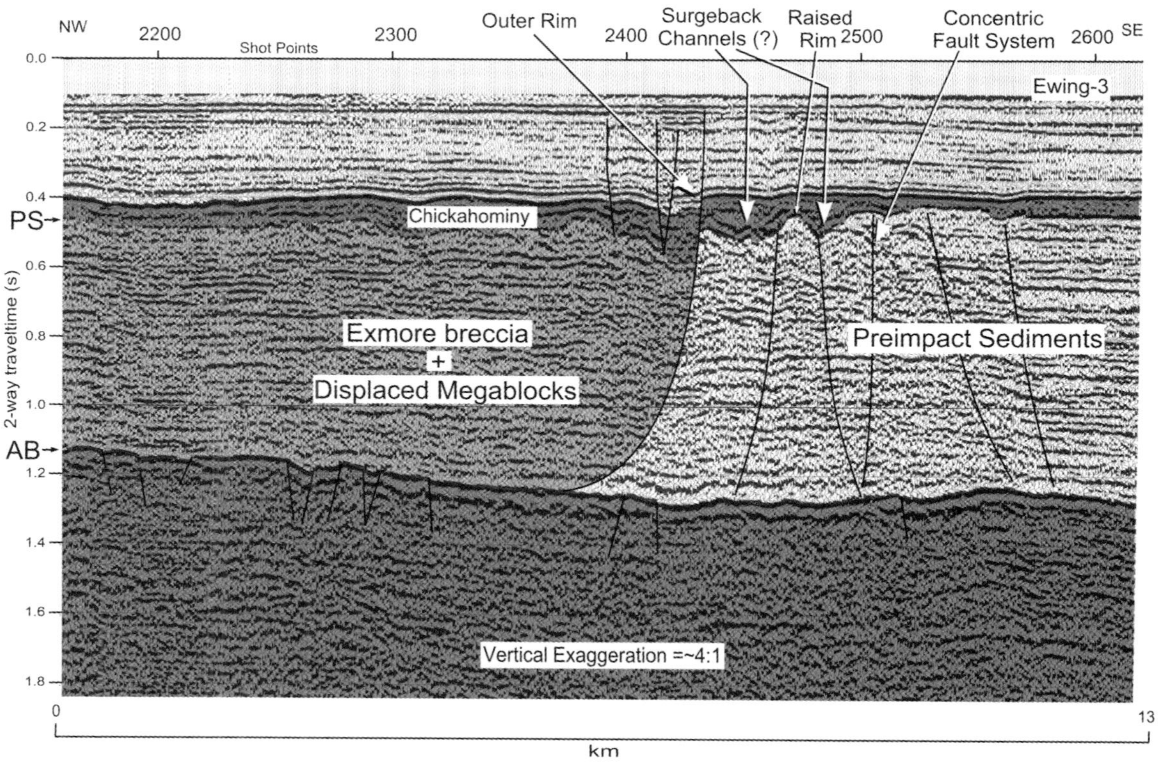

Fig. 4.13. Interpreted segment of seismic reflection profile E-3 (*Ewing*-3) crossing outer rim of Chesapeake Bay impact crater on inner continental shelf adjacent to mouth of Chesapeake Bay. See Fig. 4.12 for precise location and CD-ROM.16 for full-scale profile Reflection abbreviations as in Fig. 4.3.

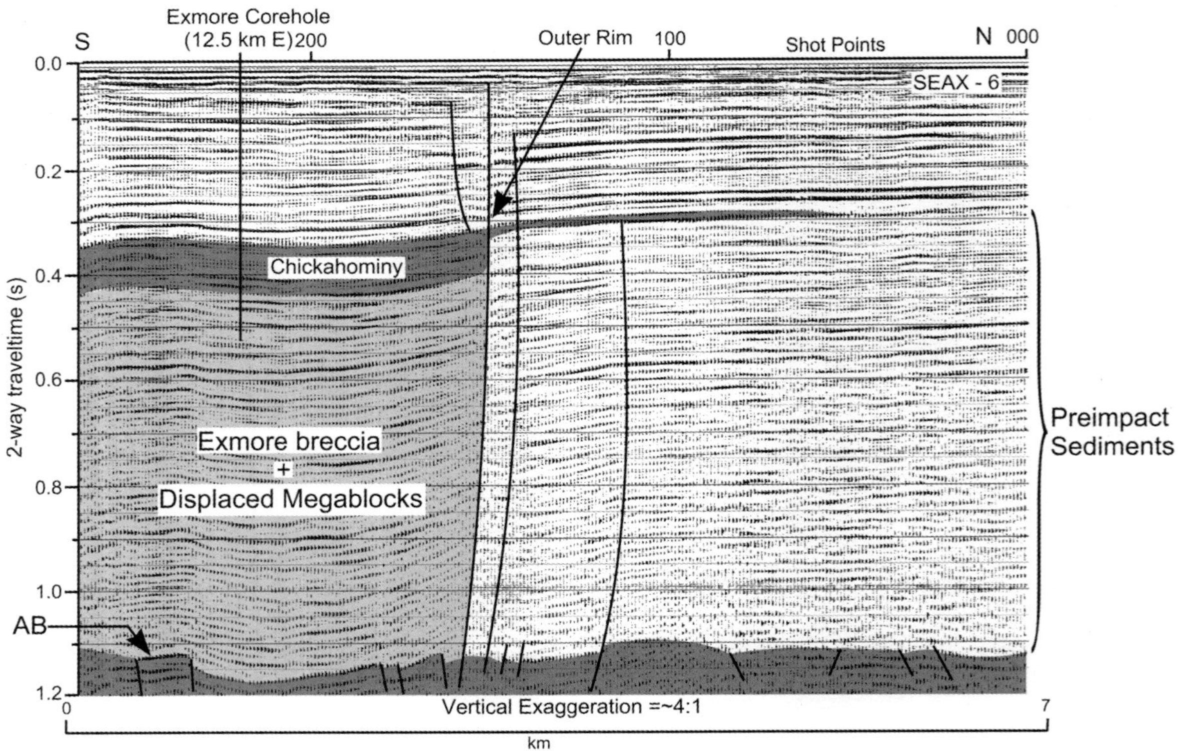

Fig. 4.14. Interpreted segment of seismic reflection profile SEAX-6 (S-6) crossing northern part of outer rim of crater near western shore of Delmarva Peninsula. See Fig. 4.4 for precise location and CD-ROM.9 for full-scale profile. Reflection abbreviations as in Fig. 4.3.

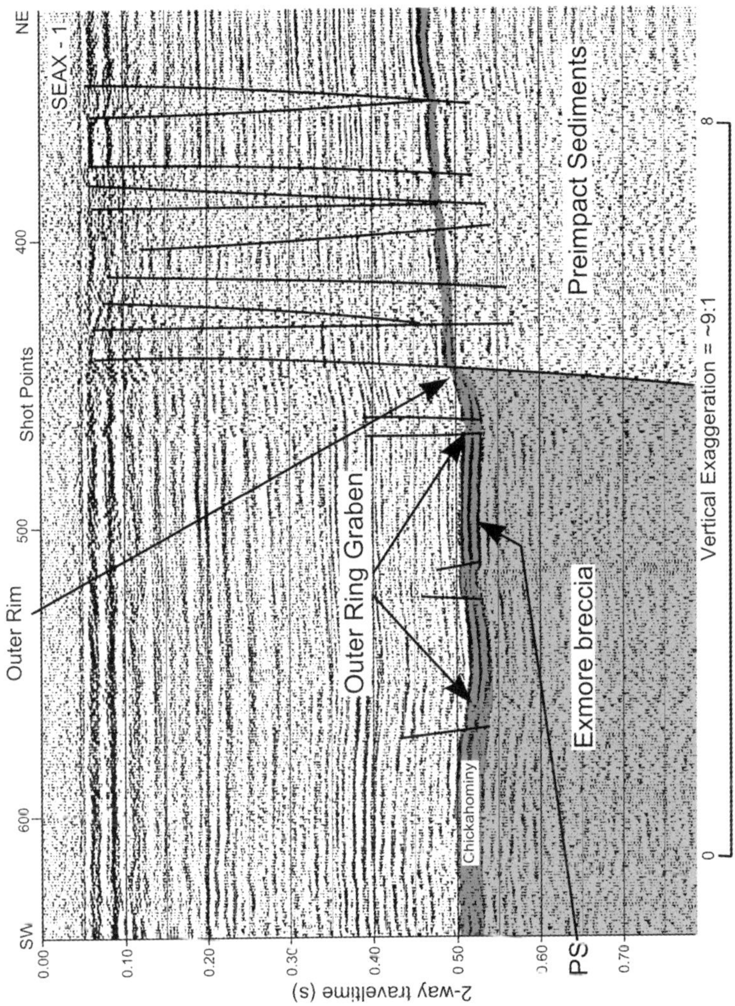

Fig. 4.15. Interpreted segment of seismic reflection profile SEAX-1 (S-1) crossing outer rim in eastern sector of Chesapeake Bay impact crater. See Fig. 4.12 for precise location. Reflection abbreviations as in Fig. 4.3.

Table 4.2. Morphometric data for structural and depositional features of outer rim of Chesapeake Bay impact crater, as determined from seismic profiles .

Profile #	Basement Sag [m]	Rim Terrace [km]	Postimpact Sediment Thickness Outside [m]	Postimpact Sediment Thickness Inside [m]	Total Thickness Increase [m]	Percent Thickness Increase [m]	[1] Lateral Distance [km]	[2] Concentric Fault	Raised Rim [m]
S-2	40	NO	110	210	100	91	2.5	?	NO
S-3	40	0.8	130	210	80	62	2.5	YES	NO
T-13-YR	40	NO	130	240	110	85	3.0	YES	NO
S-16	50	NO	150	250	100	67	3.2	YES	NO
S-17	50	NO	150	300	150	100	3.2	YES	NO
N-3	?	NO	130	240	110	85	1.7	YES	NO
S-12	30	NO	160	300	140	88	3.4	YES	NO
S-13	60	1.3	150	340	190	130	0.8	YES	NO
T-10-RR	70	0.9	200	300	100	50	1.6	YES	NO
S-5	?	NO	230	340	110	48	1.0	YES	NO
S-6	70	NO	280	380	100	36	2.7	YES	NO
S-8	NONE	1.3	240	380	140	42	10.0	YES	10

Table 4.2. (cont.)

S-9	10	0.7	290	380	90	31	4.0	YES	NO
S-10	40	NO	190	310	120	63	2.2	?	?
T-1-CB	NONE	NO	200	300	100	50	1.5	YES	NO
S-11	50	NO	190	290	100	53	0.7	NO	NO
S-19	?	?	410	540	130	32	3.5	YES	YES
S-22	?	?	420	530	110	26	2.0	YES	NO
S-25	?	?	380	540	160	42	6.0	YES	YES
S-27	?	?	380	470	90	24	1.5	YES	YES
S-1	?	?	390	550	160	41	6.4	YES	YES
Average	46	1.0	234	352	119	59	3.0		

[1] Indicates lateral distance between thinner postimpact section outside rim and thicker postimpact section inside crater.

[2] Indicates presence of a concentric normal fault or faults outside the rim scarp.

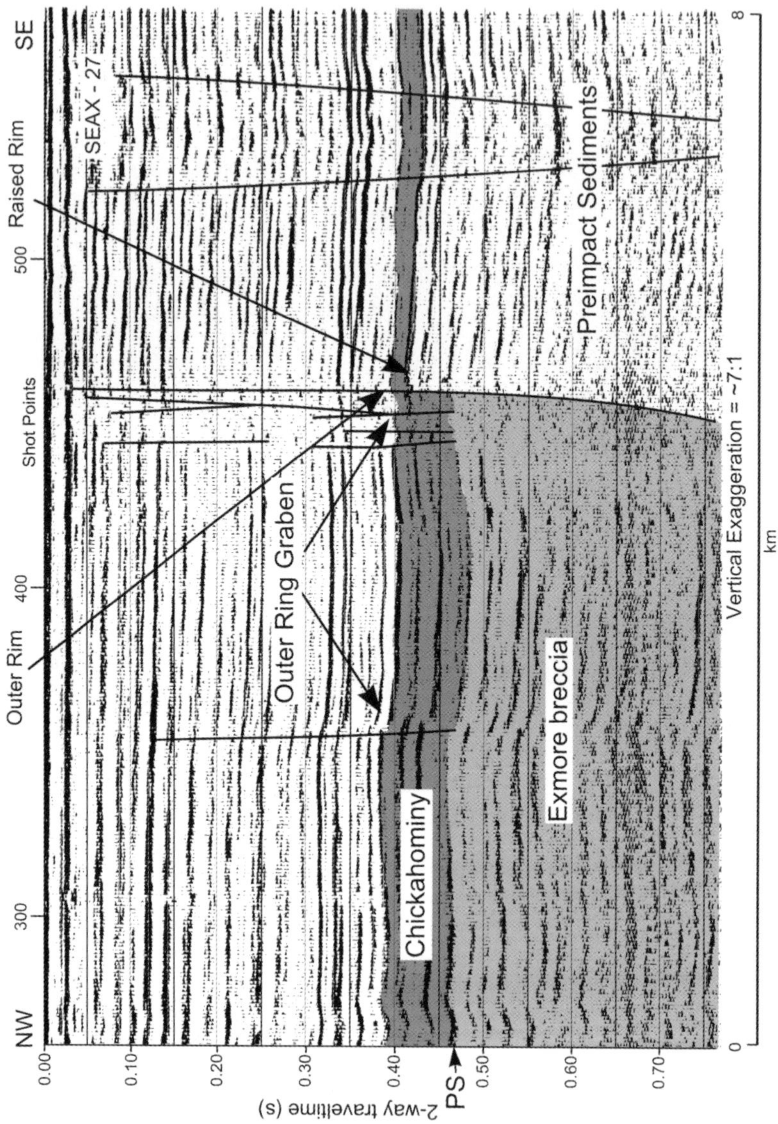

Fig. 4.16. Interpreted segment of seismic reflection profile SEAX-27 (S-27) crossing outer rim in eastern sector of Chesapeake Bay impact crater. See Fig. 4.12 for precise location. Reflection abbreviations as in Fig. 4.3.

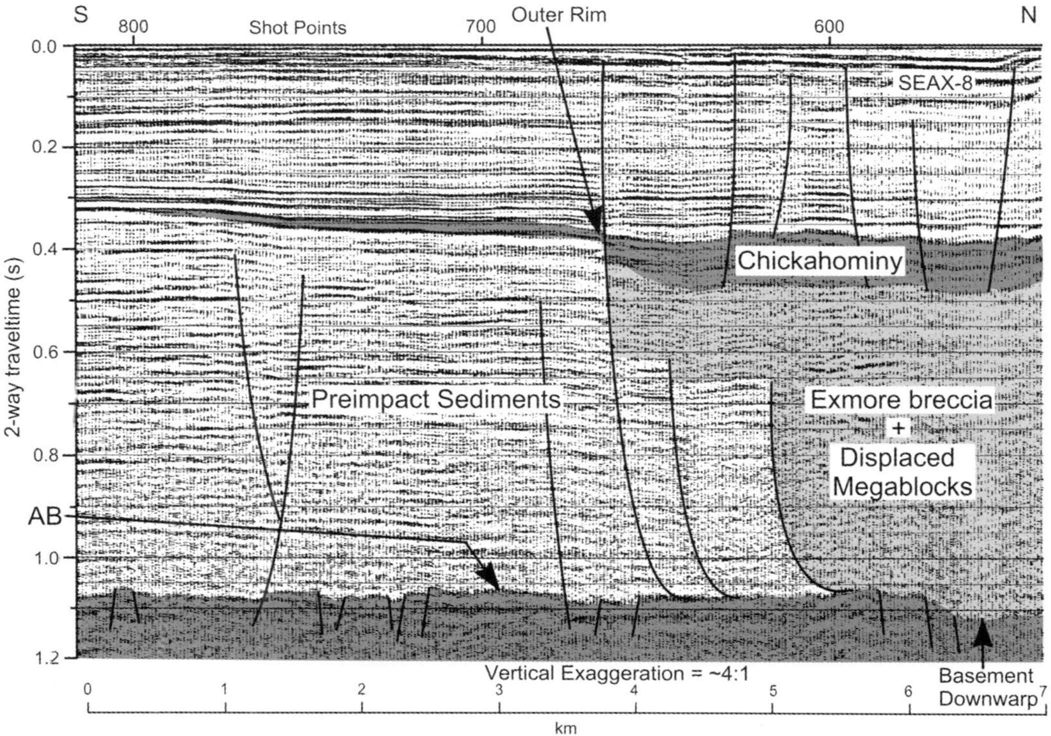

Fig. 4.17. Interpreted segment of seismic reflection profile SEAX-8 (S-8) crossing outer rim in southeastern sector of Chesapeake Bay impact crater. See Fig. 4.12 for precise location and CD-ROM.9 for full-scale profile. Reflection abbreviations as in Fig. 4.3.

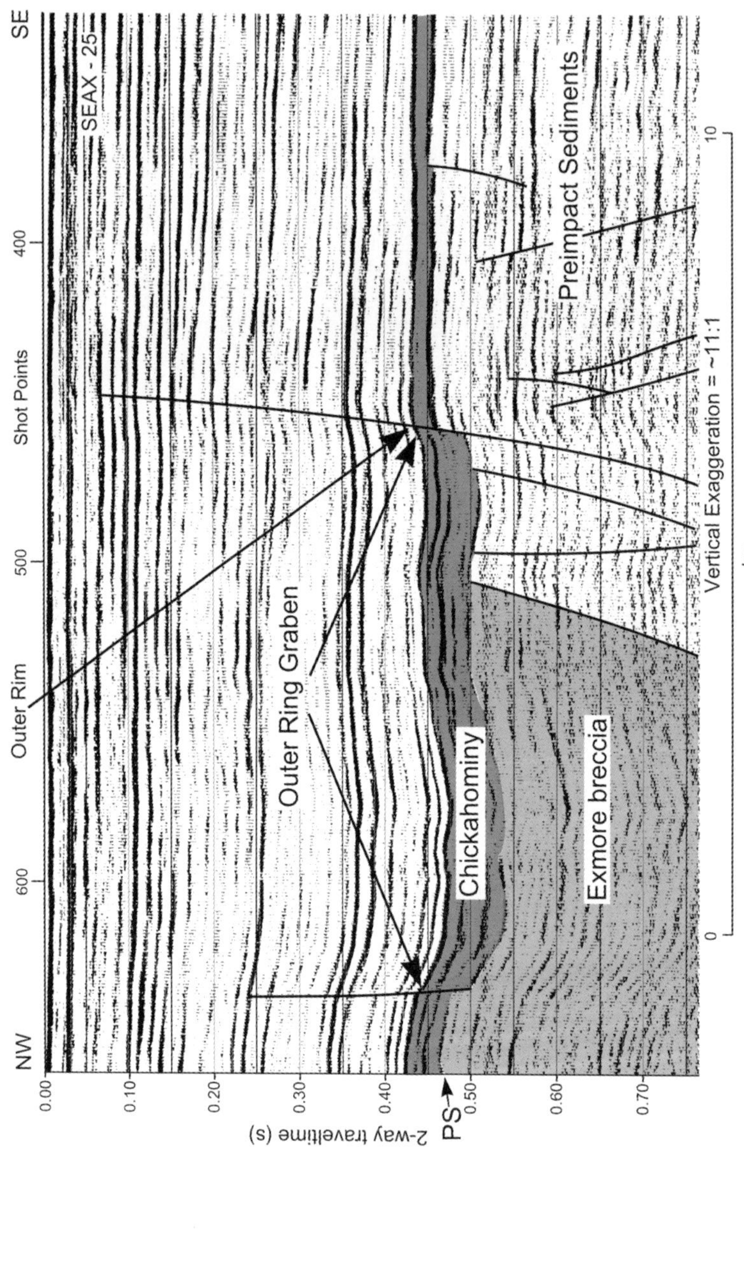

Fig. 4.18. Interpreted segment of seismic reflection profile SEAX-25 (S-25) crossing outer rim in eastern sector of Chesapeake Bay impact crater. See Fig. 4.12 for precise location. Reflection abbreviations as in Fig. 4.3.

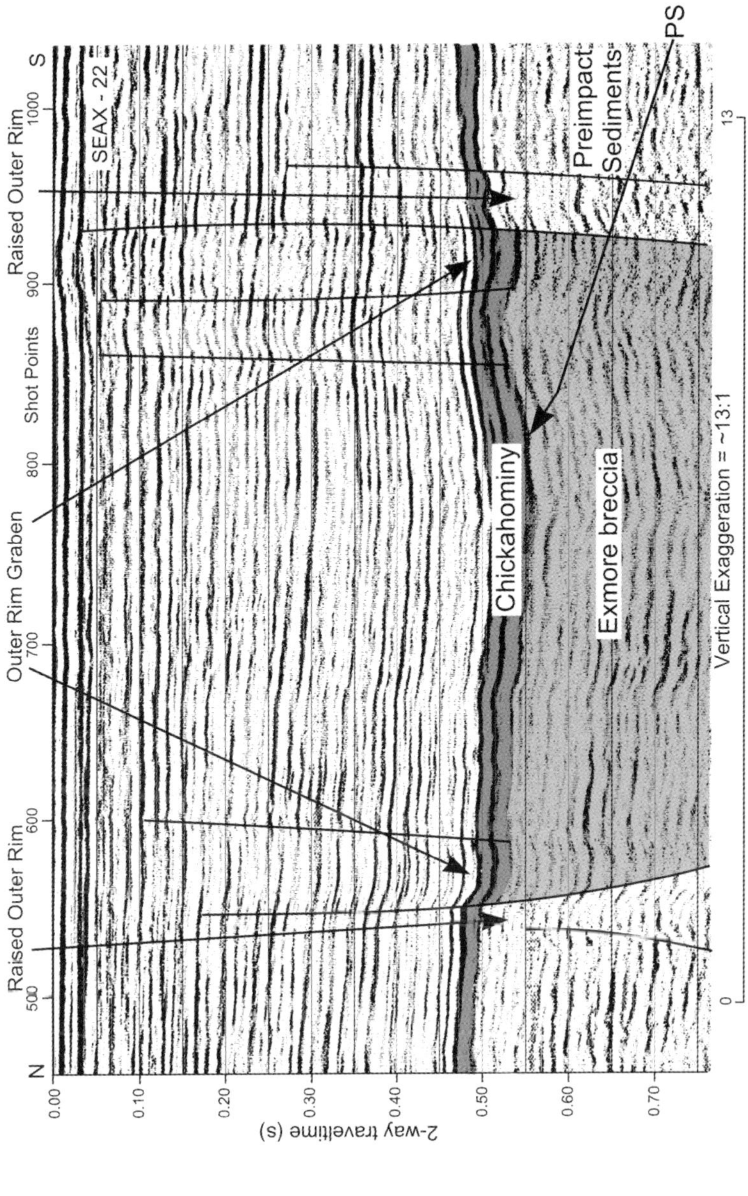

Fig. 4.19. Interpreted segment of seismic reflection profile SEAX-22 (S-22) crossing outer rim in eastern sector of Chesapeake Bay impact crater. See Fig. 4.12 for precise location. Reflection abbreviations as in Fig. 4.3.

The point at which sagging and thickening takes place provides a means to recognize the outer rim where other sedimentary features are difficult to discern on seismic profiles and where basement reflections are not visible. The eastern sector of the crater is a good example of the latter, and we have used the sagging-thickening relationship to identify the position of the outer rim in that part of the crater (Figs. 4.12, 4.14–4.19).

In several places, the crater's outer rim also is manifested on the surface of the crystalline basement. We have relatively closely spaced seismic control on the basement structure and morphology at the outer rim covering 230° (about 2/3) of the crater circumference – essentially the western and southern sectors. On most seismic profiles that cross the outer rim, the basement surface is offset by a marked downwarp into the crater and by extensive normal faulting. The lateral position of the downwarp varies from approximately 1 km outside the sedimentary rim (SEAX-16 in the James River; Fig. 4.11) to directly at the toe of the outer-rim escarpment (SEAX-3 in the York River; Fig. 4.9A), to ~1 km inside the crater rim (SEAX-8; Fig. 4.17). The relief across the downwarp varies from 10 m (SEAX-9) to 70 m (T-10-RR), and averages 48 m (Table 4.2). Six profiles on the eastern side of the crater did not image the basement surface (Figs. 4.15, 4.16, 4.18, 4.19).

4.1.1.2 Annular Trough

The floor of the annular trough laterally separates the toe of the outer rim scarp from the elevated outer flank of the peak ring (Fig. 4.2). As seen on most multichannel profiles in the western half of the crater, the floor of the trough is formed by a broad, wrinkled, faulted, eastward-dipping surface of crystalline basement. The width of the annular trough is highly variable (15–28 km), and averages ~21.4 km (Figs. 4.1, 4.2; CD-ROM.3, 5; Table 4.3). Extensional features dominate the floor of the trough, which is riddled with small-offset (<25 m) normal faults (Figs. 4.1, 4.13, 4.14), several normal faults with >25-m offsets (see Chapter 9), a few reverse faults, three principal radial fault systems, diverse linear depressions, and scattered concentric, low-relief, sometimes faulted, compression ridges (Figs. 4.7, 4.11, 4.20; Table 4.4). Many of the normal fault blocks form small grabens and horsts in the basement surface. The most prominent compression ridges are elevated 30–80 m above the floor of the annular trough, and are 0.9–3.5 km wide (Figs. 4.11, 4.20). Some compression ridges also are present beneath the outer rim (Fig. 4.7) and beneath the Rappahannock Canyon (Figs. 4.7, 4.20).

4.1.1.3 Peak Ring

The boundary between the annular trough and the peak ring is defined by a point at which the crystalline basement surface begins to rise gently in a radial direction toward the crater center. The surface gradually reaches an apex at the crest of an irregular, fractured, blocky ridge, which encircles the deep inner basin of the crater (Figs. 1.5, 4.1, 4.2, 4.21–29; CD-ROM.3–6). This subcircular ridge, or peak

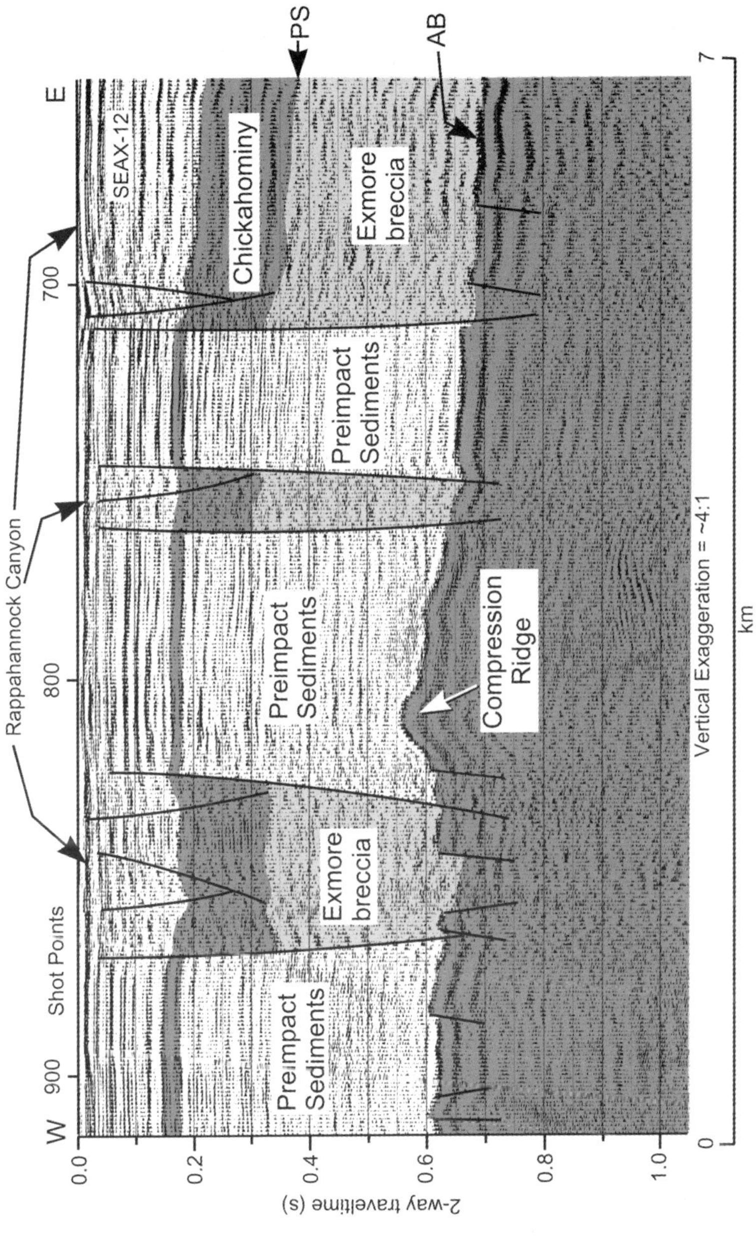

Fig. 4.20A. Interpreted segment of seismic reflection profile SEAX-12 (S-12), showing compression ridge on basement surface in annular trough of Chesapeake Bay impact crater. See Fig. 4.6 for precise location. Reflection abbreviations as in Fig. 4.3.

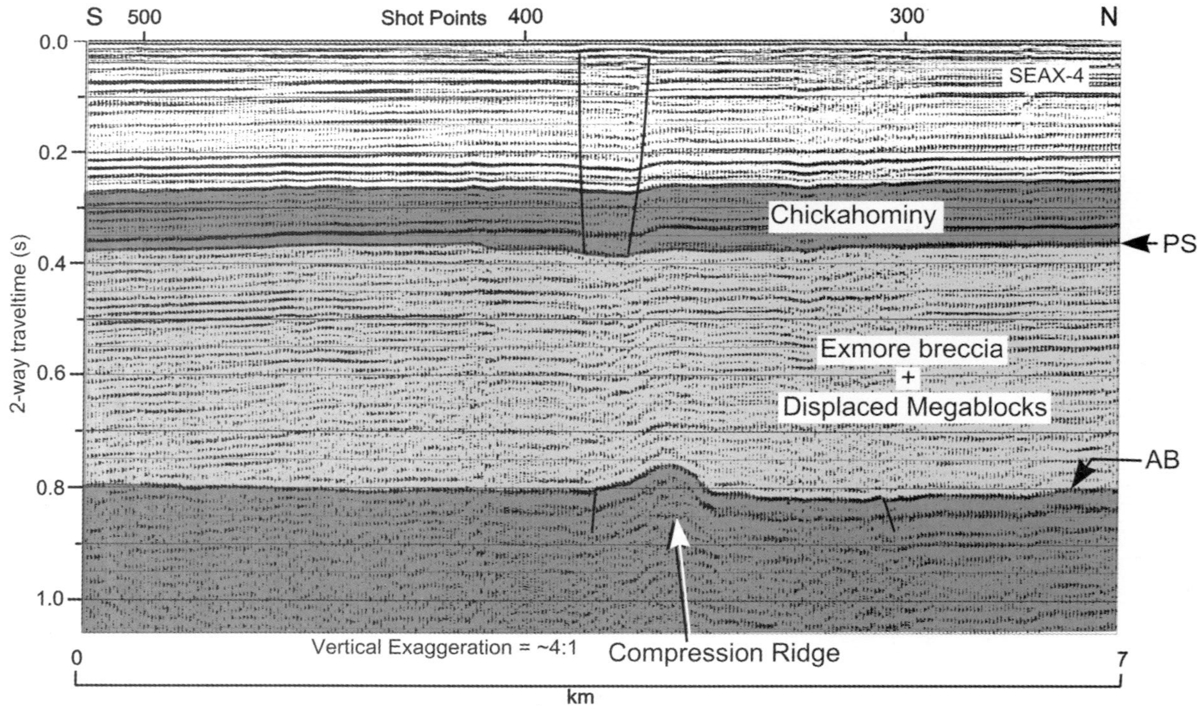

Fig. 4.20B. Interpreted segment of seismic reflection profile SEAX-4 (S-4), showing compression ridge on basement surface in annular trough of Chesapeake Bay impact crater. See Fig. 4.21 for precise location and CD-ROM.8 for full-scale profile. Reflection abbreviations as in Fig. 4.3.

Table 4.3. Morphometric data for annular trough of Chesapeake Bay impact crater.

Profile #	Width [km]	Shot Point [outer edge]	Shot Point [inner edge]
T-1-CB	27.5	2650	1550
S-4	28	1635	805
S-6	22	145	880
S-8	15	668	218
S-10	20	2590	1800
S-14	21	~800	640
S-3 (dogleg)	16	135	720
T-13-YR (dogleg)	16	2125	1375
S-16/S-14A Composite (dogleg)	24	435	55
E-3/S-8 Composite	24	668	1261
Average	21.4		

Table 4.4. Morphometric data for compression ridges in the crystalline basement surface of the Chesapeake Bay impact crater.

Profile #	Shot Point	Structural Relief [m]	Width [km]	Position
S-4A	360	40	0.9	Annular trough
S-12	810	50	1.8	Rappahannock Canyon
S-13	820	40	1.9	Annular trough
S-16	280	30	1.5	Annular trough
S-15	825	70	1.2	Annular trough
T-10-RR	1200	50	1.8	Outer rim
E-1	150	80	3.5	Annular trough
Average		51	1.8	

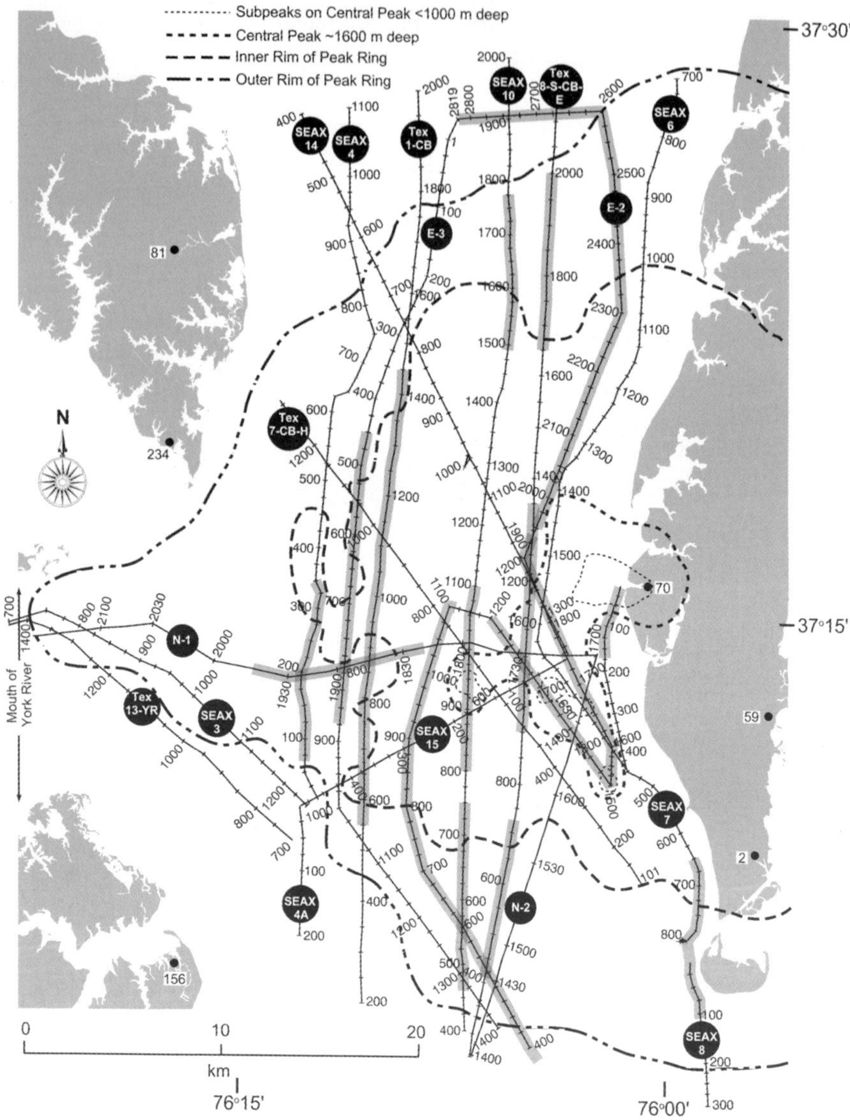

Fig. 4.21. Detailed map, showing shot points of seismic reflection profiles crossing peak ring, inner basin, and central peak of Chesapeake Bay impact crater. Shaded lines indicate segments of profiles shown in Figs. 4.22–4.30 and 4.32–4.34. See CD-ROM.2 for more details of seismic tracklines.

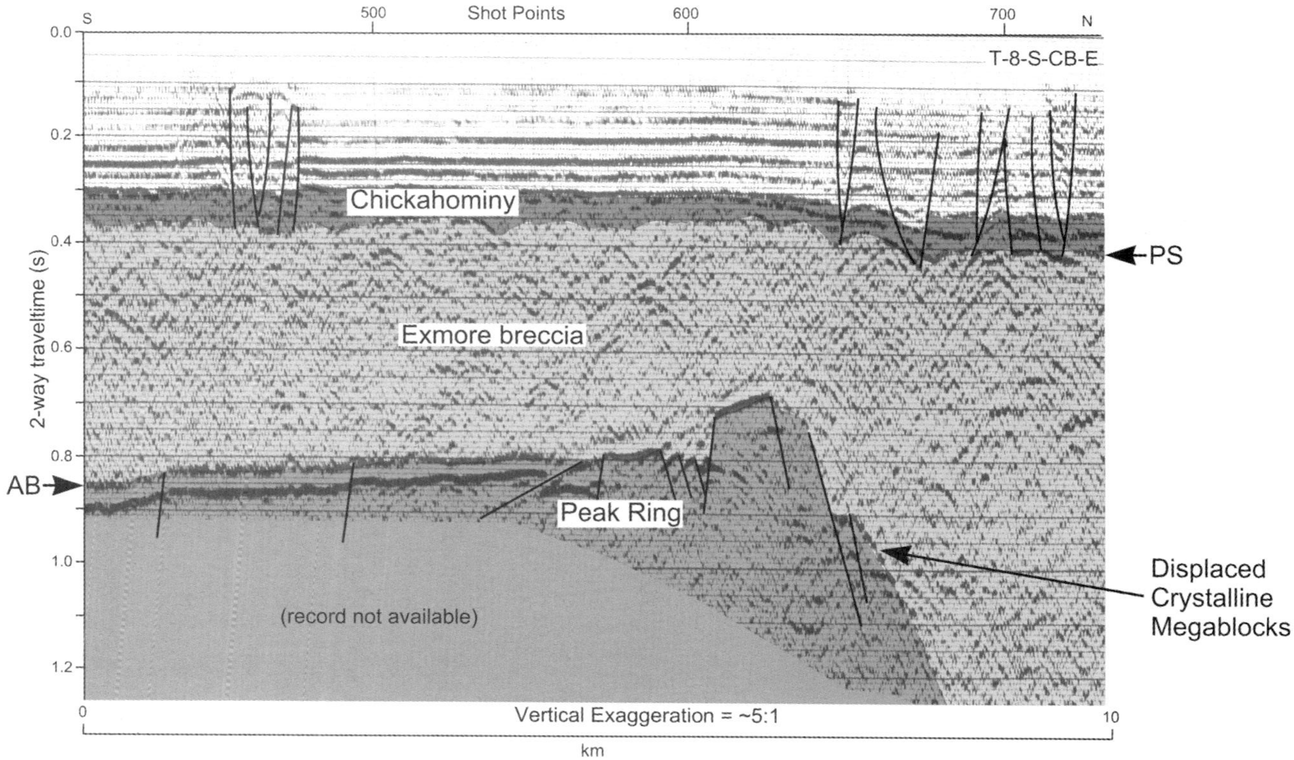

Fig. 4.22. Interpreted segment of seismic reflection profile T-8-S-CB-E crossing peak ring in southern sector of Chesapeake Bay impact crater. See Fig. 4.21 for precise location and CD-ROM.12 for full-scale profile. Reflection abbreviations as in Fig. 4.3.

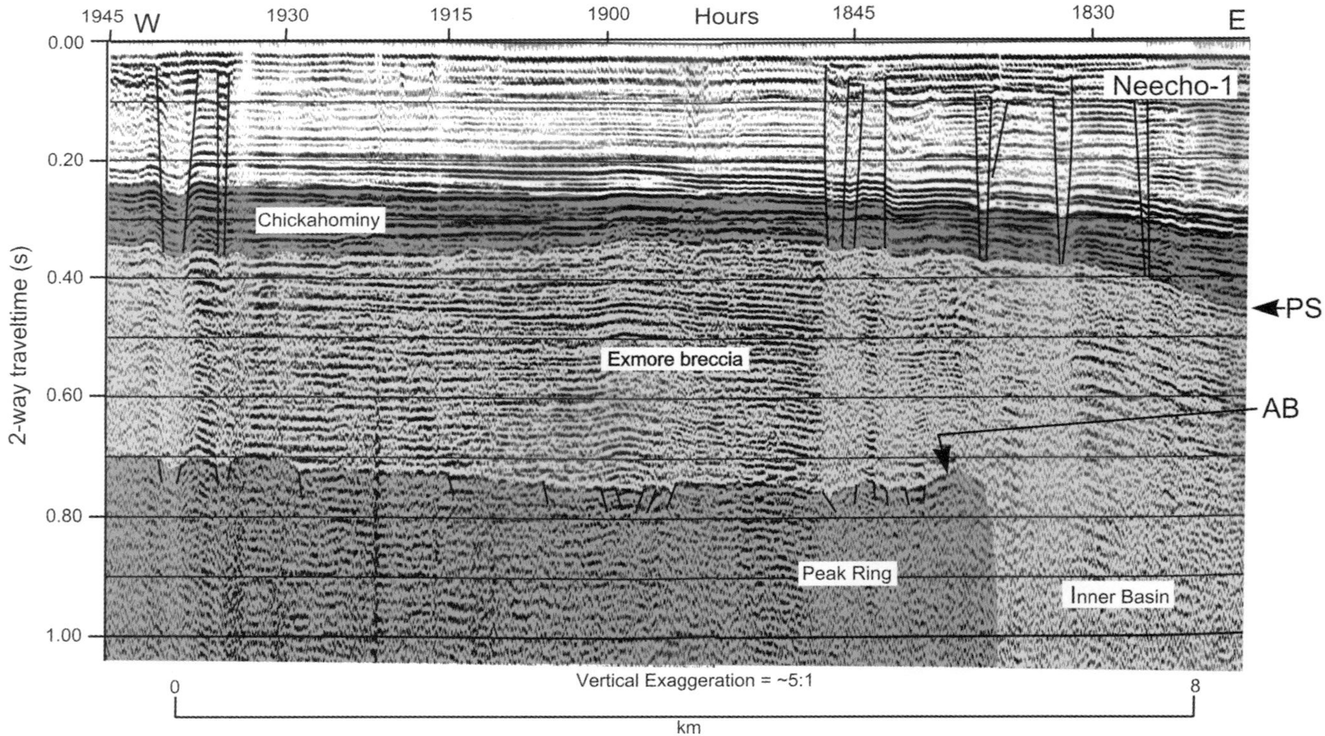

Fig. 4.23A. Interpreted segment of seismic reflection profile Neecho-1 (N-1) crossing peak ring in western sector of Chesapeake Bay impact crater. See Fig. 4.21 for precise location. Reflection abbreviations as in Fig. 4.3.

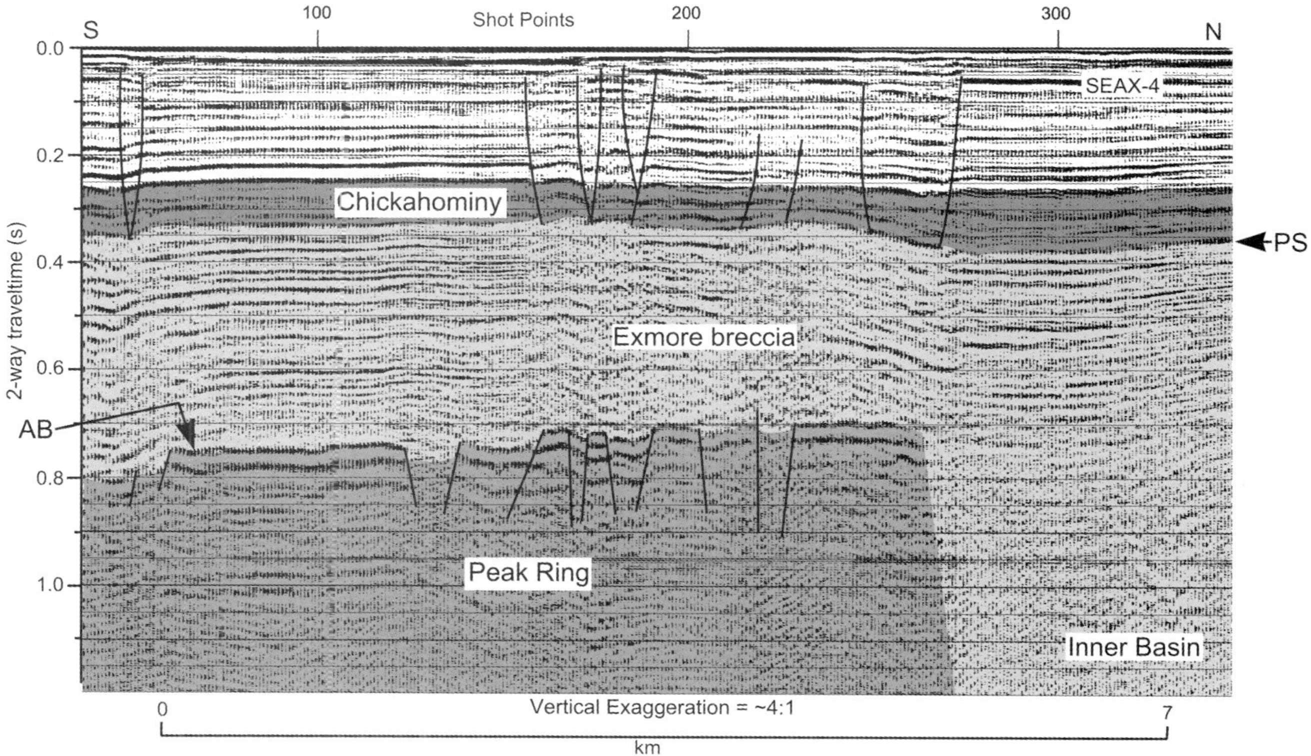

Fig. 4.23B. Interpreted segment of seismic reflection profile SEAX-4 (S-3) crossing peak ring in western sector of Chesapeake Bay impact crater. See Fig. 4.21 for precise location and CD-ROM.8 for full-scale profile. Reflection abbreviations as in Fig. 4.3.

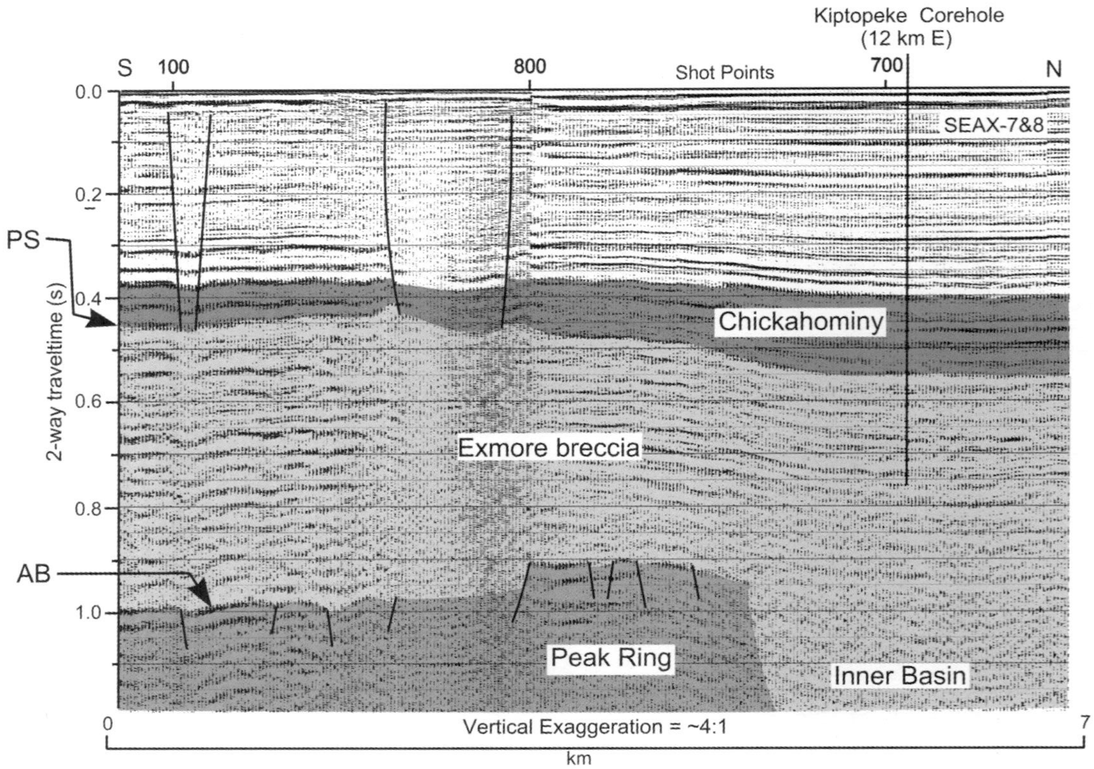

Fig. 4.24A. Interpreted segment of seismic reflection profiles SEAX-7 and 8 (part of S-1) crossing peak ring in southern sector of Chesapeake Bay crater. See Fig. 4.21 for precise location and CD-ROM.9 for full-scale profile. Reflection abbreviations as in Fig. 4.3.

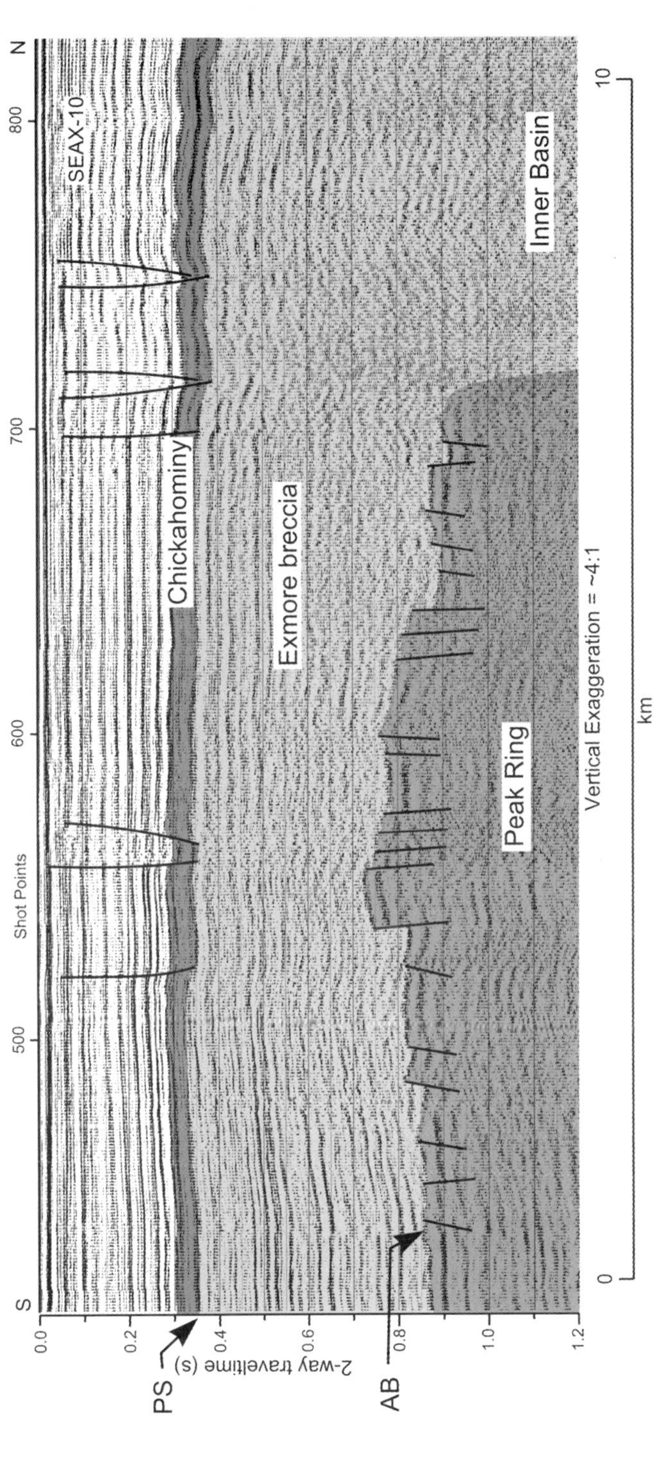

Fig. 4.24B. Interpreted segment of seismic reflection profile SEAX-10 (S-2) crossing peak ring in southern sector of Chesapeake Bay crater. See Fig. 4.21 for precise location and CD-ROM.10 for full-scale profile. Reflection abbreviations as in Fig. 4.3.

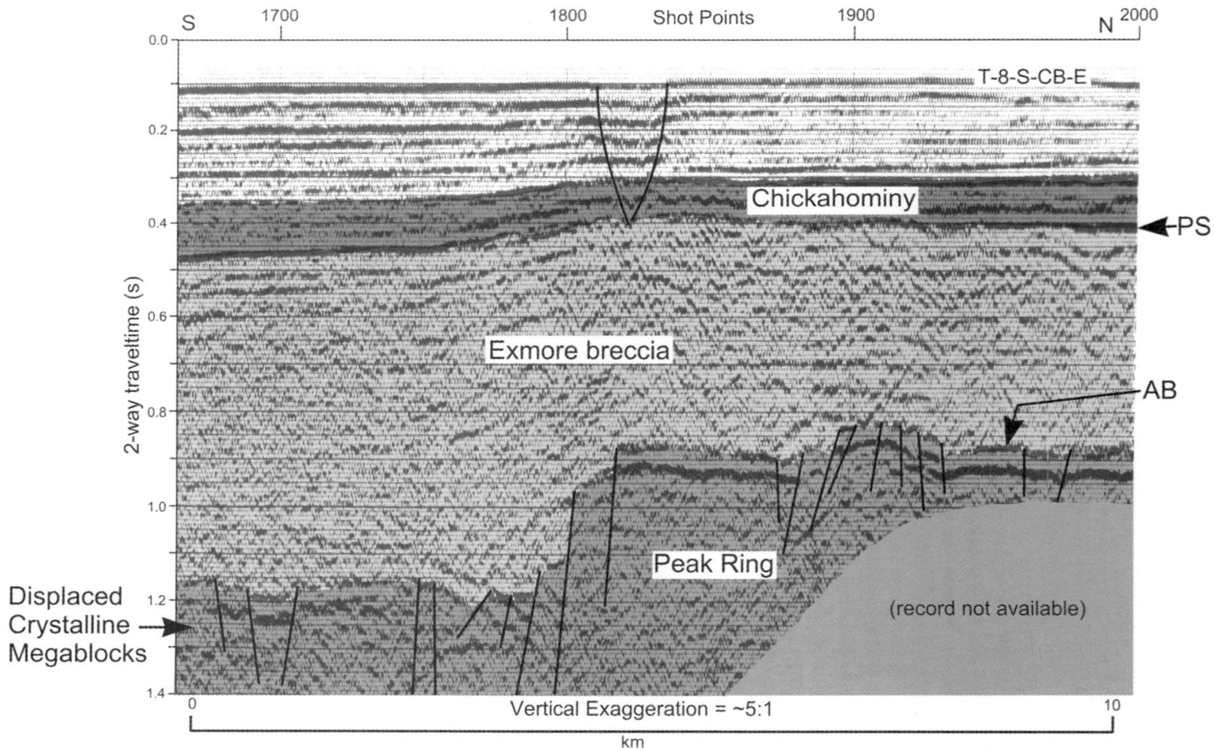

Fig. 4.25A. Interpreted segment of seismic reflection profile T-8-S-CB-E crossing peak ring in northern sector of Chesapeake Bay impact crater. See Fig. 4.21 for precise location. and CD-ROM.12 for full-scale profile. Reflection abbreviations as in Fig. 4.3.

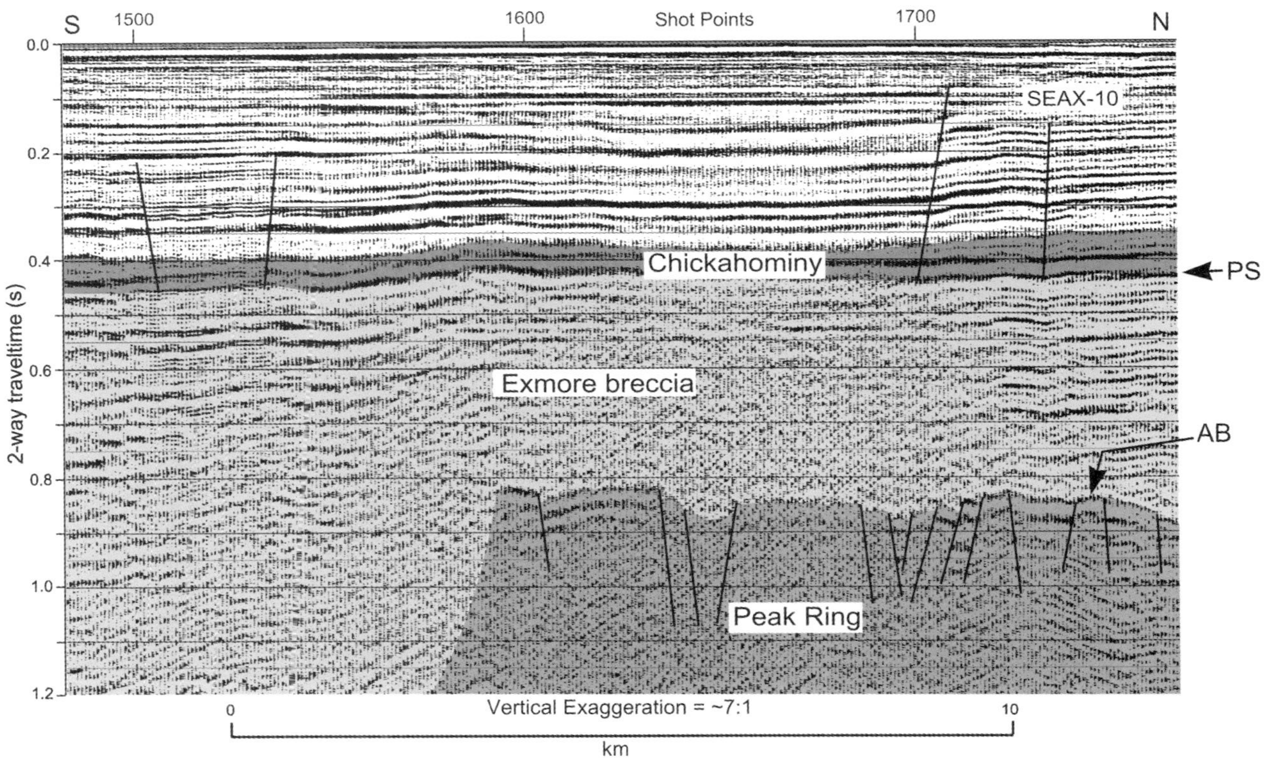

Fig. 4.25B. Interpreted segment of seismic reflection profile SEAX-10 (S-10) crossing peak ring in northern sector of Chesapeake Bay impact crater. See Fig. 4.21 for precise location and CD-ROM.10 for full-scale profile. Reflection abbreviations as in Fig. 4.3.

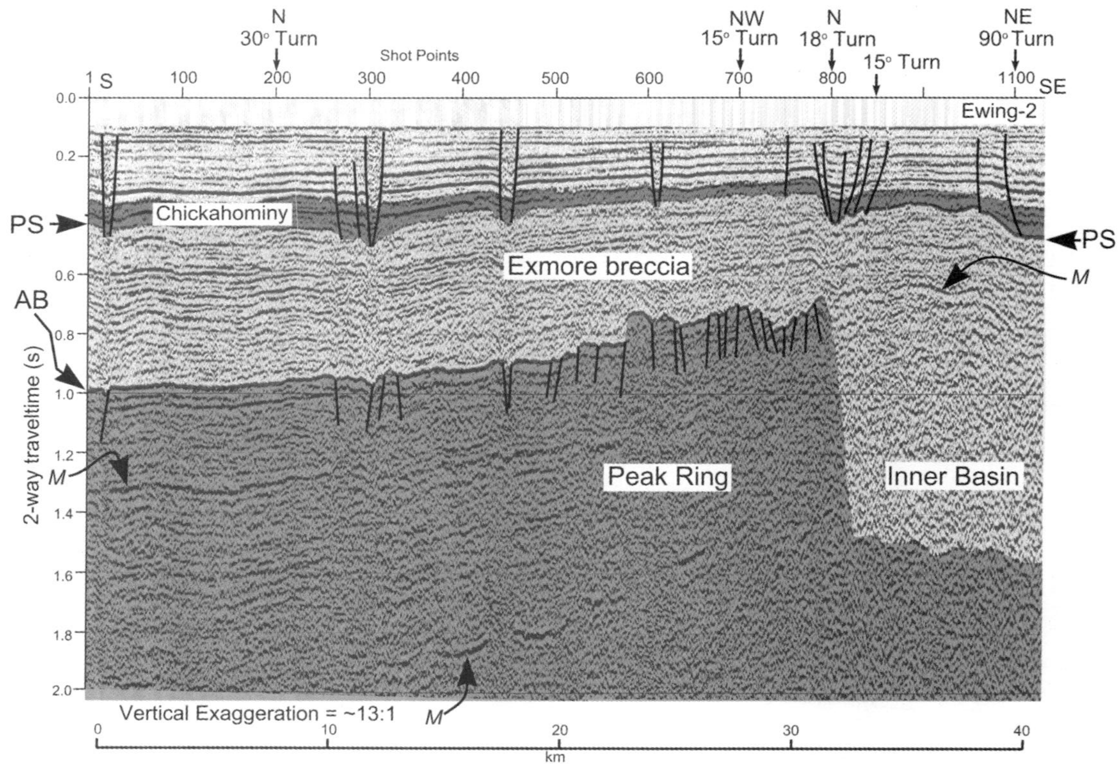

Fig. 4.26A. Interpreted segment of seismic reflection profile *Ewing*-2 (E-2) crossing peak ring in northern sector of Chesapeake Bay impact crater. See Fig. 4.21 for precise location and CD-ROM.15 for full-scale profile. *M* = multiple; other reflection abbreviations as in Fig. 4.3.

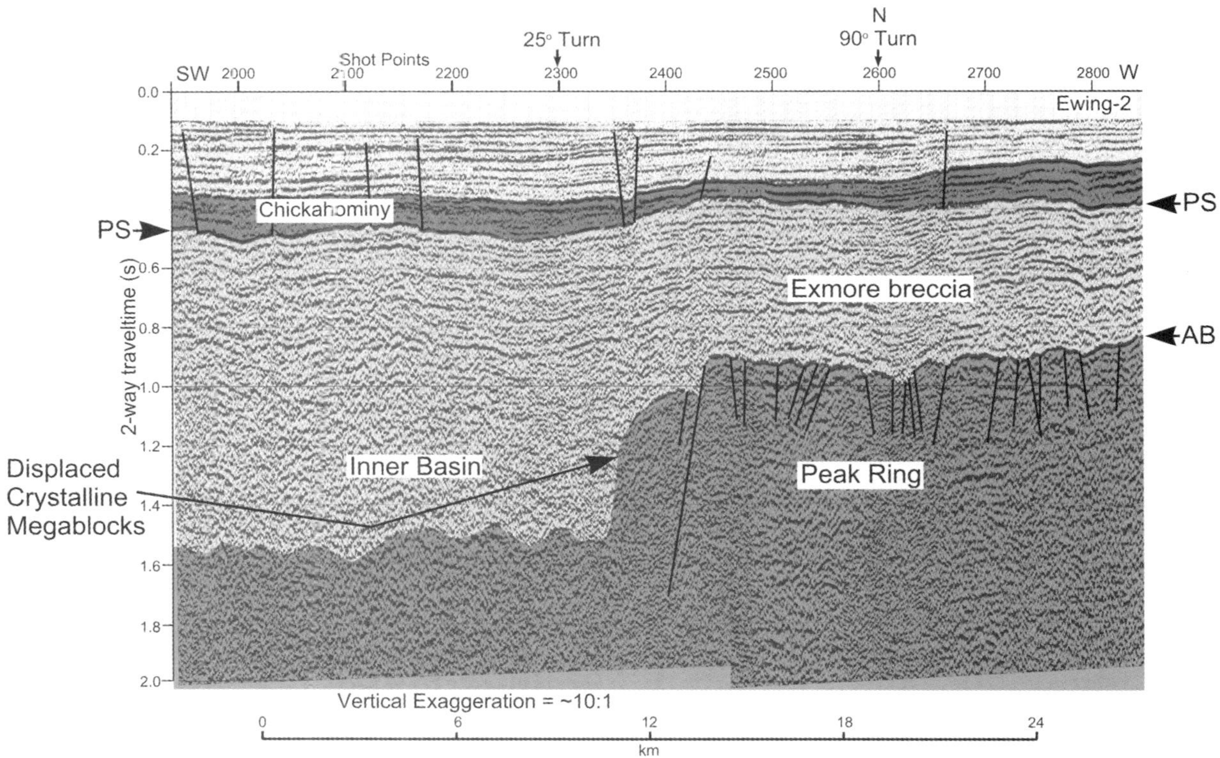

Fig. 4.26B. Interpreted segment of seismic reflection profile *Ewing*-2 (E-2) crossing peak ring in southern sector of Chesapeake Bay impact crater. See Fig. 4.21 for precise location and CD-ROM.15 for full-scale profile. Reflection abbreviations as in Fig. 4.3.

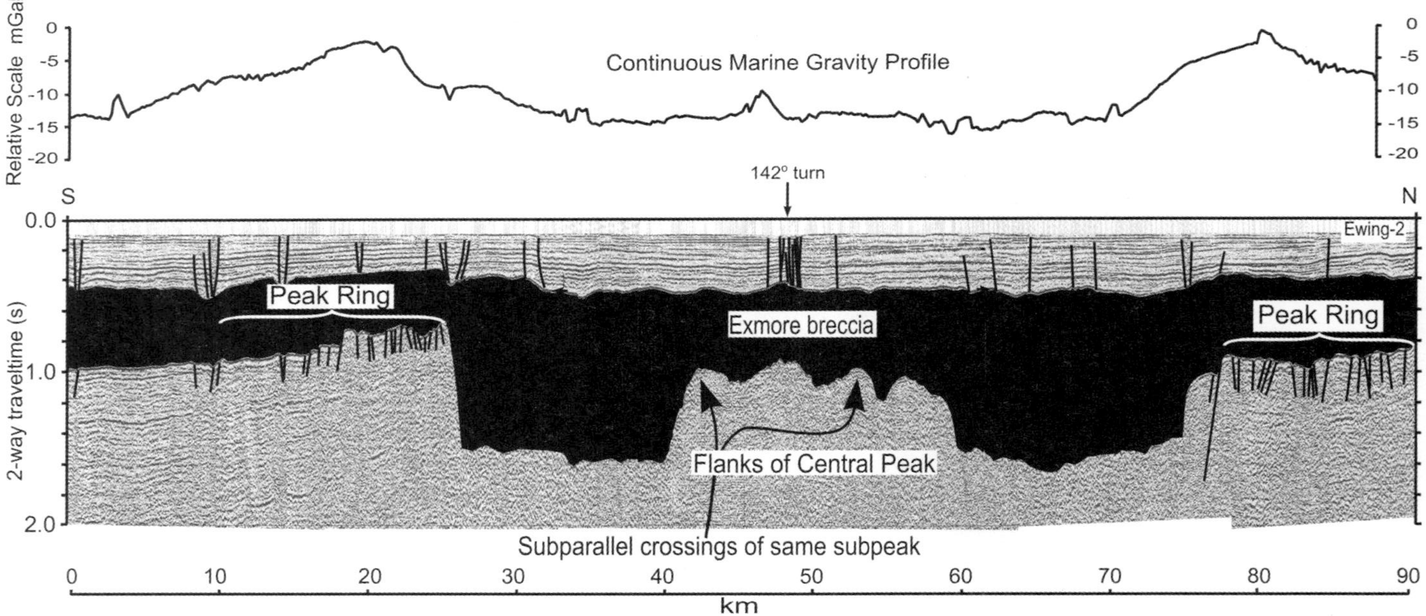

Fig. 4.27. Interpreted segment of seismic reflection profile *Ewing*-2 (E-2) and corresponding gravity profile crossing inner basin of Chesapeake Bay impact crater. See Fig. 4.21 for precise location, Fig. 4.26A,B for navigation details, and CD-ROM.15 for full-scale profile. Reflection abbreviations as in Fig. 4.3.

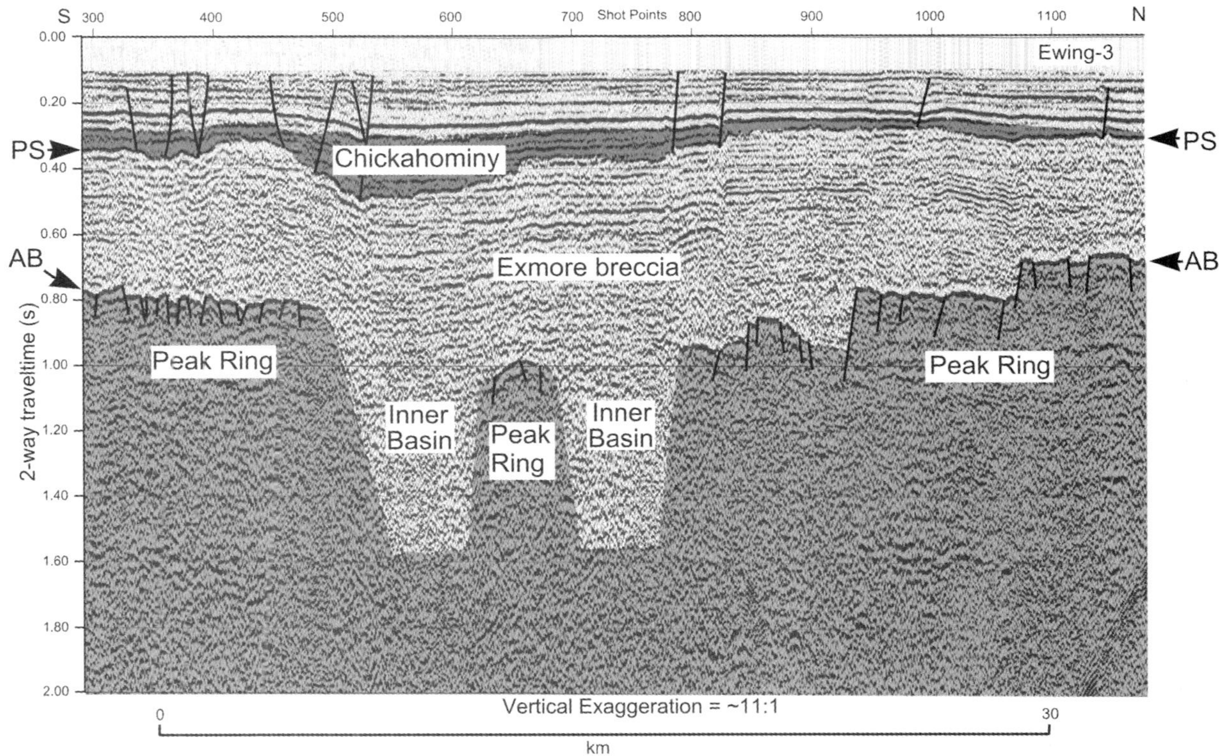

Fig. 4.28. Interpreted segment of seismic reflection profile *Ewing*-3 (E-3) crossing inner basin of Chesapeake Bay impact crater. A Narrow segment of peak ring juts out into inner basin from western wall. See Fig. 4.21 for precise location and CD-ROM.16 for full-scale profile. Reflection abbreviations as in Fig. 4.3.

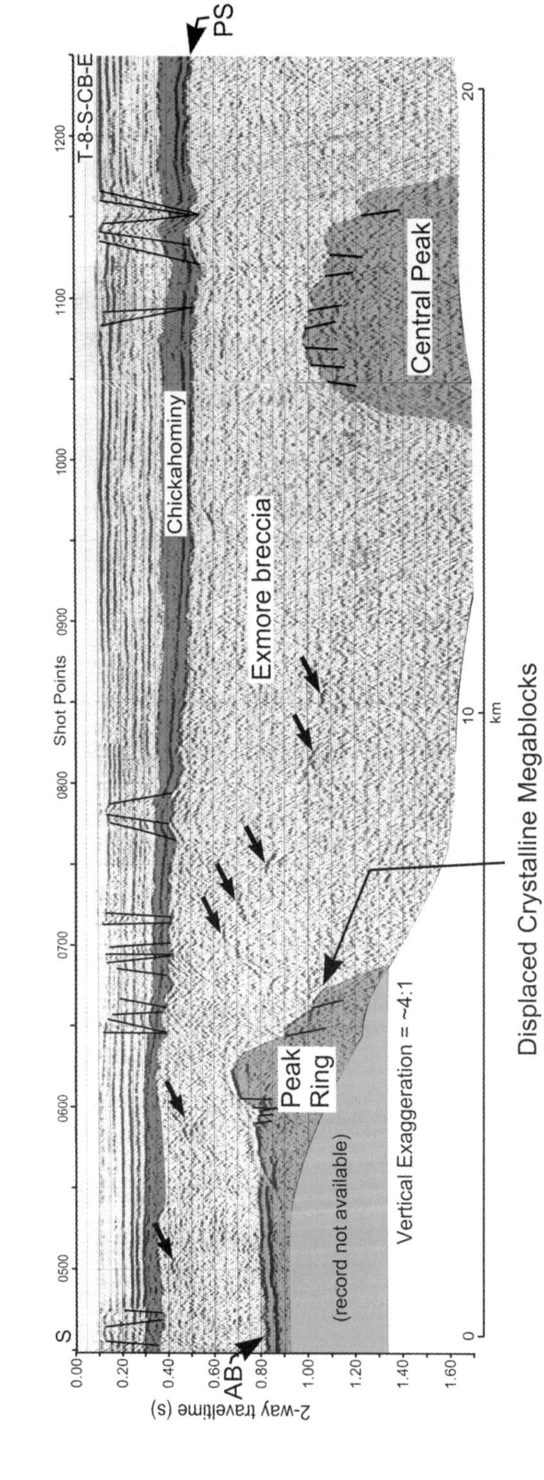

Fig. 4.29A. Interpreted southern segment of seismic reflection profile T-8-S-CB-E crossing peak ring, inner basin, and central peak of Chesapeake Bay impact crater. Arrows indicate strong (high-amplitude) reflections within Exmore breccia, which may represent megablock clasts or melt bodies. See Fig. 4.21 for precise location and CD-ROM.12 for full-scale profile. Reflection abbreviations as in Fig. 4.3.

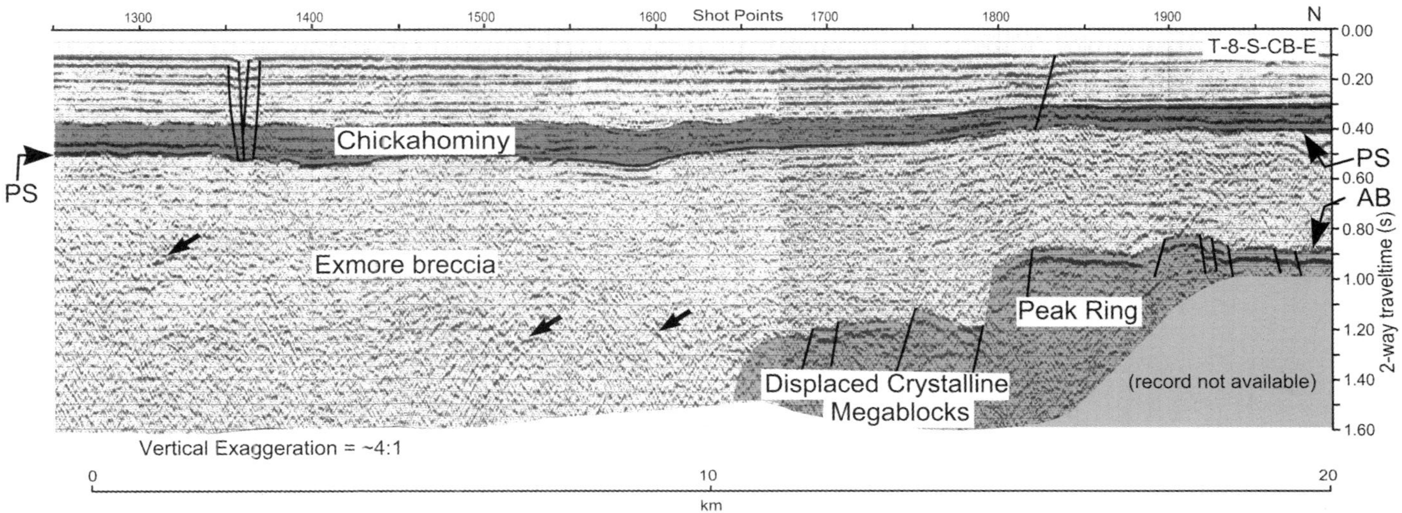

Fig. 4.29B. Interpreted northern segment of seismic reflection profile T-8-S-CB-E crossing inner basin, crystalline displaced megablocks, and peak ring of Chesapeake Bay impact crater. Arrows indicate strong (high-amplitude) reflections within Exmore breccia, which may represent megablock clasts or melt bodies. See Fig. 4.21 for precise location and CD-ROM.12 for full-scale profile. Reflection abbreviations as in Fig. 4.3.

Table 4.5. Morphometric data for peak ring of Chesapeake Bay impact crater.

Profile #	Width [km]	Relief [m]	Shot Point [outer edge]	Shot Point [inner edge]
[1]S-4 south	6.5	65	55	275
S-4 north	10.8	60	835	435
S-10 south	8.4	140	429	705
S-10 north	6.0	100	1809	1605
S-6 north	11.0	100	680	1010
S-8 south	8.5	140	218	740
S-3 west [1]	4.5	150	720	not crossed
S-3 west [2]	4.5	150	852	not crossed
S-15 west	11.0	50	not crossed	430
T-8-S-CB-E south	5.7	175	430	680
T-8-S-CB-E north	11.0	80	2020	1700
T-7-CB-H west	?	~50	not crossed	1100
N-1 west	9.75	150	[2]2050 hrs	[2]1833 hrs
N-2 south	8.0	200	[2]1400 hrs	[2]1537 hrs
Ewing-2 north	14.5	220	430	800
Ewing-2 south	>7.5	>100	2600	2350
T-1-CB north	11.5	140	1775	1335
T-1-CB south	12.5	300	450	925
Ewing-3 north	14	120	095	501
Ewing-3 south	22	220	1410	761
T-13-YR west (1)	4.25	40	1225	not crossed
T-13-YR west (2)	4.25	40	1375	not crossed
S-14	6.7	50	635	775
Average	9.2	124		

[1]Direction indicates side of peak ring shown on profile; no data for eastern half of crater
[2]Location on seismic trackline measured in clock time rather than in shot points

ring, varies considerably in all morphological and structural characteristics. Measured width is 4.25–22 km; the interpolated average is ~10 km (Table 4.5). The ring is widest on the western side, near the mouth of the York River, and narrowest about 20 km southeast of the York River mouth (Fig. 4.21; Table 4.5). The diameter of the peak ring, as measured along its crest, varies from 35 to 45 km, and averages 40 km. The elevation of the ring crest varies from 500 to 950 m below sea level; relief is 40–300 m, with an average of 123 m, where measured at 17 locations on seismic profiles (Table 4.5). Four individual peaks and a small sinuous ridge segment are prominent features of the southwest sector of the peak ring, which attains the highest ring elevations (Figs. 4.1, 4.21; CD-ROM.3–5). The largest of the four peaks is a triangular feature, whose eastern flank slopes down to the inner edge of the peak ring. This peak displays the maximum relief (300 m) of any structural or morphological feature so far identified on the peak ring or in the annular trough. The three smaller peaks jut out into the western inner edge of the inner basin and produce an irregular outline in map view (Fig. 4.1). The geometry of the peak ring in the eastern side of the crater is not as precisely known as elsewhere, because we lack basement images in that area. We have assumed that the peak-ring structure is roughly symmetrical, as indicated by the gravity anomaly data (Fig. 3.6), and have extrapolated the morphology of the peak ring on that basis. The smoother surface of the basement shown on the eastern side of the crater (Figs. 4.1; CD-ROM.3,4) is a result of the lack of control there.

4.1.1.4 Inner Basin

The inner edge of the peak ring is a steep cliff of crystalline basement rock, roughly circular in outline, which forms the outer wall of the inner basin (Figs. 1.5, 4.1, 4.21–4.29; CD-ROM.3–6; Table 4.6). On most seismic profiles (western half of crater only), the position of the crystalline wall is indicated by the abrupt truncation of the basement reflection couplet (Figs. 4.21–4.29). The exact diameter of the inner basin has not been measured, because no profile traverses the entire crater passing through its center. However, profiles SEAX-10, E-3, T-8-S-CB-E, and E-2 approximate the diameter at 29.75 km, 31 km, and 41.5 km, respectively (Fig. 4.21; Table 4.6). The inner basin radius, as measured from the approximate center along seven different profiles, ranges from 10 to 18.3 km, averaging 14.2 km, which would yield an average diameter of 28.3 km (assuming symmetry; Table 4.6). From the structure map, we estimated an average diameter of ~30 km, and used this value to extrapolate the geometry of the inner basin wall to the eastern half of the crater.

The best seismic data available to us prior to 1989 (when we acquired the Texaco multichannel profiles) were stratigraphically too shallow (<1 km) to image the floor of the inner basin. The *Ewing* profiles provide deeper data, but we observed no obvious reflections that could be interpreted as a distinct floor of the inner basin. From this we conclude that either the seismic system was not sufficient to resolve this contact through more than a kilometer of breccia, or the impedance contrast between the basal part of the crater-fill breccia and the fractured upper part of crystalline basement is not significant. Seismic studies of several other craters also have failed to distinguish a prominent reflection that would identify the floor of the inner basin (e.g., Manson crater, Schultz and Anderson 1994; Chicxulub crater, Christeson et al. 1999; Houghton crater, Scott and Hajnal 1988), which supports the hypothesis that impedance contrasts are negligible at that level. However, Poag (1997a), Poag et al. (1999), and we (herein) estimate the minimum depth of the floor of the inner basin to be at ~1.6 s 2-way traveltime (~1.6 km; Figs. 4.26–4.29), using the scaling relationship proposed by Grieve and Robertson (1979) [($d_t = 0.52D^{0.2}$), in which d_t is true depth, and D is the final diameter of the crater (85 km)].

Table 4.6. Morphometric data for the inner basin of the Chesapeake Bay impact crater.

Profile #	[1]Apparent Diameter [km]	Shot Point Interval [wall to wall]
S-4	5	275–440 (south to north)
S-10	29.75	709–1609 (south to north)
S-6-7-8	[2]37	730–1000 (south to north)
T-1-CB	12	925–1365 (south to north)
T-8-S-CB-E	31	670–1625 (south to north)
7-CB-H	>26	101–1100 (southeast to northwest)
Ewing 2	[2]41.5	800–2350 (south to north)
Average	28.3	

Profile #	Radius [km]	Shot Point [Outer wall]
N-1 + S-7	11.75	[3]1835 hrs
N-2 + S-7	10.00	[3]1550 hrs
S-15 + S-7	[2]18.30	430
S-7	16	730
S-14	14.75	775
Average	14.2	

[1]No profile transects entire inner basin directly through crater center
[2]Not a straight-line profile
[3]Location on seismic trackline measured in clock time rather than in shot points

4.1.1.5 Central Peak

Several seismic profiles offer subtle evidence that a rugged, irregular central peak is present near the center of the inner basin (Fig. 4.21; Table 4.7; CD-ROM.3–6). Profile SEAX-7, for example, located a few kilometers south of the mouth of Cape Charles harbor (Figs. 4.21, 4.30), displays a series of persistent, though faint, diagonal reflections, which Poag et al. (1999) and we interpret to be side reflections from a central peak. The top of the peak on profile SEAX-7 is not plainly obvious, but Poag et al. (1999) and we (herein) place it at roughly 575–600 m

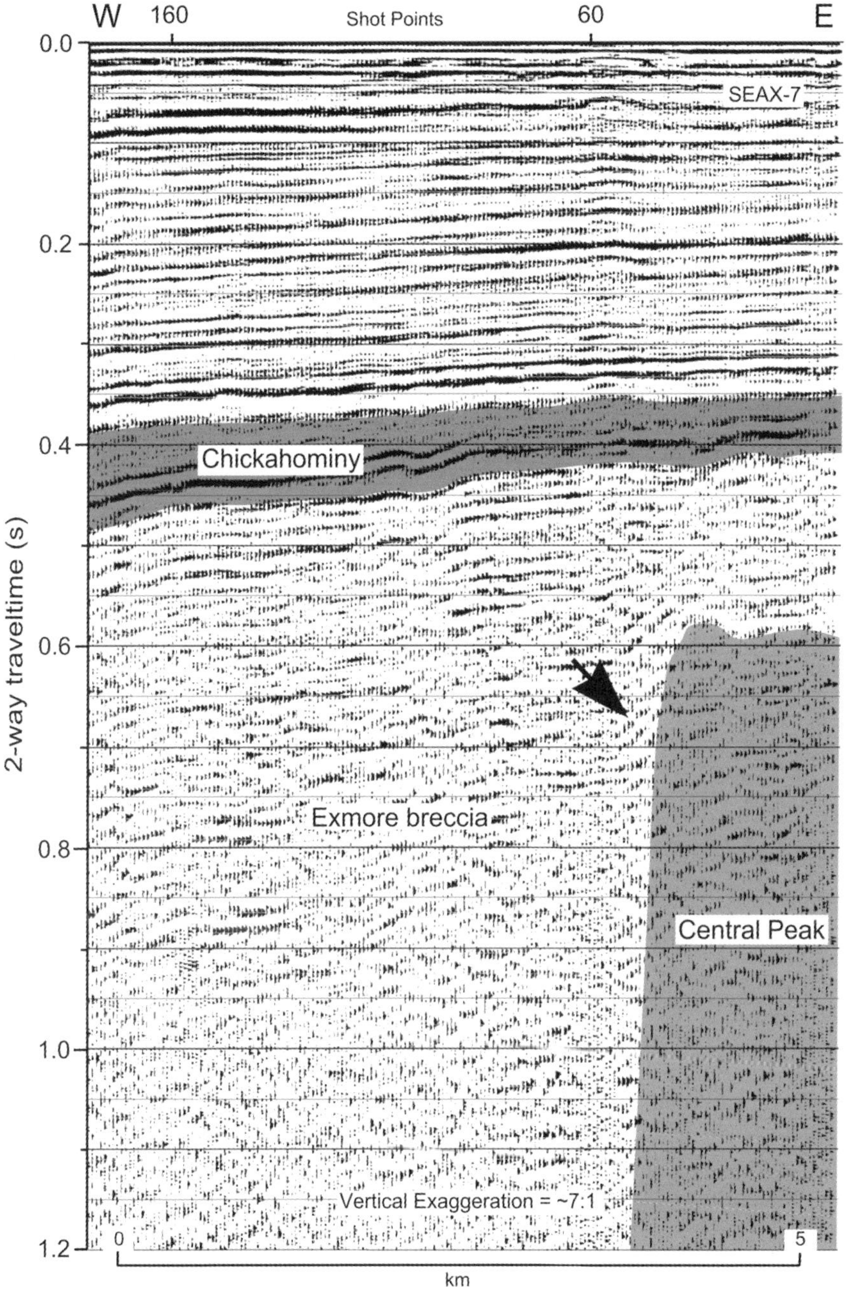

Fig. 4.30. Interpreted segment of seismic reflection profile SEAX-7 (at entrance to Cape Charles harbor, Delmarva Peninsula) crossing central peak of Chesapeake Bay impact crater. Arrow indicates inclined parallel reflections interpreted to be side echoes from central peak. See Fig. 4.21 for precise profile location.

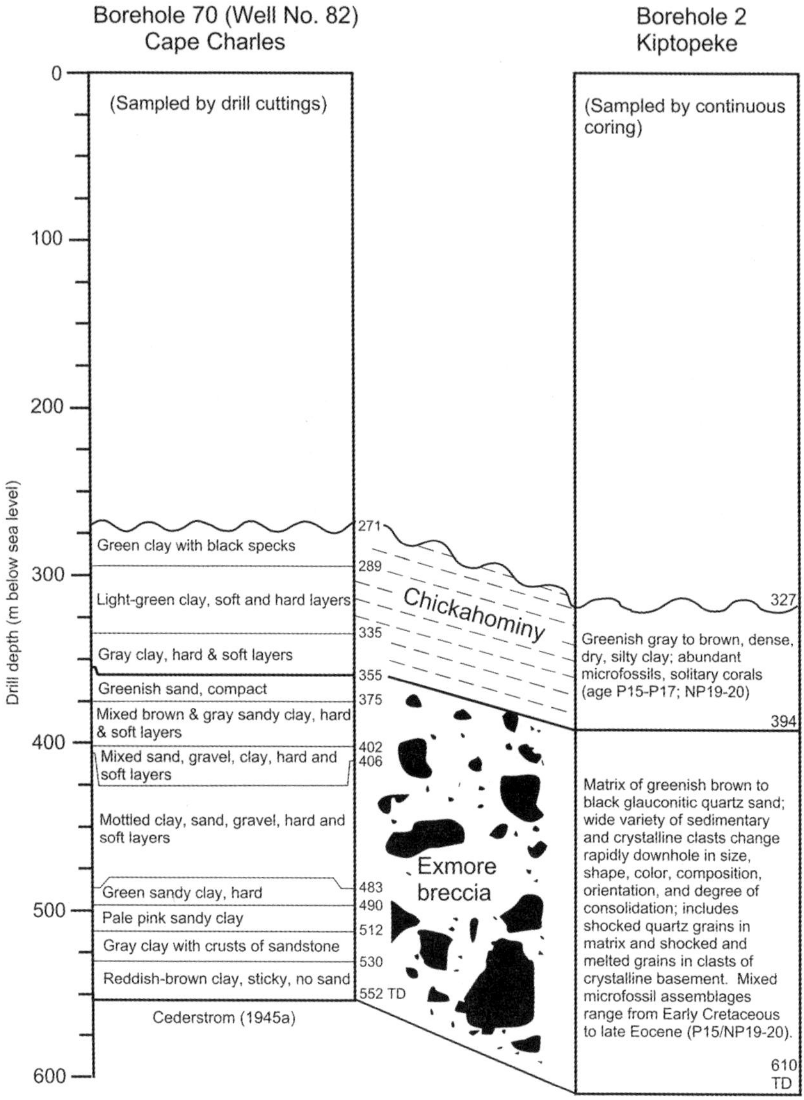

Fig. 4.31. Stratigraphic correlation between two boreholes drilled on Delmarva Peninsula into inner basin of Chesapeake Bay impact crater. See Figs. 1.2, 4.12, and CD-ROM.1 for borehole locations.

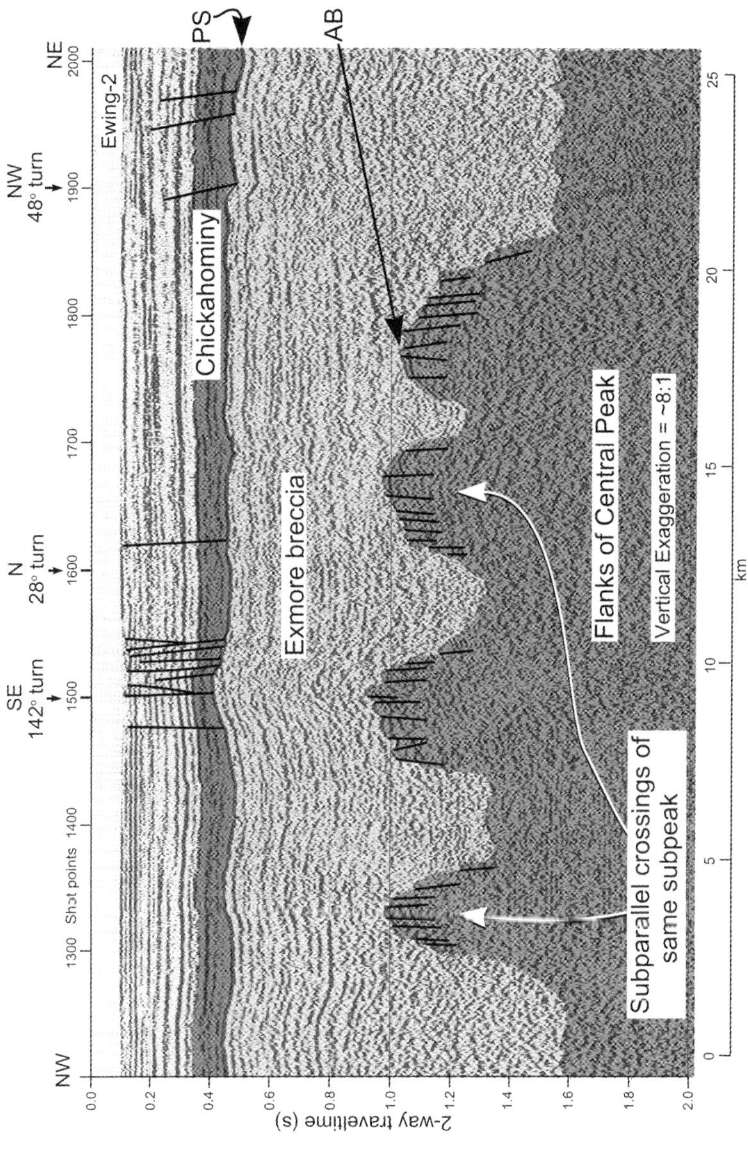

Fig. 4.32. Interpreted segment of seismic reflection profile *Ewing-2* (E-2), showing subparallel crossings of southwestern flank of central peak of Chesapeake Bay impact crater. See Fig. 4.21 for precise location and CD-ROM.15 for full-scale profile. Reflection abbreviations as in Fig. 4.3.

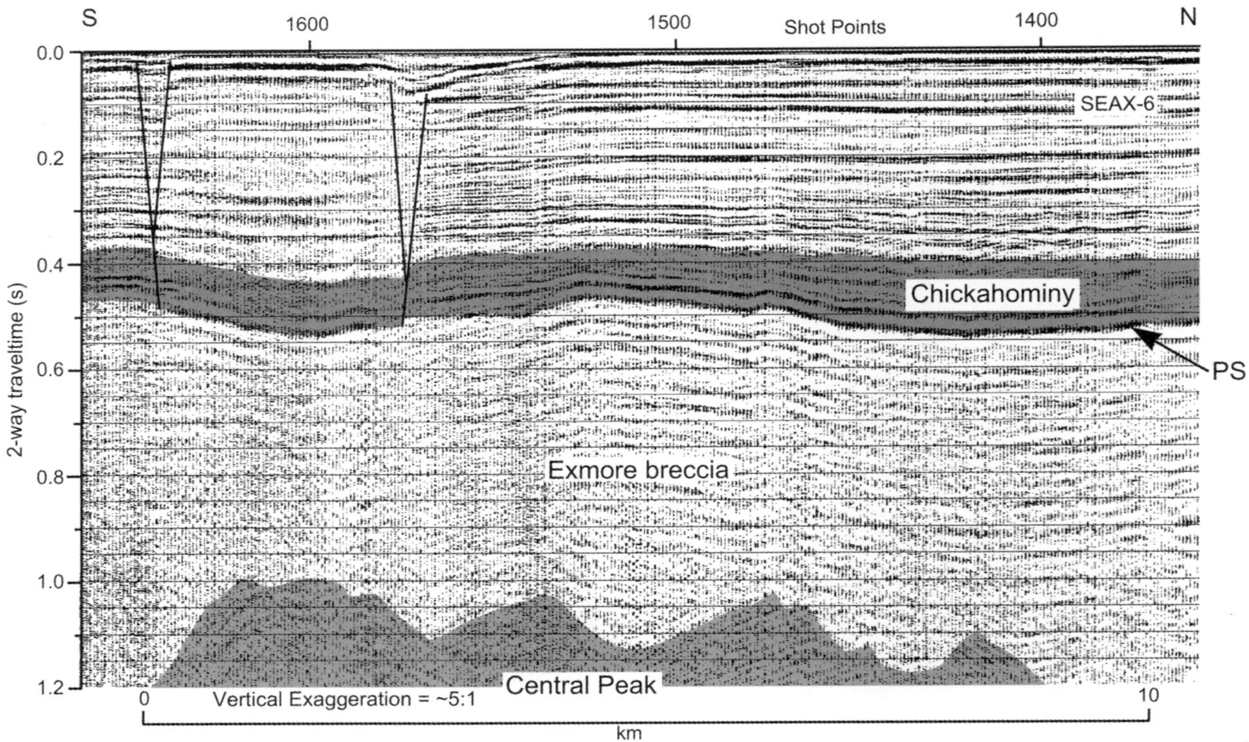

Fig. 4.33. Interpreted segment of seismic reflection profile SEAX-6 (S-6), showing inclined, parallel, side echoes from flank of central peak of Chesapeake Bay impact crater. See Fig. 4.21 for precise location and CD-ROM.9 for full-scale profile. Reflection abbreviations as in Fig. 4.3.

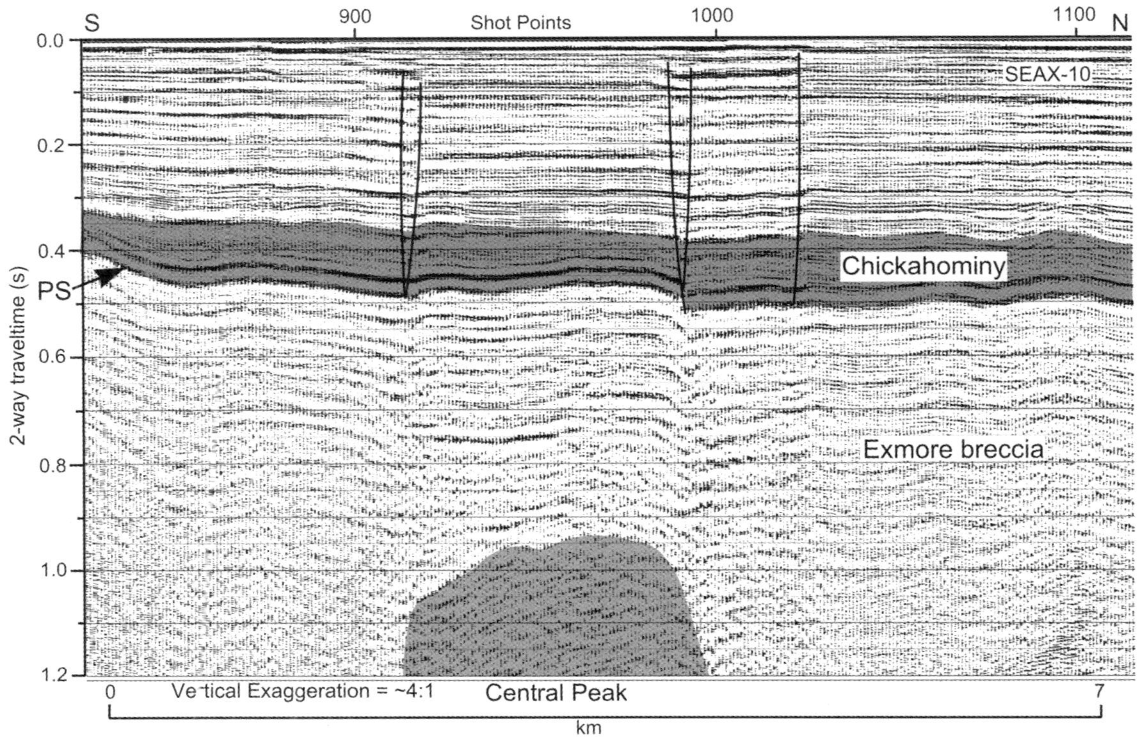

Fig. 4.34. Interpreted segment of seismic reflection profile SEAX-10 (S-10), showing inclined, parallel, side echoes from flank of central peak of Chesapeake Bay impact crater. See Fig. 4.21 for precise location and CD-ROM.10 for full-scale profile. Reflection abbreviations as in Fig. 4.3.

depth (below sea level) between shot points 0 and 31. Onshore, about 0.3 km to the east, a deep well (borehole 70, not cored) terminated in Exmore breccia at 552 m without having encountered crystalline basement (Figs. 4.21, 4.31). This provides some evidence that the peak is not more than 1 km in diameter at this location. The best evidence for the central peak, however, is provided by profile E-2 (Figs. 4.21, 4.32). This profile images the southwestern flank of the central peak on two subparallel crossings between shot points 1280 and 1850. The peak massif is approximately 12 km in diameter where intersected by E-2, and expresses irregular relief of 200–400 m. Three prominent knobs or subpeaks are shown on this profile, each approximately 2.5–3 km in diameter, and their crests vary from 950 to 1050 m below sea level. Maximum subpeak relief is 500–600 m above the inferred floor of the inner basin (~1.6 km depth). Distinctive diagonal and hyperbolic reflections at roughly 1-km depth on profiles SEAX-6 (Figs. 4.21, 4.33; shot points 1430–1730), SEAX-10 (Figs. 4.21, 4.34; shot point 1109), SEAX-15 (Fig. 4.21; shot points 1200–1350), T- 8-S-CB-E (Fig. 4.21; shot points 1025–1175), and at the intersection of N-1 and N-2 (Fig. 4.21; 1675–1710 hrs) provide evidence that, on its lower flanks, the central peak broadens out to 12 km or more in diameter, and has an average vertical relief of 620 m and an average crestal elevation of 890 m (Figs. 4.1, 4.21; Table 4.7; CD-ROM.3–6).

Table 4.7. Morphometric data for central peak of Chesapeake Bay impact crater.

Profile #	Crest Elevation [km]	Approximate Vertical Relief [km]	Width [km]	Bounding Shot Points
T-8-S-CB-E	1.00	0.60	4.5	1010–1180
T-7-CB-H	1.00	0.60	5.5	600–825
S-6	1.00	0.60	8.0	1400–1650
S-7	0.60	1.00	1.0	0–61
S-10	0.95	0.65	3.0	910–1000
S-14	0.75	0.20	2.5	1200–1350
E-2	0.90	0.70	12.0	1240–1951
S-15	Not distinguishable	?	?	
Neecho	Not distinguishable	?	?	
Average	0.89	0.62	4.8	

4.1.2 Gravity Interpretation

To test the seismically interpreted geometry of the impact structure, we compared simple Bouguer gravity (Fig. 4.35) and residual gravity (Fig. 4.36) anomaly maps (see Chapter 3), and applied 2-D geologic modeling along transects through the

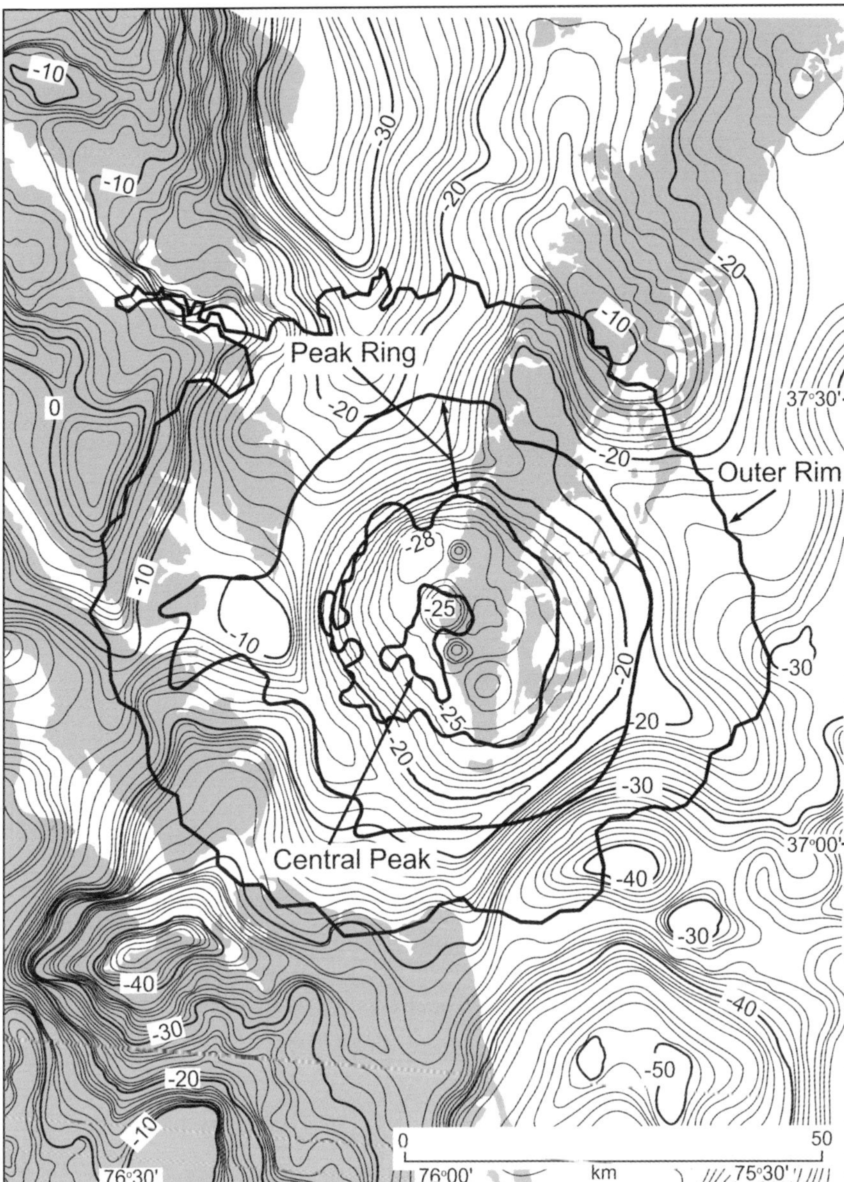

Fig. 4.35. Bouguer gravity anomaly map over Chesapeake Bay impact crater (onshore contours from simple Bouguer values; offshore contours from free air values). Contour interval 1 mGal. Modified from Poag (1997a).

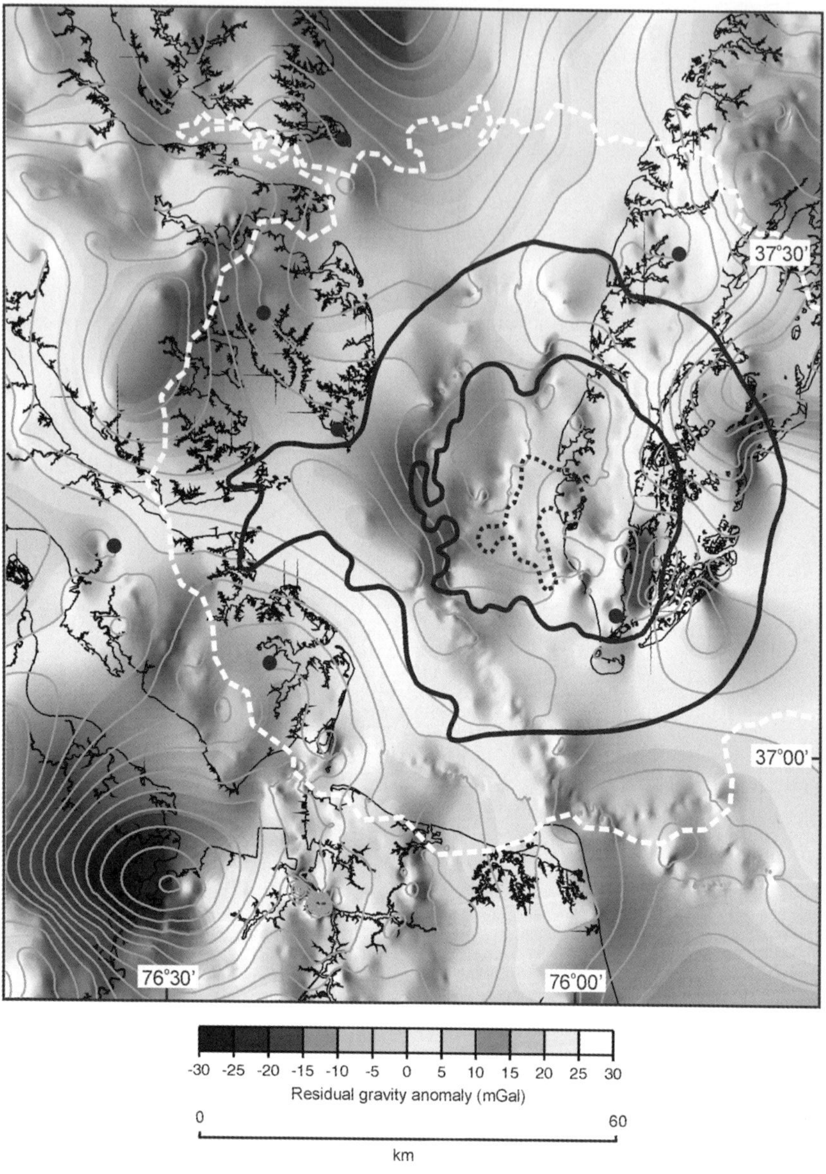

Fig. 4.36. Residual Bouguer gravity anomaly map with superimposed outlines of principal structural features derived from seismic reflection profiles. White dashed line represents outer rim of crater; two solid black lines represent outer and inner boundaries of peak ring; dashed black line represents outline of central peak. Solid black circles represent corehole locations. See text for further explanation and CD-ROM for color version of this figure.

gridded residual gravity data (Fig. 4.37A,B). In each of the three modeled sections (Fig. 4.37A), placing a low density (2.57 g/cm^3) body below the basement/sedimentary rock interface provides a large improvement from the starting model in fitting the observed residual gravity anomaly. The bodies had bottom elevations varying from -2.67 to -3.21 km, for an approximate thickness of 2 km. The positions of these bodies correlate well with seismic interpretations of the morphology of the peak ring and inner basin. An even better fit results when the basement surface is elevated by an average of 500 m at the edges of the low-density body in the basement (Fig. 4.37B). These elevated areas lie within the bounds of the zone interpreted from seismic reflection profiles to be the peak ring. Quantification of the depth of the crater and height of the peak ring depend on the assumptions made about the relative densities of the basement and crater-fill material. Although this modeling demonstrates the likely presence of an inner basin surrounded by a peak ring, their absolute elevations cannot be uniquely determined with the available data. Horizontal limits of each body are plotted in Fig. 4.37A,B. If one assumes that the residual gravity map represents a qualitative image of the inner structure of the crater, it can be used to interpret the geometry of the eastern half of the peak ring.

In general, the gravity data support the extrapolation of the seismic data across the Delmarva Peninsula. On the residual anomaly map (Fig. 4.36), a relatively symmetrical ring of positive anomalies coincides with our placement of the seismically-derived peak ring. Two broad, elongate positive anomalies on the southwest sector of the peak ring correlate with the highest peak-ring relief noted on the seismic profiles (Fig. 4.1; CD-ROM.3–6). The highest gravity-anomaly values associated with the seismically extrapolated peak ring occur on the eastern side of the crater (under the shallow eastern bays of the Delmarva Peninsula) where we lack seismic control. This suggests that the crest of the peak ring may attain its highest elevation in this area.

A broad, irregularly circular gravity low is present over the seismically imaged inner basin (Fig. 4.36). This negative depression is interrupted by several small, irregular gravity highs that suggest the presence of several individual knobs on the central peak. Furthermore, the gravity signature suggests that the highest elevation of the central peak may be ~10 km north of the Kiptopeke corehole, a location for which we have no seismic data.

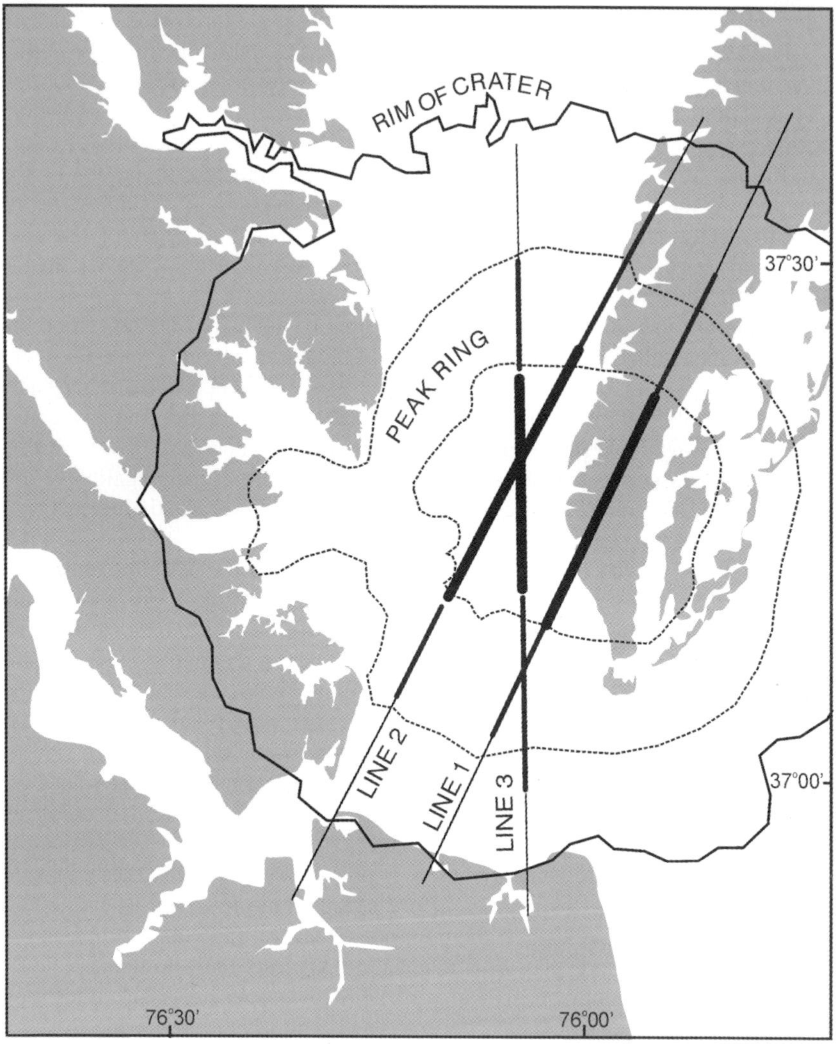

Fig. 4.37A. Location map, showing three lines of transect across Chesapeake Bay impact crater, for which gravity models are shown in Fig. 4.37B. Thickness changes along each line correspond to inner and outer boundaries of the peak ring as indicated by the gravity models. Seismic boundaries of the peak ring are indicated by dotted lines.

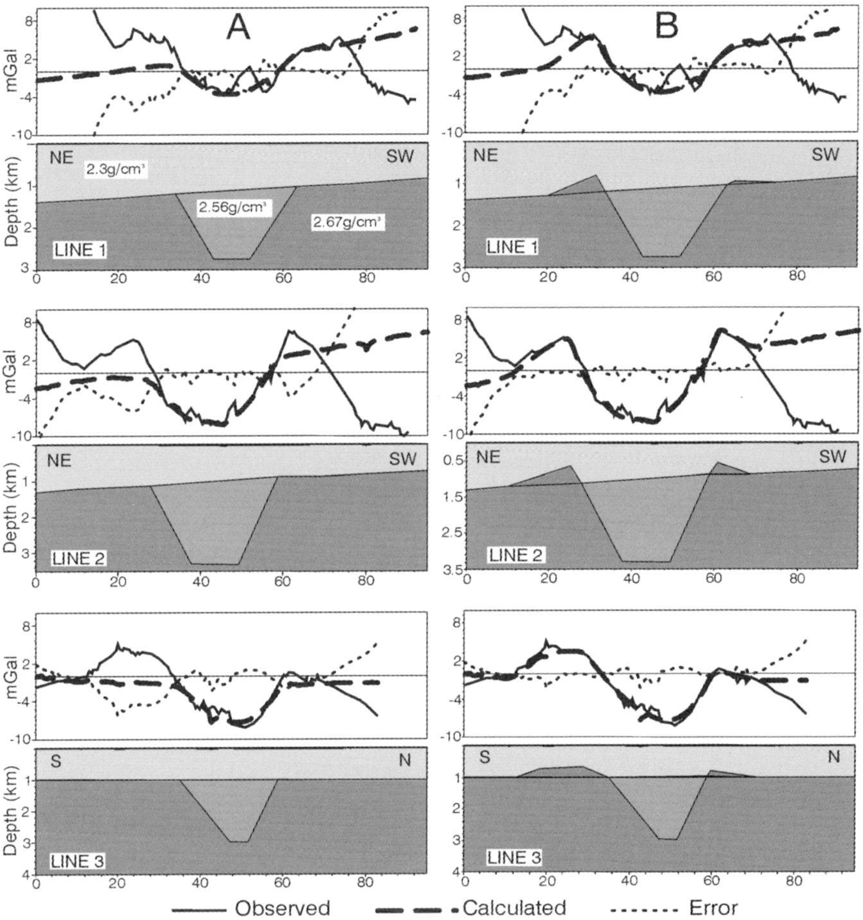

Fig. 4.37B. Two-dimensional gravity models (A,B) along three transects (see Fig. 4.37A) across Chesapeake Bay impact crater (constructed by P. Molzer). See text and Chapter 3 for further explanation.

5 Secondary Craters

5.1 Location and Identification

Telescope and satellite images of the moons and planets of our solar system reveal that large primary impact craters frequently are accompanied by smaller secondary craters of variable size, shape, and distribution (Shoemaker 1962; Melosh 1989; Spudis 1993; Greeley 1994; Fig. 5.1). Roddy (1977) showed that secondary craters also are commonly produced by large man-made explosions. The projectiles that produce secondary craters are inferred to be mainly blocks and clods derived from the target rocks, which are ejected into ballistic trajectories by the primary impact. Planetary secondaries usually are first recognizable beyond the edge of the continuous ejecta blanket, and their geographic range can extend many crater diameters from the primary crater (Melosh 1989). The maximum diameters of secondaries are roughly proportional to the diameter of their primary craters (e.g., lunar secondaries are ~4% as wide as their primaries). Besides isolated individual secondary craters, clusters (open or closed), chains, loops, gouges, and rays of secondaries are common on large and small planetary bodies. Secondary craters nearest to the rim of the primary crater may have irregular shapes because their impactors interfere with one another, and because their impact velocities are low relative to the velocity of the primary impactor. Distal secondaries usually have more regular shapes, but tend to be asymmetrical in cross section; the crater walls tend to be steepest in the direction toward the primary crater. Ejecta blocks that produce secondary craters may reach several kilometers in diameter; fragment size is inversely proportional to the ejection velocity. The extent of a secondary crater field away from the primary crater is evidently strongly controlled by gravity. Secondary craters tend to cluster closer to their primaries on larger planetary bodies than on smaller ones. According to Melosh (1989), the quantity of ejecta that produces secondary craters is small, typically one to three percent of the total ejecta derived from the primary impact. Thus, most of the large blocks and smaller clasts composing a continuous ejecta blanket do not produce well-defined secondary craters.

Despite the apparent near-ubiquity of secondary craters on other planetary bodies, secondary craters have rarely been documented on Earth. For example, though intact 1-km-long megablocks of Malm limestone have been ejected as far away as 7 km from the 24-km-diameter Ries peak-ring crater of southern Germany, no specific secondary craters have been found associated with this extensively studied primary crater (Pohl et al. 1977; Hörz et al. 1983). On the other hand, Sturkell

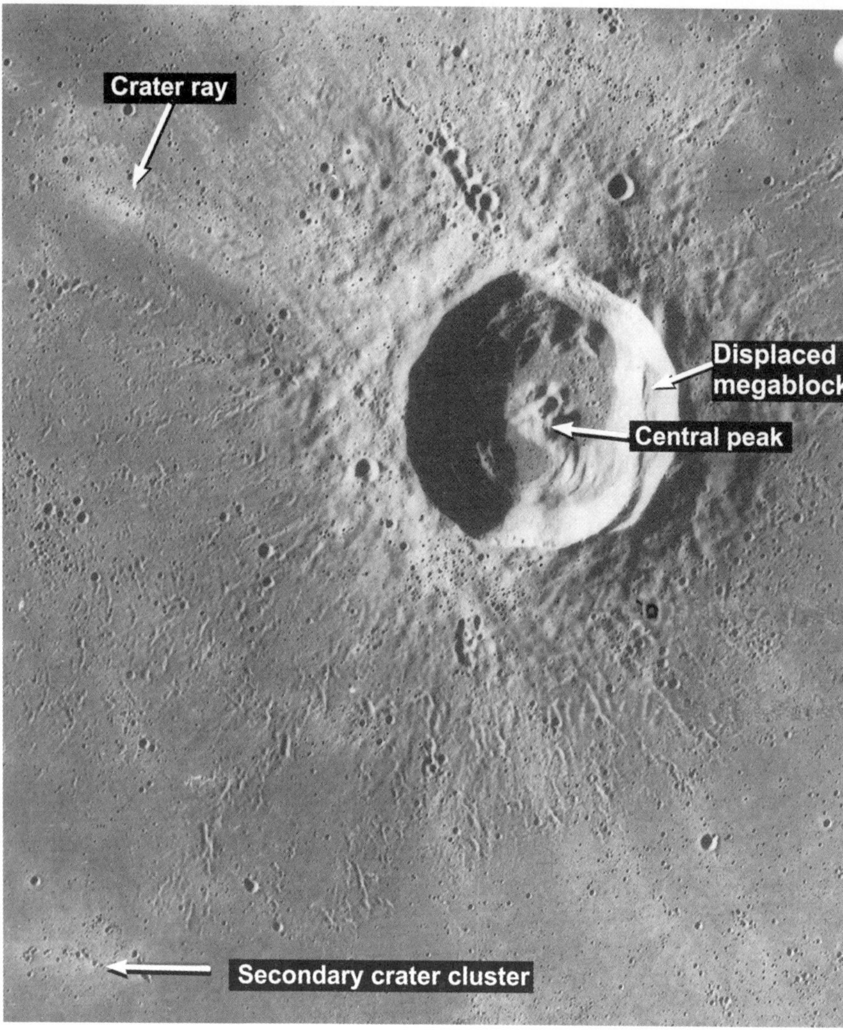

Fig. 5.1. Satellite image showing secondary craters associated with lunar crater Euler (27-km diameter; Apollo 17 image; from Greeley 1994).

(1998) reported that a 150-m-thick boulder was ejected from the Lockne crater and had excavated a 40-m-deep secondary crater.

Among the large number of small terrestrial craters (<5 km in diameter) known on our planet (Table 1.1), nearly all are interpreted to be primary craters. No secondaries have been previously reported to be associated with the larger submarine craters (*Montagnais,* Jansa and Pe-Piper 1987; *Mjølnir*, Gudlaugsson 1993; *Manson*, Koeberl and Anderson 1996a) or with the only known submarine multiring impact basin (*Chicxulub*, Hildebrand et al. 1991). Terrestrial secondary craters, like terrestrial primaries, are subject to alteration and removal by weathering, erosion, and the effects of plate subduction. At least one primary terrestrial crater, however, appears to have produced secondaries. Small secondary craters have been reported associated with the Bigach crater in Kazakhstan (Table 1.1; Kiselev and Korotuschenko 1986; Masaitis 1999).

The 85-km-diameter Chesapeake Bay primary crater is unusually well preserved, because it is relatively young (late Eocene; ~35.8 Ma; see age discussion in Chapter 8), it formed in a relatively deep marine setting (Poag et al. 1994; see also Chapter 13), and it occupies a basin characterized by relatively rapid postimpact marine sedimentation and no postimpact tectonism (Poag 1996). This advantageous setting appears also to account for the presence nearby of at least 23 smaller fault-bounded excavations, which we interpret to be secondary craters (Poag 1997a, 1998, 1999a; Fig. 5.2).

These 23 secondary craters appear north and northwest of the crater rim on two multichannel seismic profiles collected by Texaco and Exxon Exploration Company. Other segments of these same two profiles also help to define the Chesapeake Bay primary impact crater (Fig. 5.2; Poag et al. 1994; Poag 1996, 1997a, 1998). We performed a seismostratigraphic analysis of these profiles, and calibrated the profiles with lithostratigraphy and biostratigraphy from nearby outcrops and boreholes (Poag et al. 1994; Poag 1996, 1997a). Our analysis highlights the structural and stratigraphic contrasts between the normal succession of flat-lying sedimentary coastal plain rocks and the stratal disruptions characteristic of the secondary craters.

Apparent diameters (apparent because most secondaries are crossed by only a single profile) of the secondary craters range from 0.4 km to 4.7 km, and average 1.9 km; only four have apparent diameters greater than 3 km. Apparent depths of the secondaries (measured from sedimentary lip to crater floor) range from 50 to 710 m, averaging 370 m. In six of the secondaries, the entire preimpact sedimentary section appears to have been excavated and replaced by impact breccia (Table 5.1). The breccia fill ranges from 30 to 680 m in apparent thickness, and averages 266 m.

5.2 Secondary Craters on Profile T-1-CB

Five secondary craters (C-1 to C-5) are imaged by north south seismic profile T-1-CB, where it crosses Chesapeake Bay near the mouth of the Potomac River (Fig. 5.2; Table 5.1). Crater C-3 was illustrated by Poag (1997a; his Fig. 31).

The C-1 structure is the southernmost secondary crater, located at the junction of T-1-CB and T-11-PR, approximately 8 km east of Smith Point, Virginia, and 35 km north of the northern rim of the Chesapeake Bay primary crater. This location is outside the seismically identifiable periphery of the breccia apron of the primary crater (Figs. 2.14, 5.2). Undisturbed, flat-lying, coastal plain formations extend continuously (interrupted only by a few collapse structures and scattered clusters of small-offset normal faults; Fig. 5.3) for the entire 35-km distance between the

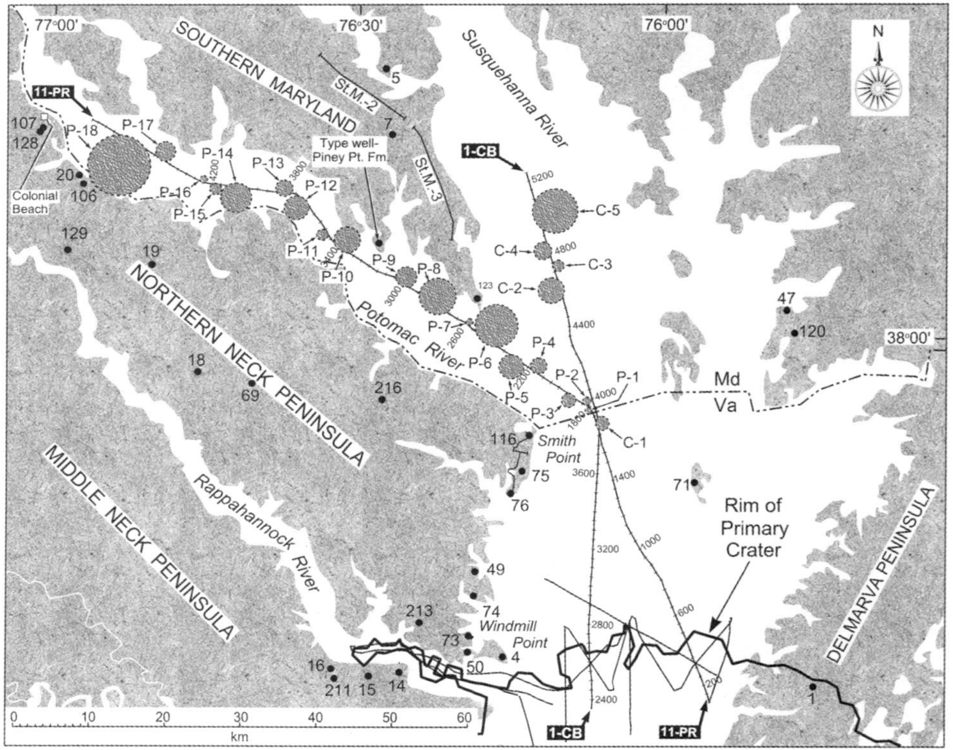

Fig. 5.2. Geographic distribution of secondary craters along seismic reflection profiles 1-CB (C-1– C-5) and 11-PR (P-1– P-18), north and northwest, respectively, of Chesapeake Bay primary crater. Solid dots indicate borehole locations; onshore seismic tracklines shown in southern Maryland (*St.M.-2*, *St.M.-3*) and at Smith Point, Virginia. *Md* = Maryland; *Va* = Virginia. See CD-ROM.14a–d for profile 11-PR.

Table 5.1. Morphometric data for 23 secondary craters inferred from reflection characteristics on seismic profiles T-1-CB (5 Chesapeake Bay secondaries) and T-11-PR (18 Potomac River secondaries).

Secondary Crater Designa-tion	Shot Point Location	Apparent Diameter [km]	Apparent Disruption Depth [1] [m]	Maximum Apparent Crater Depth [2] [m]	Maximum Apparent Breccia Thickness [m]	Raised Rim Present
CB 1	3840–3885	1.1	32	150	80	No
CB 2	4560–4660	2.7	44	310	180	No
CB 3	4720–4750	1.0	56	440	330	No
CB 4	4785–4865	2.2	74	710	[3]640	No
CB 5	4935–5085	4.2	68	700	[3]520	Yes
Potomac 1	1780–1797	0.4	18	50	30	No
Potomac 2	1808–1840	0.7	20	80	50	No
Potomac 3	1890–1945	1.5	41	340	250	No
Potomac 4	2130–2170	1.1	28	140	100	No
Potomac 5	2230–2310	2.1	60	450	270	Yes
Potomac 6	2385–2535	4.2	40	300	280	No
Potomac 7	2580–2600	0.6	40	300	120	Yes
Potomac 8	2720–2890	4.7	51	440	360	No
Potomac 9	2960–3048	2.4	66	710	[3]680	Yes
Potomac 10	3335–3425	2.4	59	640	[3]610	Yes
Potomac 11	3475–3495	0.4	22	350	110	Yes
Potomac 12	3663–3725	1.8	52	580	[3]540	No
Potomac 13	3775–3850	2.1	40	130	90	No
Potomac 14	4005–4115	3.0	48	400	300	No
Potomac 15	4165–4185	0.6	14	70	50	Yes
Potomac 16	4240–4255	0.4	20	80	50	Yes
Potomac 17	4480–4535	1.5	52	550	120	No
Potomac 18	4580–4655	2.0	48	600	[3]350	Yes
Average		1.9	43	37	266	

[1]Maximum depth of sediment disruption below sea level
[2]Depth of crater floor below crater rim
[3]Entire preimpact sedimentary section excavated and replaced by impact breccia

northern rim of the primary crater and the C-1 secondary crater.

All five secondaries on profile T-1-CB have similar general characteristics. The principal differences are in their respective apparent diameters and depths. Each crater is marked by clearly expressed rim escarpments constructed by *en echelon* (presumably concentric) down-to-the-basin normal faults (Figs. 5.3, 5.4).

The rim faults truncate horizontal, parallel, continuous to subcontinuous reflections, which represent the same preimpact target rocks disrupted by the Chesapeake Bay primary impact (Lower Cretaceous to lower Eocene siliciclastic sediments and middle Eocene bioclastic limestone; Fig. 2.4). Inside each secondary crater, the seismic reflections are chaotic or incoherent. We interpret these signatures to represent impact breccia equivalent to the Exmore breccia. A key seismic reflection horizon on profile T-1-CB is the top of the middle Eocene Piney Point Formation, represented by an essentially horizontal reflection at ~0.25 s (two-way traveltime; ~160 m) north and south of the group of secondary craters. The type section for the Piney Point Formation is ~19 km west of the profile, in a borehole drilled at Piney Point on the north bank of the Potomac River (Fig. 5.2). The Piney Point Formation also is present at 64 m below sea level in the Baltimore Gas and Electric No.1 borehole, 48 km northwest (and updip) from the T-1-CB secondary craters (Ward 1984; Gibson 1989). The same key horizon can be identified on seismic profile T-11-PR, which passes within 5 km of the Piney Point borehole at shot point 3200.

The floors of secondary craters C-1 and C-2 appear to be formed wholly within the fractured and faulted sedimentary rocks of Early Cretaceous (Potomac Formation) to Paleocene (Aquia Formation) age (Fig. 5.4), but C-3, C-4, and C-5 appear to have been excavated down to the surface of the crystalline basement (Fig. 5.5). The overlying postimpact formations (mainly middle Miocene to Quaternary sediments) thicken and sag into all five secondary craters on profile T-1-CB, just as they do where they cross the outer rim of the primary crater.

Along profile T-1-CB, the relatively flat basement reflection rises gradually northward (updip) from 0.9 s (~900 m) at the northern rim of the primary crater to ~0.925 s (~925 m) at secondary C-1. The basement reflection rises to 0.875 s (875 m) beneath C-2, then dips to 0.9 s beneath C-3 and >0.9 s (>900 m) below C-4; it then rises to 0.83 s (830 m) north of C-5. Along this profile segment, the basement reflection is gently warped and cut by numerous individual and clustered normal faults, having small vertical offsets. These faults bound scattered grabens and horsts interspersed between wide intervals that contain few or no faults (Fig. 5.6). Only a few of the crater-bounding faults can be traced into the crystalline basement. Most appear to be detached from the basement surface, similar to those at the outer rim of the primary crater (Figs. 5.4, 5.5).

5.3 Secondary Craters on Profile T-11-PR

Eighteen similar, small, collapse or excavation structures are distributed along a 110-km segment of profile T-11-PR (P-1–P-18; Figs. 5.2, CD-ROM.14a–d; Table 5.1), which extends from the northern rim of the primary crater to a location in the Potomac River, ~5 km east of the town of Colonial Beach. We infer that these structures, too, are secondary impact craters created by blocks ejected from the primary crater. Though most features of the Potomac River (P) secondaries are

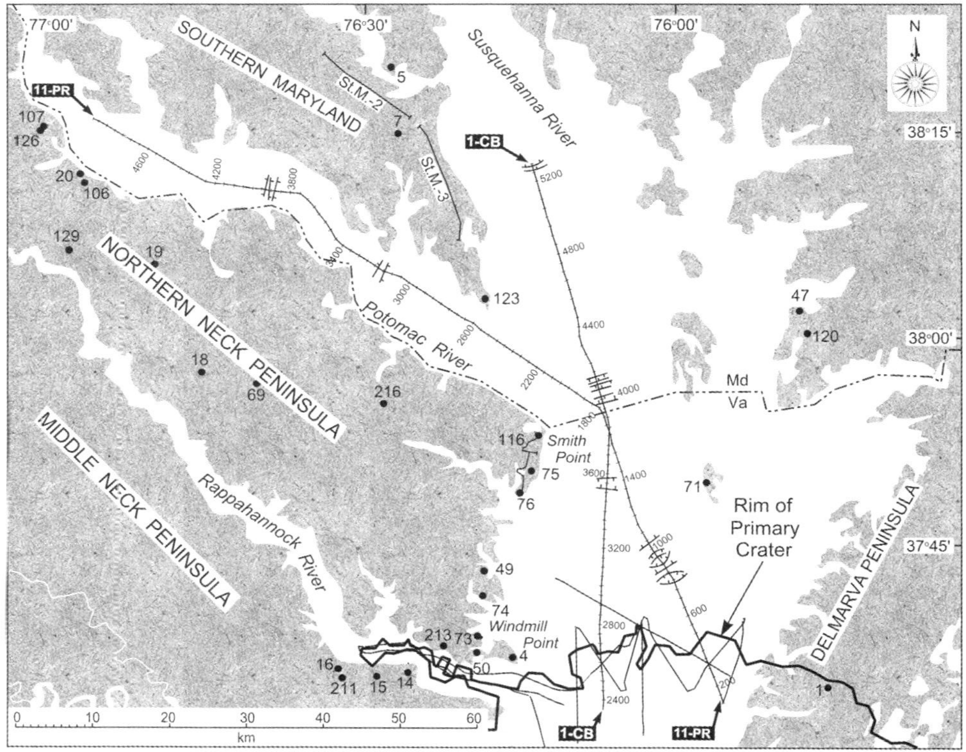

Fig. 5.3. Geographic distribution of normal faults documented in sedimentary section along seismic reflection profiles 11-PR and 1-CB, northwest and north of Chesapeake Bay primary crater. Solid dots indicate borehole locations; *Md* = Maryland; *Va* = Virginia.

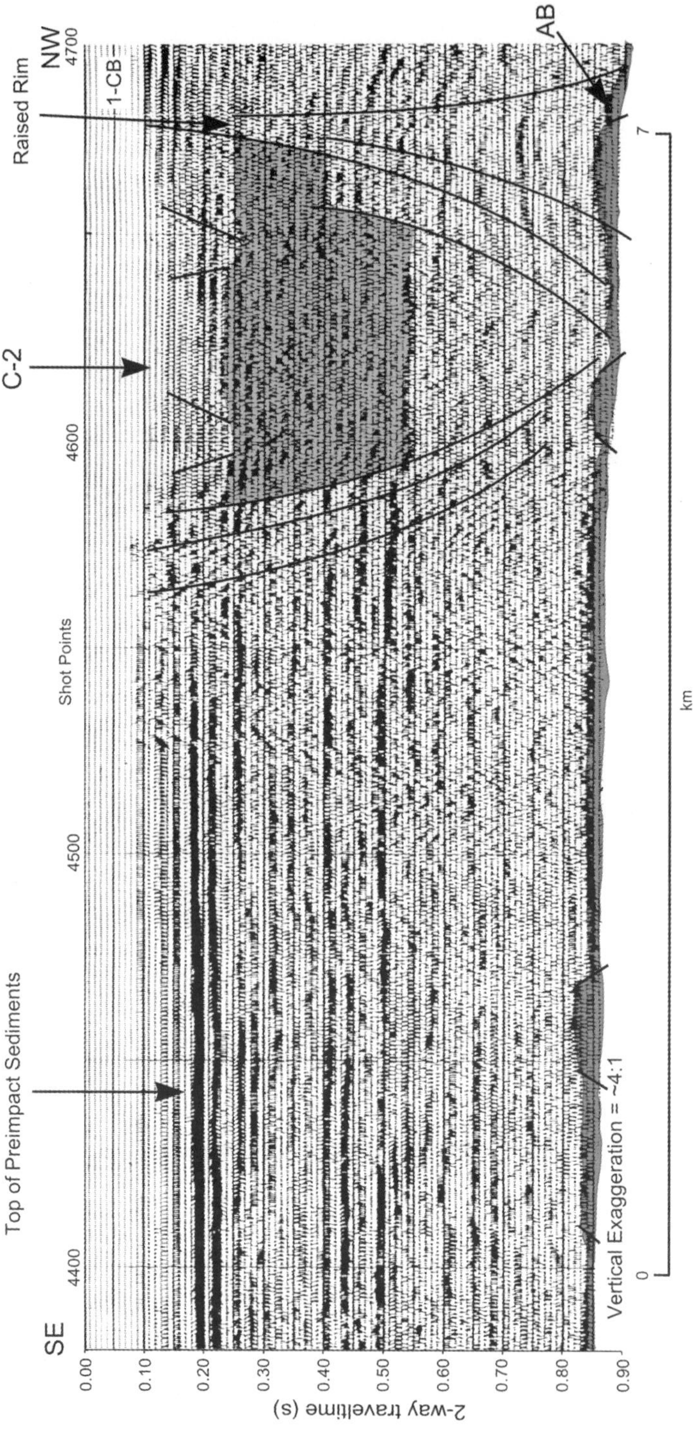

Fig. 5.4. Interpreted segment of seismic reflection profile 1-CB crossing Chesapeake Bay secondary crater C-2, showing fault traces, crater-fill breccia (lighter shading), and crystalline basement (*AB*; darker shading). See Fig. 5.2 for location.

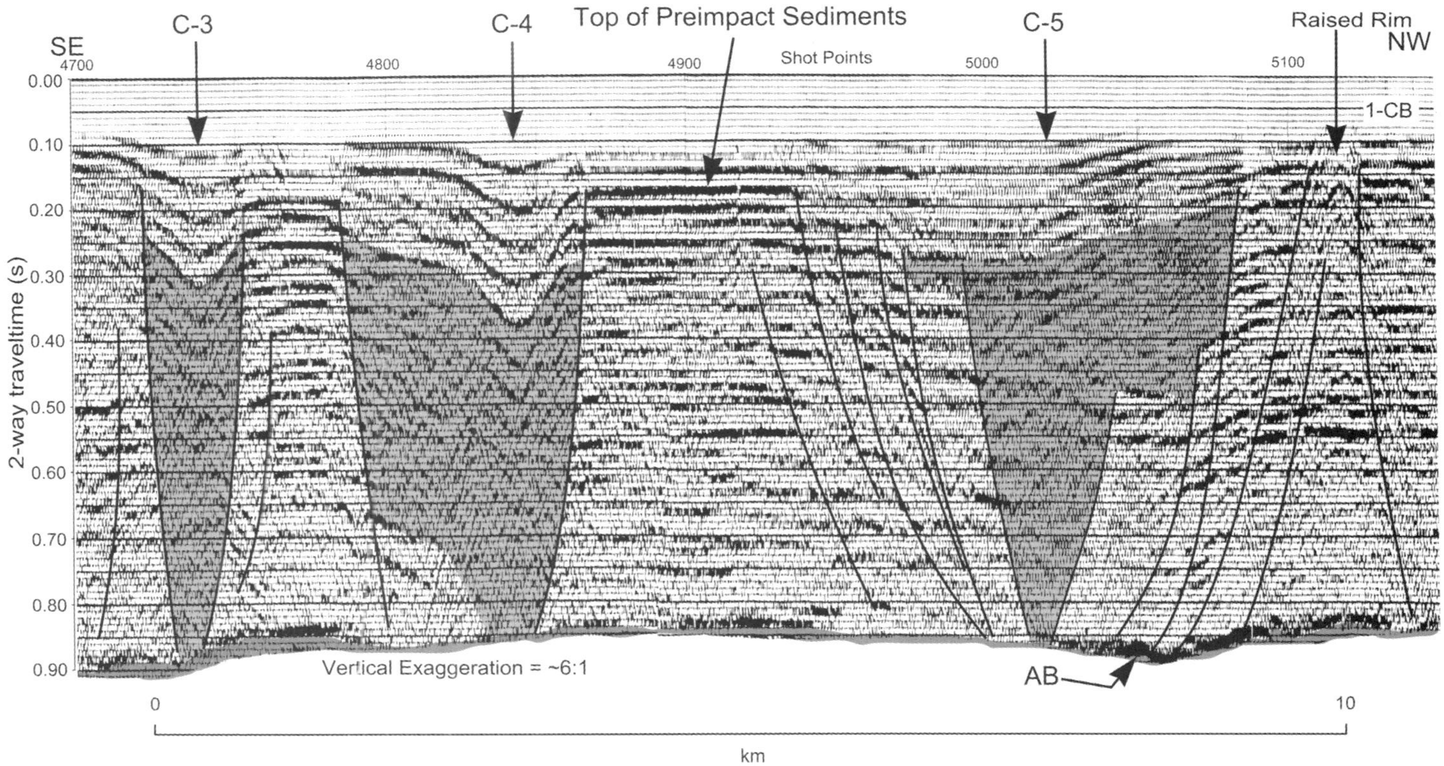

Fig. 5.5. Interpreted segment of seismic reflection profile 1-CB crossing Chesapeake Bay secondary craters C-3, C-4, and C-5, showing fault traces, crater-fill breccia (lighter shading), and crystalline basement (*AB*; darker shading). See Fig. 5.2 for locations.

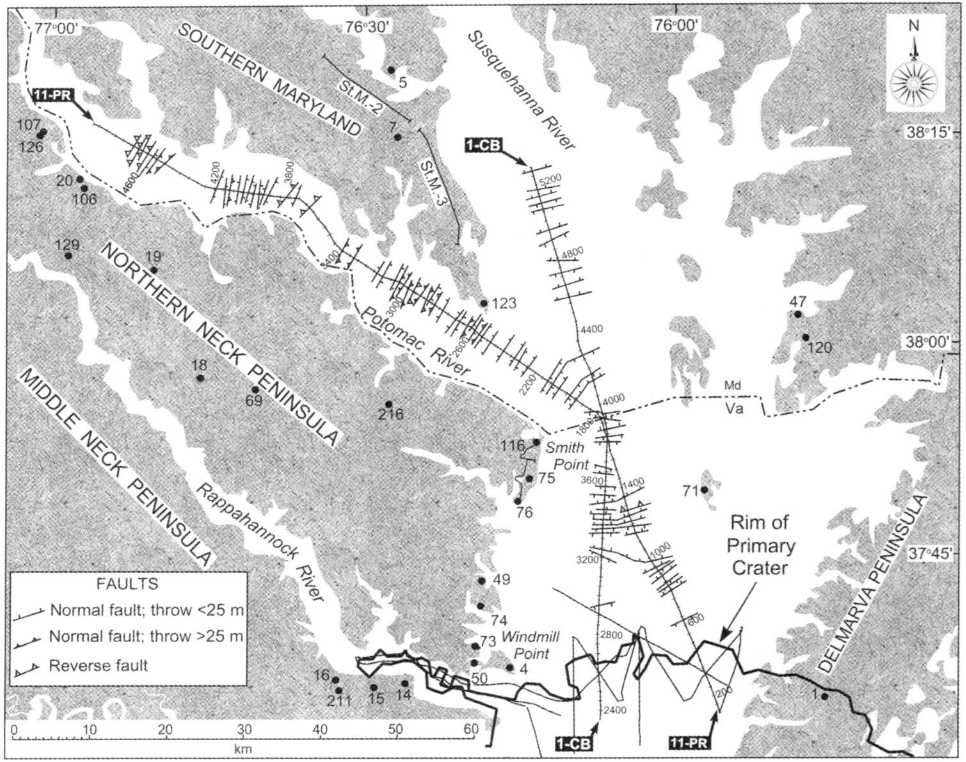

Fig. 5.6. Geographic distribution of normal and reverse faults that displace basement surface (and in some cases, overlying sedimentary section) along seismic reflection profiles 11-PR and 1-CB north and northwest of Chesapeake Bay primary impact crater. Solid dots indicate borehole locations; onshore seismic tracklines as in Fig. 5.2; *Md* = Maryland; *Va* = Virginia.

similar or identical to those of the C secondaries, some differences can be noted (Figs. 5.7–5.12). For example, eight of the P secondaries have well-developed, raised, sedimentary rims, almost identical to the few raised rim segments preserved on the primary crater.

Perhaps the most important difference, however, is that some of the normal faults on most of the P secondaries appear also to disrupt the basement reflection. This basement disruption is especially evident for P-5, P-7–P-10, P-13, P-14, P-17, and P-18 (Figs. 5.7–5.12). In this region, the basement rocks have had a particularly complex history. Along part of the same profile, the basement rocks consist of metasedimentary units within the Sussex terrane and Chesapeake block, and, in part, constitute the upper lithified sedimentary deposits of the Taylorsville rift basin (Fig. 2.1). The basement surface in the Taylorsville segment (particularly beneath P-13 and P-18) has been offset by eight reverse faults of 100 m or more vertical displacement (lateral displacement is unknown), seven of which display underthrusting to the west (Figs. 5.11, 5.12). The basement beneath P-18 (Fig. 5.12) has been thrust into two high-angle anticlinal folds. The thrust folds extend to the top of the preimpact target rocks. Thus, the thrusting coincided with the primary impact (i.e., late Eocene). We interpret these thrusts as distal products of the compressive shock wave that radiated from the primary impact site. Some of the reverse faults, however, appear to have been reactivated as normal faults, probably during the late stages of crater deformation. The original fault planes acted as zones of crustal weakness, along which normal displacement took place during the ejecta bombardment that formed the secondary craters. All this would have taken place within a few minutes of the primary impact (see Chapter 12).

The C-1 crater is the only secondary for which we can determine a true diameter, because all the other secondaries are crossed by only a single seismic profile. It is highly unlikely, of course, that a single profile would symmetrically bisect any secondary crater. Secondary C-1, however, is crossed by both profiles T-1-CB and T-11-PR (Fig. 5.2), which allows us to estimate the true diameter to be 1.1 km. The position and size of the secondaries illustrated in Figure 5.2 are diagrammatic, and intended only to show: (1) the locations where they intersect the profiles; (2) that they are not all centered on the profiles; and (3) that their diameters vary.

5.4 Implications of Secondary Crater Record

The stratigraphic and structural characteristics of the 23 structures we interpret as secondary craters match the general morphologic and structural features of simple craters (as opposed to complex craters; e.g., Melosh 1989), though the Chesapeake secondaries presumably were not formed by hypervelocity impacts. The principal difference is the apparent lack of overturned flaps (though any flap present may be too small to be resolved on our profiles). This is not surprising, however, because the impacts took place in the late Eocene ocean. There is ample evidence that even small oceanic impact craters lack these features, probably as a result of ex-

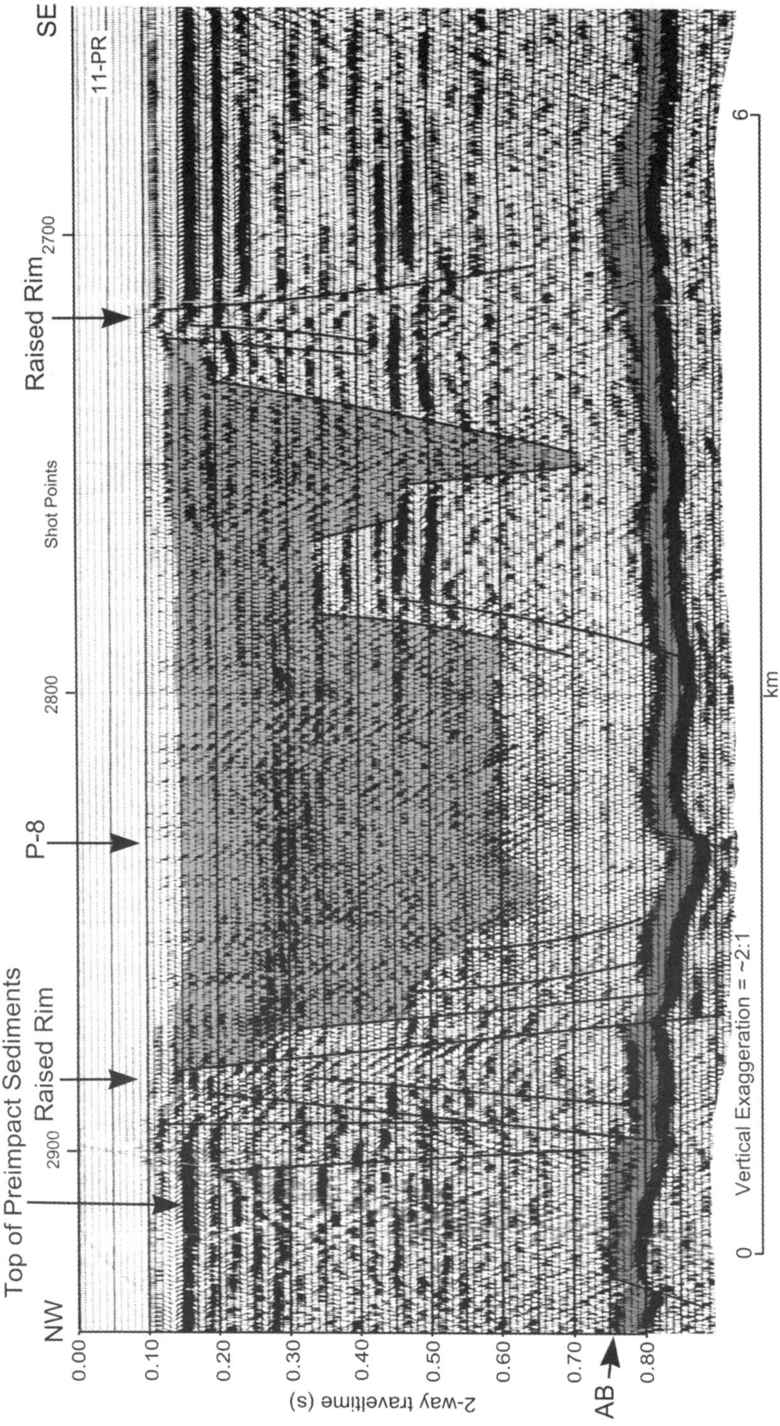

Fig. 5.7. Interpreted segment of seismic reflection profile 11-PR crossing Chesapeake Bay secondary crater P-8, showing fault traces, crater-fill breccia (lighter shading), and crystalline basement (*AB*; darker shading). See Fig. 5.2 for location and CD-ROM.14b for full-scale profile.

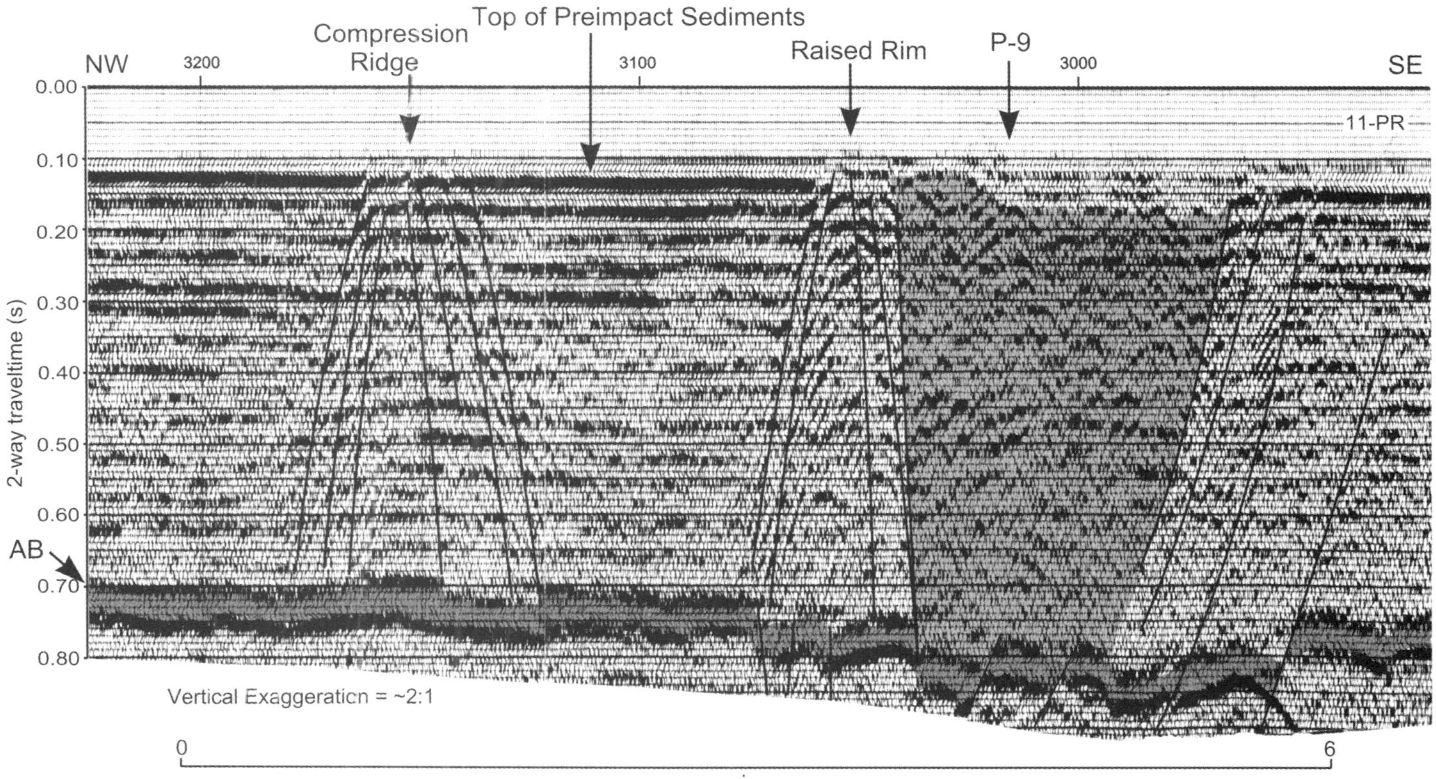

Fig. 5.8. Interpreted segment of seismic reflection profile 11-PR crossing Chesapeake Bay secondary crater P-9, showing fault traces, crater-fill breccia (lighter shading), and crystalline basement (*AB*; darker shading). See Fig. 5.2 for location and CD-ROM.14b for full-size profile.

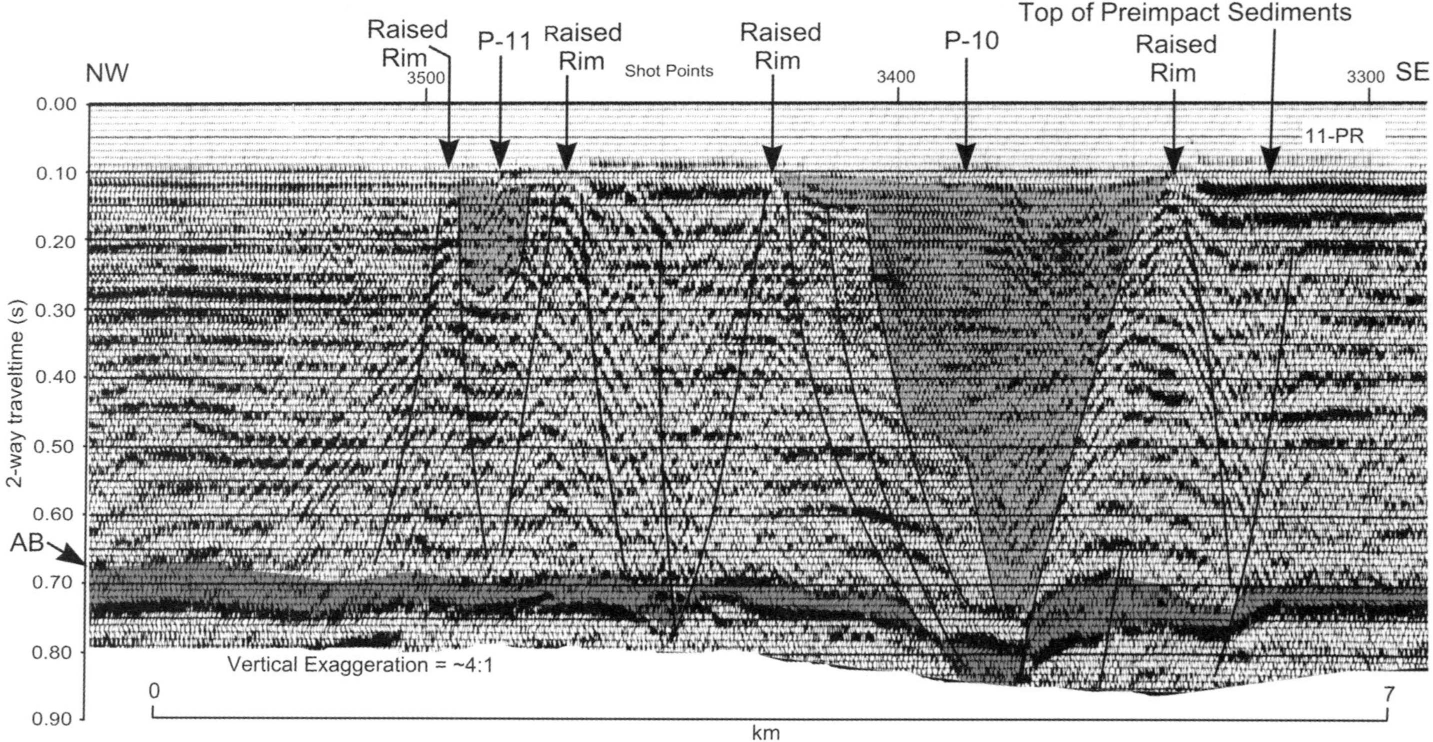

Fig. 5.9. Interpreted segment of seismic reflection profile 11-PR crossing Chesapeake Bay secondary craters P-10 and P-11, showing fault traces, crater-fill breccia (lighter shading), and crystalline basement (*AB*; darker shading). See Fig. 5.2 for locations and CD-ROM.14b for full-size profile.

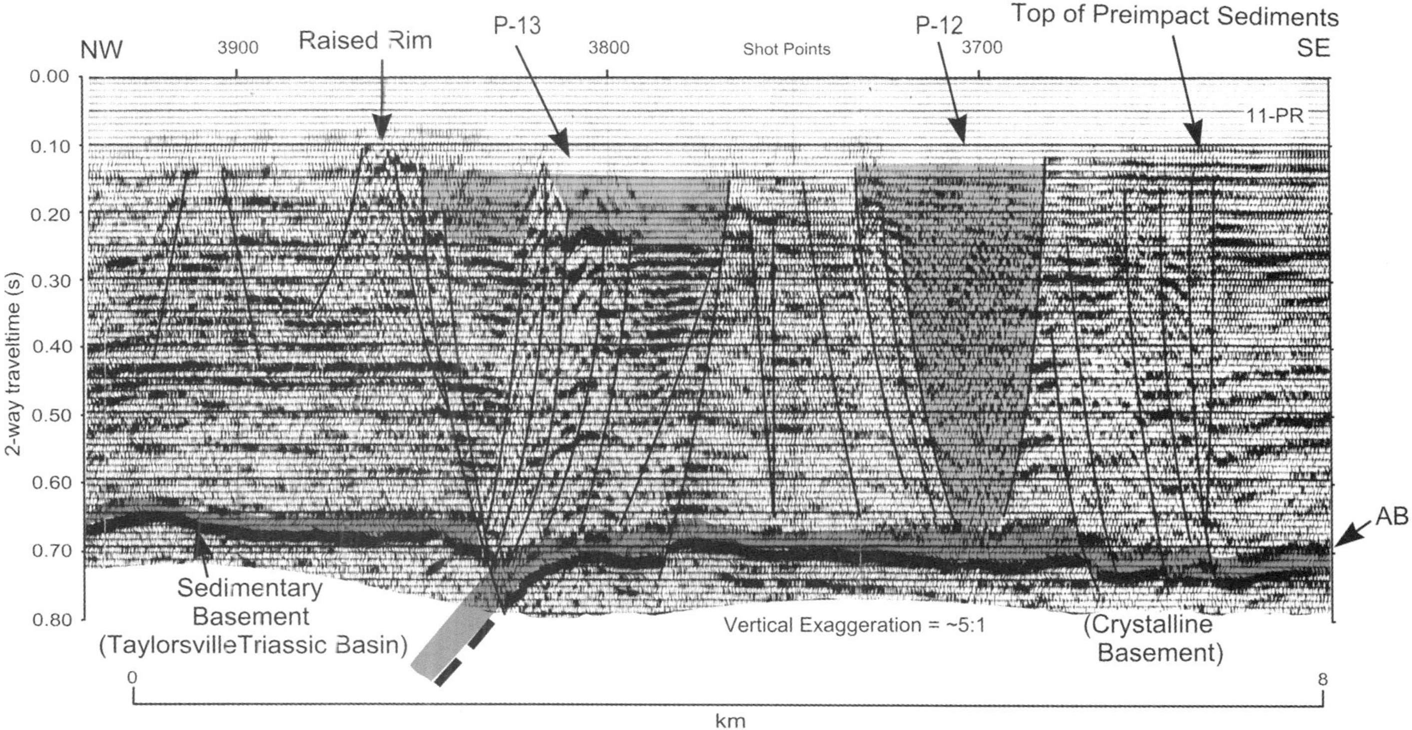

Fig. 5.10. Interpreted segment of seismic reflection profile 11-PR crossing Chesapeake Bay secondary craters P-12 and P-13, showing fault traces, crater-fill breccia (lighter shading) crystalline basement (*AB*; darker shading), and change from crystalline to lithified sedimentary basement composition at edge of Taylorsville Triassic basin. See Fig. 5.2 for locations and CD-ROM.14a,b for full-size profile.

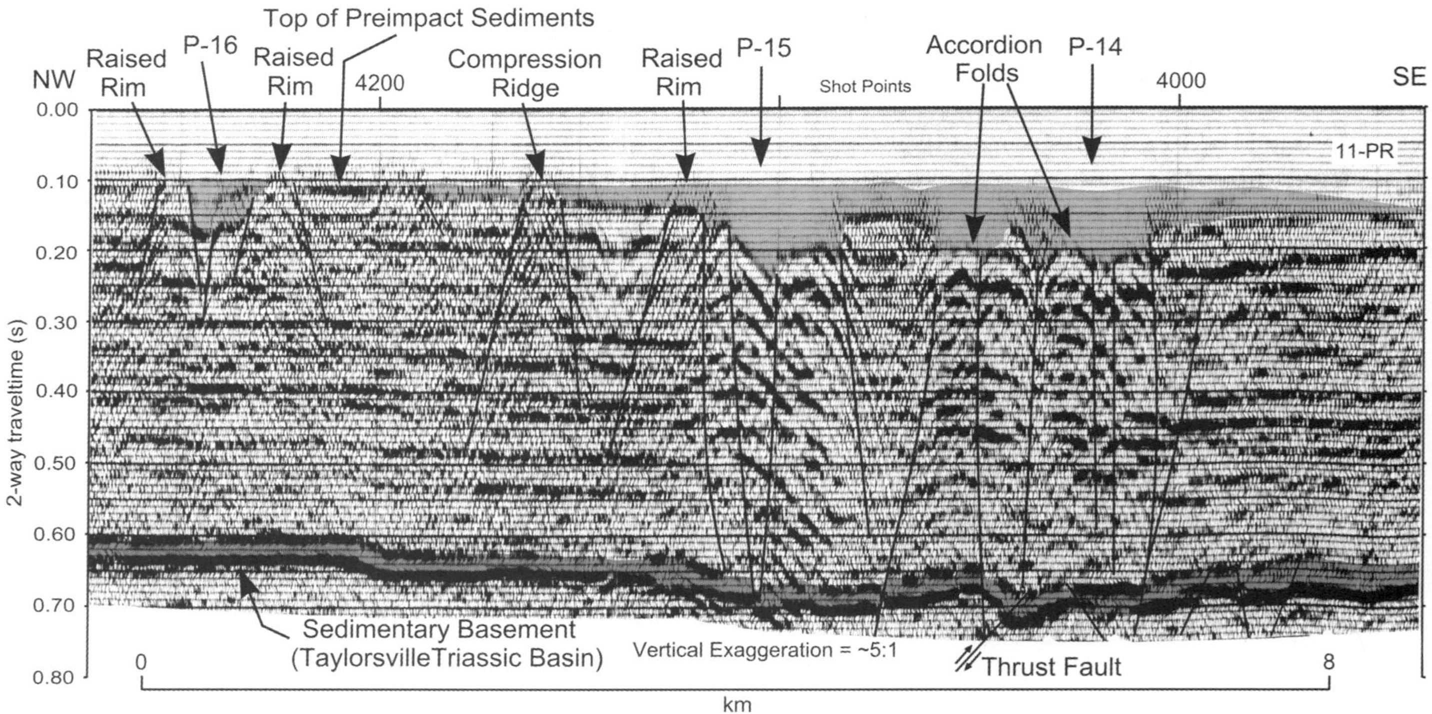

Fig. 5.11. Interpreted segment of seismic reflection profile 11-PR crossing Chesapeake Bay secondary craters P-14–P-16, showing fault traces, crater-fill breccia (lighter shading), and crystalline basement (*AB*; darker shading). See Fig. 5.2 for locations and CD-ROM.14a for full-size profile.

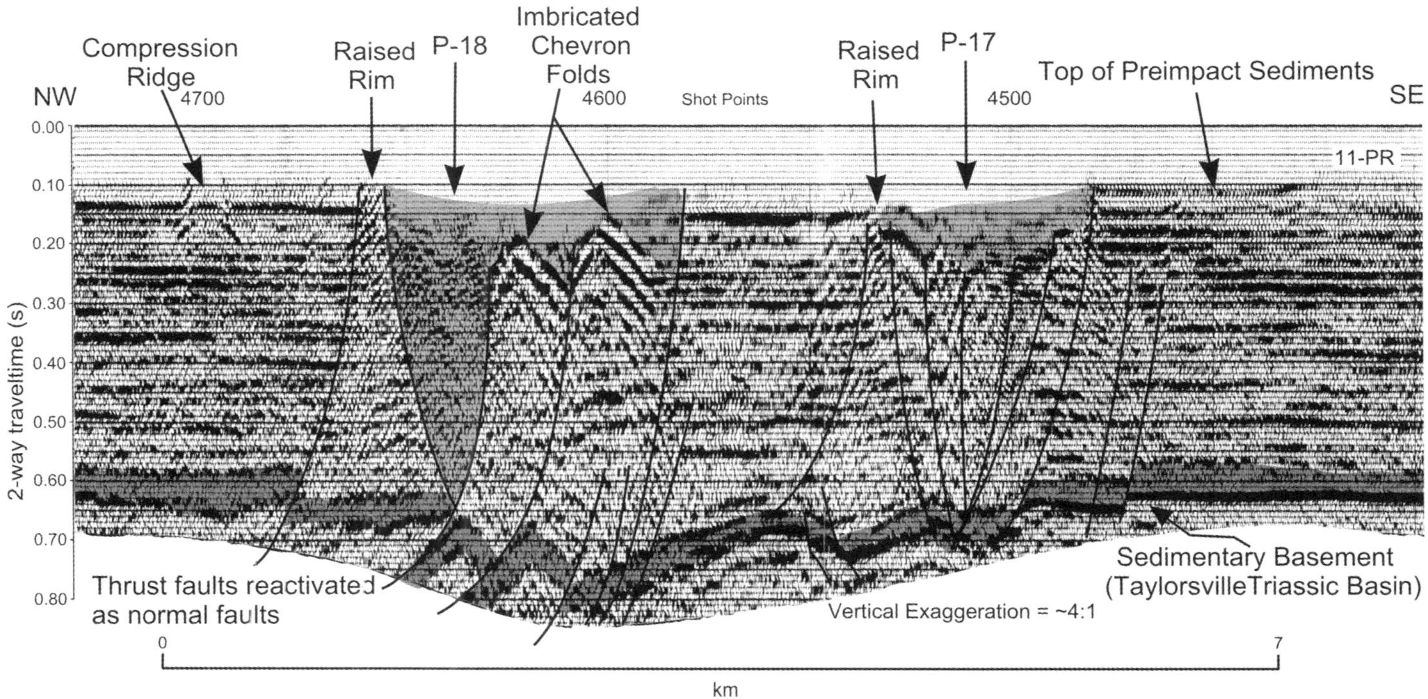

Fig. 5.12. Interpreted segment of seismic reflection profile 11-PR crossing Chesapeake Bay secondary craters P-17 and P-18, showing fault traces, crater-fill breccia (lighter shading), and crystalline basement (*AB*; darker shading). Note thrust faults reactivated as normal faults and note associated imbricated chevron folds. See Fig. 5.2 for locations and CD-ROM.14a for full-size profile.

tensive slumping of their unconsolidated, water-saturated walls, and of hydraulic erosion resulting from both surgeback of the oceanic water column, and washback from tsunami wavetrains (Higgins and Butkovich 1967; Kieffer and Simonds 1980; McKinnon and Goetz 1981; Silver 1982; Roddy et al. 1987; Jansa 1993; Poag and Poppe 1998; Ormö and Lindström 2000; see Chapter 12).

Ours is the first report of possible buried secondary craters associated with a buried primary source crater. On other planetary bodies, of course, only exposed secondaries can be identified because of limitations imposed by observation instruments currently available. The characteristics of the buried Chesapeake secondary craters are in general agreement with the features of exposed planetary secondaries. In both cases, the secondaries appear to occur in distinct clusters, or perhaps chains, and their apparent diameters range from 0.05 to 0.1 of the diameters of their respective primaries. Most of the Chesapeake secondaries, though, unlike many planetary secondaries, appear to be symmetrical rather than asymmetrical in cross section. Additional seismic surveys are necessary, however, to confirm this.

Dip reversal is notable within the postimpact sediments overlying the secondary craters. This phenomenon caused Poag (1997a) to speculate that reversed dips in outcropping postimpact coastal-plain sediments of southeastern Virginia (Ward and Strickland 1985; Johnson et al. 1998) might be caused by faulting and differential subsidence (due to breccia compaction) associated with underlying secondary craters. If this hypothesis proves to be accurate, we can expect to find numerous additional secondary craters scattered around the perimeter of the Chesapeake Bay primary crater at locations of dip reversal.

The presence of secondary craters below the bed of the Potomac River may shed light on one of the problems regarding recognition of the Mattaponi Formation as a formalized lithostratigraphic unit. As pointed out by Poag (1997a) and Powars and Bruce (1999), the presence of the Mattaponi Formation in the two boreholes that Cederstrom (1957) chose as the Mattaponi co-type sections, has been in doubt for nearly 50 years. This doubt stems from two principal sources. First, the type boreholes (106 and 107; Figs. 1.3, 2.14, 5.2; Table 1.2) are located near Colonial Beach and Washington's Birthplace, on the south bank of the Potomac River, more than 50 km northwest of the main depocenter for the Mattaponi Formation. Second, several additional boreholes, drilled within a few kilometers of the co-type boreholes, have not yielded sediments attributable to the Mattaponi Formation. If, however, the Mattaponi Formation consists of impact breccia, as we infer (see also Poag 1997a), and in the vicinity of Colonial Beach is confined to secondary impact craters, it would not be surprising to find it this far away from the main crater. It also would be normal not to find the Mattaponi in other boreholes near Colonial Beach that happen not to penetrate a secondary crater. Our inference that secondary crater P-18 extends under the south bank of the Potomac River (Fig. 5.2) rests on our assumption that Cederstrom (1957) correctly identified the Mattaponi Formation in borehole 106.

6 Synimpact Crater-Fill Deposits

6.1 Oldest Breccia Unit

Five coreholes (Exmore, Kiptopeke, NASA Langley, North, Bayside) have sampled crater-fill deposits of the Chesapeake Bay crater, and two others (Windmill Point, Newport News) have sampled partly equivalent deposits outside the outer rim (Figs. 6.1–6.3, CD-ROM.7; Table 1.2). The crater-fill deposits can be subdivided into several distinctive units, which we will discuss in chronostratigraphic order from oldest to youngest.

The stratigraphically oldest crater-fill debris, which is inferred from seismic reflection profiles to be present only in the inner basin, has not yet been sampled by drilling. On the seismic profiles, this deepest crater-fill is expressed by chaotic or incoherent reflections (Figs. 4.2, 4.26, 4.28, 4.29), which hypothetically represent impact breccia dominated by large blocks of shock-metamorphosed crystalline basement rocks (megabreccia).

The oldest crater-fill unit sampled to date was encountered just above crystalline basement in the Bayside corehole. The unit is 20.33 m thick (66.7 ft), and consists of polymict, sediment- and crystalline-clast, matrix-supported breccia. The number of pebble- and cobble-sized crystalline clasts (mainly weathered granite) is significantly greater than in any other cored section. In addition, this unit contains numerous pebbles and cobbles of hard, cemented red (Triassic?) sandstone, which are unique to this core segment. The red sandstone clasts display white, millimeter-scale rinds, which may represent shock alteration, though no petrographic analyses have been completed at this writing. We interpret this breccia unit to represent debris ejected from the inner basin (Figs. 6.2, 6.4, CD-ROM.7).

6.2 Displaced Megablocks

6.2.1 Seismic Signature and General Lithic Composition

The stratigraphically next-to-deepest crater-fill deposits sampled to date come from a layer composed of 0.5- to 2-km-sized sedimentary megaslump blocks (with extensive internal deformation) and megaslide blocks (with little or no internal deformation) (Figs. 4.3A,B, 4.7B, 4.9A,B, 6.1–6.3, CD-ROM.7). As seen on the multichannel profiles, these displaced sedimentary megablocks are extensively

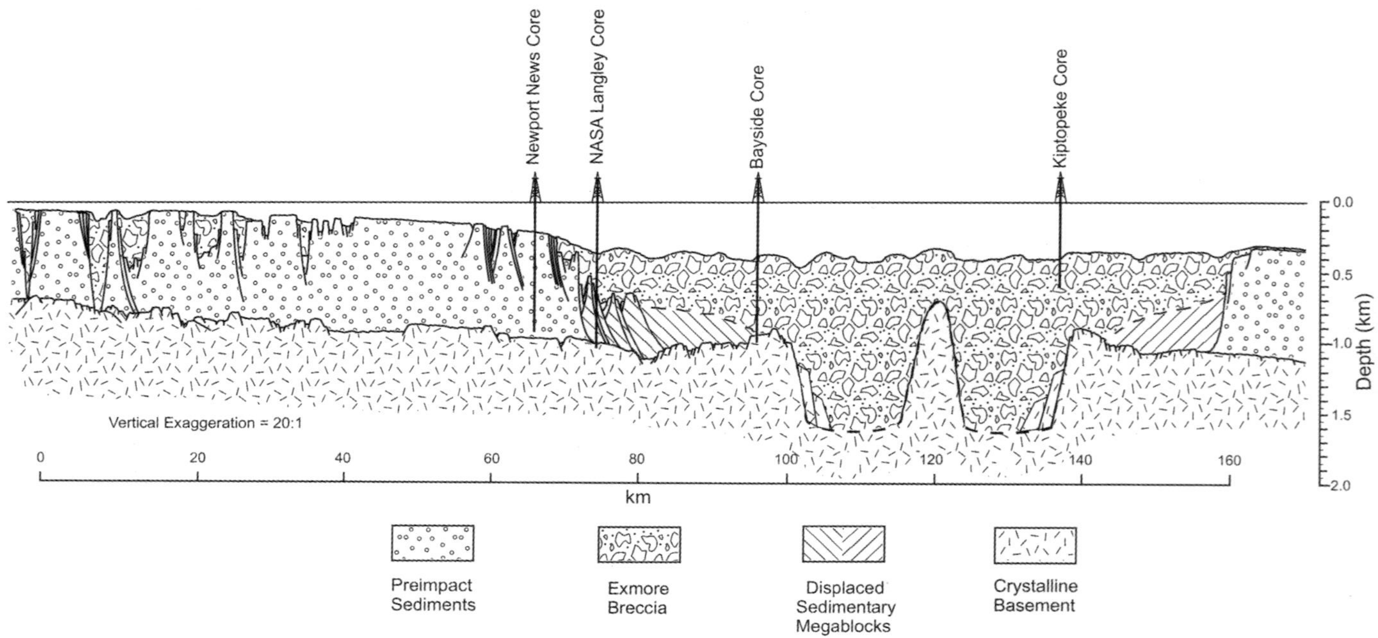

Fig. 6.1. Depth-scaled cross section of Chesapeake Bay primary and secondary impact craters along composite seismic reflection profile, showing two-dimensional geometries of, and spatial relations between, crystalline basement rocks, displaced sedimentary and crystalline megablocks, preimpact sedimentary strata, and Exmore breccia. Postimpact sediments have been removed from this section. Corehole locations are projected to this profile; their total depths are not measured from sea level, but are adjusted to show approximate depth relative to the basement surface. Compare with Fig. 4.2A.

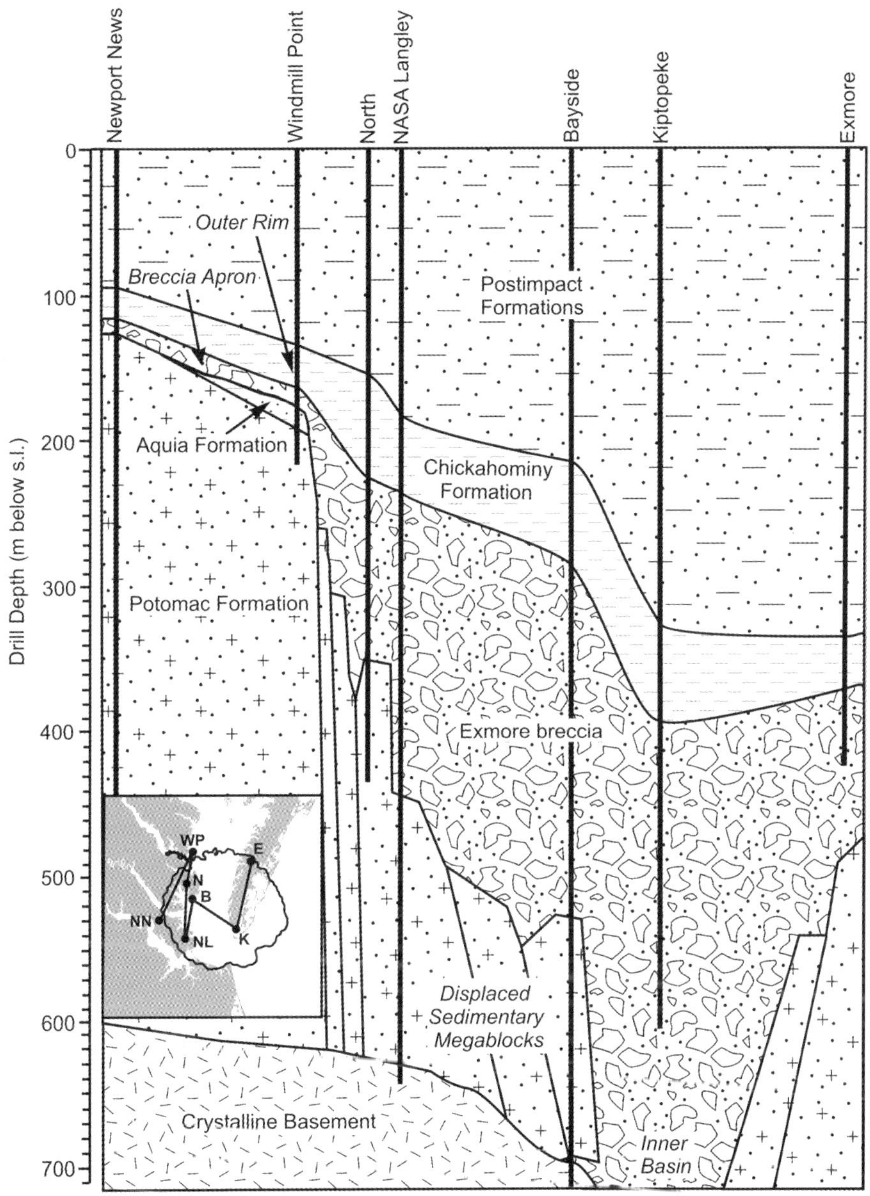

Fig. 6.2. Stratigraphic correlation of seven coreholes from which Exmore breccia has been recovered. Vertical exaggeration ~400:1. *NN* = Newport News; *WP* = Windmill Point; *N* = North; *NL* = NASA Langley; *B* = Bayside; *K* = Kiptopeke; *E* = Exmore. Inset map shows line of section. See CD-ROM.7 for more details about cored sections and correlations.

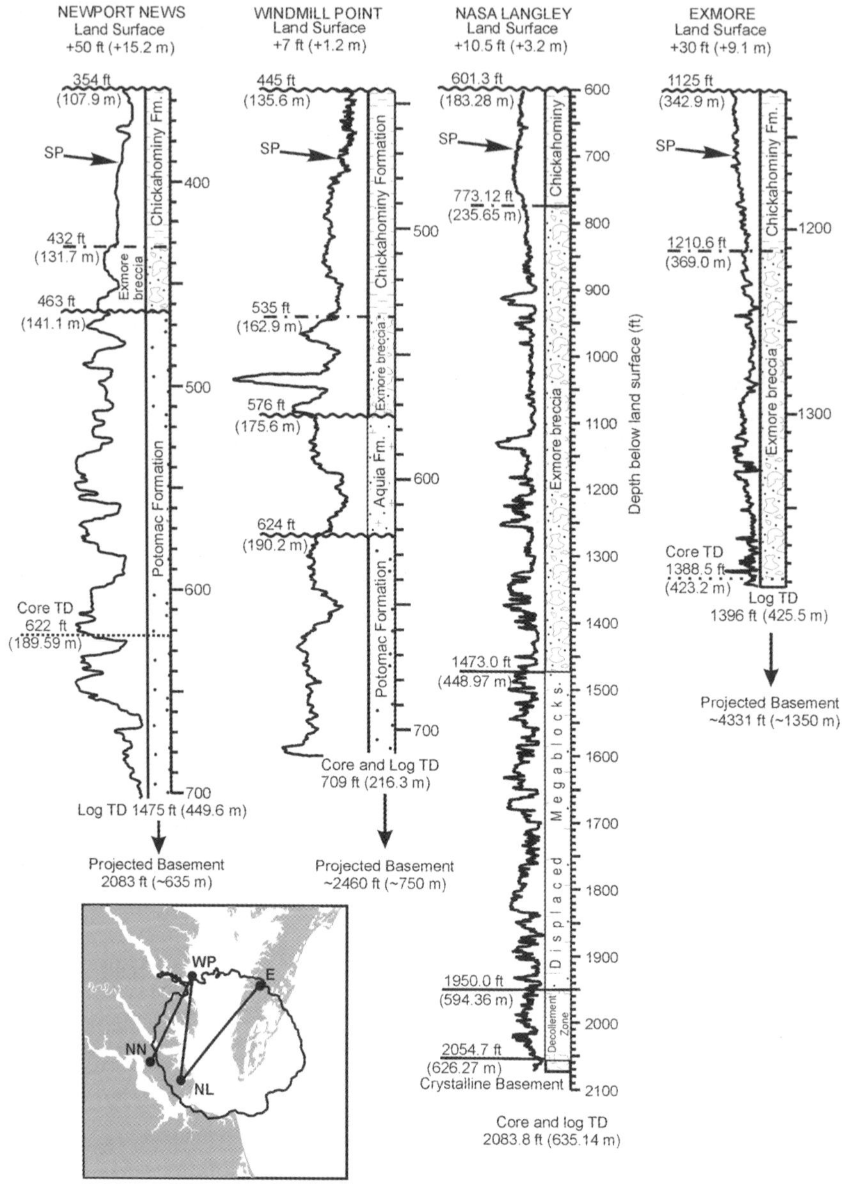

Fig. 6.3A. Sections of downhole SP logs from two coreholes outside Chesapeake Bay impact crater (Newport News and Windmill Point) and two coreholes in outer part of annular trough (NASA Langley and Exmore), showing stratigraphic succession from crystalline basement to postimpact sedimentary deposit (Chickahominy Formation). Vertical scale given in feet below land surface to match original logs and coring record. See CD-ROM.7 for more details about cored sections, SP logs, and correlations.

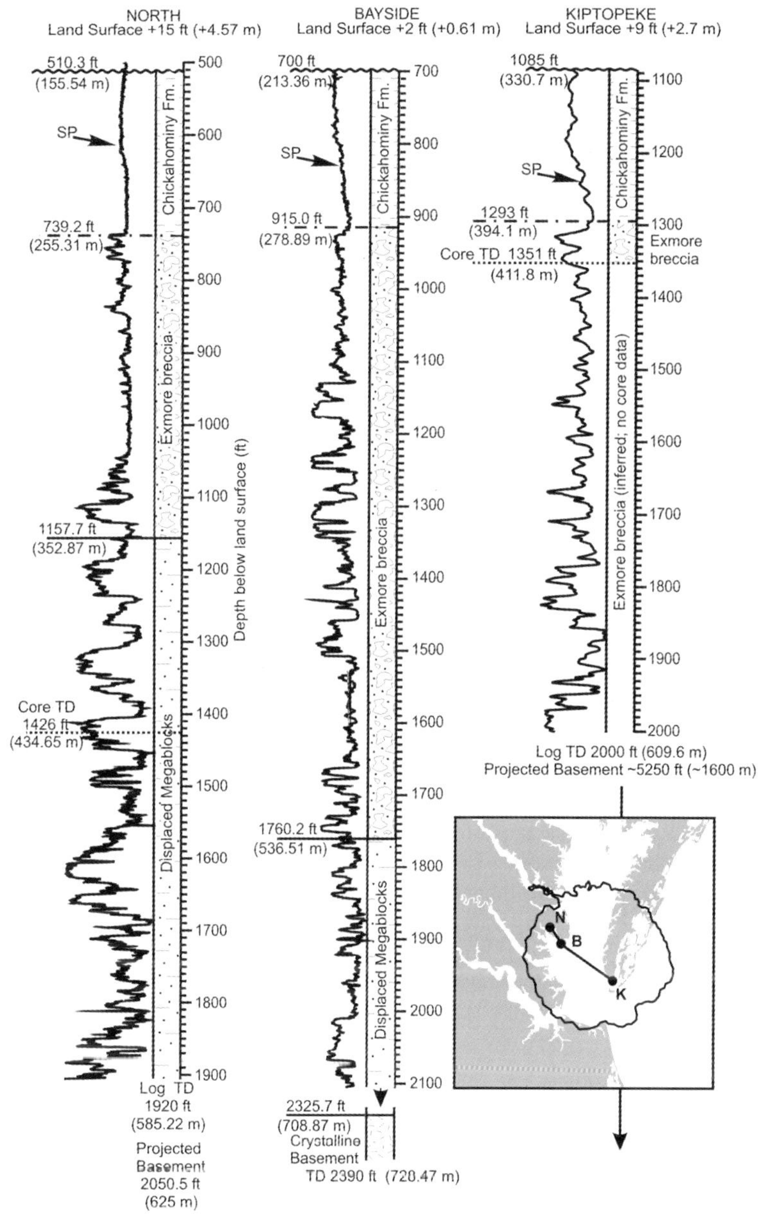

Fig. 6.3B. Sections of downhole SP logs from North corehole (in outer part of annular trough) of Chesapeake Bay impact crater, Bayside corehole (near outer flank of peak ring), and Kiptopeke corehole (inside peak ring), showing stratigraphic succession from crystalline basement to Chickahominy Formation. Vertical scale given in feet below land surface to match original logs and coring record. See CD-ROM.7 for more details.

fractured and cut by myriad compressional and extensional faults. On the single- and two-channel profiles, however, ringing interference (repetition of shallower reflections) between the basement and postimpact sediments prevents clear definition of the displaced megablocks (Figs. 4.11, 4.13, 4.14, 4.17, 4.20B). On seismic profiles, most displaced megablocks appear to be translational or collapse features that are detached from the crystalline basement surface by a zone of décollement (Fig. 4.3A,B). However, excellent examples of detached rotational megablocks also are prominent on several multichannel profiles. The best expressions of rotational megablocks are on profiles 11-PR and 9-CB-F, near the toe of the outer rim scarp (Fig. 4.3A,B). At both of these locations, the preimpact sedimentary section has been broken into megaslide blocks at least 0.75–1 km long (as measured on the seismic profiles). The blocks have rotated so that their toes are now 100–150 m lower than their heads. Additional sets of parallel, inclined reflections can be seen at numerous other locations on the multichannel profiles (at shot point 1100 on profile T-10-RR, for example; Fig. 4.7B), which indicates the widespread occurrence of both small and large rotational blocks. The layer of displaced sedimentary megablocks averages 200–300 m in thickness over most of the annular trough. Our seismic data indicate that, in contrast, sedimentary megablocks are absent over most (perhaps all) of the peak ring (Figs. 4.22–4.26), and are totally lacking in the inner basin and over the central peak (Figs. 4.32–4.34).

The upper surface of the megablock layer is highly irregular, commonly with hundreds of meters of relief (Figs. 4.3A,B, 4.7B, 4.9A,B, 6.2). Many megablocks appear to retain the seismic signature (vertical succession of reflection packages) of their undisturbed stratigraphic equivalents outside the crater. By correlating these reflection packages, we estimate that the highest points of the displaced megablocks are approximately 100–150 m lower than their stratigraphic equivalents outside the crater. In other words, the minimal vertical displacement of individual blocks within the megablock layer is ~100 m; maximum vertical displacement is ~350 m.

Displaced sedimentary megablocks have been sampled in three coreholes, NASA Langley, North, and Bayside (Figs. 6.2, 6.3, CD-ROM.7). As a rule, the general sedimentary compositions and structures of the three cored megablock sections are similar. Each is composed of a variable succession of massive nonmarine sands (1–20 m thick), coarsely to finely stratified sands (2–10 m thick), shattered dark clays (0.3–1 m thick), and sticky varicolored clays (1–7 m thick). Some of the sandy strata are inclined at 15–30 degrees from horizontal, some are cross-bedded, and others are virtually horizontal. Relatively thin intervals (<1 m) of chaotically mixed clasts, clay-injected clasts, and(or) other expressions of plastic, soft-sediment deformation punctuate the thicker stratified sections. Sediments at the top and base of the displaced megablocks tend to be massive sands, which appear to have been destabilized or fluidized by the impact shock wave.

In detail, however, each cored section of displaced megablocks is unique. The most complete stratigraphic sections through the megablock interval were cored at NASA Langley and Bayside, where the sections are 177 and 152 m thick, respectively (Figs. 6.2, 6.3, CD-ROM.7). The base of the megablock section at NASA Langley is a coarse, matrix-supported breccia, 1.4 m thick (Fig. 6.5A–C), which

Fig. 6.4. Photographs of four segments of Bayside core, showing polymict, sediment- and crystalline-clast, matrix-supported fallback breccia. Note white (fusion?) rinds on several clasts (*R*) and sulfide splotches due to possible hydrothermal mineralization (*H*). Numbers at top of each segment indicate drill depth at top of segment.

Fig. 6.5. Photographs of three segments of NASA Langley core, showing contact (dashed line) between weathered crystalline basement and basal breccia. Labeled clasts are crystalline basement (*B*) and clay (*C*). Splotches of red and yellow mineralization inferred to be possible hydrothermal deposits (*H*). Numbers at top of each segment indicate drill depth at top of segment.

Fig. 6.6. Photographs of four segments of NASA Langley core, showing different aspects of soft-sediment deformation within basal zone of décollement (possible shock fluidization): clay injection; brecciation; scaly clay; and possible hydrothermal mineralization. Numbers at top of each segment indicate drill depth at top of segment.

Fig. 6.7. Photographs of five segments of NASA Langley core, showing four different manifestations of stratified sands (*A–D*) and two examples of brecciation (*B, E*) within displaced sedimentary megablocks. Arrows indicate crossbeds in segment *A*; brecciation in segment *B*; ripple cross-lamination in segment *C*. Numbers at top of each segment indicate drill depth at top of segment.

Fig. 6.8. Photographs of four segments of NASA Langley core, composed of brightly colored sticky clays of Lower Cretaceous paleosols found within displaced sedimentary megablocks. Numbers at top of each segment indicate drill depth at top of segment.

Fig. 6.9. Photographs of four segments of NASA Langley core, showing three different manifestations of sedimentary breccias (*A, B, C*) and three examples of soft-sediment deformation (arrows on *B, C, D*) within displaced sedimentary megablocks. Numbers at top of each segment indicate drill depth at top of segment.

Fig. 6.10. Photographs of three segments of the NASA Langley core, showing steeply inclined stratification (dashed lines on *A*) in sand section and complexly distorted contact geometry (arrows on *B, C*) in upper part of displaced sedimentary megablock. Numbers at top of each segment indicate drill depth at top of segment.

rests directly on the weathered surface of crystalline basement (metagranite; Fig 6.5C). Angular to rounded pebbles and cobbles of clay dominate the clasts, but several cobbles of relatively fresh granite also are present. The matrix is gray, clayey, quartz sand.

Resting above the basal breccia at NASA Langley is a 32-m interval of shattered scaly clays and massive sands (no clear evidence of stratification; Fig. 6.6B,D). Several short (<1 m thick) sections within this interval display chaotic sedimentary features attributable to plastic deformation of soft sediment (Fig. 6.6A). This interval and the basal breccia also are notable for patches or splotches (2–10 cm in diameter) of brightly colored red, purple, and greenish-yellow mineralization, which appear to be sulfide deposits resulting from hydrothermal activity (Fig. 6.5A, 6.6A,D). Some sands in this basal megablock interval and its equivalent at Bayside are more indurated (cemented) than any other sediment yet encountered in the coreholes.

The middle section of the displaced sedimentary megablock at NASA Langley contains several excellent examples of finely and coarsely stratified sands (Fig. 6.7A–D) interspersed with sticky, varicolored, paleosol clays (Fig. 6.8A–D) and short sections (<1 m thick) of chaotically mixed sediment breccias (Fig. 6.7E, 6.9A,B).

The upper 4 m of the displaced megablock interval at NASA Langley consists of medium to coarse-grained, moderately stratified sand, with bedding planes inclined at 20–30 degrees from horizontal (Fig. 6.10A). Intense internal disruption in this upper section is particularly well displayed, however, by a 60-degree inclined contact (Fig. 6.10C). The upper boundary of this megablock interval is a nearly vertical contact with an overlying clay-clast breccia (Fig. 6.10B). Similar lithic and structural attributes characterize the megablock section at North and Bayside, but their stratigraphic distribution and lithic composition are notably less similar.

Displaced megablocks of crystalline basement rocks also can be distinguished along the walls of the inner basin of Chesapeake Bay crater on some seismic profiles (Figs. 4.22, 4.25A, 4.26B, 4.29A,B). No crystalline megablock, however, has yet been sampled by drilling.

6.2.2 Expression on Downhole Geophysical Logs

As part of our analysis of sedimentological properties of the crater-fill deposits at Chesapeake Bay, we have used the spontaneous potential (SP) logs, a standard method for identifying the lithic boundaries between subsurface units and for determining their relative permeabilities (Asquith 1982). The general expression of the displaced megablock sections on downhole spontaneous potential (SP) logs is a predominance of thick, blocky, permeable units (sands, silts) separated by distinctly less permeable intervals (mostly sticky clay paleosols; Figs. 6.3A,B, CD-ROM.7). The less permeable sections are notably thinner than the more permeable sections in the outer crater at NASA Langley, but at North and Bayside they are equal in thickness (9–18 m) to the more permeable sections. Several upward-fining sequences can be distinguished at each of these three sites (Figs. 6.3A,B,

CD-ROM.7). No correlation is obvious between individual SP units in the megablock sections of these coreholes.

6.3 The Exmore Breccia

6.3.1 Seismic Signature and General Geometry

The central peak, inner basin, peak ring, annular trough, and outer rim of the Chesapeake Bay impact crater are totally buried by an unusually thick deposit of impact breccia (Figs. 6.1, 6.2). This breccia is informally named the Exmore breccia, after the town of Exmore, Virginia, where the deposit was first cored (Powars et al. 1992; Poag 1997a). We infer that the bulk of the breccia formed from fragments (sand-sized particles to kilometer-sized megablocks) of sedimentary rocks and crystalline basement rocks violently disrupted initially by the impact and further deformed and turbulently mixed together by subsequent water-column collapse and surgeback, followed by runup and washback of the resultant tsunami wave train (see Chapters 11 and 12 for further discussion). Gohn (in press) refers to the breccia by the more general, nongenetic, term *diamicton*, in deference to the abundance of both rounded and angular clasts within the debris.

The upper surface of the Exmore breccia is manifest on most seismic reflection profiles as a distinctive high-amplitude, nearly continuous reflection (reflection PS; Figs. 4.3A,B, 4.5A,B, 4.7B, 4.9A,B, 4.13, 4.15, 4.16, 4.18–4.20, 4.22, 4.24–4.26, 4.28, 4.29A,B, 4.32–4.34). This upper surface reflection is horizontal to subhorizontal and relatively smooth on a kilometer scale, but is quite irregular on a scale of tens to hundreds of meters. The upper surface of the breccia is easily traceable over the entire crater, and characteristically sags into the crater along radial transects across the outer rim (Figs. 6.2, 6.11). Relief on the breccia surface at this outer-rim sag varies from 30 to 190 m, averaging 110 m (Table 6.1). Where measured inside the crater, the upper surface of the breccia varies from 210 m below sea level under the York River (profile S-3), to more than 550 m below sea level in the eastern sector of the crater (profile S-1; Fig. 6.11). Average elevation of the breccia inside the rim (relative to mean sea level) is -347 m, versus -238 m, on average, outside the rim.

The upper surface of the breccia also sags as it crosses the peak ring into the inner basin (Figs. 4.22–29; Table 6.2). The sag relief ranges from 40 to 230 m and averages 117 m. Average elevation outside the inner basin is -298 m, versus -416 m inside the basin. An equivalent sag takes place across the central peak (Figs. 4.27–30; Table 6.3). Sag there ranges from 30 to 90 m and averages 47 m. Average elevation of the breccia on the central peak is -372 m, whereas it averages -417 m in the inner basin.

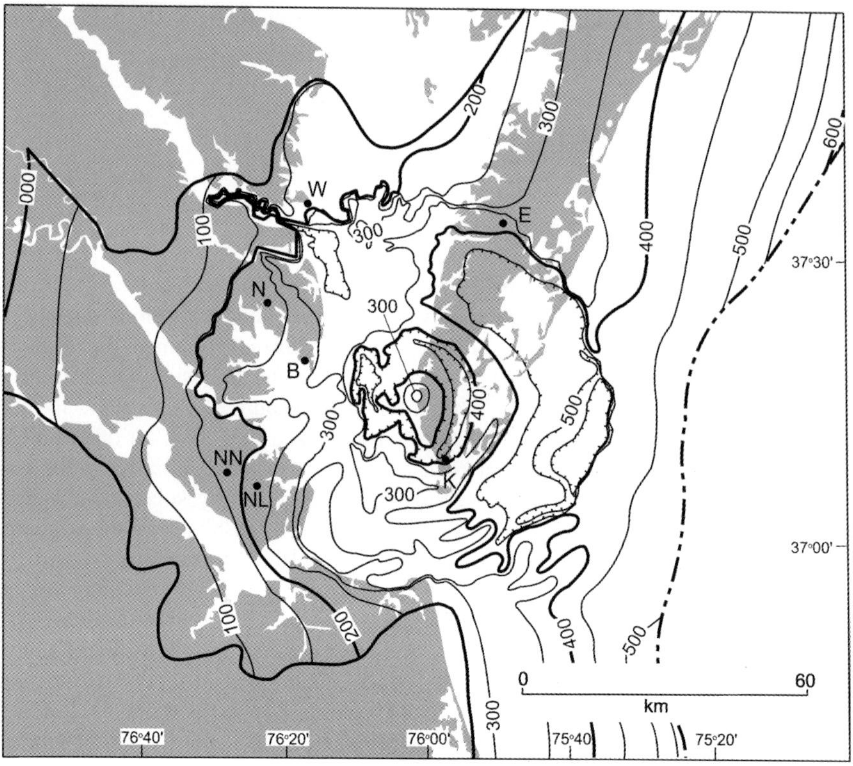

Fig. 6.11. Structural map of upper surface of Exmore breccia (depths in meters below sea level; contour interval 50 m). *NN*, Newport News corehole; *L*, NASA Langley corehole; *K*, Kiptopeke corehole; *E*, Exmore corehole; *W,* Windmill Point corehole, *N*, North corehole; *B*, Bayside corehole. Heavy dashed line approximates seaward extent of breccia distribution.

In contrast, the lower surface of the Exmore breccia is difficult to determine with precision, except where it rests directly on crystalline basement (on the peak ring, for example). Over most of the annular trough, the Exmore breccia lies upon the kilometer-scale sedimentary megablocks (Figs. 1.5, 4.3A,B, 4.7B, 4.9A,B, 6.1, 6.2). The breccia-megablock contact can be approximated as an irregular boundary of high relief (200–300 m) on some of the Texaco multichannel profiles (e.g., Fig. 4.3A,B). Seismic ringing and the presence of extensive intrabed multiples, however, obscure the megablock morphology and structure on nearly all single-channel and 2-channel profiles (e.g., Figs. 4.11, 4.13).

Between its upper and lower surfaces, the internal seismic signature of the Exmore breccia is variable, depending on the type of data analyzed (multichannel,

Table 6.1. Vertical sag of upper surface of Exmore breccia across outer rim of Chesapeake Bay crater as measured on 22 seismic profiles.

Profile #	Elevation Outside Rim [2][mbsl]	Elevation Inside Rim [2][mbsl]	[1]Sag [m]
S-2	110	210	100
S-3	130	210	80
T-13-YR	130	240	110
S-16	150	250	100
S-17	150	300	150
N-3	130	240	110
S-12	160	220	60
S-13	150	340	190
T-10-RR	200	300	100
T-1-CB	200	280	80
S-5	230	340	110
S-6	280	380	100
S-8	240	380	140
S-9	290	380	90
S-10	190	310	120
S-11	200	290	90
S-19	410	540	130
S-22	420	530	110
S-25	380	540	160
S-27	380	470	90
S-1	390	550	160
E-3	310	340	30
Average	238	347	110

[1]Sag is elevation difference from outside to inside crater
[2]meters below sea level

single-, or two-channel), and on the thickness of the breccia and its overlying sedimentary cover. In its clearest expression, the breccia appears on many of the Texaco multichannel profiles as a zone of incoherent hyperbolic reflections, which contrasts markedly with the high-amplitude, horizontal reflections of the crystalline basement surface and of the postimpact sedimentary section (Figs. 4.3A,B, 4.7A,B, 4.9A,B, 4.22, 4.25A). On almost all single-channel and two-channel profiles (and older multichannel profiles, e.g., *Neecho*), however, acoustic ringing and intrabed multiples completely mask the internal breccia signature (Figs. 4.5B, 4.11, 4.13, 4.14, 4.20B, 4.23A,B, 4.24A,B, 4.26A,B, 4.32).

Table 6.2. Vertical sag of upper surface of Exmore breccia across peak ring of Chesapeake Bay impact crater as measured at 18 crossings by seismic reflection profiles.

Profile #	Elevation Outside Basin [[2]mbsl]	Elevation Inside Basin [[2]mbsl]	[1]Sag [m]
S-6	400	440	40
[3]S-10 (N)	330	420	90
[4]S-10 (S)	270	370	100
S-4 (N)	300	350	50
S-4 (S)	290	350	60
S-7	340	400	60
S-14	330	430	100
S-15	360	440	80
T-1-CB (N)	300	475	175
T-1-CB (S)	250	475	225
T-8-S-CB-E (N)	300	420	120
T-8-S-CB-E (S)	280	380	100
7-CB-H	220	450	230
E-2 (N)	320	460	140
E-2 (S)	280	420	140
E-3	220	380	160
N-1	280	420	140
N-2	300	400	100
Average	298	416	117

[1]sag is elevation difference from outside to inside crater
[2]meters below sea level
[3]north end of given profile
[4]south end of given profile

6.3.2 Distribution and Thickness

We determined the structure, distribution, and thickness of the Exmore breccia (Figs. 6.11, 6.12) within the crater by integrating multichannel, single-, and two-channel seismic profiles, with seven continuously cored boreholes. We supplemented these data with 124 additional non-cored boreholes (Figs. 2.12, CD-ROM.1; Table 1.2). The Exmore breccia covers the entire crater, completely fills the inner basin and the annular trough, and overlaps the outer-rim escarpment. Lithostratigraphic successions in the Exmore breccia sections cored at NASA Langley, Exmore, Bayside, and Kiptopeke are easily correlatable with seismostratigraphic interpretations of the nearest reflection profiles (SEAX-16, S-6, S-4, and S-7, respectively; Figs. 4.10, 4.21). The thickest section of Exmore breccia cored inside the crater was 249 m at the Bayside site (Figs. 1.3, 2.14, 6.2, 6.3, CD-

Table 6.3. Vertical sag of upper surface of Exmore breccia across central peak of Chesapeake Bay impact crater as measured on 10 seismic reflection profiles.

Profile #	Elevation above peak crest [2][mbsl]	Elevation in adjacent inner basin [2][mbsl]	[1]Sag [m]
S-6	290	330	40
S-7	330	390	60
S-10	370	400	30
S-14	420	460	40
S-15	410	440	50
T-8-S-CB-E	390	440	50
T-7-CB-H	370	420	50
N-1	380	410	30
N-2	380	410	30
E-2	380	470	90
Average	372	417	47

[1]sag is elevation difference from outside to inside crater
[2]meters below sea level

ROM.7; Table 1.2). At the Kiptopeke site, the projected thickness of this breccia (derived from the seismic profiles) is on the order of 1.2 km, but only the upper 17.7 m was cored there (prior to identification of the crater).

Over the annular trough, the Exmore breccia is generally ~200 m thick in the western half of the crater, but thickens eastward to ~400 m (rough estimate) in the eastern half, where seismic evidence is poorest. The breccia thickens markedly to more than a kilometer (~1.2 km) in the inner basin (Figs. 4.27–4.29, 6.12), but the elevation of its basal surface there cannot be determined directly from our current seismic data.

Just outside the crater rim, the Exmore breccia makes up an irregularly distributed band of impact debris, averaging ~30 km in width (Figs. 6.2, 6.12). At most impact craters, such a debris band would be referred to as an ejecta blanket. However, though the debris band at Chesapeake Bay contains some ejected materials, it cannot be treated as an ejecta blanket if, as we propose, it was deposited in large part by washback from the impact-generated tsunami wave train. We prefer to call this band a breccia apron, a neutral term that does not imply a particular depositional origin. The breccia apron has been cored at Newport News (9.4 m thick) and at Windmill Point (12.7 m thick; Figs. 6.2, 6.3, CD-ROM.7). At Windmill Point, the breccia apron is too thin to distinguish on the nearby seismic profiles, but the rest of the lithostratigraphic succession is unambiguously correlative with the seismostratigraphy displayed on profile S-12, which is only 1.5 km south of the core site (Figs. 4.6, 4.7A,B).

One hundred-twenty four additional boreholes have sampled the breccia apron on the west side of the crater (Figs. 2.14, CD-ROM.1). Many of these additional boreholes are quite old, having been completed in the late 1800s and early 1900s,

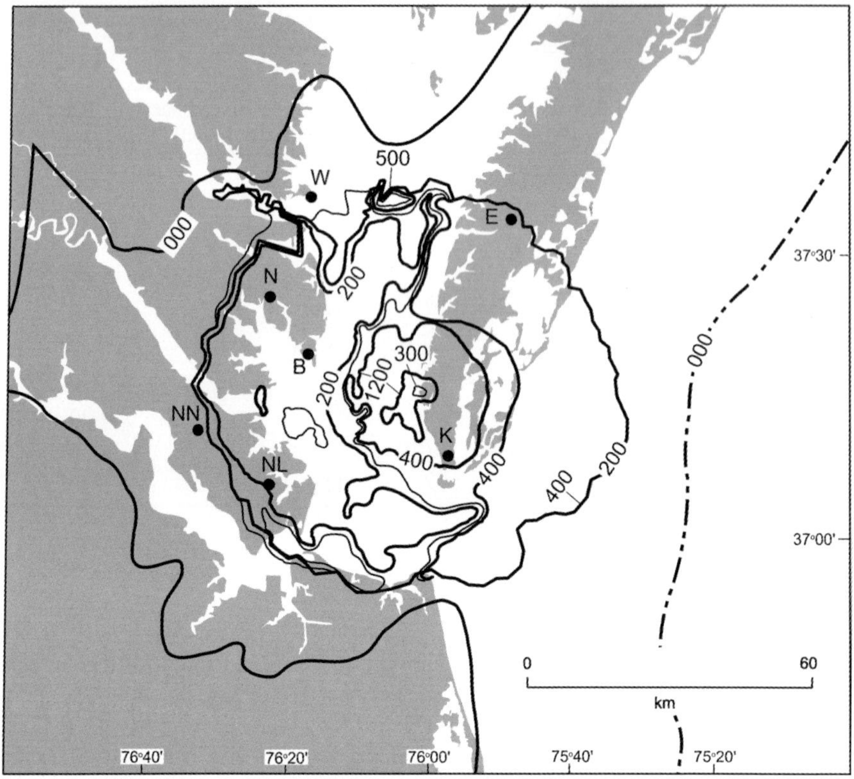

Fig. 6.12. Isopach map of Exmore breccia (thickness measured in meters; contour interval irregular). Abbreviations as in Fig. 6.11.

and were sampled from cable-tool cuttings or rotary cuttings, but no cores were taken. Our interpretation of the lithostratigraphy from these cuttings depends, in part, on 50-year-old drillers' logs, and upon interpretations of the distribution of the Mattaponi Formation (Cederstrom 1945a,b, 1957), which we infer to be equivalent to the Exmore breccia. Close to the crater (within ~10 km), there is good agreement between our interpretation and those of Cederstrom (1957) and Powars and Bruce (1999), regarding the distribution of the breccia apron. Farther west on the York-James Peninsula, however, our viewpoint diverges from that of Powars and Bruce (1999). We accept Cederstrom's contention that the Mattaponi Formation spreads westward of Williamsburg, which creates a broad tongue of breccia between the Mattaponi and Pamunkey Rivers (Fig. 2.14). Using Cederstrom's interpretations, we constructed three structural cross sections perpendicular to the inferred direction of tsunami washback (Fig. 6.13). The cross sections show that the breccia fills a distinct, broad channel. Because we tentatively infer that this channel was eroded by the runup and washback of the impact tsunamis, we tentatively call it a washback channel.

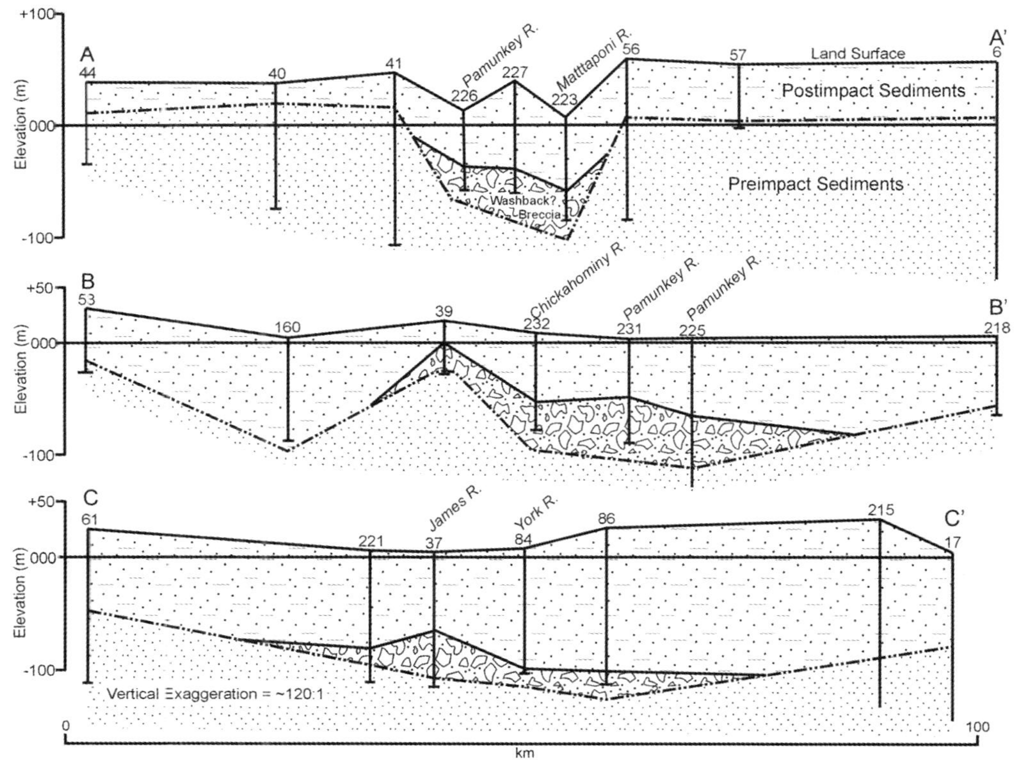

Fig. 6.13. Three cross sections through inferred Chesapeake Bay impact-tsunami washback channel filled with washback breccia. See Fig. 2.14 for locations of cross sections and CD-ROM.1 for borehole locations.

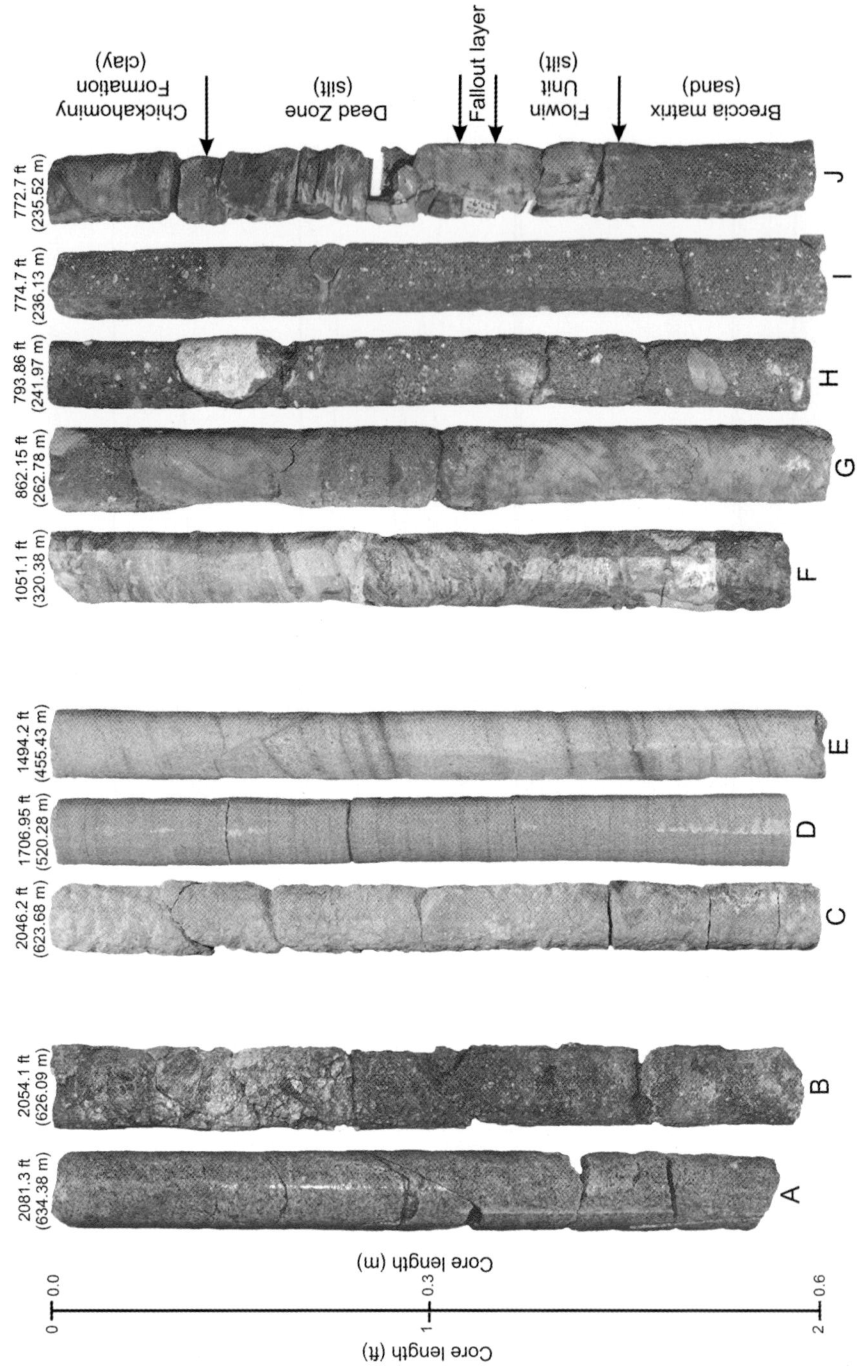
Chickahominy Formation (clay)
Dead Zone (silt)
Fallout layer
Flowin Unit (silt)
Breccia matrix (sand)
772.7 ft (235.52 m)
J
774.7 ft (236.13 m)
I
793.86 ft (241.97 m)
H
862.15 ft (262.78 m)
G
1051.1 ft (320.38 m)
F
1494.2 ft (455.43 m)
E
1706.95 ft (520.28 m)
D
2046.2 ft (623.68 m)
C
2054.1 ft (626.09 m)
B
2081.3 ft (634.38 m)
A
Core length (m)
0.0
0.3
0.6
Core length (ft)
0
1
2

The position of the limit of the breccia apron around the eastern half of the crater (Fig. 6.12) is a rough estimate. The ejecta layer here is too thin to identify on the seismic profiles. Poag (1997a), however, reported evidence of correlative impact ejecta from coastal wells in North Carolina, Maryland, and New Jersey. Miller et al. (1998) also reported evidence of late Eocene impact-generated subsurface deposits in the New Jersey Coastal Plain.

6.3.3 General Lithology

Several lithic units, or lithofacies, can be recognized within the Exmore breccia on the basis of cores and downhole geophysical logs. These finer subdivisions are discussed in detail in Chapters 11 and 12. In this section, we limit our descriptions to the broader lithic aspects of the breccia body.

In a general sense, the Exmore breccia is polymictic, variably matrix-supported to clast-supported, and upward fining, and the entrained clasts consist mainly of poorly consolidated sediments. At the six sites where core recovery was most complete (NASA Langley, North, Bayside, Exmore, Windmill Point, Newport News; CD-ROM.7), the upward-fining trend resembles that of a coarse-grained megaturbidite or debris-flow deposit (Powars et al. 1992; Poag and Aubry 1995; Poag 1997a; Fig. 6.14). Microfossils generally are common to dominant within the Exmore matrix and constitute a stratigraphically mixed assemblage of marine and nonmarine taxa (ostracodes, planktonic and benthic foraminifera, calcareous nannofossils, dinoflagellate cysts, spores, and pollen) ranging in age from Albian to late Eocene (Poag and Aubry 1995; Poag 1997a; Gohn in press; Fig. 6.15A,B; see also Chapter 13). The mixed assemblages are derived from the entire succession of eight preimpact sedimentary formations present in the target area (Fig. 2.4). Individual lithic fragments ripped from each of these formations make up the large majority of clasts within the breccia, as determined by their characteristic lithologies and individual microfossil suites (Poag et al. 1992; Poag and Aubry 1995; Gohn in press).

In the lower part of each cored breccia section, the percentage of matrix in the breccia averages ~10% (by volume; Fig. 6.14F), and clasts (mainly of sedimentary rocks) reach cored thicknesses of up to 20 m (CD-ROM.7). Small quantities of crystalline basement clasts also are present in the lower sections, but dominantly in millimeter to centimeter sizes, with occasional lengths of 10–30 cm. Seismic

Fig. 6.14. (Opposite page) Photographs of ten core segments recovered from NASA Langley corehole showing: *A*, slightly weathered crystalline basement rocks; *B*, strongly weathered crystalline basement rocks overlain by basal breccia; *C–E*, silts and sands of displaced sedimentary megablocks (*C*, massive, fluidized? sand from décollement zone; *D*, horizontal stratification; *E*, inclined stratification); *F–J*, succession of upward-fining breccias (*F*, clast-supported sediment-clast breccia with no matrix; *G*, matrix-supported breccia with moderately large sedimentary clasts; *H*, matrix-supported breccia with small sedimentary clasts; *I*, nearly 100% matrix of glauconitic quartz sand; *J*, continuation of matrix interval, overlain in succession by flowin unit, fallout layer and laminated dead zone, and capped by Chickahominy Formation (marine clay). Numbers at top of each segment indicate drill depth at top of segment. See CD-ROM for color version of this figure.

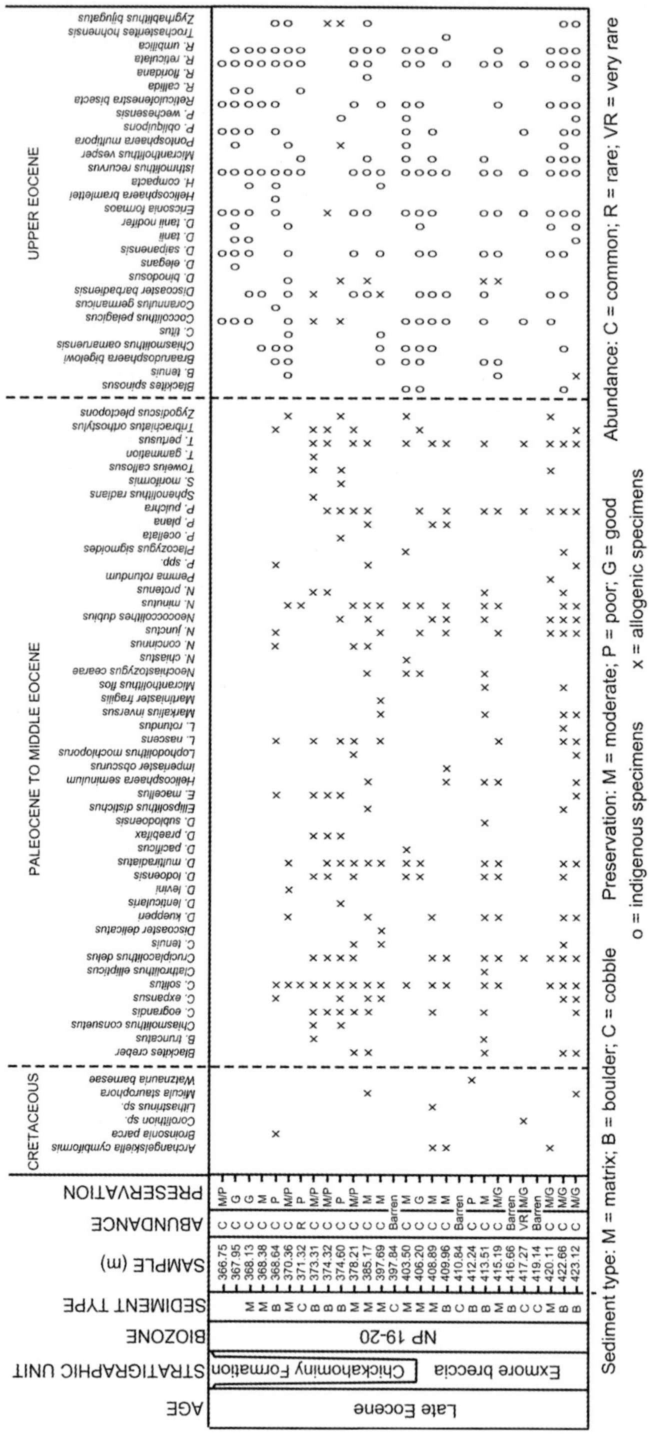

Fig. 6.15A. Distribution of stratigraphically mixed assemblages of calcareous nannofossils in Exmore breccia (samples taken from Exmore core-hole). Figure modified from Poag and Aubry (1995).

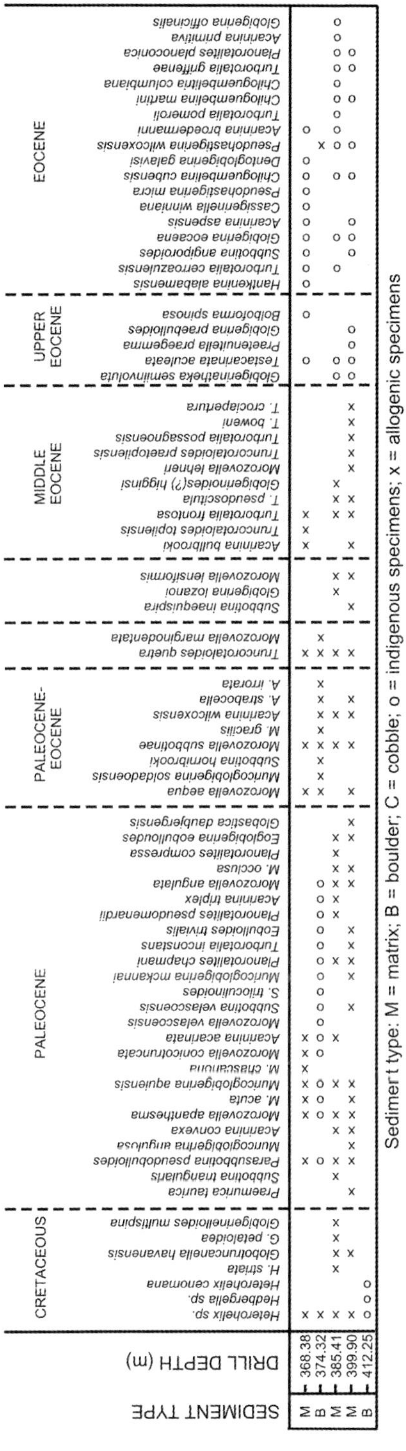

Fig. 6.15B. Distribution of stratigraphically mixed assemblages of planktonic foraminifera and bolboformids in Exmore breccia (samples taken from Exmore corehole). Figure modified from Poag and Aubry (1995).

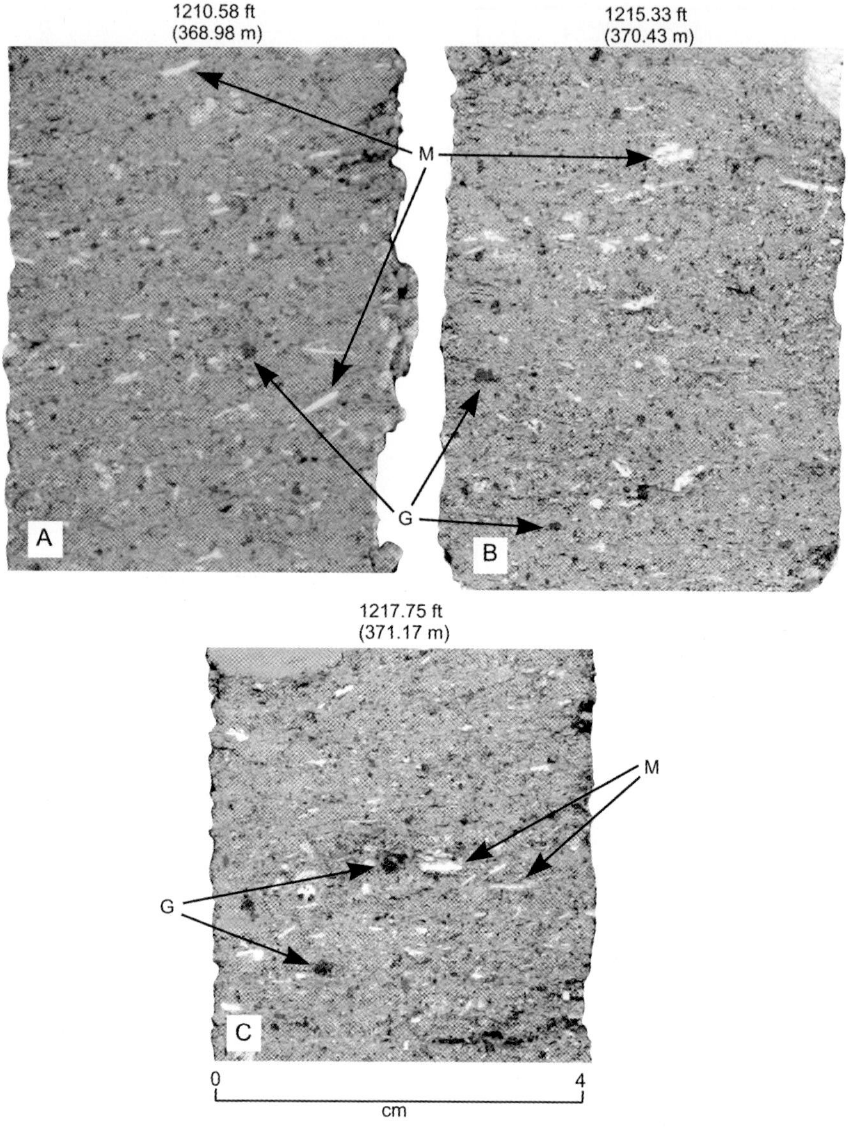

Fig. 6.16. Photographs of three segments of split cores from matrix of Exmore breccia (from Exmore corehole). Note scattered dark glauconite grains (*G*), variable orientations of mollusc shells (*M*; white streaks), and dominance of clayey quartz sand (clasts rarely larger than a few millimeters). Numbers at top of each segment indicate drill depth at top of segment. See CD-ROM for color version of this figure.

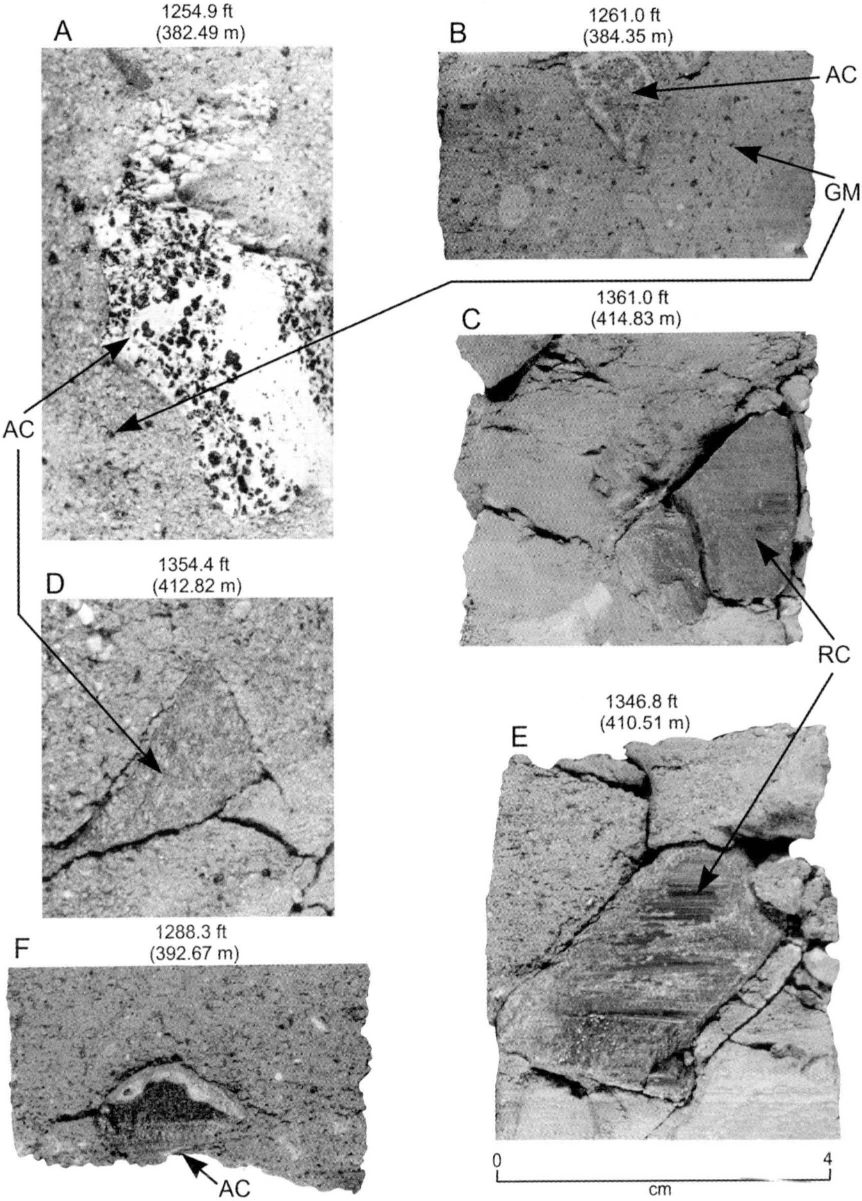

Fig. 6.17. Photographs of whole (*A, C, D*) and split (*B, E, F*) sections of core from Exmore corehole, showing angular clasts (*AC*) and rounded clasts (*RC*) within glauconite/quartz matrix (*GM*) of Exmore breccia. Numbers at top of each segment indicate drill depth at top of segment. See CD-ROM for color version of this figure.

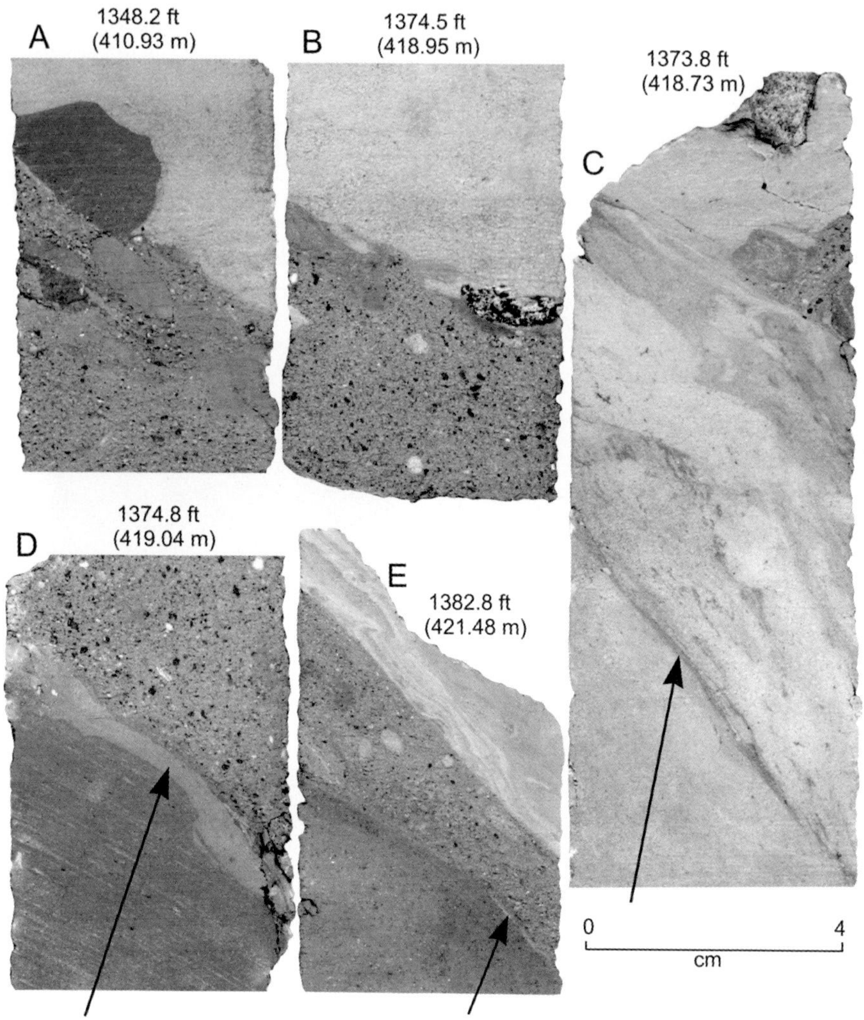

Fig. 6.18. Photographs of split sections of core from Exmore corehole, showing inclined contacts between clasts and matrix within Exmore breccia and mud rims on clasts (arrows on *C, D, E*). Numbers at top of each segment indicate drill depth at top of segment. See CD-ROM for color version of this figure.

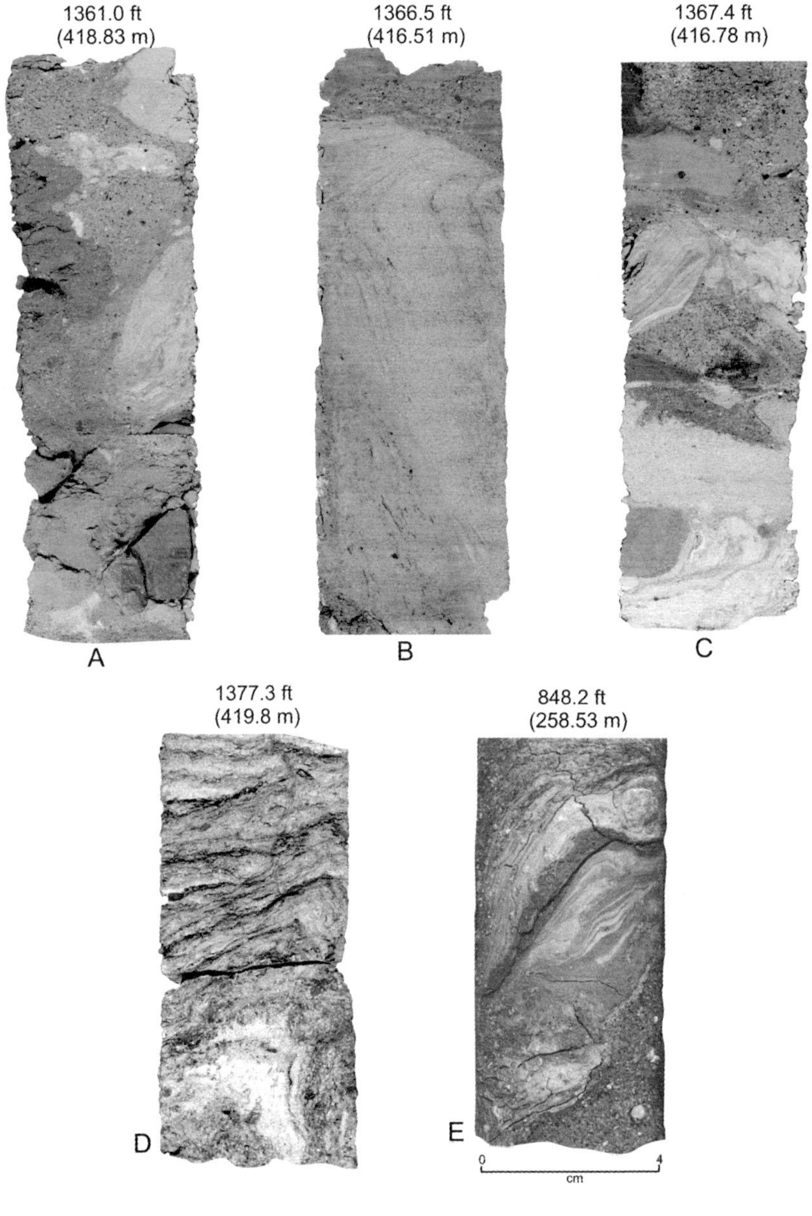

Fig. 6.19. Photographs of split sections of core from Exmore corehole (*A, B, C, D*) and whole section from North corehole (*E*), showing complex plastic deformation of soft-sediment clasts within Exmore breccia. Numbers at top of each segment indicate drill depth at top of segment. See CD-ROM for color version of this figure.

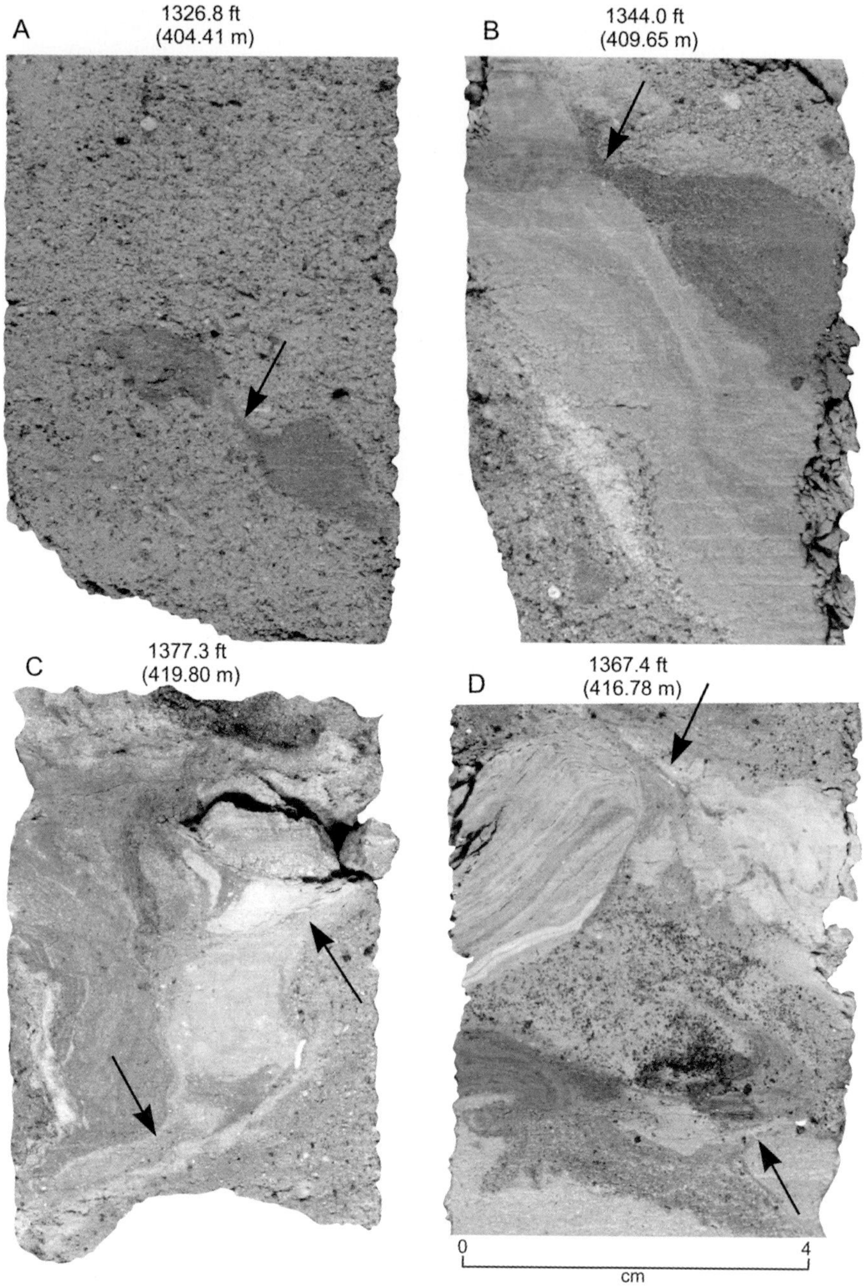

Fig. 6.20. Photographs of split sections of core from Exmore corehole, showing squeeze-outs (arrows) in clay clasts within Exmore breccia. Numbers at top of each segment indicate drill depth at top of segment. See CD-ROM for color version of this figure.

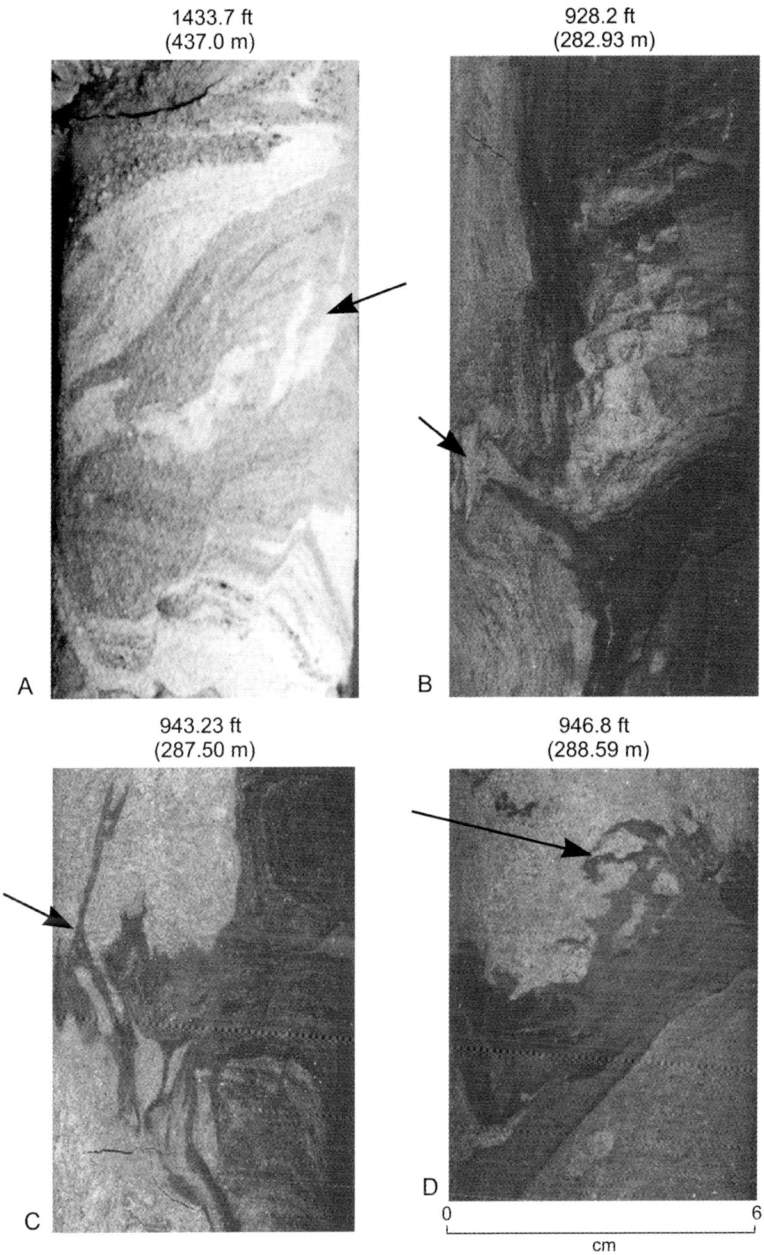

Fig. 6.21. Photographs of whole sections of core from Exmore breccia showing flame structures (arrows), in which one lithic unit has been injected into another lithic unit. *A* Bayside core; *B, C, D* North core. Numbers at top of each segment indicate drill depth at top of segment. See CD-ROM for color version of this figure.

data suggest that kilometer-sized blocks also may be supported by this matrix at some uncored locations (Fig. 4.29A,B).

In the middle part of the Exmore breccia, the size and frequency of clasts (mostly sedimentary pebbles, cobbles, and boulders) decreases, and the percentage of matrix consequently increases (Fig. 6.14G).

The upper part of the Exmore breccia consists of 50–95 percent matrix (Fig. 6.14H,I). Macroscopically, the matrix is a medium gray to brownish gray, calcareous, fossiliferous (mainly microfossils), clayey, glauconitic, medium- to coarse-grained, angular to rounded, quartz sand (Figs. 6.16–6.22). In the upper 1–4 m, few clasts are larger than 1–2 cm.

At the very top of the breccia, the glauconite-quartz sand is replaced by dark greenish-gray, clayey silt, with centimeter-scale laminae of fine to very fine sand. This silty interval also contains numerous nodular concentrations of framboidal pyrite (Figs. 6.14, 6.22). This silt-rich facies contains no indigenous foraminifera, but foraminifera reworked from deeper in the Exmore breccia (chalky, leached specimens) are concentrated in the thin, white, horizontal sand laminae, which also contain concentrations of muscovite flakes (Fig. 6.22C). At NASA Langley, this silt-rich layer is in sharp contact with the underlying glauconitic quartz sand of the Exmore breccia (Fig. 6.22A, D). At Bayside and North, however, the silt layer is in transitional contact with the underlying sandy breccia matrix. Also, the upper part of the silty layer at Bayside and North is more obviously stratified with white sandy, micaceous laminae and burrow casts, whose spatial orientations change markedly along the core (Fig. 6.23; flowin lithofacies of Chapter 11). Some laminae are horizontal, but others are inclined at variable angles and in different directions. A few rippled laminae and occasional centimeter-sized clasts are present in the silt at Bayside.

In the upper ~3 cm of the laminated silt-rich interval at NASA Langley, is a concentration of millimeter-sized, porous lattices of framboidal pyrite (Figs. 6.22A, 6.24A,B). The key impact-related feature of the pyrite lattices is their pore structure. Each pore is nearly perfectly spherical, of uniform ~1-mm diameter, and spatially arranged as if the lattice originally had enveloped a layer of micro-

Fig. 6.22. (Opposite page) *A* whole segment of NASA Langley core, showing lithic transition from Exmore breccia to dead zone. Dashed line is contact between sand matrix of Exmore breccia and silt-rich layer, which contains coarse-grained burrow-fills, pyrite lattices with spherical pores, nodular concentrations of framboidal pyrite, and reworked specimens of foraminifera. Dotted straight lines are boundaries of fallout layer. Black rectangles are sample locations. *B* vertically split segment of NASA Langley core, showing concentrations of framboidal pyrite in upper part of silt-rich layer (arrows indicate position of this segment in *A*). *C* horizontally split segment of NASA Langley core showing complex microlithologies (micaceous silt, dark clay band, coarse-grained burrow-fill, nodular pyrite concentrations) of silt-rich layer just above contact with Exmore breccia; thin, micaceous, white laminae contain benthic foraminifera reworked from Exmore breccia. *D* vertically split segment of NASA Langley core, showing irregular contact between sand matrix of Exmore breccia and overlying silt-rich layer. Datum for vertical scale is top of fallout layer. See Chapters 5, 13 for further discussion of dead zone; see CD-ROM for color version of this figure.

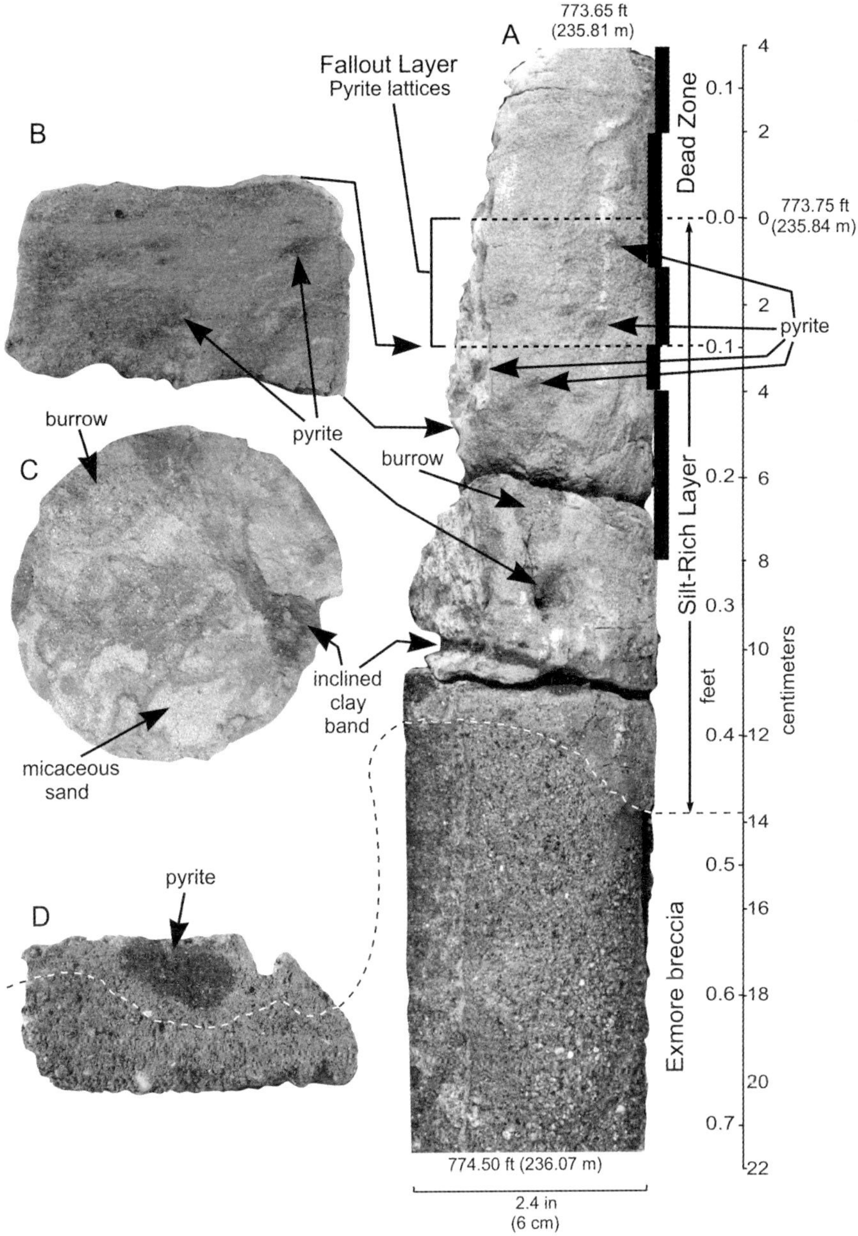
A
773.65 ft
(235.81 m)
Fallout Layer
Pyrite lattices
B
Dead Zone
773.75 ft
(235.84 m)
pyrite
burrow
C
pyrite
burrow
Silt-Rich Layer
inclined
clay
band
micaceous
sand
feet
centimeters
D
pyrite
Exmore breccia
774.50 ft (236.07 m)
2.4 in
(6 cm)

spherules at least three-microspherules-thick (~3–4 mm). These properties are quite similar to those of impact-derived layers of glass (or glass altered to clay) microspherules (microtektites) reported from other fallout ejecta deposits (Bohor 1990; Olsson et al. 1997; French 1998). Poag (2000, 2002b), Poag et al. (2001), and Poag and Norris (in press) inferred that the pores in the pyrite lattices originally contained glass microspherules ejected from the Chesapeake Bay crater. They speculated that after the spherules settled out as sedimentary particles, the framboidal pyrite encompassed them, like foam rubber around ball bearings. Over time, the microspherule glass dissolved, or altered to clay, which was inadvertently washed away during sample preparation. (See Chapter 13 for further discussion of this layer of pyrite lattices). Though a stratigraphically equivalent silt-rich interval is present at the top of the Exmore breccia at the other core sites (where it is inferred to be the final impact-generated deposit), the pyrite lattices have been found only at NASA Langley.

6.3.4 Sedimentary Structures

Whole-core segments and split-core sections of the Exmore breccia reveal an immense variety of excellently preserved sedimentary flow structures and soft-sediment deformation features indicative of deposition in a turbulent, even violent, fluid milieu. Clear examples of brittle deformation also are present, but are relatively scarce (Figs. 6.14, 6.18–6.21). Both rounded and angular clast boundaries are common (Fig. 6.14, 6.21). Inclined clast-to-matrix contacts dominate (Figs. 6.18–6.21). Squeezing, folding, stretching, twisting, rotation, truncation, shearing, faulting, and fracturing are manifest in various clasts ranging from massive clay to laminated silt and clay, to fine, medium, and coarse sand, to indurated bioclastic limestone, and granitic crystalline basement (Figs. 6.14, 6.16–6.21). On both a geological and historical time scale, all these deformational processes operated essentially simultaneously (within a few minutes to hours) in various parts of the extremely turbulent incipient breccia body, but the detailed succession of deformation (on a scale of hours) depended upon distance from the point of impact and the progression of shock compression, rarefaction, ejection, fallback, slumping, sliding, surgeback, washback, flowin turbulence, and fallout, produced by the impact event.

Perhaps the most noticeable sedimentological feature of the Exmore breccia, on a macroscopic scale, is the scarcity of horizontal contacts, horizontal laminae, or horizontal bedding planes. Nearly all clast-to-matrix contacts (for pebbles and larger clasts) are inclined, often at 45 degrees or more from horizontal (Figs. 6.18–6.21). Sixteen of the most pronounced inclinations in the Exmore core ranged from 18° to 90°, and averaged 43°. Drag and shear folds, fractures, faults, truncated laminae, and complexly convoluted flow bands give evidence of diverse differential motions having taken place along these inclined contacts. Laminations and(or) bedding planes within individual clasts also are frequently inclined, attesting to rotational motion during the depositional process (Figs. 6.18–6.21). Many

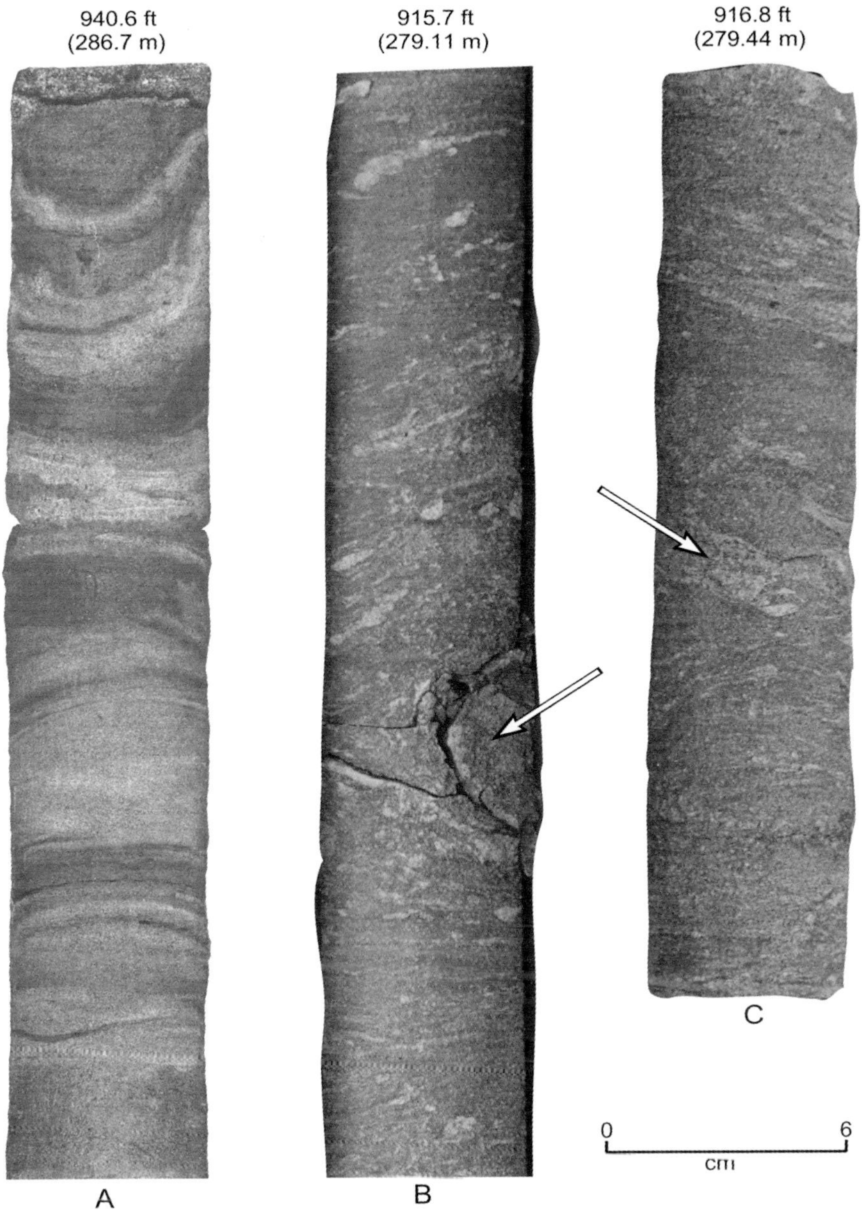

Fig. 6.23. Photographs of one whole-core section from North corehole (*A*), showing uphole changes in stratal geometry in silt-rich layer and two whole-core sections from Bayside corehole (*B, C*), showing multidirectional, inclined stratification in silt-rich layer. Sparsely scattered clasts shown by arrows. These silt-rich layers represent flowin depositional facies discussed in Chapter 11. See CD-ROM for color version of this figure.

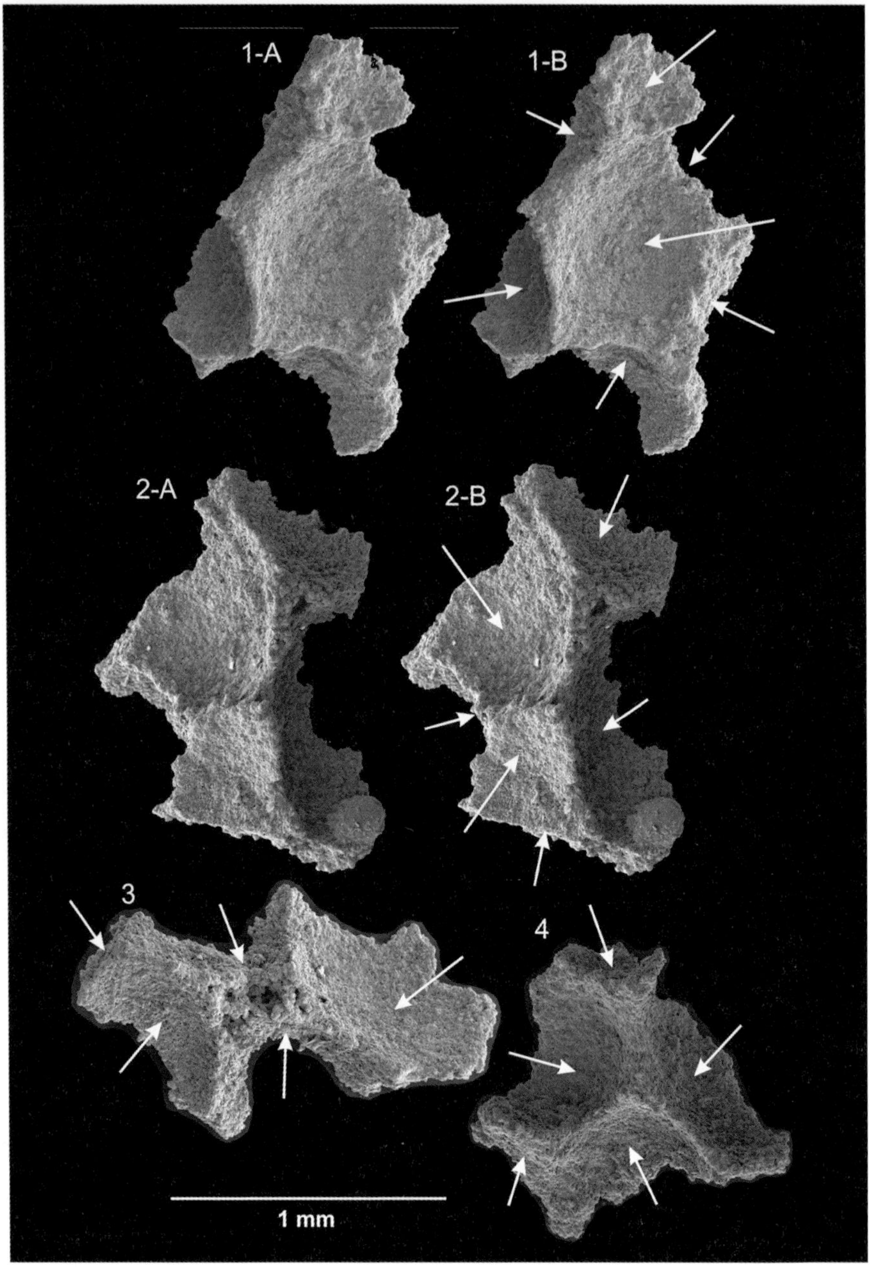

Fig. 6.24A. Scanning electron micrographs showing fragments of framboidal pyrite lattices extracted from fallout silt layer at NASA Langley core site. Note similar diameters of hemispherical depressions in lattice fragments. Arrows indicate smoothly concave depressions in which glass microspherules are inferred to have originally rested. *1-AB* and *2-AB* are stereopairs.

Fig. 6.24B. Conceptual reconstruction of pyrite lattices with glass microspherules restored. Numbered clusters refer to pyrite-lattice fragments shown in Fig. 6.24A. Modified from Poag (2002b).

clasts are surrounded by thin (millimeter-scale) rims of clay (Fig. 6.18), from which we infer that these clasts moved through the fluid as individual fragments, rather than being merely entrained within a larger slump block or debris flow.

Folds are frequent among the less consolidated silt, sand, and clay clasts. The folds vary from gentle, open folds, to tight, 180° recumbent folds (Figs. 6.18–6.21). The limbs of some primary folds have undergone additional wavy or crenulate folding (Fig. 6.25). Fold axes display a wide variety of spatial orientations.

Small faults and fractures are present in abundance, even in some of the less competent silt and clay clasts (Figs. 6.18–6.21, 6.25–6.27). Many of the faults and fractures display clay fillings, even though the lateral offsets are measured only in millimeters (Fig. 6.18).

Some clasts, clays in particular, have been stretched or squeezed into thin necks that connect adjacent thick blebs of the same lithology (Figs. 6.19–6.21). The margins of some clasts are drawn out into fine, wispy peaks or "tails" (Figs. 6.20, 6.21). Less frequently, clasts have been stretched by shear forces, so that wispy tails extend in opposite directions away from the parent clast. A few spectacular examples of flame structures are present, in which one lithic component has been injected into another. It is not uncommon to observe flame structures in which the compressive forces that formed them appear to have been directed laterally rather than vertically (Fig. 6.21).

In some cores, centimeter-scale mollusk fragments are abundant within the matrix of the Exmore breccia, and stand out as thin white streaks (Fig. 6.16). The shell fragments appear to have no preferred orientations, though no statistical measurements of them have yet been carried out.

Angular clasts are abundant in the breccia, and are composed of lithified and unlithified sands and clays, as well as tightly cemented bioclastic limestones and fragments of crystalline basement rocks (Figs. 6.17–6.20, 6.27). Cement-filled fractures are common in the more competent clasts, but macroscopic examination does not reveal whether or not the fractures formed as a result of impact shock (Fig. 6.27). However, petrographic studies of basement clasts (see section 6.3.6) reveal microscopic shock-deformation features and incipient shock melting produced at impact pressures as high as 60 GPa (Koeberl et al. 1996).

One of the most interesting lithologic expressions of high shear strain in sedimentary rocks is *scaly clay* (Maltman 1994a,b; Lash 1989; Moore et al. 1986). This unusual fabric has been described as "... subparallel undulations of shiny surfaces anastomosing around narrow lenticles of less fissile material" (Maltman 1994b, p 275). In macroscopic perspective, scaly clay appears to be intensely fractured on a centimeter to millimeter scale. This "shattered" fabric has been reported from deep-sea accretionary prisms, in stratal disruption zones of folded mountains, and even in drumlins and landslides. The highly fractured fabric of large Paleocene clay clasts incorporated within the Exmore breccia (Fig. 6.26) is similar to scaly clay, and may represent a similar physical response to impact shock. However, Moore et al. (1986) concluded that scaly clays in the Nankai Trough formed at much lower strain rates ($10^{-13}s^{-1}$) than those typically developed in bolide impacts (10^4s^{-1} to 10^6s^{-1}; French 1998). This difference may indicate that the scaly clays sampled at Exmore formed at localities distal from ground zero.

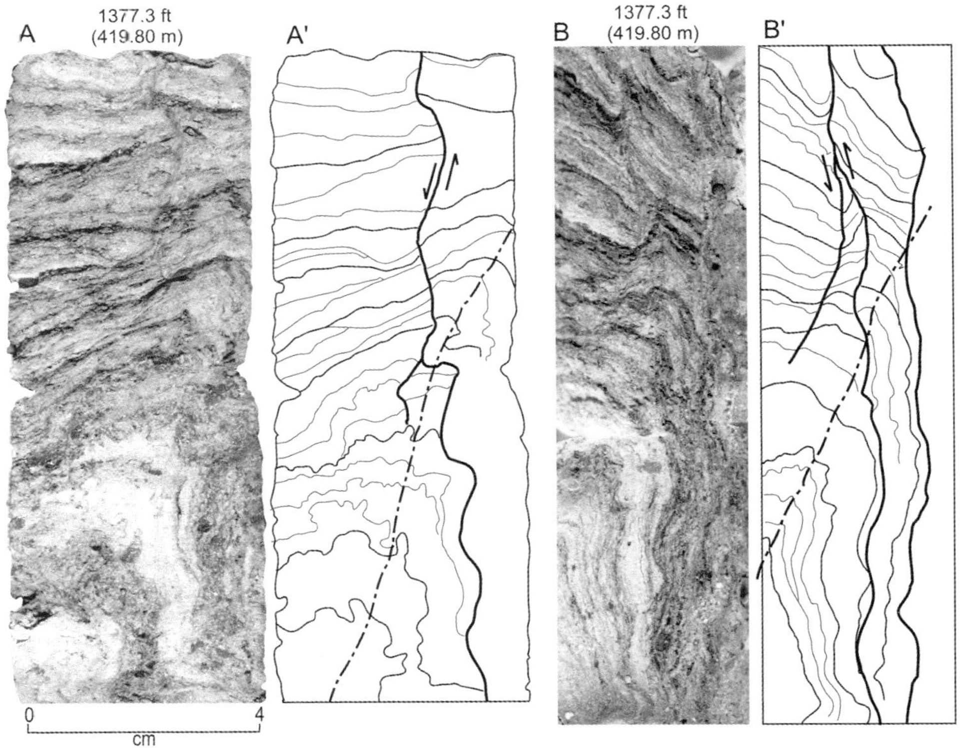

Fig. 6.25. Split (*A*) and whole (*B*) core sections from Exmore corehole, showing vertical squeeze deformation and drag folds in clast of Lower Cretaceous laminated, organic-rich silt within Exmore breccia. Tracings (*A'*, *B'*) show interpreted sense of motion on principal faults (solid lines) and trace of main fold axis (dashed lines). Numbers at top of each segment indicate drill depth at top of segment. See CD-ROM for color version.

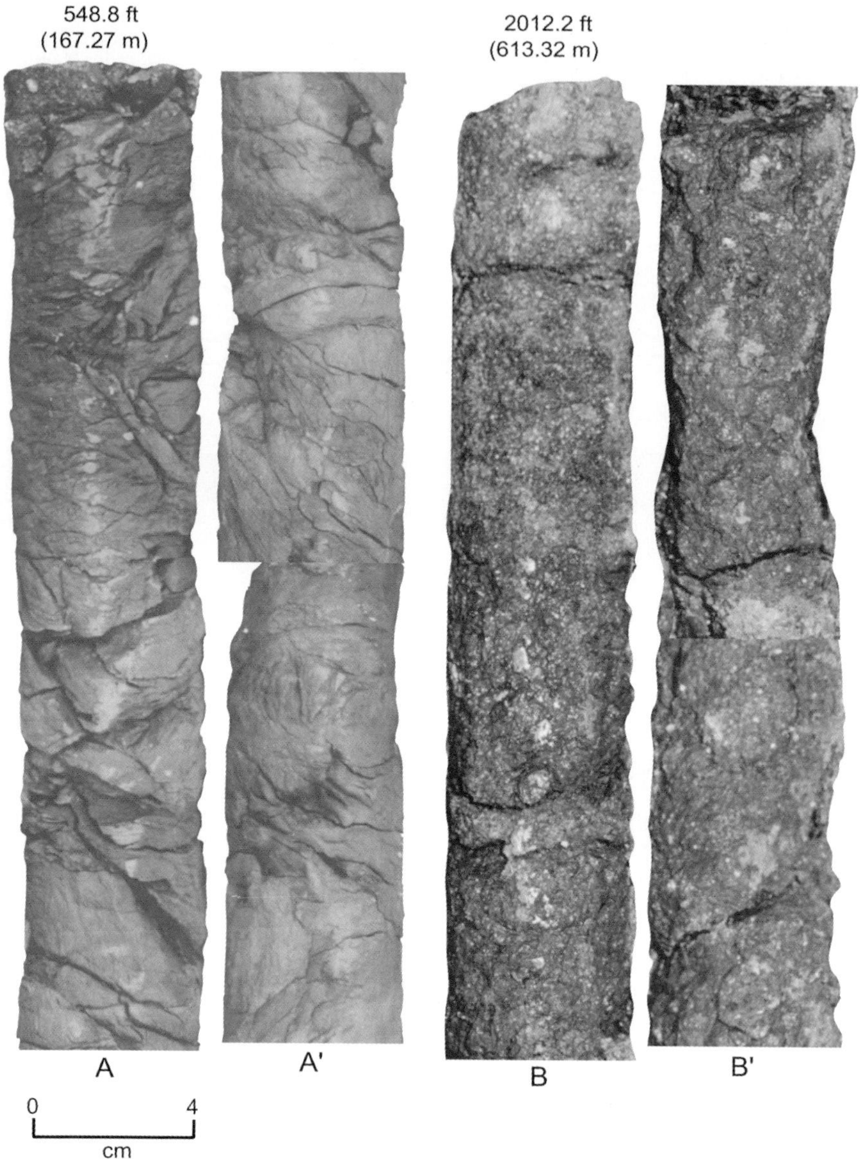

Fig. 6.26. Photographs of whole core sections, showing boulders (continuous core) of scaly clay within Exmore breccia in Windmill Point core [*A* 0.76 m (2.5 ft) thick] and NASA Langley core [*B* 0.37 m (1.2 ft) thick]. Numbers at top of segments indicate drill depth at top of segments. See CD-ROM for color version of this figure.

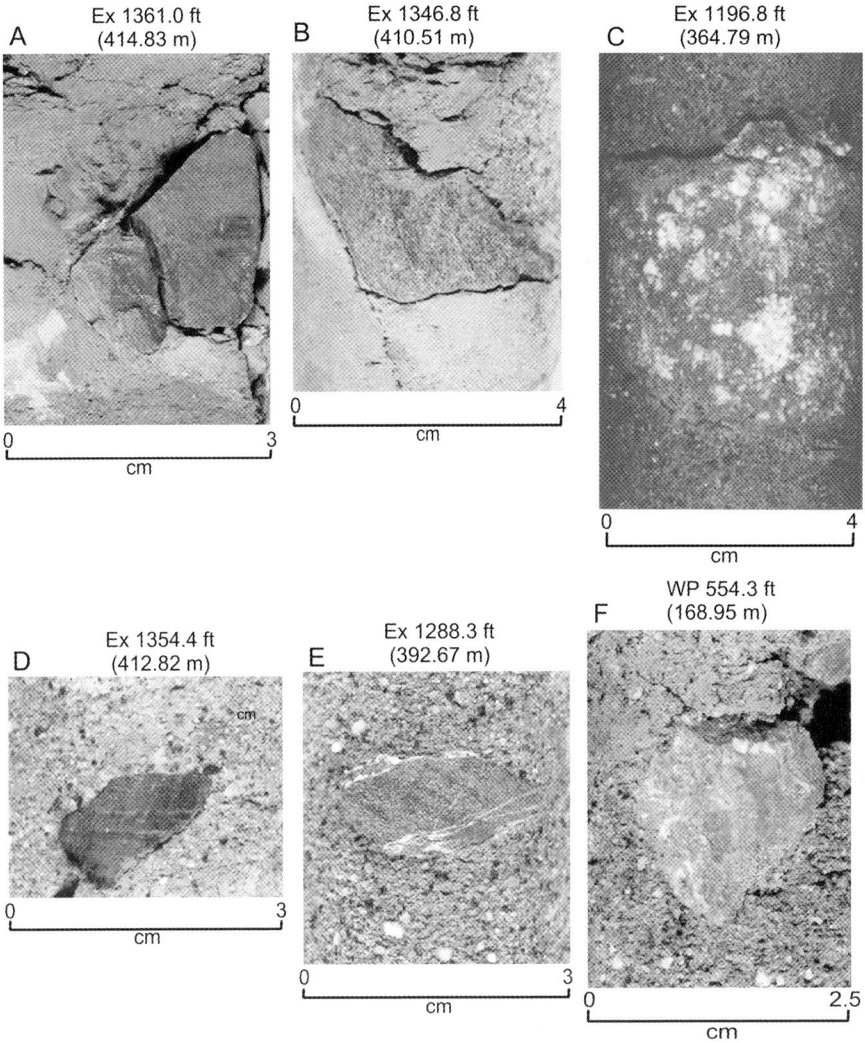

Fig. 6.27. Photographs of five whole sections (*B, C, D, E, F*) and one split section (*A*) of core, showing variety of small, angular, crystalline basement clasts within Exmore breccia from Exmore corehole (*Ex*) and Windmill Point corehole (*WP*). Numbers at top of segments indicate drill depth at top of segments. See CD-ROM for color version of this figure.

6.3.5 Expression on Downhole Geophysical Logs

A comparison of downhole geophysical logs from each of the coreholes reveals considerable variability in the expression of various matrix-rich and boulder-rich sections. For example, though some of the matrix-rich intervals produce strong positive deflections in the SP curve, indicative of high permeability, other matrix-rich intervals produce negative deflections, which indicate low permeability (Figs. 6.3A,B, CD-ROM.7). Likewise, in some cases, boulders produce more positive SP deflections than matrix-rich intervals, but other boulders produce more negative deflections than matrix-rich intervals (Figs. 6.3A,B, CD-ROM.7). In other words, because of the random spatial distribution of different clasts within the Exmore breccia, and their great lithic variability, one cannot confidently use the SP curves to differentiate sands from clays in the traditional manner of electric-log interpretation. These relationships create severe difficulties in trying to correlate different parts of the breccia column from corehole to corehole. It also means that any attempt to use downhole logs to interpret lithologic successions in drilled (but not cored) sections of crater-fill deposits is likely to be unreliable.

On the other hand, the SP logs are quite useful in understanding the general stratigraphic succession relative to overlying and underlying strata. The logs also provide important information regarding the relative permeability of units and the nature of their interstitial fluids. In this chapter we show the principal general relationships between the rock types and their log expressions in individual coreholes. We have found it more instructive in this section to discuss the log characteristics from the top to the base of the breccia, rather than in stratigraphic order of deposition.

6.3.5.1 Windmill Point Corehole

In six of the seven coreholes (NASA Langley is the exception), in the upper part of the Exmore breccia, the SP curve on downhole geophysical logs deflects to more positive values relative to those of the overlying Chickahominy Formation, though the relative increase is variable (Figs. 6.3A,B, CD-ROM.7). The Windmill Point log (location outside the crater) is a particularly good example of this positive deflection (Figs. 6.3A, CD-ROM.7). In this corehole, the Exmore breccia displays three intervals of positive SP (upper, middle, basal), each of whose values are higher (greater permeability) than any other in this corehole, except for the 1.5-m-thick (5-ft-thick) Potomac sand recorded at total depth. The upper permeable section is 2.7 m thick (9 ft), and corresponds to a matrix-dominated (>95% by volume), friable, glauconitic, quartz sand. Clasts larger than small pebbles are rare.

The next lower lithic unit displays the lowest SP values in the Windmill Point corehole (Figs. 6.3A; CD-ROM.7). This interval corresponds to a 1.5-m-thick (5 ft) boulder of scaly clay.

Below the scaly clay boulder, SP values increase gradually downward in a 1.8-m-thick (6 ft) interval, which contains three ~0.3-m-thick (1 ft), silty, Cretaceous boulders encompassed by glauconitic quartz-sand matrix (Fig. CD-ROM.7).

Next downhole is a 1.5-m-thick (5 ft) interval that displays the highest SP values recorded in the borehole. But rather than representing a matrix-rich interval, this deflection corresponds to a boulder of indurated, bioclastic, middle Eocene limestone. The limestone must be highly fractured to produce such high SP values (Fig. CD-ROM.7). In normally stratified sediments, this log pattern would indicate a sand, but in this case, we drilled a boulder, not a stratified bed. Next below the limestone interval is a 0.3-m-thick (1-ft-thick) interval of low SP, which corresponds to a hard, silty, Cretaceous boulder.

The basal interval displaying high SP values in the Exmore breccia at this site is 3.4 m thick (11 ft; Fig. CD-ROM.7). The SP increases gradually downward through a section of mainly glauconitic quartz-sand matrix, which encompasses a few cobbles of silt and limestone. No core was recovered in the lower 1.5-m (5-ft) interval, which displays the next-to-highest SP in the corehole, so its lithology is not definitely known. By analogy with the overlying log-core relationships, this basal breccia interval could represent either a permeable, matrix-dominated core section, or a fractured limestone (Fig. CD-ROM.7).

6.3.5.2 Newport News Corehole

On the Newport News geophysical log (this core site also is outside the crater), the Exmore breccia again shows a marked positive shift of SP values relative to those of the Chickahominy Formation. Here, the breccia displays particularly high SP values in the highest interval (4.9 m thick; 16 ft) and lowest interval (3.1 m thick; (10 ft)(Fig. 6.3A). The upper interval corresponds to a sandy matrix-dominated core section containing abundant pebbles and cobbles of hard, white limestone (Fig. CD-ROM.7).

The upper 2.6 m (8.5 ft) of the basal 3.1-m (10-ft) interval corresponds to a highly fragmented section of core, composed of short sections (3.75–5 cm; 1.5–2 in) of sandy matrix alternating with similar thicknesses of small glauconite-sand cobbles and pebbles (Fig. CD-ROM.7). The lower 0.5 m (1.5 ft) of this cored interval was not recovered, but a collection of loose quartz and limestone pebbles at the bottom of the core is consistent with good permeability indicated by high SP values (Fig. 6.3A).

Sandwiched between the two permeable sections is a 2.7-m (9 ft) interval of low SP values, which corresponds to a core section dominated by boulders of 0.5–0.6 m (1.5–2 ft) apparent thickness (Fig. CD-ROM.7).

6.3.5.3 NASA Langley Corehole

Geophysical logs from the NASA Langley corehole (~5 km inside outer rim of crater) and the North corehole (also ~5 km inside outer rim) are the only downhole records of the complete synimpact crater-fill succession documented in the outer part of the crater's annular trough (the Exmore core recovered only the top of the Exmore breccia). Each of these two coreholes records a different manifestation of the transition from the Exmore breccia to the Chickahominy Formation (Figs. 6.3A, CD-ROM.7). In particular, the SP values of the uppermost 38.7 m (127 ft) of the Exmore breccia at NASA Langley are lower than those of the overlying

Chickahominy Formation (rather than higher, as at other sites). In fact, with the exception of ~10.7 m (35 ft), SP values in the entire upper 105.5 m (346 ft) of Exmore breccia at NASA Langley are more negative (lower permeability) than those of the Chickahominy Formation. These relatively low values arise from a section of highly variable lithologies, ranging from nearly 100% clayey sand matrix in the upper 38.7 m (127 ft) to a 16.5 m-thick (54 ft) clast of highly fractured sandy, silty clay near the base of this section.

The most permeable interval in the upper 105.5 m (346 ft) of breccia at NASA Langley, as indicated by high SP values at 275.8–278.3 m (905–913 ft), is only 5.5 m (18 ft) thick. This permeable interval is not a cohesive sand body, as the log signature might suggest, but is composed of numerous matrix-supported, 15–20-cm-thick (6–8 in) clasts, which vary in composition from weathered granite to silty clay (Figs. 6.3A, CD-ROM.7).

The NASA Langley SP log from 341.4 to 442 m (1120–1450 ft) shows a series of alternating permeable and impermeable intervals, each ~9.2–24.4 m thick (30–80 ft). Below 442 m (1450 ft), which approximates the boundary between the Exmore breccia and the displaced megablocks, permeable sand intervals dominate (Figs. 6.3A, CD-ROM.7). The unconsolidated nature of these sands is reflected in a significant decrease in core-recovery in the megablock interval.

6.3.5.4 Exmore Corehole

At Exmore, ~5 km inside the outer rim of the crater, 54.2 m (177.8 ft) of Exmore breccia was cored and logged (Figs. 6.3A, CD-ROM.7). The SP differential between the lower Chickahominy Formation and the upper Exmore breccia is not nearly as marked at this locality as at those outside the crater rim (Windmill Point and Newport News). Moreover, the SP values at Exmore are higher in the upper Chickahominy than in the upper breccia.

The upper log unit at Exmore is a 4.0-m-thick (13 ft) interval of intermediate SP values, which corresponds to a cored section almost entirely composed of clayey, glauconitic, quartz-sand matrix (clasts larger than small pebbles are rare; Fig. CD-ROM.7). The lowest 0.6 m (2 ft) of core in this section contains increased glauconite and a few cobbles of crystalline basement. This basal section corresponds to a modest increase in SP.

Below this upper permeable unit, the SP decreases in a 1.8-m-thick (6 ft) interval, which correlates with a boulder of scaly clay in the core (note that a similar scaly clay boulder is present at about the same distance below the top of the Exmore breccia at Windmill Point).

At 375.2-m depth (1231 ft), a 1.2-m-thick (4 ft) interval of higher SP values corresponds to another section of glauconitic quartz-sand matrix in the core (Fig. CD-ROM.7).

The 16.2-m-thick (53 ft) interval from 377 to 393.2 m (1237–1290 ft) displays relatively low SP. This interval corresponds to another core section dominated by glauconitic quartz-sand matrix. Two boulders [~0.3-m (1 ft) apparent thickness] are present in the core, but are too small to record a recognizable log signature.

The 13.7-m-thick (45 ft) section from 393.2 to 406.9 m (1290–1335 ft) displays slightly elevated SP values. This section maintains the matrix-dominated lithology.

The SP values decrease from 406.9 m (1335 ft) to total depth. Lowest values occur between 416.7 m (1367 ft) and TD (Fig. CD-ROM.7). This interval correlates with a core section in which sedimentary boulders dominate over matrix. Some boulders reach ~1 m (3 ft) in apparent thickness.

6.3.5.5 North Corehole

The position of the North corehole relative to the morphology of Chesapeake Bay crater is analogous to that of the NASA Langley and Exmore coreholes – it is situated in the annular trough ~5 km inside the outer rim (Fig. 6.2). The North site, however, is about half way between the NASA Langley and Exmore coreholes, as measured along the circumference of the crater, and is farther updip (west) than any of the other intracrater coreholes.

The upper ~10.7 m (35 ft) of breccia at North displays elevated permeabilities, but the next deeper ~84.4 m (277 ft) between 236.2 and 320.7 m (775–1052 ft) are notably impermeable (Figs. 6.3B, CD-ROM.7). The section consists of a variety of rotated, parallel-bedded blocks of sand and clay (~19.8-m-section; 65 ft), plus intervals of highly fractured, sticky, clay-rich paleosols (also ~19.8-m section; 65-ft). The basal 28 m (92 ft) of the Exmore breccia at North (320.7–348.7 m; 1052–1144 ft) regains significantly more permeability than most of the overlying section, before decreasing again near the top of the underlying displaced megablocks. This basal section consists mainly of tilted blocks of sand and silt.

6.3.5.6 Bayside Corehole

The Bayside corehole is located in the outer part of the annular trough, a few kilometers from the outer flank of the peak ring (Fig. 6.2). At Bayside, the upper ~9.8 m (32 ft) of section shows the typical increased permeability relative to the overlying Chickahominy Formation. The next lower 51.8 m (170 ft), from 292.0 to 343.8 m (958–1128 ft), shows moderate, but rapid downhole shifts in permeability, in a section dominated by cobble-size sedimentary clasts (Fig. 6.3B, CD-ROM.7). Between 292.0 and 460.3 m (1128–1510 ft) is a 116.4-m (382 ft) section of thick (6.1–15.2 m; 20–50 ft), blocky, high-permeability intervals, separated by equally thick low-permeability intervals. Most of the high-permeability intervals consist of bedded and massive sands, whereas the low-permeability intervals comprise highly fractured, clay-rich paleosols, with lesser amounts of matrix-supported, cobble-rich breccias (Figs. 6.3B, CD-ROM.7). The 45.7-m (150 ft) interval from 460.3 to 506 m (1510–1660 ft) is dominated by low permeability arising from a succession of highly fractured, clay-rich paleosols. A 30.5-m (100 ft) basal section of dominantly high permeability (506–536.5 m; 1660–1760 ft) separates the Exmore breccia from the underlying section of displaced megablocks (Figs. 6.3B, CD-ROM.7). Tilted blocks of bedded sand with internal soft-sediment deformation characterize this basal section.

6.3.5.7 Kiptopeke Corehole

The Kiptopeke corehole is the only site drilled to date inside the peak ring (Fig. 6.2). It is particularly unfortunate that only a 17.7-m (58 ft) interval was cored there (394.1–411.8 m; 1293–1351 ft). Moreover, core recovery was poor in the breccia interval, which precludes direct correlation of the logs with downhole rock types. Based on analogies with the other logged coreholes, however, we have interpreted the two uppermost intervals of relatively high SP values (combined thickness of 12.2 m; 40 ft) to be permeable sands of the Exmore matrix, whereas the intervening low SP values we interpret to be a less permeable interval of undetermined lithic composition (Figs. 6.3B, CD-ROM.7). We have not attempted to interpret the detailed succession of lithic units below the cored interval. We note, however, that the upper 12.2 m (40 ft) of relatively permeable section is underlain by 54.9 m (180 ft) of section (413–467.9 m; 1355–1535 ft) that displays relatively low permeabilities, like the upper ~100-m sections at NASA Langley, North, and Bayside.

6.3.6 Petrography

In order to firmly establish the impact origin of the Exmore breccia, and to assess the types of impact metamorphism brought about by the impact, we performed petrographic analyses on individual quartz grains and on small clasts (mm- to cm-sized) of crystalline basement rocks extracted from the breccia (Tables 6.4–6.7). We examined samples mainly from the two core sites outside the crater (Windmill Point and Newport News), two sites drilled in the outer part of the annular trough (Exmore and NASA Langley), and from the only site drilled inside the peak ring (Kiptopeke). We also examined thin sections from basement cores outside the crater (Table 6.8) for comparison with future analyses of basement rocks inside the crater at NASA Langley and Bayside. Our analyses corroborate the findings of Poag et al. (1992), Koeberl et al. (1996), Powars et al. (2001), and Horton et al (2001, 2002), that the Exmore breccia contains abundant evidence of shock metamorphism. The shock-metamorphic features most common in the Exmore breccia samples fall into four categories: (1) shock fractures; (2) multiple sets of PDFs (planar deformation features); (3) shock melt; and (4) glass microspherules. We also documented the lack of shock metamorphic features in basement rocks outside the crater.

6.3.6.1 Shock Fractures

Typical shock fractures in quartz, indicative of relatively low shock pressures (≤8 GPa), are common in clasts of crystalline basement extracted from the Exmore breccia (Fig. 6.28).

6.3.6.2 Planar Deformation Features (PDFs)

Higher shock pressures produce planar deformation features, and these are common in crystalline basement fragments from the Exmore breccia, where they are expressed mainly in quartz and feldspar grains (Fig. 6.29). On the other hand, PDFs in individual quartz grains of the breccia matrix (i.e., grains not incorporated in basement clasts; Fig. 6.29) are quite rare, constituting <1% (by volume) of the matrix grains (Poag et al. 1992). In the basement clasts, multiple sets of PDFs (as many as six sets) are common in both quartz and feldspars. In order to verify the shock origin of the PDFs, Koeberl et al. (1996) measured their crystallographic orientations in selected quartz grains (Fig. 6.30). The resultant orientations [(0001), $\{10\bar{1}3\}$, $\{10\bar{1}2\}$, $\{11\bar{2}2\}$, $\{10\bar{1}1\}$, $\{11\bar{2}1\}$, $\{51\bar{6}1\}$] (c, ω, π, ξ, r, z, s, x, respectively) are characteristic of impact shock. The relative frequencies of these crystallographic orientations indicate initial shock pressures of 20–30 GPa (Koeberl et al. 1996).

We have measured the crystallographic orientations of PDFs in quartz within an additional eight thin sections of Exmore breccia from the Exmore corehole (Table 6.4; samples 1235.43–.67 ft, 1237 ft, 1280.78–1291.0 ft, 1290.6–.76 ft, 1312.0–.14 ft, 1329.20 ft, 1341.5–.67 ft, 1356.8 ft). A total of 22 quartz grains with 45 sets of PDFs could be detected in these eight sections (Table 6.9). Two grains contained a single set of PDFs each. Seventeen grains exhibited two sets of different crystallographic orientations, and in 3 cases, three sets of PDFs per grain could be measured. Thus, a total of 45 orientation measurements were carried out (Table 6.9; Fig. 6.30). Measurements in 14 grains (63.6% of all data) could be properly indexed with {hkl}. The results clearly show that most measurements correspond to the orientation (angle relative to the c-axis) of the $\{10\bar{1}3\}$ zone, but also, a few measurements correspond to other shock-characteristic crystallographic orientations, including the $\{10\bar{1}2\}$, $\{10\bar{1}1\}$, and $\{21\bar{3}1\}$ zones. This predominance of the $\{10\bar{1}3\}$ orientation can be interpreted to indicate that most of the shock-metamorphosed quartz grains we identified originated from a moderately shocked crater regime (~16 GPa shock pressure). However, after comparing these results with those of Koeberl et al. (1996; Figure 6.30), derived from a different set of Exmore breccia samples, we conclude that the individual statistics, though independently confirming the presence of impact-characteristic shock-metamorphic effects in quartz grains, do not allow us to determine a statistically valid average shock pressure for the combined sets of data. Instead, we infer that these shocked clastic components were derived from different parts of the evolving impact crater and were subsequently mixed together by turbulent depositional processes that produced the Exmore breccia.

Table 6.4. Petrographic analyses of individual clasts in Exmore breccia, taken from Exmore borehole inside Chesapeake Bay impact crater.

Sample Depth [ft (m)]	Description
1210.2 (368.9)	Coherent silt with abundant fine-grained quartz, calcite, glauconite, magnetite, biotite, feldspar, and muscovite. Matrix consists of calcite and silica; no foliation present. No shock deformation noted in fine-grained minerals.
1210.575 (369.0)	Coherent sample of interbanded (at several mm spacing) medium-grained sand/silt; fossiliferous. No shock deformation in quartz or other minerals.
1220.0 (371.9)	Particulate sample: chert with partially shock-melted quartz; also fragment of reddish breccia that contains several shocked quartz clasts (shock fractures, partial isotropization). Fig.6.31A shows a nearly completely melted and annealed quartz grain. Fig. 6.28A shows an example of typical shock fracturing (product of relatively low shock pressure, ca. 8 GPa) in quartz of granitoid or vein quartz origin.
1220.625 (372.05)	Breccia pocket in granitoid fragment, with some evidence of melting (isotropic matrix; Fig. 6.31B).
1228.0 (374.3)	Coherent sample of medium- to fine-grained silt; larger grains consist of glauconite, calcite, quartz, muscovite, and magnetite. Grains well-rounded to angular; generally cemented by calcite. Calcite also is dominant mineral in matrix; no foliation. No evidence of shock deformation.
1231.35 (375.3)	Coherent sample of glauconite cemented by phyllosilicate matrix; also a few spherulitic grains of quartz and carbonate ooliths; limited fracturing is not shock-specific.
1232.1 (375.5)	Coherent sample of glauconitic sand; almost entirely composed of well-rounded, moderately sorted, coarse-grained glauconite grains; some quartz in interstices. Most glauconite grains have brownish rims – presumably of limonite. The cement is formed mainly by limonite, besides some minor silica. No shock deformation.
1235.43 (376.6)	Particulate sample; glauconite in carbonate matrix; contains a cryptocrystalline (devitrified glass?) particle, shale, shocked quartz fragments with fracture patterns typical of Hospital Hill quartzite shocked to 8 GPa (Reimold 1988; Huffman and Reimold 1996) or single sets of PDFs; silt, granite-derived fragments with shocked quartz; shocked quartz (1 set of PDFs) in a carbonate clast adhering to strongly shocked K-feldspar.
1237.0 (377.0)	Grain mount; 1 clay, 1 silt, 1 cm-sized K-feldspar grain with small quartz poikiloblasts (possibly derived from vein quartz or pegmatoid). Both K-feldspar and quartz contain up to two sets of PDFs per crystal. Fig. 6.29B shows K-feldspar grain with PDFs).

Table 6.4. (cont.)

1238.0 (377.3)	Coherent sample of fine-grained, slightly deformed glauconitic sand (elongation of nodules).
1239.98 (377.9)	Coherent sample of clay with very fine-grained quartz fragments; no shock deformation detected.
1240.85 (378.2)	Particulate of granite fragments, silt, microcline, sand, a shocked granite fragment containing quartz with shock fractures and one set of PDFs, a piece of a reddish breccia with annealed quartz clasts and shocked granite-derived clasts (quartz + feldspar with shock fracturing and/or one set of PDFs).
1244.01 (379.2)	Coherent medium-grained feldspathic litharenite; poorly sorted and immature (> 5 vol% phyllosilicates). In order of decreasing abundance, this sample consists of quartz, glauconite, microcline, some other feldspar, hematite and magnetite, and muscovite. Glauconite is generally well-rounded; other minerals also angular. Cement mainly quartz and calcite, besides phyllosilicates. No foliation; no shock deformation.
1248.0 (380.4)	Coherent sample of medium- to coarse-grained litharenite; poorly sorted, rounded to angular grains. Quartz is most abundant, besides glauconite, muscovite, magnetite, feldspar, calcite, carbonate-shell fragments (brachiopods?), and chert. All these phases are set into a fine- grained matrix of predominantly calcite; locally stained by hematite. No evidence of shock deformation.
1249.5 (380.8)	Coherent sample; matrix-dominated (ca. 50 vol%), medium-grained, glauconitic sand; mostly angular grains, except glauconite pellets mostly rounded. A significant feldspar clast component (dominantly K-feldspar). No shock deformation; only limited intragranular fracturing.
1250.85 (381.3)	Particulate sample comprising clay, silt, a sericite-schist clast, medium-grained arkose (unshocked), medium-grained and strained carbonate, mylonite, a few granite-derived particles, partially melted sandstone (or quartzite); also one fragment of fine-grained melt breccia (containing partially annealed quartz clasts).
1261.05 (384.4)	Coherent sample of glauconitic sand; has significant feldspar component and small component of carbonate and pyroxene (presumably volcanic-derived). No shock deformation.
1262.87 (384.9)	Coherent glauconitic sand; medium-grained, unshocked.
1263.0 (385.0)	Particulate sample. One strongly altered, heavily fractured, locally brecciated, granitoid fragment, with partial annealing of breccia zones, as well as shock features (planar fractures in feldspar, shock fracturing, mosaicism, and local isotropization in quartz); one clay particle and one 0.5-cm microcline fragment (unshocked).
1264.48 (385.4)	Coherent, medium-grained glauconitic sand; no visible shock effects.

Table 6.4. (cont.)

1269.11 (386.8)	A 1-cm quartz pebble, unshocked and barely fractured, in glauconitic sand similar to that from 1249.5 ft depth. Unshocked.
1272.6 (387.9)	Glauconitic sand, very similar to material from 1349.5 ft depth, but clast size, on average, somewhat larger. Matrix is dominantly sericite/brownish phyllosilicate. Clasts mostly angular; strong feldspar clast component (ca. 15 vol%), mainly K-feldspar; some carbonate clasts. No shock deformation. A second specimen from this depth has up to 3-mm-wide pellets of glauconite. Carbonate could be of fossil origin. No shock evidence.
1280.78 (390.4)	Particulate sample: nine silt fragments, one clay (all 10 are well-rounded), and one quartz fragment with shock-derived planar fracturing. Fig. 6.29A shows densely spaced PDFs in K-feldspar. Typical, low-shock "shock fracturing" visible in Fig. 6.28.D.
1280.78 (390.4)	Particulate sample comprising silt, shocked quartz, unshocked microcline, clay, and glauconitic sand fragments.
1286.7 (392.2)	Polymict lithic and mineral breccia. Contains red breccia clasts in contact with glauconitic sand. Medium-grained, angular mineral and lithic clasts often show minerals with reduced birefringence (mostly quartz and feldspar). A part of two sections is hematite-stained. The contact to this zone is irregular, but well-defined. Large mica clasts are kinkbanded. Enhanced cleavage present in some feldspar grains, but no unambiguous PDFs or other shock deformation. This sample represents either a clast of brecciated granite in the sandstone or a thin injection vein of polymict allogenic breccia.
1288.36 (392.7)	Glauconitic sand, similar to sample at 1349.5 ft. Angular clasts, some carbonate, and a lot of feldspar. No shock deformation.
1290.6 (393.4)	Particulate sample: one microcline fragment, one quartz fragment with possible shock fractures (Fig. 6.28C), four clay particles, two silt fragments, and one fragment of K-feldspar with a dense PDF pattern (multiple sets of different crystallographic orientation (Fig. 6.29B).
1290.6 (393.4)	Particulate sample: a breccia particle containing shocked quartz, unshocked granite, and granite-derived fragments, pegmatite- or vein quartz-derived quartz (this seems to be the precursor material for many of the shocked quartz particles; also observed in bands across granite-fragments), clay, sand, and a single shocked granite clast. Generally, the shocked mineral and lithic fragments are derived from crystalline basement lithologies; no bona fide shocked sediment particles were observed. Some rare shocked quartz grains are present in carbonate, but there are carbonate veins cutting across granite as well.
1291.37 (393.6)	Coherent, medium-grained arkose (abundant microcline, some orthoclase and plagioclase). No shock effects, though some frayed and kinked muscovite.
1294.6 (394.6)	Coarse-grained vein quartz clast in glauconitic sand (ca. 25 vol% matrix); chert, carbonate, clay, and oxide components in clast population. Much granitoid-derived feldspar (similar to the other sandstone samples); also an iron-oxide-cemented nodule. No shock deformation.

Table 6.4. (cont.)

1298.54 (395.8)	Coherent, fine-grained clay/biotite-schist; banded. Too fine to recognize shock effects.
1302.01 (396.9)	Silt comprised of quartz, magnetite, calcite, glauconite, muscovite, and biotite fragments, cemented by calcite. Secondary quartz and calcite form veins and fill interstices. No foliation; no shock deformation.
1307.95 (398.7)	Coherent sample of interbedded (at a 3-5 mm scale) clay/glauconitic clay, which contains some carbonate fragments.
312.0 (399.9)	Particulate of unshocked quartz, brecciated quartz with hematite-rich material adhering to it (equivalent of reddish breccia?), clay, weakly shocked/strained sand (undulous extinction of quartz grains), a shocked granite fragment with quartz showing local shock fracturing, brecciation, and melting (strong shock heterogeneity at the grain scale), silt, and banded carbonate.
1313.0 (400.2)	Coherent silt. Abundant, fine-grained quartz grains, less calcite and muscovite, and minor feldspar and magnetite; set into a phyllosilicate-dominated, locally ferruginous matrix, and cemented by calcite. No foliation, no shock deformation.
1314.75 (400.7)	Coherent sample of well-banded clay/sand (interlayered quartz-rich/quartz-poor layers in pattern reminiscent of cross-banding). Abundant medium-grained quartz; no shock deformation.
1323.82 (403.5)	Particulate of silt, shocked (mosaicism and planar fractures, or shock fracturing) and unshocked granitic fragments, shale, clay with a shocked quartz grain (mosaicism), partially shocked (shock fracturing or one set of PDFs in quartz) and locally melted granite, unshocked microcline, weakly shocked (fractured) and fine-grained chert, unshocked sandstone, a carbonate plus opaque mineral fragment. Fig. 6.29C shows PDFs in K-feldspar.
1329.20 (405.1)	Particulate of carbonate, chert, granite with fractures and undulous extinction in quartz; shocked granite with mosaicism and numerous sets of PDFs; unshocked granite fragments.
1332.0 (406.0)	Four clay and two quartzite fragments; one quartzite fragment has shock fractures.
1332.35 (406.1)	Particulate containing silt, brecciated leucogranite, strongly altered and unshocked granite, and sand with carbonate lenses. Besides local cataclasis in a leucogranite fragment, no likely shock metamorphic effects.
1339.03 (408.1)	Sand (apparently layered); locally glauconitic, locally sheared. Some large (up to 1 cm) quartz clasts. Variable grain size in different bands ranging from fine- to medium-grained. No shock deformation.
1341.5 (408.9)	Particulate containing clay, unshocked microcline, partially altered impact melt (aphanitic, with K-feldspar clasts; Fig. 6.31.D), silt, sand, shocked (shock fracturing, rare PDFs; local melting along fractures and grain boundaries) quartzite (annealed granite- or vein-derived?), shocked sand (local isotropization in quartz).

Table 6.4. (cont.)

1345.0 (410.0) A	Seven clay and four silt fragments (too fine-grained to identify possible shock deformation).
1345.0 (410.0) B	Particulate of sand, arkose, clay, and schist fragments.
1346.0 (410.3) A	Silt, banded with Fe-oxide-rich and Fe-oxide-poor layers.
1346.0 (410.3) B	Glauconitic sand with 1.5-cm nodule of protomylonitic, arkosic sand (hardly any fine-grained matrix). Large (several mm) concentric nodules of glauconite. No shock deformation.
1346.15 (410.3)	Coherent sample of fine-grained silt. Some opaque globular fragments in clastic matrix (fragmental impact breccia?) composed of mostly feldspar grains.
1347.0 (410.6)	Arkosic sand; poorly sorted with >15 vol% fine-grained matrix. Clasts include (in order of decreasing abundance) quartz (angular to subangular), glauconite (rounded grains), magnetite, calcite, and muscovite. Many glauconite pellets have brownish margins, apparently as a result of oxidation. Feldspar grains also appear brownish due to oxidation of Fe. Brown staining in the matrix can be attributed to oxidation of fine-grained glauconite particles.
1349.5 (411.1)	Glauconitic sand with largely sericitic matrix; a significant component of feldspar (ca. 10 vol%; mostly microcline) clasts. Most clasts are angular. No shock deformation. A second section contains a few (up to 0.5 cm, ovoid) silt clasts.
1356.8 (413.6)	Shocked quartz (Fig. 6.29.D) in a granitoid fragment.
1358.0 (413.9) A	Completely weathered (oxides, chert, some carbonate) granitoid clast. Seems to have mafic patches that could be relics of primary mafic minerals.
1358.0 (413.9) B	Silt, grading into clay; sheared; tiny flakes of mica.
1358.0 (413.9) C	Similar to sample B. Possibly containing some pollen. None of these three samples shows shock deformation.
1359.4 (414.3) A	Silt layer in slightly coarser (still fine-grained) sand. No glauconite.
1359.4 (414.3) B	Section of the sand only. Rather mature (<20 vol% matrix, rather quartz-rich). Contains some feldspar clasts (about 5 vol%). No shock deformation.
1362.18 (404.2) A	Particulate with many clay and silt clasts. One silt fragment with a relatively large, unshocked granitoid fragment; two perthitic, unshocked K-feldspar particles.

Table 6.4. (cont.)

1362.18 (404.2) B	Particulate composed of schist; shocked and unshocked, fine-grained sand; clay; shocked (shock fracturing, mosaicism) and unshocked quartz particles.
1365.9 (416.3)	Coherent, fine-grained clay.
1366.6 (416.5)	Well-laminated (on a mm-scale) silt to sand sequence; slightly sheared. Contains a few feldspar clasts. No shock deformation.
1371.2 (417.9) A	Particulate of two quartzite fragments, one spherule remnant, and one weathered, monomict granitic breccia (cataclastite). The latter two of likely impact origin.
1371.2 (417.9) B	Medium-grained glauconitic sand; unshocked.
1374.0 (418.8)	Contact between fine-grained, glauconitic sand (ca. 35 vol% phyllosilicate in the matrix) and clay, containing sand lenses. Few feldspar clasts. No shock deformation.
1375.3 (419.2)	Silt consisting of quartz, calcite, muscovite, feldspar, biotite, minor magnetite, and glauconite; all within a matrix of phyllosilicates, clacite, and minor quartz. No foliation; no shock deformation.
1377.3 (419.8)	Folded and micro-boudinaged ferruginous clay band in sheared sand. Abundant tiny mica flakes. Some quartzitic nodules. The sand is very fine-grained, nearly a silt. No shock deformation.
1377.6 (419.9) A	Medium- to coarse-grained litharenite. Poorly sorted, with rounded to angular fragments. Calcite and quartz are the most abundant clast phases, besides glauconite, muscovite, microcline and plagioclase. Fragments within a fine-grained matrix of similar mineralogical composition, but with calcite as most abundant mineral. In places, magnetite alteration has caused Fe-oxide staining of the matrix. No foliation; no shock deformation.
1377.6 (419.9) B	Fine-grained sand, dominated by subangular quartz and chert fragments, with minor glauconite, muscovite, biotite, and magnetite. The cement consists of microcrystalline calcite and silica. Coherent sample; no shock deformation.
1378.33 (420.1)	Particulate of one fine-grained sand particle, one clay fragment, and one silt fragment.
1387.4 (422.9)	Medium-grained glauconitic sand. In order of decreasing abundance, the constituents are: glauconite, quartz, calcite, carbonate fragments (including intraclasts, brachiopod shell fragments, and mollusc shell fragments), magnetite, and limonite. The groundmass is almost entirely opaque, but also contains some carbonate. No foliation; no shock deformation.
1388.2 (423.1)	Particulate. One fine-grained meta-quartzite particle (in places chert-like), attached to which is a glass spherule (Figs. 6.32D,E).

Table 6.5. Petrographic analyses of bulk samples of Exmore breccia taken from Kiptopeke corehole, inside peak ring of Chesapeake Bay impact crater.

Sample Depth [ft (m)]	Description
1329.2 (405.1)	Quartz, microcline, some plagioclase fragments, some mudstone and siltstone, orthoclase, glauconite, glauconitic sand, some micro-oolitic (foraminifera?) carbonate, some other carbonate fossils; >90 vol% granite-derived material, ~10 vol% sediment. Some of the fine-grained quartzitic fragments could be annealed granite-derived clasts. No shock deformation.
1331.0 (405.7)	Three sections from this sample. Quartz, microcline, some plagioclase fragments, some mudstone and siltstone, orthoclase, glauconite, glauconitic sand, some micro-oolitic (foraminifera?) carbonate, some other carbonate fossils; somewhat lower proportion of sediment than sample at 1331.0 ft, especially less fine-grained sand/quartzite. No shock deformation.
1331.2 (405.8)	Clay, aplite, and quartz fragments. No shock deformation.
1332.25 (406.1)	Two sections of this material, which is similar to sample at 1331ft. Some quartz crystals contain subplanar fluid inclusion trails, as well as some undulatory extinction, but no definitive shock deformation.

6.3.6.3 Impact Melt Rocks

At impact pressures greater than 45 GPa, target rocks begin to melt. Typical shock-melted minerals or mineral assemblages are present in many crystalline basement clasts within the Exmore breccia (Fig. 6.31). We observed partly or wholly melted fragments of granite and feldspar, breccia clasts enclosed by possible suevite, melted/annealed quartz grains, annealed melt veins in granite, aphanitic impact melt with K-feldspar clasts, and local melt zones around fractures and grain boundaries. This group of melt products indicates that shock pressures as high as 55–60 GPa are recorded in basement clasts within the Exmore breccia (see also Koeberl et al. 1996).

6.3.6.4 Glassy Microspherules

Spherical bodies of impact melt in the 1-mm or smaller size range, are found in proximal impact deposits of only a few terrestrial impact craters [Barringer (Meteor Crater, Arizona; Mark, 1987), Wabar (Saudi Arabia; Krinov 1966; See et al. 1989), Lonar (India; Murali et al. 1987), Ries (Engelhardt 1997)]. In contrast, glass microspherules (microtektites) are widely distributed in the four documented distal ejecta (tektite) strewn fields (Koeberl 1994). The Chesapeake Bay crater contains the first known occurrence of proximal, glassy, impact-melt microspherules in a submarine impact crater (Poag 2002b; Fig. 6.32). Three microspherules

Table 6.6. Petrographic analyses of bulk samples and individual clasts from Exmore breccia, taken from Newport News corehole, outside Chesapeake Bay impact crater.

Sample Depth [ft (m)]	Description
426.0 (129.8)	One apparent breccia fragment composed of medium-grained quartz fragments in a silica-phyllosilicate matrix; quartz clasts show undulatory extinction and local annealing, but no unequivocal shock deformation; three granitic clasts are altered but apparently unshocked.
427.15 (130.2)	One quartz-rich schist particle (greywacke?), one silt fragment, one large quartz grain with shock fracturing (≤8 GPa) and strong undulatory extinction, but no PDFs.
431.2 (131.4)	One altered (much secondary carbonate) microgranite fragment, one granitic fragment (quartz plus altered perthitic K-feldspar), two chert fragments, and one metaquartzite fragment. No shock deformation.
432.25 (131.7)	A coherent piece of unshocked silt/clay in contact with greywacke. Minerals present include quartz, glauconite, magnetite, calcite, feldspar, and muscovite. Poorly sorted sample with grain shapes ranging from angular to well rounded; matrix consists mainly of calcite but also some fine-grained fragments of the other listed mineral phases. Slightly foliated rock with banding on a 0.5-cm scale.
433.8 (132.2)	One piece of fractured and locally brecciated granite, one fragment of a fine-grained melt rock that could represent impact melt rock (its clast population comprises a number of unshocked feldspar clasts), one unshocked granite clast, and a piece of fine-grained melt rock with angular as well as well-rounded quartz clasts and heavily altered matrix. Whether this melt rock represents impact melting or endogenous deformation is not clear.
438.2 (133.6) A	One glauconite-quartz fragment with some carbonate clasts, one piece of fossiliferous carbonate, and one piece of fossiliferous carbonate breccia.
438.2 (133.6) B	Three fragments of cherty breccia with granite-derived clasts, two fresh, unshocked granite fragments, one of which has a granophyric component.
441.95 (134.7)	One fragment of fine-grained melt rock with an angular, unshocked plagioclase clast, two ca. 0.5-cm unshocked pieces of perthitic K-feldspar, and one brecciated and locally melted granitoid.
444.72 (135.6)	One glass spherule attached to a strongly altered fragment with silicic matrix and granite-derived clasts (Fig. 6.32C); one chert fragment with a spherule indenting this chert and a second, chloritized fragment (Figs. 6.32A,B); one chert particle with a single, angular, quartzitic clast.
446.7 (136.2)	Several completely altered granitoid fragments; one piece of a silicic breccia after granitoid-derived material.
449.25 (136.9)	One chert fragment, one medium-grained and unshocked granite fragment, one fragment of metasediment with a cherty matrix and several granite-derived clasts, and one fragment of metasediment with phyllosilicate matrix and very small granitoid-derived clasts.

Table 6.7. Petrographic analyses of bulk samples and individual clasts from Exmore breccia, taken from Windmill Point corehole, outside Chesapeake Bay impact crater.

Sample Depth [ft (m)]	Description
539.8 (164.5)	Particulate sample: three silt fragments and remnants of a number of clay fragments; one plagioclase fragment nearly isotropic, heavily fractured and showing, in places, mosaic extinction.
544.03 (165.8)	Glauconitic sand with probably shell-derived fragments of carbonate. Generally similar to the other Windmill Point samples in this series. Coherent; no shock deformation.
552.11 (168.3)	Medium-grained glauconitic sand with ca. 20 vol% matrix. Many internally very fine-grained carbonate clasts (possibly shell-derived) show a slight alignment; some folded, phyllonitic (white mica) clasts. Fossiliferous carbonate is prominent, in contrast to most of the glauconitic sands from the Exmore corehole. Coherent; no shock deformation.
553.7 (168.8)	Particulate: One coarse-grained, strongly altered granitoid fragment, fractured but lacking characteristic shock effects; one fine-grained fragment of a feldspathic melt rock containing unshocked quartz and K-feldspar clasts; one piece of metasediment with cherty matrix and granite-derived minerals (feldspar and quartz) and lithic clasts.
555.6 (169.3) A	Matrix-dominated glauconitic sand with silty matrix. A few sand clasts in the glauconitic sand and a significant carbonate component. Coherent; no shock deformation.
555.6 (169.3) B	Straight contact between glauconitic sand (as in 555.6 A) and a relatively finer-grained, clast-dominated sand with distinct micro-laths of muscovite. Coherent; no shock deformation.
555.6 (169.3) C	Glauconite-rich sand layer grading into a less-glauconitic variety that is similar to 555.6 A and B, and then grading into a thin, dark-brown clay layer and silt (finer-grained than the fine-grained sand in 555.6 B). Coherent; no shock deformation.
563.7 (171.8)	Medium-grained litharenite; poorly sorted and immature sand composed of angular quartz, rounded glauconite, >5 vol% of phyllosilicate. Other grains include muscovite, magnetite, feldspar, and calcite. Several carbonate clasts are present, including intraclasts (fragments of sediment eroded from older strata and redeposited) and some bioclasts (bryozoans?). The cement consists largely of carbonate. Coherent; no shock deformation.
564.55 (172.1)	Particulate: one fragment of vein quartz (or quartz-pegmatoid)), partially annealed, especially along grain boundaries and fractures; only irregular fracturing noted. Two unshocked fragments of metaquartzite.
565.05 (172.2)	Similar to sample 552.11, but with prominent clay nodules. Coherent; no shock deformation.
566.4 (172.6)	Particulate: one chert particle with small, angular quartz and feldspar clasts; one fragment of fine-grained quartzite; one piece of silicic breccia of granitic material (could represent a monomict fragmental breccia), and 1 fragment, ca. 0.5 cm wide, of quartz with irregular fracturing and undulatory extinction.

Table 6.8. Petrographic analyses of crystalline basement samples derived from coreholes outside Chesapeake Bay impact crater.

Corehole [Name/ Number]	Sample Depth [ft (m)]	Description
C25/111	1869.75 (569.9)	Coarse-grained microcline-granite with chloritized amphibole and sericitized as well as saussuritized plagioclase. No shock deformation.
C25/111	1960.4 (597.5)	Magnetite-bearing, muscovite-plagioclase granite with a few irregular fractures. No shock deformation.
C26/112	1498.6 (454.0)	Coarse-grained microcline-granite, similar to C25/1869.75 ft, but more strongly altered. No shock deformation.
C26/112	1500.0 (457.2)	Similar to C26/1489.6 ft, but more strongly altered; quartz strongly annealed; contains secondary biotite; plagioclase completely altered. No shock deformation.
C26/112	1524.0 (464.5)	Biotite-granite; some annealing and deformation in the form of relatively large subgrain domains, the formation of which would have required a significant time. No shock deformation.

Table 6.9. New measurements of planar deformation features in grains from Exmore breccia of Chesapeake Bay impact crater.

Sets of planes	Orientations	Number of grains
1	$112\bar{1}$, $(0001\text{-}10\bar{1}3)$	2
2	$52\bar{6}1$, $10\bar{1}2$	1
2	$(0001\text{-}10\bar{1}3)$, $10\bar{1}1$	1
2	$11\bar{2}2$, $21\bar{3}1$	1
2	$10\bar{1}3$, $10\bar{1}3$	2
2	$21\bar{3}1$, $21\bar{3}1$	1
2	$10\bar{1}3$, $(10\bar{1}2\text{-}11\bar{2}2)$	4
2	$10\bar{1}3$, $10\bar{1}1$	2
2	$11\bar{2}1$, $22\bar{4}1$	1
2	$(000\bar{1}\text{-}10\bar{1}3)$, $10\bar{1}3$	1
2	$10\bar{1}3$, $10\bar{1}2$	1
2	$10\bar{1}3$, $11\bar{2}2$	1
2	$10\bar{1}3$, $11\bar{2}1$	1
3	$21\bar{3}1$, $10\bar{1}2$, $10\bar{1}3$	1
3	$(000\bar{1}\text{-}101\bar{3})$, $(000\bar{1}\text{-}101\bar{3})$, $10\bar{1}1$	1
3	$10\bar{1}1$, $11\bar{2}2$, $10\bar{1}2$	1

$(0001\text{-}10\bar{1}3)$ means between 0° and 23°

$(10\bar{1}2\text{-}11\bar{2}2)$ means between 24° and 48°

Of 22 grains examined, data for 14 (63.6%) could be indexed (see Grieve et al. 1996)

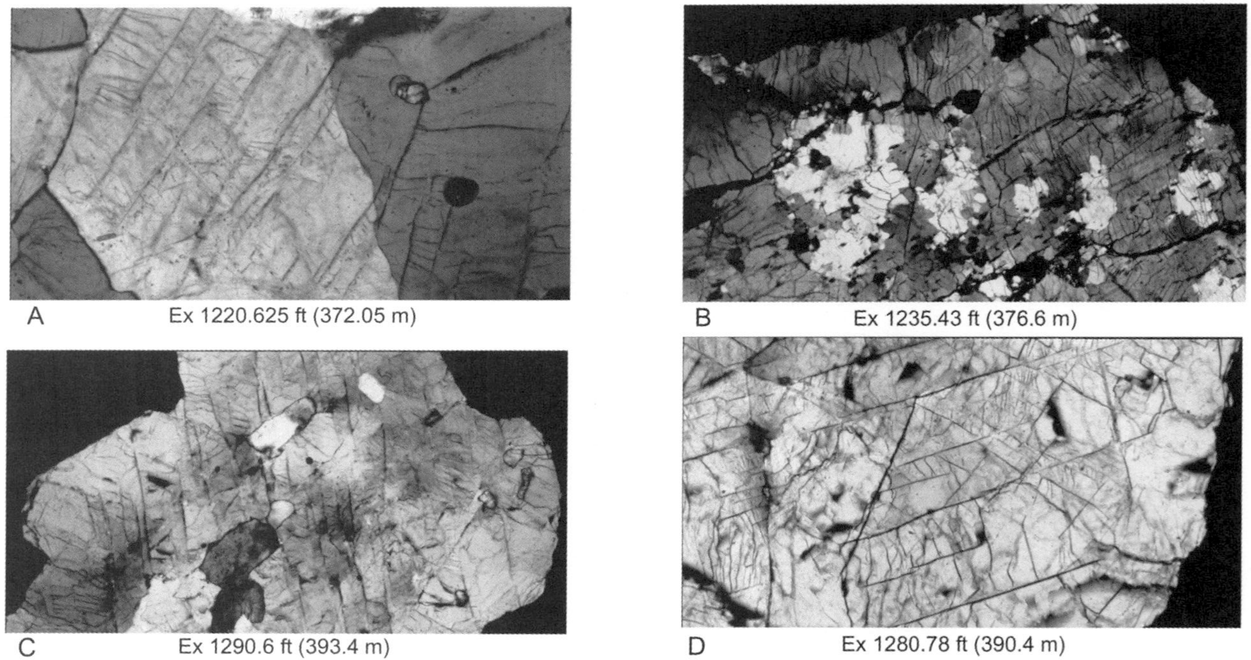

Fig. 6.28. Photomicrographs of thin sections (*A, D* plane-polarized light; *B, C* crossed polarizers), showing typical fracture patterns resulting from low-pressure (~8 GPa) shock metamorphism in clasts of crystalline basement extracted from Exmore breccia (Exmore corehole). *A* typical shock fractures in quartz (shocked to ≤8 GPa) within granitoid fragment or vein, width of field 1.1 mm; *B* shocked quartz fragment with fracture patterns similar to those of Hospital Hill quartzite (South Africa; shocked to ≤8 GPa; see Reimold 1988; Huffman and Reimold 1996), width of field 3.4 mm; *C* quartz with possible shock fractures, width of field 2.75 mm; *D* typical shock fractures in feldspar, width of field 2.2 mm. See CD-ROM for color version of this figure.

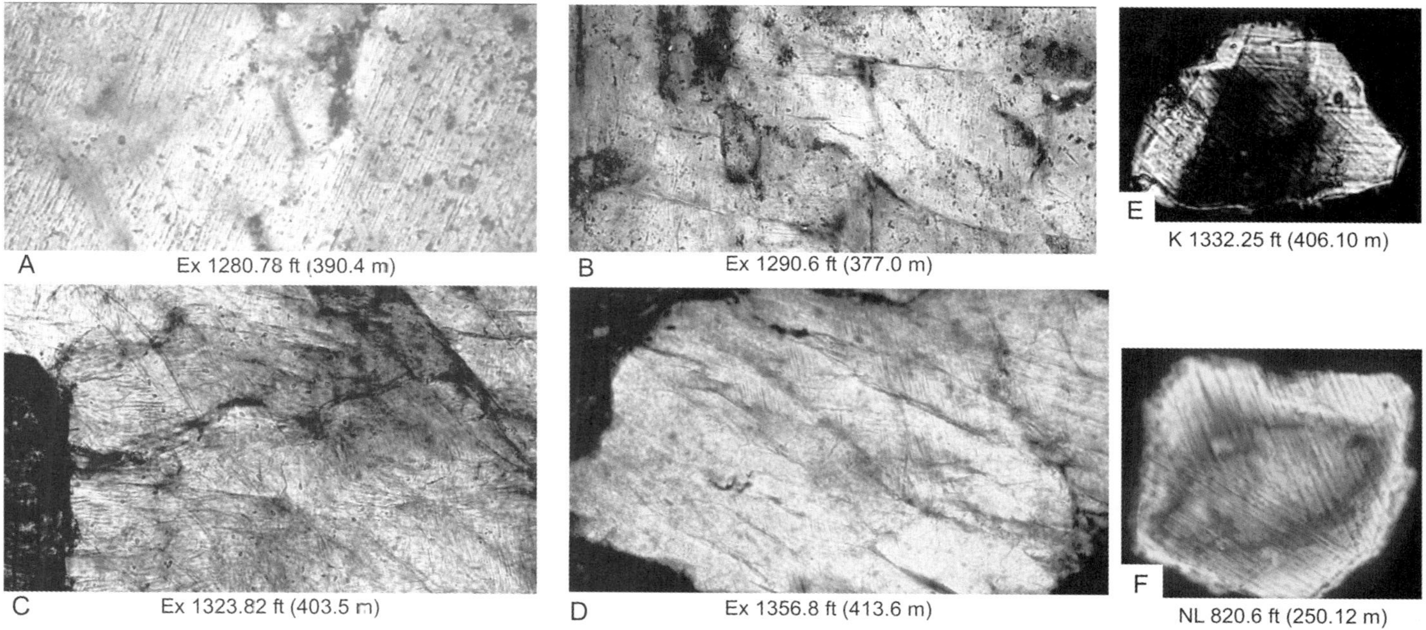

Fig. 6.29. Photomicrographs of thin sections (plane-polarized light), showing PDFs (planar deformation features) resulting from shock metamorphism (20–30 GPa) in clasts of crystalline basement (*A–D*) and in individual quartz grains (*E, F*) extracted from Exmore breccia. *A* densely spaced multiple sets of PDFs in K-feldspar grain, width of field 220 µm; *B* K-feldspar grain with dense pattern of multiple PDFs, width of field 355 µm; *C* multiple sets of PDFs in K-feldspar grain, width of field 335 µm; *D* multiple sets of PDFs in quartz grain from granitoid fragment, width of field 565 µm; *E* and *F* individual quartz grains from matrix of Exmore breccia (cross-polarized light), each showing two sets of PDFs, width of field ~0.2 mm. *Ex* = Exmore corehole; *NL* = NASA Langley corehole; *K* = Kiptopeke corehole. See CD-ROM for color version of this figure.

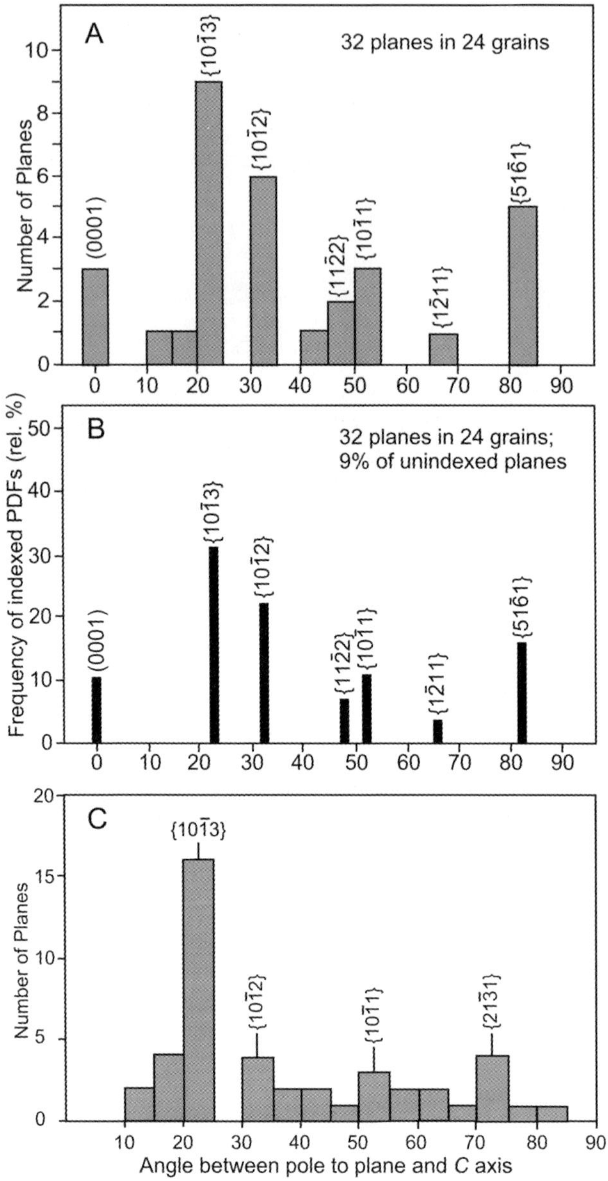

Fig. 6.30. Crystallographic orientation of PDFs in quartz from clasts of crystalline basement extracted from Exmore breccia. *A* standard histogram plot (after Engelhardt and Bertsch 1969) showing all measured data; *B* histogram showing frequency of indexed PDFs (after Grieve and Stöffler 1996; Grieve et al. 1996) versus angle between c axis and the poles of PDFs after transformation of the optic axis into the center of a standard stereographic projection; from Koeberl et al. (1996). *C* additional measurements made by us for this volume.

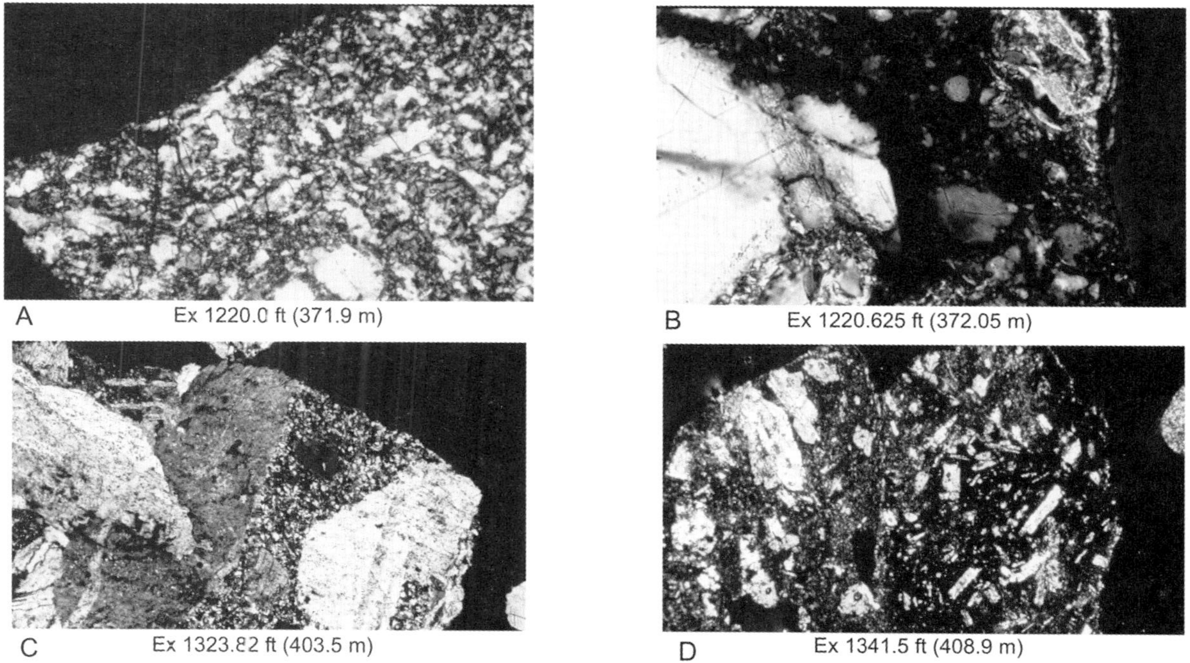

Fig. 6.31. Photomicrographs of thin sections (all with crossed polarizers), showing melt features resulting from shock metamorphism (35–60 GPa) in clasts of crystalline basement extracted from Exmore breccia. *A* nearly completely melted/annealed quartz grain; width of field 3.4 mm; *B* breccia pocket with apparent presence of melt matrix; width of field 3.6 mm; *C* granite fragment with annealed melt vein; width of field 3.4 mm; *D* aphanitic impact melt with K-feldspar clasts; width of field 3.4 mm. *Ex* = from Exmore corehole. See CD-ROM for color version of this figure.

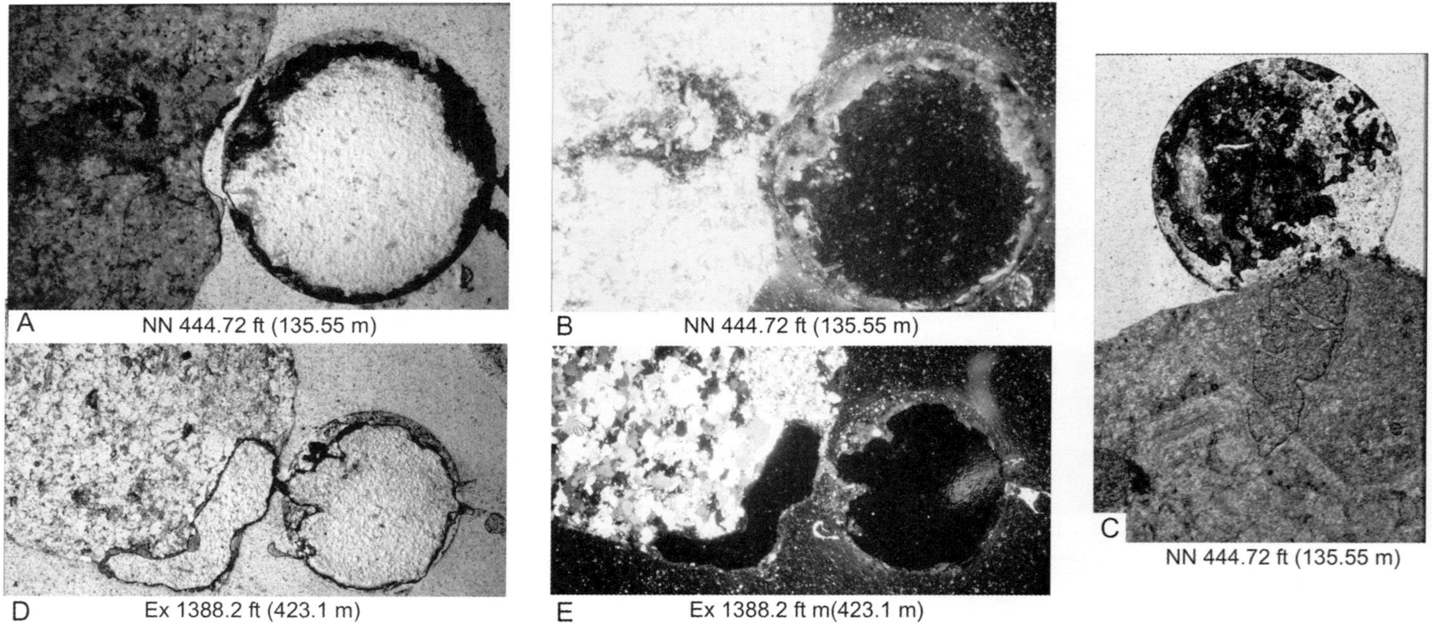

Fig. 6.32. Photomicrographs of thin sections, showing glass microspherules from Exmore breccia. Microspherules indicate shock-melting of sedimentary target rocks at Chesapeake Bay crater. *A* chert fragment and chloritized fragment, both indented by glass microspherule; parallel polarizers; width of field 2.75 mm; *B* same glass microspherule as in *A* (crossed polarizers); *C* glass microspherule attached to and indenting strongly altered fragment with silicic matrix and granite-derived clasts; parallel polarizers; height of field 2.5 mm; *D* glass microspherule firmly attached to glass splash on clast of fine-grained particle of meta-quartzite; parallel polarizers; width of field 4.5 mm; *E* Same glass microspherule as in *D* (crossed polarizers). *NN* = Newport News corehole; *Ex* = Exmore corehole. See CD-ROM for color version of this figure.

have been identified to date, one from the Exmore corehole inside the crater, and two from the Newport News corehole, outside the crater. Current geochemical evidence indicates that such microspherules form when near-surface target rocks (in this case, unconsolidated, siliciclastic, shallow-marine sediments) are melted prior to the impactor's contact with the crystalline basement rocks (Koeberl 1994; Engelhardt 1997; French 1998).

Our petrographic observations can be summarized as follows: (1) The suites of particulate samples we investigated contain populations of sedimentary clasts, as well as a range of granitoid materials. Clast populations in the Exmore, Kiptopeke, and Newport News samples are not distinctly different from each other. Some individual samples are dominated by sedimentary clasts, whereas others have a significant granitoid portion. Generally, however, all samples have an important or dominant glauconite-rich fraction. In contrast, all the Windmill Point samples we studied are sediment-dominated; (2) All granitoid samples we studied from outside the crater are unshocked; (3) The proportion of shock-metamorphosed clasts is consistently small (<1%), and most are derived from the crystalline basement; (4) Mafic minerals other than biotite are rare, and we encountered no obvious volcanic rock fragments; (5) We detected no systematic variation in clast population within the entire suite of samples we analyzed. Nor was there any stratigraphically significant variation in the mean degree of shock metamorphism; clasts with low and high grades of shock metamorphism appear to have been thoroughly mixed; (6) Shock effects cover the entire range of shock metamorphism from <5 GPa to >45 (~60) GPa. This involves (in order of increasing shock pressure) shock fracturing, single and multiple sets of PDFs per grain, beginning isotropization of quartz or feldspar, diaplectic glass of quartz or feldspar, mineral melting, and, finally, bulk rock melting (particles of impact melt rock); (7) We found impact glass microspherules in two samples from the lower part of the Exmore breccia, as sampled in the Exmore corehole, and in one sample from the Newport News corehole; (8) Clasts of suspected impact melt breccia are present in three different samples from Newport News and Windmill Point; (9) The PDF orientation statistics of Koeberl et al. (1966) are significantly different from our data, because of the different numbers of grains with only one set, or more than one set, of PDFs that could be measured; (10) Considering the fact that we measured grains with one to >3 sets of PDFs, one could conclude that these grains were shocked at pressures of ~10–30 GPa. However, we discourage calculation of average shock pressures on the basis of relative abundance of specific orientations, because different results would be obtained for these different samples (suites of measurements).

6.3.7 Geochemistry

For this study, we carried out a large number of new chemical analyses for 40 clasts (85 samples and subsamples) of different lithic types (quartz sand, glauconitic sand, clayey sand, silt, clay, cherty breccia,) separated from the Exmore breccia (Tables 6.10, 6.11, 6.12A–C, 6.13, 6.14) together with corresponding CIPW normative mineral compositions. The first objective of these analyses was to investigate whether it was possible to chemically discriminate the different lithological components that constitute the Exmore breccia. The second objective was to carry out mixing calculations, in order to better characterize the indigenous siderophile elements in the breccia. The third objective was to determine whether the Exmore breccia carries a meteoritic component. In addition, we attempted to determine the likely protolith lithologies for the sedimentary components in the Exmore breccia.

We analyzed the samples in two different forms. Some fragments were quite small (millimeter-sized) and were analyzed as they are. For most samples, however, several grams of drillcore material were available. These samples were carefully broken (in a few cases with the help of a tabletop-sized alumina (ceramic) jaw crusher), and then powdered in either a boron carbide rapid mill (Retsch) in the case of small samples ($\leq$2 g), or in an automatic alumina ball mill. Up to 5 g of the sample powder were then used for major and trace element analysis by x-ray fluorescence (XRF) analysis. Trace elements analyzed by XRF are V, Cr, Co, Ni, Cu, Rb, Sr, Y, Zr, Nb, and Ba. Details (especially on precision and accuracy) are reported by Reimold et al. (1994). Some major and minor elements, and the trace elements were analyzed by instrumental neutron activation analysis (INAA). Of the trace elements analyzed by XRF, Cr, Co, Ni, Rb, Sr, Zr, and Ba were also analyzed by INAA, and, if the results agreed within errors, they were averaged; if not, the data with the better precision were selected. Information on irradiation, counting, instrumentation, standards, accuracy, and precision is given by Koeberl (1993). The only differences between our methods and those described by Koeberl (1993) are that we used a larger volume (more sensitive) germanium detector with 48% relative efficiency for all counts of the sedimentary samples, and we used four international standard reference materials (granites AC-E and G2; Govindaraju 1993; WMG1, and Allende meteorite; Jarosewich et al. 1987) for standardization and internal accuracy check in all counting batches (instead of the synthetic standards described by Koeberl 1993).

The results of the major and trace element analyses of all Exmore breccia samples are given in Tables 6.10 and 6.11. The samples were all of sedimentary origin and included mainly glauconitic sands, and some silt and clay. The petrographic descriptions, including the results of our search for evidence of shock metamorphic effects, were given in Tables 6.4 to 6.7. Samples are ordered by borehole and depth within the boreholes. A large compositional range for the major and trace element contents of the various samples is evident from the data in Tables 6.10 and 6.11. For example, the SiO_2 content of the Exmore breccia in the Exmore core samples varies from 32.6 to 85.3 wt%, and the other major elements show similar or even wider ranges. No systematic compositional difference between the different sediment types (i.e., silt versus glauconitic sandstone) is evident from the data in these tables.

To check for any systematic changes with depth in the borehole, we plotted the contents of some of the major elements with depth in Fig. 6.33. None of the three oxides plotted there (CaO, Al_2O_3, Fe_2O_3) show any such systematic variation. In fact, the diagram clearly shows a fairly constant average content for at least two of

Table 6.10. Chemical composition of samples from Exmore breccia (Exmore corehole).

Depth [ft]	1208.2	1210.2	1215.4	1217.5	1228.0	1232.1	1234.1	1234.5
Depth [m]	368.3	368.9	370.5	371.1	374.3	375.5	376.2	376.3
	Silt	Sd	Glc.Sd	Glc.Sd	Silt	Sd	Glc.Sd	Glc.Sd
SiO_2	49.12	55.43	64.04	60.53	46.98	43.11	68.36	68.82
TiO_2	0.47	0.62	0.61	0.51	0.60	0.14	0.02	0.57
Al_2O_3	8.82	9.63	7.85	7.74	9.58	5.96	6.54	8.96
Fe_2O_3	6.41	4.55	5.88	6.10	4.21	18.61	8.61	5.19
MnO	0.028	0.032	0.050	0.060	0.035	0.022	0.010	0.070
MgO	1.67	1.28	1.31	1.25	2.12	3.82	1.45	1.04
CaO	12.19	8.86	6.70	8.56	14.06	6.60	3.97	3.81
Na_2O	1.17	1.76	1.20	1.04	1.25	0.85	0.12	1.18
K_2O	2.28	2.38	2.64	2.66	2.21	7.24	3.23	3.06
P_2O_5	0.21	0.13	0.27	0.30	0.13	2.93	0.63	0.29
LOI	17.42	14.01	8.54	10.09	18.75	9.95	5.45	5.86
Total	99.78	98.67	99.09	98.84	99.92	99.22	98.39	98.85
Sc	8.87	9.21	6.43	7.35	10.6	4.99	5.44	n.d.
V	n.d.	n.d.	105	138	n.d.	n.d.	20	83
Cr	133	96.4	81.3	88.7	128	147	101	79
Co	7.11	11.9	6.97	8.22	13.1	16.5	5.95	11
Ni	18	18	18	25	33	21	8	17
Cu	<2	<2	<2	<2	<2	<2	<2	<2
Zn	116	118	53	72	335	125	90	44
Ga	15	13	8	9	14	21	11	8
As	21.8	42.3	8.57	10.4	5.95	13.5	21.2	n.d.
Se	0.96	0.61	0.3	0.5	1.01	0.26	0.1	n.d.
Br	1.23	0.45	1.5	1.1	0.45	0.26	0.8	n.d.
Rb	87.1	83.2	90.3	84.2	98.1	207	105	83
Sr	513	365	326	382	414	247	430	196
Y	n.d.	n.d.	19	19	n.d.	n.d.	5	20
Zr	235	322	122	129	286	271	80	189
Nb	n.d.	n.d.	11	10	n.d.	n.d.	4	11
Ag	0.04	0.04	0.01	0.02	0.06	0.05	0.02	n.d.
Sb	0.83	1.3	0.45	0.42	0.48	0.96	0.74	n.d.
Cs	4.19	3.33	2.45	3.22	5.94	2.89	2.23	n.d.
Ba	211	299	249	259	116	51	250	305
La	24.9	28.6	21.9	23.1	27.2	81.8	22.4	n.d.
Ce	52.2	60.0	50.2	53.5	55.6	221	53.1	n.d.
Nd	21.9	26.1	25.1	26.3	24.4	76.7	25.1	n.d.
Sm	4.94	5.62	4.77	4.61	4.74	16.5	4.49	n.d.
Eu	0.95	0.27	1.02	1.05	0.91	3.4	1.12	n.d.
Gd	4.35	4.72	5.2	4.5	4.81	14.3	5.1	n.d.
Tb	0.61	0.74	0.68	0.65	0.64	2.21	0.68	n.d.
Dy	3.8	4.5	3.9	4.2	4.1	9.3	4	n.d.
Tm	0.29	0.39	0.33	0.37	0.35	0.69	0.31	n.d.
Yb	2.07	2.74	1.81	2.38	2.32	4.88	1.71	n.d.
Lu	0.29	0.36	0.27	0.31	0.31	0.60	0.25	n.d.
Hf	3.59	5.95	5.97	6.52	6.05	0.89	4.93	n.d.
Ta	0.67	0.85	0.73	0.65	0.78	0.17	0.45	n.d.
W	0.4	0.87	5.2	3.1	0.97	0.37	1.3	n.d.
Ir (ppb)	<0.8	<0.7	<1	<1.3	<0.7	<0.7	<0.8	n.d.
Au (ppb)	1.4	0.9	0.8	0.3	1.1	1.8	0.6	n.d.
Hg	0.12	0.11	n.d.	n.d.	0.13	0.041	n.d.	n.d.
Th	7.26	8.51	5.86	6.66	9.28	2.62	4.01	n.d.
U	4.42	2.80	1.85	2.12	2.64	5.04	1.51	n.d.
K/U	5152	8493	11892	10456	8371	14365	17826	--
Zr/Hf	65.5	54.1	20.4	19.8	47.3	304	16.2	--
La/Th	3.43	3.36	3.74	3.47	2.93	31.22	5.59	--
Hf/Ta	5.36	7.00	8.18	10.0	7.76	5.24	11.0	--
Th/U	1.64	3.04	3.17	3.14	3.52	0.52	2.66	--
La_N/Yb_N	8.13	7.05	8.18	6.56	7.92	11.3	8.85	--
Eu/Eu*	0.63	0.16	0.63	0.70	0.58	0.68	0.72	--

Major elements in wt%, trace elements in ppm, except as noted. All Fe as Fe_2O_3. See Table 6.4 for petrography of samples. Glc. = glauconite; Sd = sand; n.d. = not determined.

Table 6.10. (cont).

Depth [ft]	1248.0	1249.5	1254.2	1261.2	1269.5	1272.5	1279.4	1283.1
Depth [m]	380.4	380.8	382.3	384.4	386.9	387.9	390.0	391.1
	Arenite	Glc.Sd	Glc.Sd	Glc.Sd	Glc.Sd	Glc.Sd	Silt	Glc.Sd
SiO_2	64.66		64.77	67.12	64.54	69.05	70.11	67.78
TiO_2	0.44		0.53	0.50	0.52	0.55	0.78	0.55
Al_2O_3	7.35		7.24	8.01	8.19	8.92	13.02	8.19
Fe_2O_3	6.02	4.81	5.89	4.99	4.42	6.67	4.21	7.81
MnO	0.027	0.016	0.050	0.035	0.028	0.040	0.034	0.020
MgO	1.03		1.12	0.96	0.90	1.16	0.76	1.30
CaO	6.49		7.27	5.79	7.67	2.65	1.26	3.42
Na_2O	1.44	1.13	0.99	1.11	1.08	1.11	2.05	1.01
K_2O	2.99	3.01	2.72	2.79	2.52	3.10	2.61	3.30
P_2O_5	0.22		0.33	0.30	0.24	0.36	0.05	0.44
LOI	8.92		8.52	7.49	8.73	5.48	4.32	5.94
Total	99.58		99.43	99.10	98.84	99.09	99.21	99.76
Sc	6.34	7.29	6.41	7.31	7.37	7.60	11.9	7.31
V	n.d.	n.d.	119	90	72	116	n.d.	125
Cr	101	61.3	77.1	68.3	62.1	93.7	67.6	102
Co	8.95	8.22	7.32	7.36	7.68	8.46	23.2	8.51
Ni	15	45	19	15	14	18	21	26
Cu	<2	<2	<2	<2	<2	<2	<2	<2
Zn	87	63	59	54	55	70	367	63
Ga	20	9	10	12	9	11	22	8
As	12.8	5.69	10.4	5.44	5.54	10.1	2.15	14.1
Se	0.28	0.2	0.1	0.2	0.1	0.3	0.31	0.3
Br	0.23	0.4	1.3	0.9	0.7	1.1	0.15	1.5
Rb	85.5	89.1	88.6	78.3	73.4	105	77.1	106
Sr	276	150	250	214	200	177	201	200
Y	n.d.	n.d.	19	21	21	24	n.d.	22
Zr	236	195	162	149	159	118	333	135
Nb	n.d.	n.d.	11	11	10	10	n.d.	9
Ag	0.07	0.02	0.02	0.02	0.01	0.02	0.04	0.03
Sb	0.69	0.41	0.51	0.28	0.36	0.46	0.39	0.51
Cs	2.14	2.59	2.65	2.26	2.24	3.18	2.14	3.09
Ba	363	400	247	296	284	292	403	273
La	24.9	22.7	22.5	22.3	21.5	28.6	26.1	24.8
Ce	54.3	53.2	50.7	50.5	48.2	62.5	53.1	54.6
Nd	25.5	22.8	25.2	20.3	23.3	30.7	25.7	27.2
Sm	5.08	4.15	4.42	4.23	3.82	5.63	5.16	4.79
Eu	0.75	1.15	1.11	1.09	1.11	1.34	0.5	1.09
Gd	4.64	4.9	4.8	4.1	4.2	5.8	4.32	4.7
Tb	0.70	0.71	0.69	0.67	0.71	0.81	0.63	0.75
Dy	4.1	4.3	4.2	3.9	4.1	4.9	4.0	4.1
Tm	0.29	0.36	0.38	0.33	0.34	0.41	0.37	0.34
Yb	2.05	2.02	2.11	2.07	2.04	2.31	2.30	1.94
Lu	0.25	0.28	0.31	0.31	0.31	0.34	0.32	0.31
Hf	4.32	4.05	5.45	6.31	5.72	4.34	7.11	4.77
Ta	0.51	0.68	0.63	0.58	0.62	0.63	0.75	0.58
W	0.77	1.7	1.2	1.8	1.3	0.4	0.58	1.5
Ir (ppb)	<0.7	<1	<0.7	<0.5	<0.6	<0.8	<0.7	<1
Au (ppb)	1.2	0.4	0.4	0.7	0.3	0.8	1.2	0.4
Hg	0.04	0.02	0.03	0.04	0.02	0.05	0.028	0.05
Th	6.33	5.29	4.98	5.39	5.15	6.13	7.37	5.79
U	1.94	1.32	1.61	1.78	1.61	1.77	2.20	1.97
K/U	15412	22803	14079	13062	13043	14595	11864	13959
Zr/Hf	54.6	48.1	29.7	23.6	27.8	27.2	46.8	28.3
La/Th	3.93	4.29	4.52	4.14	4.17	4.67	3.54	4.28
Hf/Ta	8.47	5.96	8.65	10.9	9.23	6.89	9.48	8.22
Th/U	3.26	4.01	3.09	3.03	3.20	3.46	3.35	2.94
La_N/Yb_N	8.21	7.59	7.21	7.28	7.12	8.37	7.67	8.64
Eu/Eu*	0.47	0.78	0.74	0.80	0.85	0.72	0.32	0.70

Major elements in wt%, trace elements in ppm, except as noted. All Fe as Fe_2O_3. See Table 6.4 for petrography of samples. Glc. = glauconite; Sd = sand; n.d. = not determined.

Table 6.10. (cont.)

Depth [ft]	1284.0	1288.4	1294.5	1302.0	1305.3	1313.0	1317.5	1330.5
Depth [m]	391.4	392.7	394.6	396.8	397.9	400.2	401.6	405.5
	Arenite	Glc.Sd	Glc.Sd	Silt	Glc.Sd	Gr.Breccia	Glc.Sd	Glc.Sd
SiO_2	67.95	68.06	68.52	56.93	85.31	69.82	70.69	69.84
TiO_2	0.55	0.54	0.57	0.91	0.40	0.81	0.57	0.55
Al_2O_3	9.99	8.85	8.98	15.43	6.78	12.02	8.51	8.73
Fe_2O_3	6.59	7.17	6.78	5.51	0.76	5.88	6.36	6.41
MnO	0.049	0.048	0.020	0.034	0.090	0.021	0.010	0.019
MgO	1.28	1.28	1.23	1.88	0.06	0.76	1.04	1.06
CaO	2.35	3.10	2.69	2.86	0.58	0.47	2.73	2.47
Na_2O	1.45	1.08	1.22	1.33	2.14	1.65	1.12	1.24
K_2O	3.26	3.32	3.28	2.54	2.62	1.95	3.11	3.27
P_2O_5	0.25	0.41	0.37	0.06	0.06	0.09	0.34	0.32
LOI	6.00	5.62	5.28	12.01	0.43	5.91	5.20	5.09
Total	99.71	99.48	98.94	99.50	99.23	99.38	99.68	99.00
Sc	8.52	7.22	6.67	15.6	2.02	13.6	6.79	7.42
V	n.d.	122	123	n.d.	<15	n.d.	119	117
Cr	108	87.1	79.2	151	11.4	65.2	76.8	80.7
Co	11.7	8.11	7.84	17.5	1.75	10.8	7.86	7.77
Ni	17	22	21	21	8	18	22	23
Cu	<2	<2	<2	<2	<2	<2	<2	<2
Zn	510	71	64	110	11	272	54	52
Ga	9	11	14	21	9	14	11	12
As	8.59	11.1	1.03	8.97	10.9	1.81	8.7	8.72
Se	0.23	0.1	0.2	0.75	0.03	0.31	0.2	0.3
Br	0.19	1.2	1.1	0.52	0.1	0.14	0.7	1.1
Rb	91.4	116	83.2	135	60.9	57.5	89.1	100
Sr	197	203	192	176	168	104	174	185
Y	n.d.	21	21	n.d.	8	n.d.	21	21
Zr	275	142	131	338	134	401	130	135
Nb	n.d.	11	11	n.d.	9	n.d.	12	12
Ag	0.09	0.02	0.03	0.06	0.02	0.06	0.03	0.02
Sb	0.59	0.43	0.41	0.64	0.29	0.53	0.38	0.46
Cs	2.41	2.84	2.81	7.84	0.81	3.52	2.72	2.75
Ba	296	300	312	269	554	239	315	312
La	22.7	20.9	21.9	36.8	11.3	28.6	21.9	26.1
Ce	48.9	49.2	50.6	71.4	20.2	61.5	50.7	58.9
Nd	23.1	22.5	24.6	27.3	10.2	25.3	24.4	28.5
Sm	4.49	4.37	4.21	5.79	1.78	5.28	4.11	5.25
Eu	0.97	1.08	1.15	0.26	0.47	1.03	1.09	1.22
Gd	3.84	4.4	4.4	4.95	1.9	4.86	4.1	5.1
Tb	0.62	0.63	0.65	0.68	0.27	0.62	0.64	0.69
Dy	3.6	3.8	3.9	4.5	1.7	3.8	3.9	4.0
Tm	0.29	0.31	0.31	0.43	0.16	0.36	0.29	0.32
Yb	2.06	1.76	1.91	2.95	0.99	2.65	1.79	2.01
Lu	0.28	0.28	0.28	0.38	0.14	0.36	0.27	0.29
Hf	5.10	4.05	4.81	7.27	4.59	9.02	4.21	4.81
Ta	0.65	0.57	0.57	1.13	0.56	0.72	0.64	0.66
W	0.69	2.2	1.7	0.49	3.0	0.82	1.1	0.1
Ir (ppb)	0.7	0.2	<0.8	<0.6	<0.3	<0.7	<1	<0.8
Au (ppb)	0.8	0.3	0.7	0.7	0.5	0.6	1.1	0.3
Hg	0.071	0.03	0.05	0.13	0.04	0.031	0.03	0.05
Th	5.38	5.28	4.56	13.2	3.21	8.61	4.78	9.21
U	1.94	1.95	1.54	2.39	0.86	1.65	1.74	1.88
K/U	16804	14188	17749	10628	25388	11818	14895	14495
Zr/Hf	53.9	35.1	27.2	46.5	29.2	44.5	30.9	28.1
La/Th	4.22	3.96	4.80	2.79	3.52	3.32	4.58	2.83
Hf/Ta	7.85	7.11	8.44	6.43	8.20	12.5	6.58	7.29
Th/U	2.77	2.71	2.96	5.52	3.73	5.22	2.75	4.90
La_N/Yb_N	7.45	8.02	7.75	8.43	7.71	7.29	8.27	8.77
Eu/Eu*	0.71	0.75	0.82	0.15	0.78	0.62	0.81	0.72

Major elements in wt%, trace elements in ppm, except as noted. All Fe as Fe_2O_3. See Table 6.4 for petrography of samples. Glc. = glauconite; Sd = sand ; n.d. = not determined.

Table 6.10. (cont.)

Depth [ft]	1334.4	1337.1	1347.0	1361.9	1366.5	1375.3	1377.6	1387.4
Depth [m]	406.7	407.5	410.6	415.1	416.5	419.2	419.9	422.9
	Glc.Sd	Glc.Sd	Greywacke	Schist,SS	Silt/Sd	Silt	Silt	Glc.Sd
SiO_2	69.97	72.05	69.51	64.37	71.43	63.56	73.57	32.65
TiO_2	0.60	0.61	0.90	0.88	0.85	0.92	0.77	0.15
Al_2O_3	9.11	9.17	12.74	16.13	13.00	14.98	9.10	3.12
Fe_2O_3	6.35	4.99	4.12	4.75	3.07	5.54	4.42	6.52
MnO	0.020	0.027	0.044	0.017	0.029	0.056	0.043	0.087
MgO	1.08	0.95	0.92	1.03	0.65	1.35	0.92	2.02
CaO	2.21	2.23	0.95	0.34	0.43	1.04	1.73	26.78
Na_2O	1.27	1.32	1.87	1.65	1.64	1.78	1.29	0.44
K_2O	3.39	3.09	2.54	2.20	1.65	2.52	2.87	2.95
P_2O_5	0.36	0.35	0.05	0.09	0.04	0.12	0.22	1.67
LOI	4.77	4.27	5.76	7.80	6.23	7.82	4.40	22.78
Total	99.13	99.06	99.41	99.25	99.02	99.67	99.34	99.17
Sc	6.83	6.89	13.9	17.3	12.2	16.5	8.34	7.32
V	113	109	n.d.	n.d.	109	n.d.	n.d.	n.d.
Cr	72.9	68.8	76.6	77.5	56.3	88.7	79.7	180
Co	7.91	7.81	12	14	14.2	21.9	10.9	4.93
Ni	24	16	19	56	22	21	23	33
Cu	<2	<2	<2	<2	<2	<2	<2	<2
Zn	59	49	321	193	90	124	453	397
Ga	12	9	11	12	8	18	8	6
As	10.3	6.59	1.58	0.93	1.96	5.95	4.67	6.01
Se	0.1	0.2	0.73	0.23	0.2	0.51	0.29	0.12
Br	1.3	0.7	0.24	0.02	0.3	0.19	0.11	0.15
Rb	88.9	90.1	75.4	98.1	101	93	73.4	81.4
Sr	199	152	189	152	149	180	250	244
Y	22	17	n.d.	n.d.	24	n.d.	n.d.	n.d.
Zr	152	121	542	776	309	418	417	275
Nb	11	10	n.d.	n.d.	15	n.d.	n.d.	n.d.
Ag	0.02	0.02	0.05	0.08	0.01	0.04	0.04	0.05
Sb	0.39	0.36	0.51	0.34	0.39	0.59	0.39	0.77
Cs	2.71	2.41	2.92	4.81	3.65	3.63	1.83	1.14
Ba	334	365	353	284	250	367	383	92
La	22.2	30.6	36.9	173	29.7	35.4	32.7	47.5
Ce	51.7	73.9	76.6	527	64.5	69.1	68.5	95.9
Nd	24.4	35.8	32.8	212	29.9	35.3	33.8	51.8
Sm	4.26	5.74	6.88	33.5	5.32	7.02	7.14	12.2
Eu	1.21	1.27	1.33	8.17	1.24	1.4	1.44	2.77
Gd	4.7	5.4	6.59	28.1	5.4	6.26	6.61	12.4
Tb	0.71	0.73	0.89	3.85	0.85	1.04	1.08	1.93
Dy	4.1	4.2	5.3	20.9	5.4	6.2	6.5	10.9
Tm	0.35	0.33	0.46	0.79	0.48	0.49	0.49	0.68
Yb	2.08	1.98	3.28	5.21	3.22	3.57	3.23	4.77
Lu	0.31	0.26	0.45	0.59	0.48	0.48	0.40	0.55
Hf	3.43	3.36	12.2	6.74	8.31	8.2	8.13	1.51
Ta	0.47	0.56	0.89	0.85	1.09	0.81	0.81	0.13
W	1.3	0.2	0.8	0.78	1.8	1.24	0.42	0.53
Ir (ppb)	<0.8	0.2	<0.7	<0.7	0.1	0.7	<0.7	<0.7
Au (ppb)	0.1	0.3	0.4	2.4	0.3	1.1	0.9	1
Hg	0.08	0.03	0.11	0.36	0.03	0.09	0.045	0.057
Th	4.37	21.8	10.6	10.1	9.11	10.1	8.62	2.68
U	1.61	2.04	2.28	2.24	1.92	2.23	2.08	4.34
K/U	17547	12623	11140	9830	7161	11278	13808	6797
Zr/Hf	44.3	36.0	44.4	115	37.2	51.0	51.3	182
La/Th	5.08	1.40	3.48	17.1	3.26	3.50	3.79	17.7
Hf/Ta	7.30	6.00	13.7	7.93	7.62	10.1	10.0	11.6
Th/U	2.71	10.7	4.65	4.51	4.74	4.53	4.14	0.62
La_N/Yb_N	7.21	10.4	7.60	22.4	6.23	6.70	6.84	6.73
Eu/Eu*	0.83	0.70	0.60	0.81	0.71	0.65	0.64	0.69

Major elements in wt%, trace elements in ppm, except as noted. All Fe as Fe_2O_3. See Table 6.4 for petrography of samples. Glc. = glauconite; Sd = sand; SS = sandstone; n.d. = not determined.

Table 6.11. Chemical composition of Exmore breccia samples from Newport News (NN) and Windmill Point (WP) coreholes.

[ft]	NN 432.3	NN 438.2	WP 544.1	WP 544.5	WP 552.5	WP 563.7	WP 565.2	WP 566.5	WP 570.7
[m]	131.7	133.6	165.8	166.0	168.4	171.8	172.3	172.7	173.9
	Shale	Calc./Chert	Glc.Sd	Glc.Sd	Glc.Sd	Sd	Glc.Sd	Glc.Sd	Calc. Sd
SiO_2	55.09	2.79	59.03	64.10	59.87	53.87	58.14	55.14	30.85
TiO_2	0.58	0.03	0.54	0.45	0.52	0.43	0.44	0.45	0.10
Al_2O_3	11.62	0.46	7.36	7.43	7.35	7.95	7.27	8.41	2.82
Fe_2O_3	10.23	1.12	7.08	6.66	6.96	8.89	7.01	11.50	8.29
MnO	0.021	0.040	0.030	0.040	0.060	0.063	0.090	0.010	0.157
MgO	2.35	1.03	1.37	1.38	1.41	1.85	1.44	2.43	2.33
CaO	3.24	51.19	10.02	6.82	8.59	8.87	10.01	6.27	26.49
Na_2O	0.58	0.91	0.90	0.79	0.79	0.76	0.77	0.69	0.35
K_2O	4.35	0.09	3.07	3.25	3.15	3.35	3.01	4.33	3.61
P_2O_5	0.36	0.08	0.42	0.49	0.64	0.64	0.47	0.84	1.59
LOI	10.79	42.50	9.32	8.12	9.69	12.51	10.59	9.53	22.85
Total	99.20	100.24	99.14	99.53	99.03	99.18	99.24	99.60	99.42
Sc	10.5	1.85	6.58	7.19	7.09	7.64	6.44	10.7	3.53
V	n.d.	211	136	123	120	n.d.	121	141	n.d.
Cr	207	6.20	97.6	101	102	145	104	145	184
Co	11.9	0.47	7.05	7.62	8.14	16.3	7.64	10.5	3.52
Ni	15	21	17	26	26	22	25	38	19
Cu	<2	<2	<2	<2	<2	<2	<2	<2	<2
Zn	239	60	56	61	71	235	74	88	99.7
Ga	16	10	9	12	8	25	8	9	4
As	16.2	2.98	14.9	9.95	12.9	19.1	12.6	15.2	4.12
Se	0.92	0.03	0.2	0.3	0.1	0.46	0.05	0.4	0.067
Br	0.29	0.4	1.5	0.7	1.3	0.22	1.1	0.6	0.077
Rb	140	5	103	101	101	110	86.1	152	95.4
Sr	158	228	211	270	299	293	325	268	185
Y	n.d.	24	16	22	24	n.d.	21	46	n.d.
Zr	266	15	77	95	98	253	109	133	165
Nb	n.d.	11	8	9	9	n.d.	9	10	n.d.
Ag	0.09	0.01	0.04	0.02	0.03	0.09	0.03	0.02	0.07
Sb	1.10	0.11	0.88	0.42	0.65	1.10	0.55	1.34	0.67
Cs	4.56	0.089	2.34	2.33	2.54	2.65	2.23	5.12	0.97
Ba	153	415	300	288	280	228	274	258	74
La	32.1	11.5	24.7	21.8	33.8	30.2	22.8	39.3	37.5
Ce	76.1	23.3	60.3	54.3	65.6	69.3	53.5	105	68.3
Nd	30.3	11.6	30.1	26.8	28.1	28.2	23.4	56.4	34.3
Sm	6.8	1.79	5.49	4.81	5.27	6.72	4.49	9.98	8.39
Eu	1.28	0.39	1.40	1.28	1.22	1.31	1.16	2.51	1.69
Gd	9.14	1.9	5.3	4.5	4.8	5.96	4.1	10.2	9.85
Tb	0.84	0.23	0.78	0.66	0.72	0.98	0.66	1.46	1.32
Dy	4.9	1.4	4.7	3.9	4.3	5.9	3.8	7.3	7.2
Tm	0.45	0.12	0.39	0.35	0.36	0.41	0.29	0.59	0.45
Yb	2.41	0.77	2.23	1.88	1.93	2.52	1.64	3.37	2.73
Lu	0.32	0.11	0.32	0.28	0.28	0.32	0.25	0.47	0.36
Hf	4.9	0.11	3.52	2.41	2.38	3.64	3.18	3.27	1.06
Ta	0.72	0.02	0.42	0.51	0.59	0.52	0.42	0.48	0.11
W	0.79	0.6	2.2	1.3	1	0.47	0.9	2.1	0.08
Ir (ppb)	<0.8	<0.4	<1	<0.8	<1	<1	<1	<1.2	<0.7
Au (ppb)	0.9	0.4	0.3	0.4	0.3	1	0.6	4	0.7
Hg	0.072	0.01	0.02	0.04	0.04	0.12	0.01	0.03	0.02
Th	8.55	1.03	4.52	4.18	5.56	4.66	4.87	5.89	1.35
U	2.87	0.71	1.98	1.51	1.92	3.34	1.41	2.24	3.98
K/U	15157	1056	12921	17936	13672	10030	17790	16109	9070
Zr/Hf	54.3	136	21.9	39.4	41.2	69.5	34.3	40.7	156
La/Th	3.75	11.2	5.46	5.22	6.08	6.48	4.68	6.67	27.8
Hf/Ta	6.81	5.50	8.38	4.73	4.03	7.00	7.57	6.81	9.64
Th/U	2.98	1.45	2.28	2.77	2.90	1.40	3.45	2.63	0.34
La_N/Yb_N	9.00	10.1	7.48	7.84	11.83	8.10	9.39	7.88	9.28
Eu/Eu*	0.50	0.65	0.79	0.84	0.74	0.63	0.83	0.76	0.57

Major elements in wt%, trace elements in ppm, except as noted. All Fe as Fe_2O_3. See Table 6.6 for petrography of NN samples; Table 6.7 for WP samples. Glc. = glauconite; Sd = sand; Calc = calcareous.

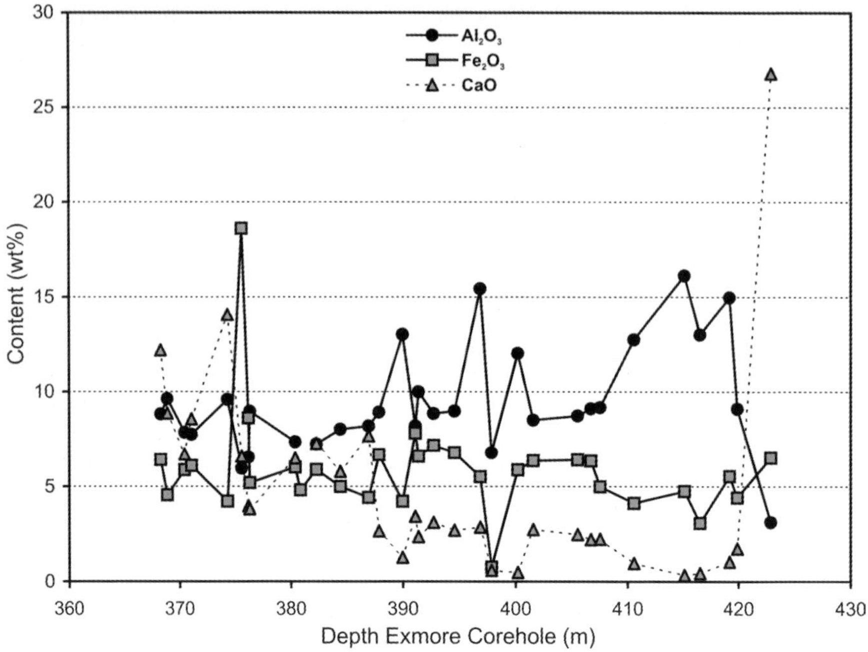

Fig. 6.33. Content (wt%) of major elements in samples of Exmore breccia plotted against depth (m) in Exmore corehole.

the oxides, with occasional wide variations. The only oxide that shows a hint of a decrease in content with depth is Fe_2O_3, but the wide variations make this trend not very clear. Similar plots (not shown here) were constructed for the other major element oxides and for a number of trace elements (Sc, Cr, Co, Rb, Sr, the REE, Hf, Th), but they expressed the same behavior (no clear systematic trend).

One of the most important groups of trace elements for provenance studies is the rare earth elements (REE). For comparison with each other, and with a variety of other reference materials, the abundances of the REEs are normalized to the abundances in C1-type carbonaceous chondrites (normalization values from Taylor and McLennan 1985). The normalized abundance patterns of four samples that represent some of the widely different patterns are shown in Fig. 6.34, and the total range of all Exmore breccia samples, as well as the overall average, are shown in Fig. 6.35. The average REE composition is similar to that of typical post-Archean upper crustal composition (cf. Taylor and McLennan 1985), with a modest slope of the light REE (LREE) pattern, a moderate negative Eu anomaly, and an almost flat heavy REE (HREE) pattern. The range mostly reflects this pattern, just with higher and lower absolute abundances. The REE patterns of individual samples show some more variety, as is indicated in Fig. 6.34. Europium anomalies of all samples are negative (see Tables 6.10 and 6.11); no positive Eu anomalies or values close to unity are recorded in our samples. Again, this is typical of sedimentary samples. The extent of the Eu anomaly (Eu/Eu*, where Eu

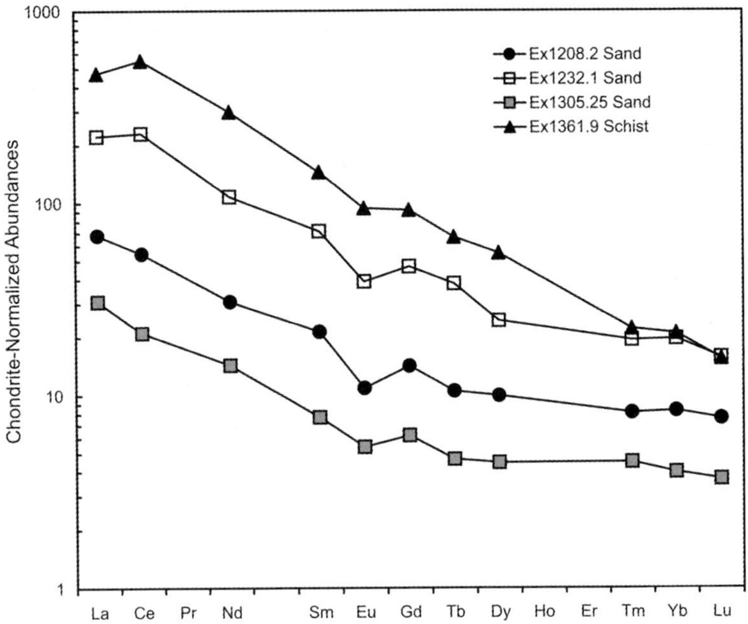

Fig. 6.34. Abundances of rare earth elements in samples of Exmore breccia from four different depths in Exmore corehole. Normalization values from Taylor and McLennan (1985).

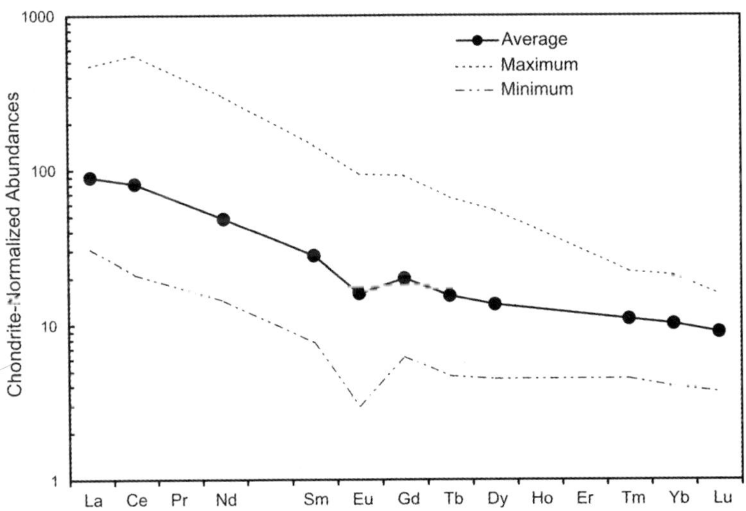

Fig. 6.35. Abundances (average, range) of rare earth elements in samples of Exmore breccia from Exmore, Windmill Point, and Newport News coreholes. Normalization values from Taylor and McLennan (1985).

is the measured value and Eu* is the calculated value if there were no Eu anomaly) ranges from 0.15 (strong negative anomaly) to 0.85 (very small negative Eu anomaly).

Some of the REE patterns are much steeper than that of the average Exmore breccia. In most cases, the slope of the LREE part of the patterns is much steeper than that of the HREE part of the pattern. A number of samples show distinct positive Ce anomalies, especially those samples that have high absolute REE contents. For some samples this could be an indication of the inclusion of a minor marine sedimentary component. For example, sample Ex1232.1, with a pronounced Ce anomaly, also has a very high phosphorus content, which might be indicative of the presence of some organic marine detritus. The situation for sample Ex1387.4 is similar. In the case of a few other samples (e.g., Ex1208.2) the Ce anomaly might simply indicate an advanced state of weathering of the sediment. The diagram of the extent of the Eu anomaly versus the slope of the REE pattern (Eu/Eu* vs. La_N/Yb_N; Fig. 6.36) shows no correlation between these two values, in agreement with the sedimentary origin of the materials. Any differentiation due to melting would appear as a correlation in this diagram. There is also no correlation between the absolute amount (sum) of the REE and the degree of the Eu anomaly.

In addition to using the elemental and oxide abundances of the Exmore breccia samples, we also calculated CIPW normative compositions (using the MinPet program, version 2.0; see Rollinson 1993 for details on the computation of the CIPW norm and for components), as well as the chemical index of weathering (CIW) and the chemical index of alteration (CIA) (Rollinson 1993). The results of the calculations are reported in Table 6.12 for the same sample suite for which the compositions are given in Tables 6.10 and 6.11. Samples are identified by their number (= depth in the core in feet). The CIW and CIA values vary widely, but, on average, are on the order of 30 to 50, which indicates fairly weathered sediments. Some of the CIPW normative data, together with the major element compositions, are plotted in ternary diagrams (Fig. 6.37).

The ternary diagrams in Figure 6.37 represent some of the multiple attempts to use major-element chemical data to determine the sedimentary lithologies forming most of the clast components in the Exmore breccia. It is obvious that neither absolute abundances nor CIPW normative abundances allow unambiguous discrimination of these rock types. On the other hand, Discriminant Function analysis (Table 6.13, Fig. 6.38), based on the equations given by Rollinson (1993) provides a strong indication that the vast majority of analyzed sedimentary clasts are derived from felsic to intermediate igneous rocks that are known to form the crystalline basement below the sedimentary section in the Chesapeake Bay region. However, a quartzose sedimentary provenance also must be considered (Fig. 6.38B).

Due to the lack of obvious macroscopic impact breccia and impact melt rock samples among our suite of Exmore breccia samples, and due to the small size of basement rock clasts among these breccia samples, we measured no chemical compositions of such clasts. We used all of those clast samples for petrographic

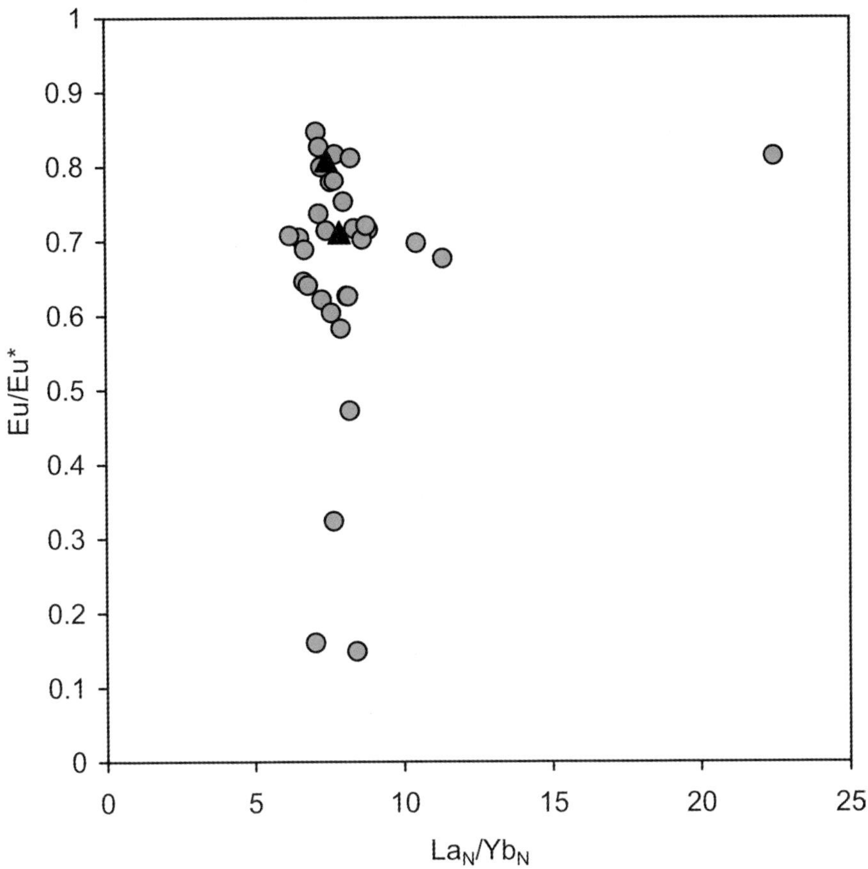

Fig. 6.36. Plot of Eu/Eu* versus La_N/Yb_N in samples of Exmore breccia from Exmore, Windmill Point, and Newport News coreholes. Solid triangles indicate values for bediasite and georgiaite tektites (cf. Table 6.14).

studies (i.e., mainly the search for shock metamorphic features; cf. Koeberl et al. 1996). Thus, it was not possible (nor necessary) to try to reproduce the composition of the Exmore breccia from target rock compositions in mixing calculations (which, given the weathered state of most of those samples, may not have yielded any meaningful results anyway). Analysis of the crystalline basement rock types will have to await data from the deep cores that are currently being drilled by the USGS (Gohn in press).

It is difficult to assess the presence of a meteoritic component in the Exmore breccia samples. Commonly the siderophile trace elements, including Cr, Co, Ni, and the platinum group elements, especially iridium (Ir), are used for such identification (cf. Koeberl 1998). This procedure requires a number of prior conditions to be fulfilled. First, breccias used for such comparison are commonly melt- sample suite. Second, the compositions of a complete set of target rocks that were in-

Table 6.12. Geochemical results from samples of Exmore breccia from Exmore (Ex), Windmill Point (WP), and Newport News (NN) coreholes.

Sample	Ex	Ex	Ex	Ex	Ex	Ex	Ex	Ex	Ex	Ex	Ex	Ex	Ex	Ex	Ex	Ex	Ex	Ex	Ex	Ex	Ex	Ex	Ex
Depth [ft]	1208.2	1210.2	1215.3	1217.1	1228.0	1232.1	1234.1	1234.4	1248.0	1254.0	1261.0	1269.1	1272.1	1279.4	1283.0	1284.0	1288.3	1294.1	1302.0	1305.2	1313.0	1317.1	1330.1
CIA	24.9	30.9	31.6	27.7	24.2	21.9	37.5	42.4	29.7	28.9	34.3	30.7	47.1	60.5	41.7	49.5	44.6	46.2	60.4	47.8	67.9	45.6	46.4
CIW	26.8	33.6	35.7	30.9	25.8	30.8	46.9	50.3	34.2	32.8	39.3	34.3	57.3	69.7	51.0	60.0	54.4	56.6	67.6	59.7	77.1	55.6	57.2
CIPW norm																							
Q	18.36	25.28	39.05	34.84	11.91	9.96	51.24	44.69	38.11	40.76	43.38	40.03	46.31	44.92	44.38	42.17	44.04	44.33	33.49	60.44	52.75	47.84	46.12
or	16.99	17.12	18.05	18.57	16.49	38.86	21.95	20.37	20.40	18.60	18.87	17.29	20.60	16.81	21.95	21.49	22.01	21.74	17.67	16.03	12.90	20.48	21.63
ab	13.25	19.25	12.47	11.03	14.17		1.24	11.94	14.93	10.29	11.41	11.26	11.21	20.06	10.21	14.53	10.88	12.29	14.06	20.24	16.59	11.21	12.47
an	15.24	13.82	9.54	10.16	17.69		8.94	11.40	5.50	8.42	9.89	11.68	11.47	6.46	9.09	11.19	10.66	10.48	16.25	1.22	1.65	10.05	9.63
C														5.53		0.49			6.99		8.01		
di	11.63	8.61	8.37	8.15	14.78	12.61	7.13	4.67	6.57	7.16	6.07				3.77		1.19			0.24		0.33	
hy						6.29	1.04	0.90				5.77	3.60	2.29	2.15	3.94	3.37	3.81	6.11	0.05	2.35	3.04	3.28
wo	17.14	10.18	5.68	10.28	19.37				8.08	7.99	4.47	8.55											
ac						7.29																	
il	0.06	0.06	0.09	0.11	0.08	0.04		0.12	0.05	0.09			0.07	0.05	0.04	0.09		0.04	0.06	0.15	0.04	0.02	
hem	5.63	3.86	4.74	5.02	3.71	13.67	6.90	4.07	4.85	4.75	3.98	3.58	5.23	3.20	6.13	5.13	5.61	5.30	4.52	0.56	4.59	4.94	5.00
ti	1.15	1.49	1.34	1.09	1.46	0.29	0.05	1.16	0.98	1.15	1.20	1.26	0.40		1.24		1.27	1.07		0.66		1.30	1.05
ap	0.55	0.33	0.65	0.74	0.34	7.32	1.52	0.68	0.53	0.80	0.72	0.58	0.85	0.11	1.04	0.58	0.96	0.87	0.15	0.13	0.21	0.79	0.75
ru													0.26	0.57		0.38		0.07	0.72		0.61		0.08
KMS						3.67																	

Table 6.12. (cont.)

Sample Depth [ft]	Ex 1334.3	Ex 1337.0	Ex 1347.0	Ex 1361.9	Ex 1366.1	Ex 1375.3	Ex 1377.6	Ex 1387.4	WP 552.6	WP 554.0	WP 554.3	WP 563.7	WP 565.0	WP 566.6	WP 570.7	NN 432.3	NN 438.2
CIA	48.2	48.9	62.8	73.9	71.2	66.5	52.1	5.6	26.6	24.2	30.1	27.5	24.2	32.8	5.1	50.1	0.5
CIW	59.9	59.6	72.6	82.9	78.9	75.6	63.3	5.9	30.3	27.2	35.2	31.4	27.2	40.2	5.5	62.9	0.5
CIPW norm																	
Q	54.64	48.46	46.75	46.37	55.78	41.30	51.76		34.12	30.74	39.35	26.40	31.12	25.96		28.21	
or	22.29	20.18	16.60	14.68	10.91	16.74	18.72		21.93	21.23	22.13	24.05	21.10	30.03		30.53	
ab	12.69	13.10	18.57	16.73	16.48	17.97	12.79		8.36	9.46	8.18	8.29	8.20	7.27		6.19	
an	9.59	9.70	4.85	1.24	2.10	4.92	7.89		8.49	8.17	8.22	10.19	8.89	8.29		16.30	
lc								18.58							16.81		
ne																	2.05
kal																	0.55
C	0.24	0.53	6.41	13.11	9.57	0.11	1.51									1.20	
di								15.21	9.18	8.85	8.78	12.41	9.43	15.75	17.56		
hy	3.32	2.90	2.81	3.21	2.01	4.19	2.80									7.71	
wo								29.71	9.39	13.42	5.73	8.46	13.28	0.12	19.85		
ol																	-129.7
ac								4.31							3.43		5.39
mt															0.23		
il	0.03		0.07	0.04		0.11	0.07	0.19	0.11			0.11	0.17	0.02	0.19	0.04	0.07
hem	4.93	3.84	3.18	3.74	2.40	4.34	3.40	5.12	5.72	5.78	5.35	7.53	5.80	9.41	6.88	8.47	
ti									1.11	1.32	1.08	0.92	0.84	1.08			
ap	0.84	0.81	0.12	0.21	0.09	0.28	0.51	4.76	1.58	1.03	1.18	1.63	1.17	2.06	4.54	0.89	88.96
pero								0.09									
ru	0.45	0.47	0.66	0.67	0.66	0.67	0.56									0.46	
KMS								0.16							2.43		
NMS																	1.18
COS								21.85							28.10		131.2

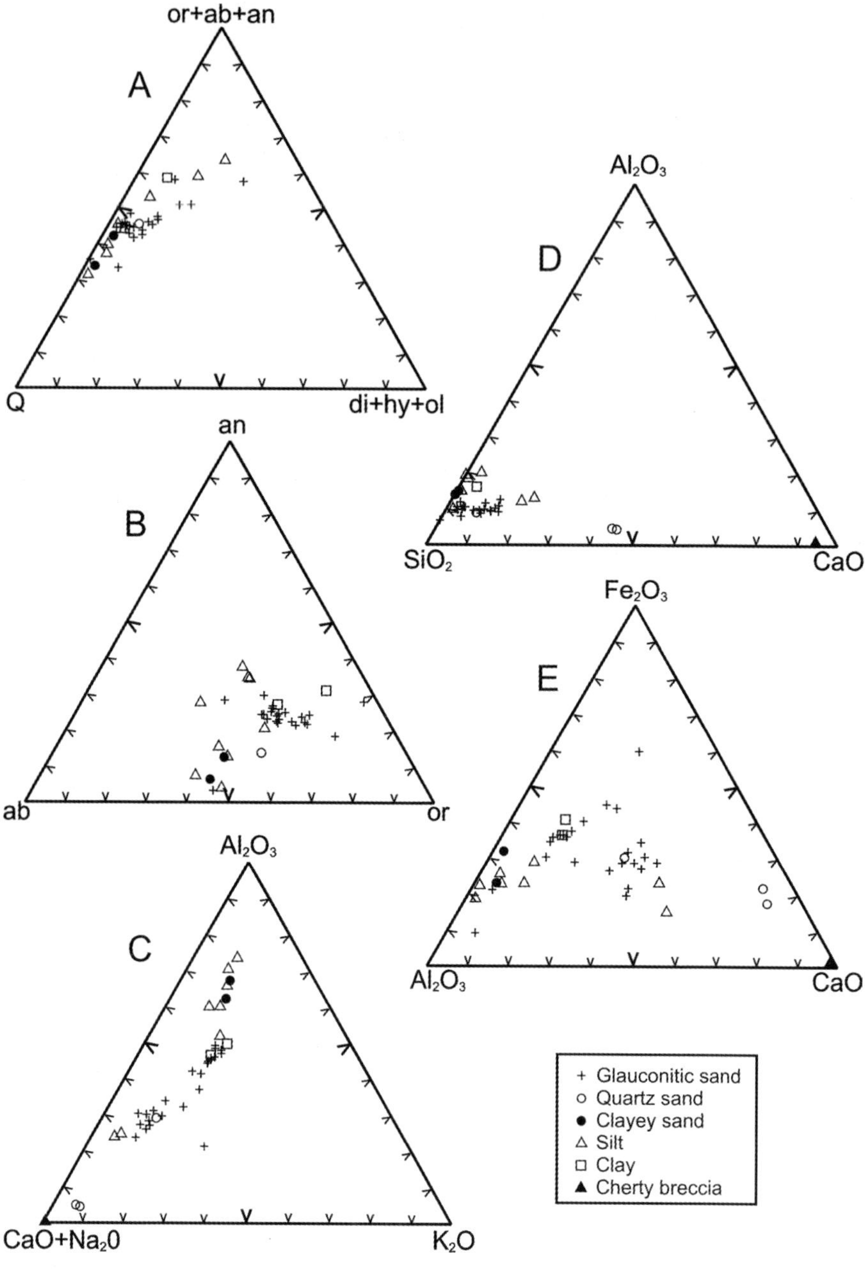

Fig. 6.37. Mineralogical and chemical classification of sedimentary clasts from Exmore breccia. *A, B* compositional variation based on CIPW normative proportions for orthoclase (*or*), anorthite (*an*), quartz (*Q*), diopside (*di*), hypersthene (*hy*), and olivine (*ol*), calculated from analyses listed in Tables 6.10–6.12. *C–E* chemical variation within samples illustrated by ternary diagrams based on different combinations of major elements.

Table 6.13. Samples for Discriminant Function analyses of Exmore breccia in this study.

Sample Drill Depth [ft]	Lithology	Subsamples	Discriminant Function 1	Discriminant Function 2
[1]Ex 1208.2	Clay	3	3.11088	0.77216
Ex 1210.2	Silt	1	0.63463	1.3603
Ex 1215.3	Glauconitic. sand	1	-1.39614	-0.53683
Ex 1217.1	Glauconitic sand	1	0.01143	0.03169
Ex 1228.0	Silt	3	2.27303	1.75655
Ex 1232.1	Glauconitic sand	1	-1.67041	-0.92765
Ex 1234.1	Quartz sand	1	-2.234	-3.68086
Ex 1234.4	Quartz sand	1	-3.07535	-0.69294
Ex 1248.0	Quartz sand	5	-1.30743	0.3984
Ex 1254.0	Glauconitic sand	1	-1.18566	-0.34666
Ex 1261.0	Glauconitic sand	1	-2.04536	-0.26981
Ex 1269.1	Quartz sand	1	-0.84147	0.35688
Ex 1272.1	Glauconitic sand	2	-2.91852	-1.76552
Ex 1279.4	Silt	3	-1.88523	-0.22543
Ex 1283.0	Glauconitic sand	1	-2.52661	-1.78654
Ex 1284.0	Glauconitic sand	1	-2.71741	-1.2094
Ex 1288.3	Glauconitic sand	1	-2.75063	-1.57034
Ex 1294.1	Glauconitic sand	1	-3.07865	-1.42341
Ex 1302.0	Silt	3	-0.64005	-2.06394
Ex 1305.2	Silt	1	-6.95648	0.67974
Ex 1313.0	Silt	6	-1.15862	-2.57676
Ex 1317.1	Glauconitic sand	1	-3.21632	-1.50673
Ex 1330.1	Glauconitic sand	1	-3.33424	-1.24429
Ex 1334.3	Glauconitic sand	1	-3.5596	-1.10179
Ex 1337.0	Glauconitic sand	1	-3.97454	-0.94997
Ex 1347.0	Quartz sand	6	-2.77325	-0.85295
Ex 1361.9	Silt	3	-0.43784	-1.98419
Ex 1366.1	Silt	3	-2.26781	-2.12231
Ex 1375.3	Silt	3	-0.98072	-1.65516
Ex 1377.6	Quartz sand	3	-4.7429	-1.28388
Ex 1387.4	Glauconitic sand	5	7.57273	6.07265
[2]WP 552.6	Clay	1	-0.53796	-0.04575
WP 554.0	Glauconitic sand	1	0.61864	0.65404
WP 554.3	Glauconitic sand	1	-1.76101	-0.5947
WP 563.7	Quartz sand	1	0.70502	-0.66519
WP 565.0	Glauconitic sand	1	0.58422	0.25911
WP 566.6	Glauconitic sand	1	-0.77437	-1.78352
WP 570.7	Glauconitic sand	5	7.32719	5.91427
[3]NN 432.3	Silt and sand	2	-1.84854	-2.55298
NN 438.2	Glauconitic sand	4	22.3283	15.6201

[1]Exmore corehole inside crater
[2]Windmill Point corehole outside crater
[3]Newport News corehole outside crater
See Table 6.4 for more complete sample descriptions
See Rollinson (1993, p. 21) for equations

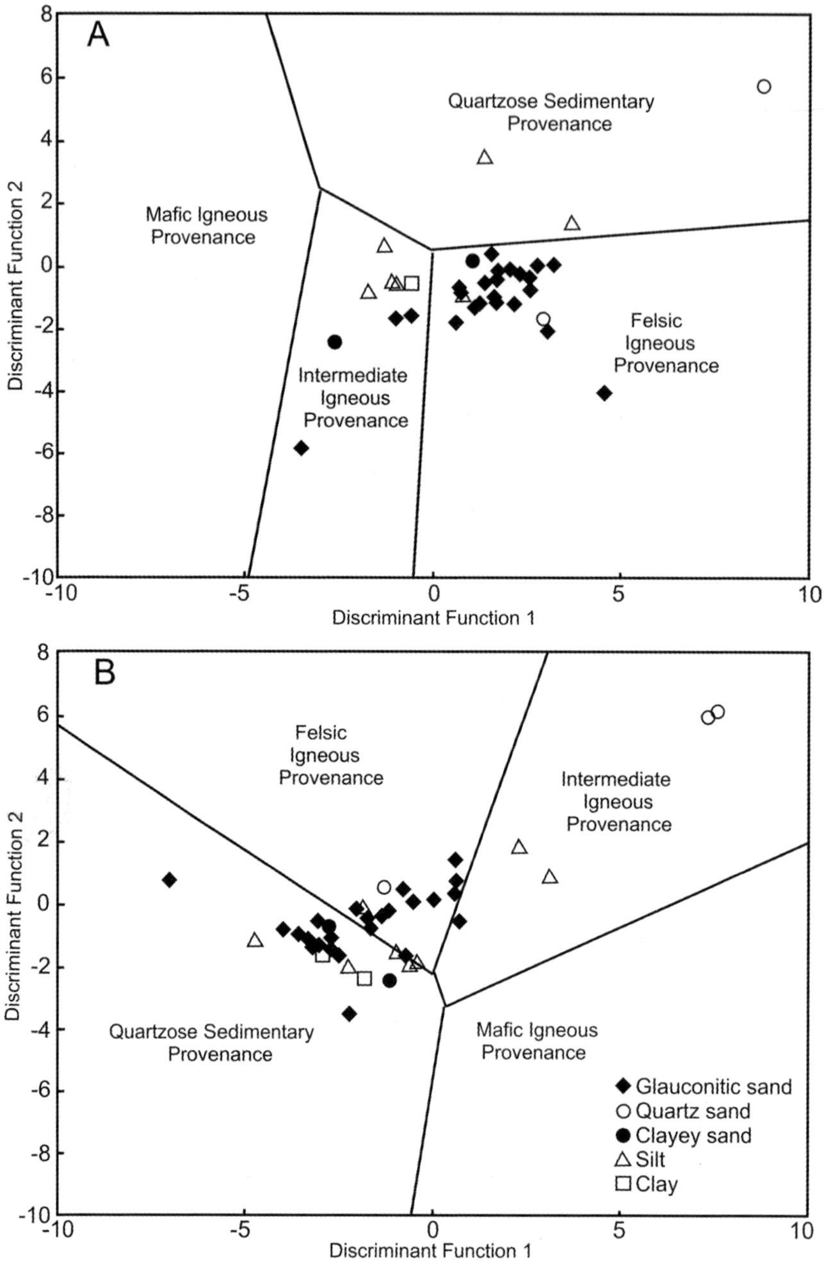

Fig. 6.38. Discriminant Function analysis for chemical compositions listed in Tables 6.10–6.13. Most samples represent felsic igneous provenance, with minor sources from intermediate igneous and quartzose sedimentary provenances. For details on diagrams see Rollinson (1993, p. 210–211).

volved in the production of the breccia need to be known. This condition is not fulfilled either. Thus, the following statements are only of qualitative nature. The values for Ir are not very reliable, because most values that did appear to give a positive signal are right at the detection limit for Ir by INAA (depending on sample type and composition, about 0.1 to 1 ppb for our procedure). A few samples seem to indicate Ir contents of 0.1 to 0.7 ppb, but we believe that these values are not reliable. No correlation between these values and the contents of the other siderophile trace elements, such as Co or Ni, are apparent. Chromium contents are fairly high in most samples, but this seems to be a characteristic of the weathered sediments, and does not indicate any meteoritic component. Cobalt and Ni contents vary somewhat, but the abundances are fairly similar to typical crustal values (Taylor and McLennan 1985). Also, the Ni and Co, or the Ni and Cr and the Co and Cr contents do not correlate with each other, which would be the case if a significant proportion were of meteoritic origin. Thus, we have not yet been able to identify a meteoritic component in these Exmore breccia samples. As most impact-derived breccias have only minor meteoritic contaminations anyway, this result was more or less expected. It will be necessary to identify suevitic-type breccia, with a significant melt rock or glass component, and analyze these, as well as crystalline and sedimentary target rocks, with a more sensitive technique than INAA (e.g., radiochemistry or ICP-MS after preconcentration) for the contents of the platinum group elements at the parts per trillion level.

Another interesting topic is the comparison of the compositions of Exmore breccia samples with that of the North American tektites. Based on age and location arguments, Poag et al. (1994) suggested that the Chesapeake Bay crater is the long-sought source crater of the North American tektites, the bediasites, and the georgiaites. In order to determine whether the Chesapeake Bay impact crater is the source of the North American tektite strewn field, Koeberl et al. (1996) analyzed major and trace elements from small samples of the Exmore breccia. They found that most breccia samples contained 32–73 weight % SiO_2. A few carbonate-rich samples yielded low SiO_2 values. The compositions of some of the high-SiO_2 samples agree well with the compositions of average North American tektites for mostly nonvolatile elements (Fig. 6.39).

To identify the exact source beds for the North American microtektites, Koeberl et al. (2001) analyzed several samples of Cenozoic sedimentary formations (Aquia, Nanjemoy, Piney Point) that would have constituted part of the uppermost target rocks at the Chesapeake Bay crater (Fig. 2.4). They found no geochemical matches, however, which indicates either that the tektites must have been formed from the underlying Cretaceous sediments, or that the North American strewn field was not produced by the Chesapeake Bay impact.

We made some further attempts to compare the samples from the present suite of Exmore breccia with North American tektites. Table 6.14 reports the average, minimum, and maximum oxide and element contents in Exmore breccia, compared to average compositions of the bediasite and georgiaite tektites. A variety of differences in composition is clearly apparent, even if the Exmore breccia composition is recalculated on a volatile-free basis. First, silica contents are simply too low, and, especially, CaO contents are too high in the breccia to provide a reas-

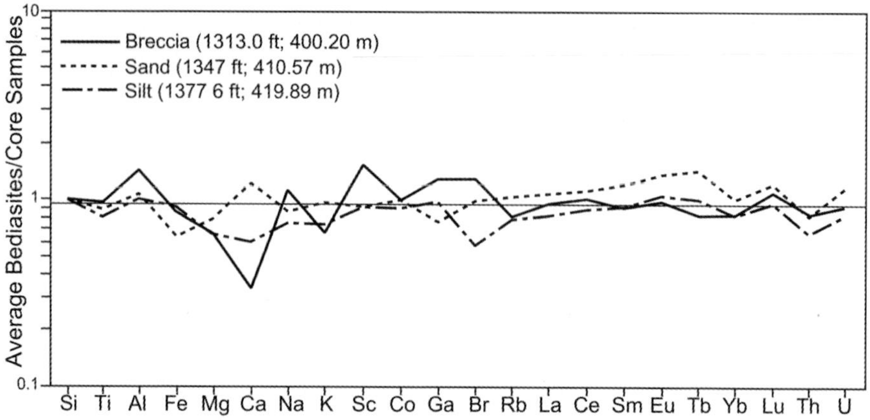

Fig. 6.39. Average composition of 32 bediasite samples compared with composition of three high-silica samples of Exmore breccia from Exmore corehole. Abundances in core samples recalculated on volatile-free basis. Sample numbers are drill depth (ft and m). From Koeberl et al. (1996).

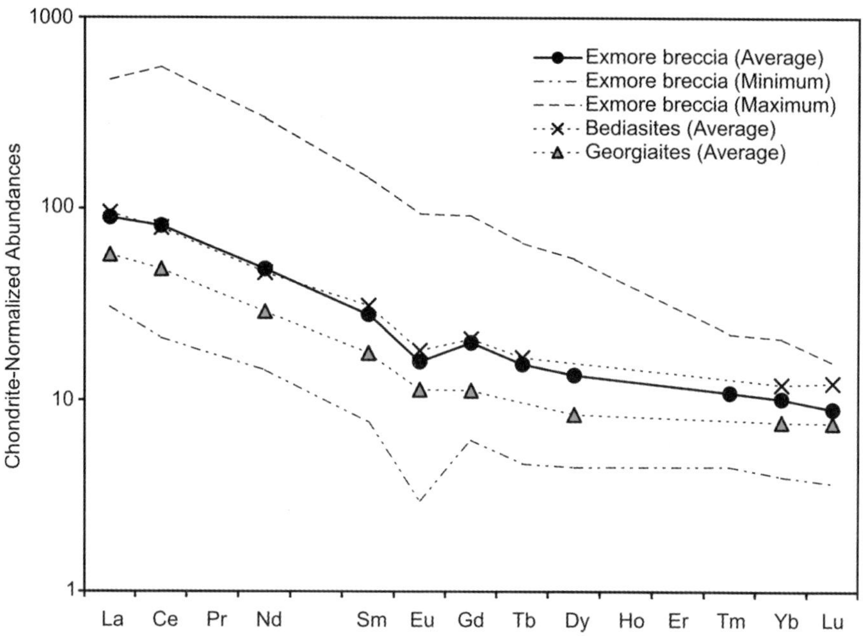

Fig. 6.40. Abundances (average, range) of rare earth elements in samples of Exmore breccia plotted against average abundances in North American tektites (bediasites, georgiaites). Normalization values from Taylor and McLennan (1985).

onable match. This could indicate that: (1) a material similar to that which was involved in forming the breccia also was involved in the tektite production, but did not contain any carbonate; or (2) the rock had to be totally decarbonated and dewatered; or (3) some near-surface rock units, which were not sampled by the breccia, were involved in the tektite production. The third possibility seems to be most likely, based on considerations from other tektite strewn fields (Koeberl 1994). The interelement ratios shown in Table 6.14 indicate that simple devolatilization would not work, as many of these ratios are simply too different. Some lithophile element distributions, though, such as the REE patterns (Fig. 6.40), indicate a general similarity between the breccia and tektite compositions. However, this general similarity is true for almost all upper crustal rocks. The diagram shown in Fig. 6.41, which was used by Albin et al. (2000) to show the relation between North American tektites and various other rock types, can be used to compare these rocks with the Exmore breccia, as well. The Th/Sc and La/Th ratios of the tektites fit relatively well with the breccia values (cf. Table 6.14), but the La/Sc ratios would be skewed to lower values, and do not provide a good fit. This relationship reinforces the conclusion made above that we have not yet identified the exact rocks from which the tektites were produced. The observations of Poag et al. (1994) and Koeberl et al. (1996), however, still stand, and the Chesapeake Bay crater remains the most likely source of the North American tektites.

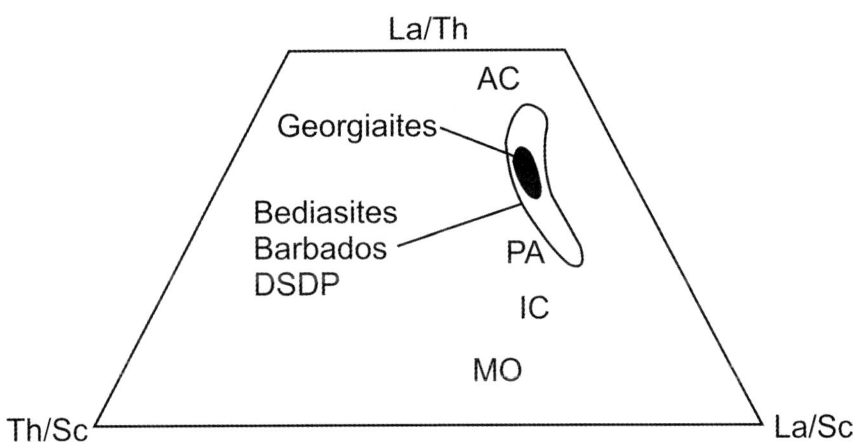

Fig. 6.41. Plot of Th/Sc, La/Th, and La/Sc for North American tektites (georgiaites, bediasites, Barbados, *DSDP*) compared to Archean (*AC*) and post-Archean (*PA*) crust (Taylor and McLennan 1985) and to moldavite (MO) and indochinite (IC). *DSDP* = Deep Sea Drilling Project cores. From Albin et al. (2000).

Table 6.14. Averages and ranges of compositions of Exmore breccia samples compared to North American tektites.

	Average Exmore Breccia	Exmore Std. Deviation	Min. Exmore Breccia	Max. Exmore Breccia	Average All Breccias	Std. Dev. All	Bediasite Average (21)	Georgiaite Average (24)
SiO_2	64.47	10.14	32.65	85.31	60.94	14.23	76.37	81.8
TiO_2	0.58	0.22	0.02	0.92	0.54	0.22	0.76	0.51
Al_2O_3	9.44	2.86	3.12	16.13	8.83	3.12	13.78	11.2
Fe_2O_3	5.92	2.73	0.76	18.61	6.28	2.82	4.21	2.79
MnO	0.04	0.02	0.01	0.09	0.04	0.03	0.03	0.03
MgO	1.25	0.62	0.06	3.82	1.36	0.63	0.63	0.61
CaO	4.91	5.33	0.34	26.78	7.09	9.28	0.65	0.45
Na_2O	1.28	0.41	0.12	2.14	1.16	0.44	1.54	0.94
K_2O	2.92	0.90	1.65	7.24	2.96	0.97	2.08	2.44
P_2O_5	0.37	0.56	0.04	2.93	0.43	0.53		
LOI	8.03	4.69	0.43	22.78	9.62	7.18	<0.01	<0.01
Total	99.21				99.25		100.05	100.77
Sc	8.76	3.54	2.02	17.3	8.33	3.46	13	8.7
V	105	28.1	20	138	115.1	33.8		45
Cr	89.0	31.9	11.4	180	96.1	40.7	49	
Co	10.2	4.60	1.75	23.2	9.77	4.62	13.5	7.5
Ni	21.8	9.3	8.0	56	22.1	8.8	8	7.4
Cu	<2				<2			
Zn	144	136	11	510	136	125		
Ga	12.0	4.3	6.0	22.0	11.9	4.7	11	9
As	9.28	7.99	0.93	42.30	9.89	7.51	1	
Se	0.32	0.25	0.03	1.01	0.31	0.26		
Br	0.65	0.48	0.02	1.50	0.66	0.48	0.1	
Rb	93.0	25.7	57.5	207	94.4	29.3	66	76
Sr	233	94	104	513	236	87	125	163
Y	19.1	5.1	5.0	24.0	20.8	7.2	25	18.2
Zr	244	148	80	776	220	143	230	187
Nb	10.5	2.2	4	15	10.2	2.0		8.1
Ag	0.04	0.02	0.01	0.09	0.04	0.02		
Sb	0.52	0.22	0.28	1.30	0.58	0.28	0.05	
Cs	3.01	1.32	0.81	7.84	2.90	1.37	1.8	1.74
Ba	291	94	51	554	283	94	470	572
La	33.0	28.6	11.3	173	31.9	25.5	35	21.1
Ce	77.8	89.3	20.2	527	74.7	79.1	76	46.2
Nd	34.5	34.8	10.2	212	33.4	31.0	33	20.6
Sm	6.46	5.66	1.78	33.5	6.35	5.08	7.2	4.07
Eu	1.39	1.39	0.26	8.17	1.38	1.25	1.58	0.99
Gd	6.11	4.69	1.90	28.1	6.13	4.32	6.4	3.44
Tb	0.90	0.66	0.27	3.85	0.89	0.60	0.97	
Dy	5.16	3.37	1.70	20.9	5.09	3.08		3.22
Tm	0.39	0.13	0.16	0.79	0.39	0.13		
Yb	2.52	0.97	0.99	5.21	2.44	0.93	3	1.91
Lu	0.34	0.10	0.14	0.60	0.33	0.10	0.47	0.29
Hf	5.54	2.24	0.89	12.2	4.90	2.39	6.7	4.64
Ta	0.66	0.21	0.13	1.13	0.60	0.23	0.6	0.57
W	1.25	1.04	0.10	5.20	1.20	0.97		
Ir (ppb)	0.4	0.3	0.1	0.7	0.4	0.3	<0.5	<0.5
Au (ppb)	0.8	0.5	0.1	2.4	0.8	0.7	<0.5	<0.5
Hg	0.07	0.07	0.02	0.36	0.06	0.06		
Th	7.17	3.71	2.62	21.8	6.57	3.59	7.6	5.81
U	2.17	0.90	0.86	5.04	2.18	0.91	2	1.46

Table 6.14. (cont.) Geochemically important ratios for Exmore breccia samples compared to ratios from North American tektites.

	Average Exmore Breccia	Std. Dev. Exmore	Min. Exmore Breccia	Max. Exmore Breccia	Average All Breccias	Std. Dev. All	Bediasite Average (21)	Georgiaite Average (24)
K/U	13275	4309	5152	25388	13132	4494	10400	16712
Zr/Hf	53.9	56.1	16.2	304.5	56.6	53.9	34.3	40.3
La/Th	5.57	5.90	1.40	31.22	6.25	6.32	4.61	3.63
Hf/Ta	8.42	2.05	5.24	13.71	8.04	2.10	11.17	8.14
Th/U	3.59	1.75	0.52	10.69	3.28	1.69	3.80	3.98
La_N/Yb_N	8.32	2.82	6.23	22.44	8.47	2.57	7.88	7.47
Eu/Eu*	0.66	0.17	0.15	0.85	0.67	0.16	0.71	0.81
La/Sc	3.95	2.77	2.10	16.39	4.12	2.68	2.69	2.43
Th/Sc	0.87	0.48	0.37	3.16	0.81	0.44	0.58	0.67
Rb/Cs	35.2	14.6	16.3	75.2	37.8	16.7	36.7	43.7

Data for tektites from Koeberl (1986), Koeberl and Glass (1988), Glass et al. (1995), Albin et al. (2000)

In summary, our geochemical studies revealed that: (1) the major element and trace element compositions of representative samples of Exmore breccia from three cores (Exmore, Newport News, and Windmill Point) show a wide variation, in agreement with their variable carbonate and clay contents and weathering states; (2) crystalline rock samples were not present in sufficient quantity for complete chemical analyses, (the few small clasts recovered from the breccia were used mainly for shock petrographic studies); (3) trace element compositions, especially of the REEs, show some variation, but are in agreement with a sedimentary, upper crustal, post-Archean source for the Exmore breccia; (4) The detection limit of the present analyses was not good enough to clearly discern the presence of a meteoritic component in the Exmore breccia, which would have to be smaller than about 0.5%; and (5) in order to better constrain the source rocks for the North American tektites, it will be necessary to analyze a wider variety of pre-impact target rocks, especially those that were close to the surface, and, especially, to carry out detailed Sr and Nd isotopic studies. This work is in progress.

7 Initial Postimpact Deposits

7.1 Depositional Setting

Postimpact deposits over and near the Chesapeake Bay impact crater consist of a thick (200–>500 m) succession of fossiliferous siliciclastic formations, which range stratigraphically from regionally distributed barren silt and richly fossiliferous late Eocene clay deposits, to a variety of locally distributed Quaternary and Holocene sand units (see Chapter 2; Figs. 2.4, 2.5, 6.2, 7.1, 7.2). Because the impact took place in moderately deep water (~300 m) on the middle continental shelf (Chapter 13; Poag 1997a), normal marine sedimentation resumed immediately after the bolide-generated atmospheric and oceanic perturbations ceased. The first (oldest) postimpact deposit is a *dead zone* at the top of the Exmore breccia; a thin (19–49 cm) layer of dominantly clayey silt, which appears to lack indigenous microfossils (Poag 2002b). This dead zone is succeeded conformably by as much as 220 m of typical clay-dominant Chickahominy Formation, which began to fill in the bathymetric lows on the irregular upper surface of the Exmore breccia (Fig.7.1). Chickahominy deposition continued for ~2.1 myr to the end of the Eocene. Thus, the nature and history of the dead zone and the Chickahominy Formation yield crucial information regarding the age of the impact (see Chapter 8) and late Eocene paleoenvironments and biota (see Chapter 13) that evolved in its aftermath.

7.2 Dead Zone

The initial postimpact deposit is a 19- to 49-cm-thick dead zone composed mainly of fine, horizontal, parallel laminae of fine to very-fine sand, silt, and clay, which constitutes the stratigraphically youngest unit above the Exmore breccia (Fig. 6.22; Poag 2002b). The dead zone directly underlies the Chickahominy Formation. The dead zone is best represented at the NASA Langley core site (Fig. 7.3), where its lower boundary can be determined by the presence of the presumed fallout layer (Fig. 6.22). There, the base of the dead zone rests conformably on the layer of pyrite lattices (Fig. 6.24A,B). The fine, horizontal, clay and silt laminae of the dead zone are disturbed in places by horizontal, vertical, and inclined burrows, which are filled with medium to coarse sand and reworked microfossils derived from the underlying Exmore breccia matrix (Fig. 7.3A). Additional reworked microfossils comprise much of the micaceous white sand concentrated in

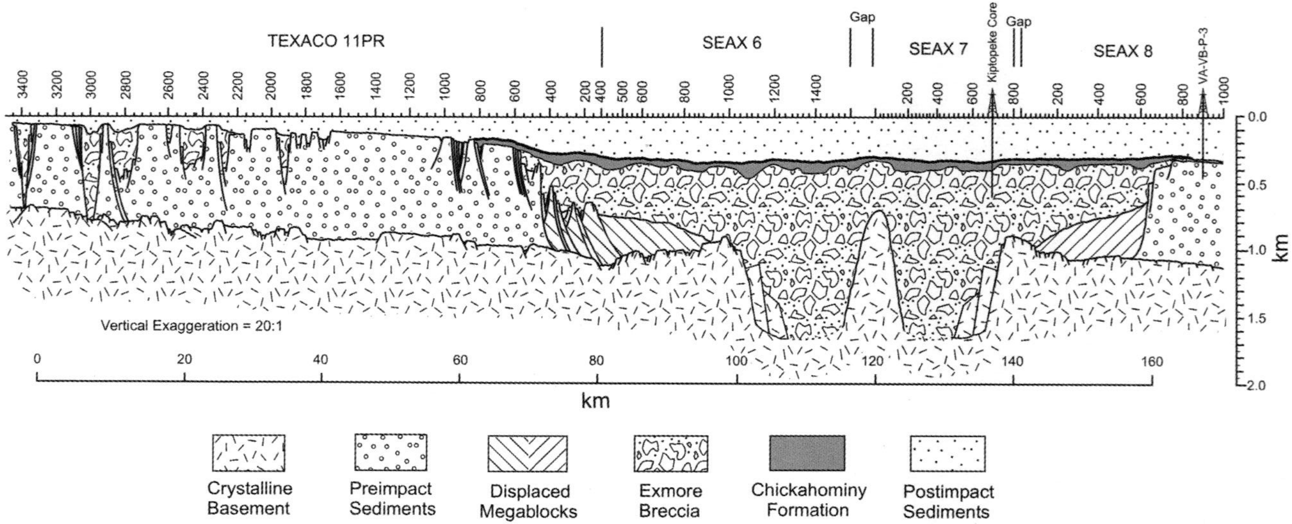

Fig. 7.1. Depth-scaled cross section of Chesapeake Bay primary and secondary craters along composite seismic reflection profile, showing complete stratigraphic succession of preimpact, synimpact, and postimpact deposits. See Figs. 3.3 and 4.21 for profile locations. Compare with Figs. 4.2 and 5.1, and CD-ROM.6.

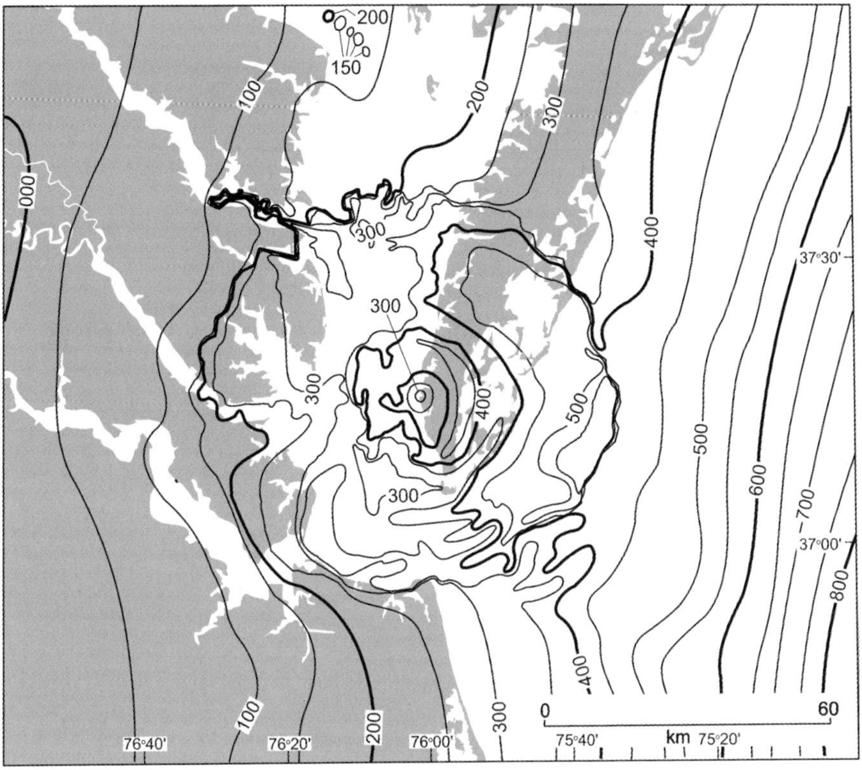

Fig. 7.2. Isopach map of postimpact sedimentary deposits associated with Chesapeake Bay impact crater (contour interval 50 m). Note that thickness variations reflect morphology of basement structure under crater (deposits thin over basement highs, such as peak ring and central peak).

horizontal laminae and lenses (Fig. 7.3B,C). At NASA Langley, the dead zone differs from the silt-rich layer below the pyrite lattices primarily in the more uniform, horizontal, parallel distribution of laminae, and the lack of nodular concentrations of pyrite (pyrite is concentrated instead in the horizontal laminae). In the Bayside core, the dead zone reaches its maximum known thickness (~49 cm). No pyrite-lattice layer has been identified at Bayside, however, so the base of the dead zone can only be approximated by the upward change from multidirectionally inclined, moderately thick lamination to horizontal, thin lamination (see further discussion of the dead zone in Chapter 13).

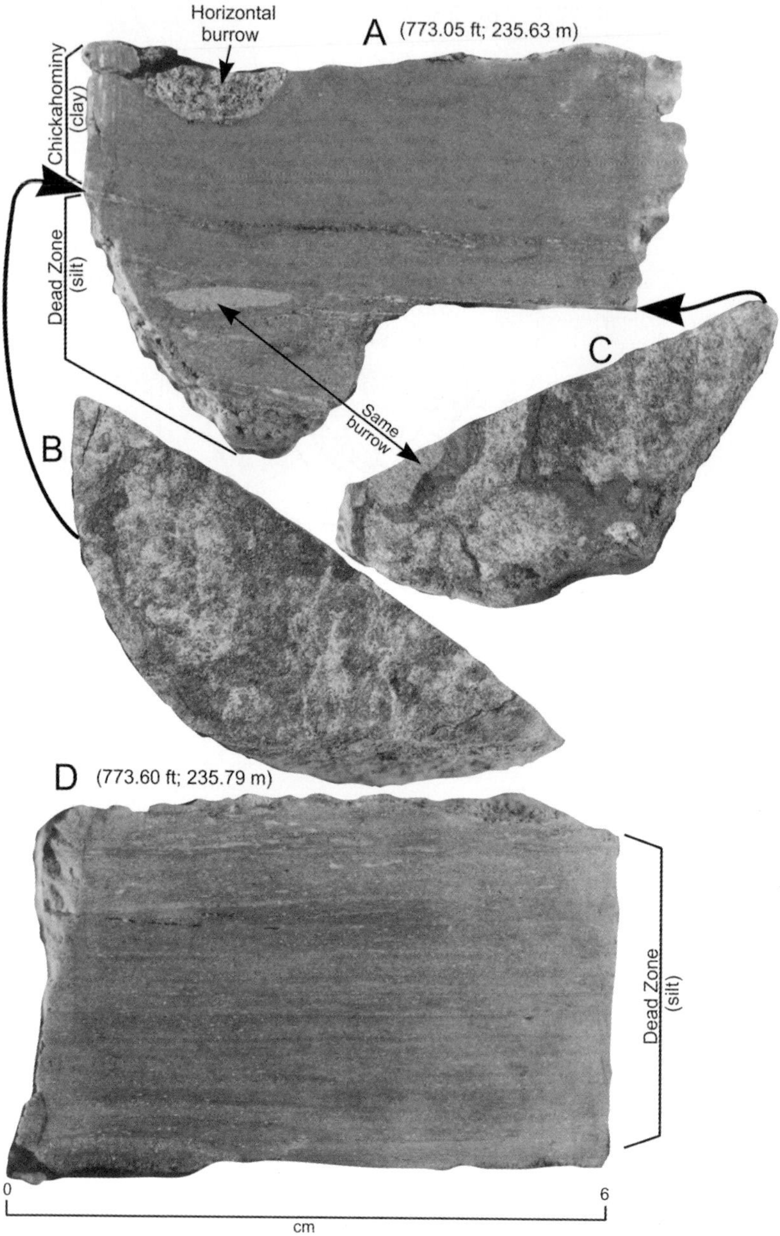

Fig. 7.3. Core segments in dead zone from NASA Langley corehole. *A*, vertically split segment, showing contact with overlying dense clay of Chickahominy Formation; *B*, horizontal split of segment *A*, showing complex laminar lithologies at contact with Chickahominy; *C*, horizontal split of segment *A* taken 0.6 cm below *B*, showing laminae and burrow; *D*, vertically split segment, showing uniform distribution of repetitious, horizontal laminae (sand, silt, clay). Approximate depths shown at tops of segments. See CD-ROM for color version.

7.3 Chickahominy Formation

7.3.1 Lithology of Cores

The most complete lithic records of the Chickahominy Formation come from four sites inside the crater: the North corehole (69.49 m; 228 ft); the Bayside corehole (65.53 m; 215 ft); the Kiptopeke corehole (63.4m; 208 ft); and the NASA Langley corehole (52.70 m; 172.9 ft; Figs. 6.3A,B; CD-ROM.7). At all core sites, fresh cores of the Chickahominy Formation are typically gray-green clay (Fig. 7.4A,B; see CD-ROM for color version) that weathers to yellowish olive brown, and contains variable amounts of finely comminuted glauconite and muscovite (Fig. 7.4C). The clay is silty to sandy, richly fossiliferous, and commonly displays fine to coarse (often faint) lamination (Fig. 7.4C). The biota are mainly marine microfossils (benthic and planktonic foraminifera, calcareous nannofossils, bolboformids, ostracodes, dinoflagellates, radiolarians), but also include common to abundant remains or evidence of invertebrates (echinoid spines, solitary corals, thin bivalves, scaphopods, pyritized burrow casts), and vertebrates (fish skeletal debris and teeth; Fig. 7.4A,B). Sediments subjacent to the upper boundary of the Chickahominy Formation are usually intensely burrowed; those near the lower boundary are moderately burrowed. Larger burrows are filled with coarser material (sand) than the Chickahominy itself (clay), and can be identified as far as 2 m into the Chickahominy. Burrows at the top of the Chickahominy are filled with glauconitic quartz sand and microfossils reworked downward from the overlying Oligocene Delmarva unit or Old Church Formation (Fig. 7.5A). At the base of the Chickahominy Formation, the smallest, most abundant burrows are filled with framboidal pyrite. The largest burrows in this basal interval are filled with quartz sand and mixed microfossil assemblages reworked upward from the Exmore breccia (Fig. 7.5B,C). The presence of the sand-filled burrows causes the upper and lower sediments in the Chickahominy section to fracture and crumble upon drying (Fig. 7.6C,D), in contrast to the dense, massive character maintained throughout most of the remainder of the unit (Figs. 7.4A,B; 7.5C; 7.6A,B).

7.3.2 Expression on Downhole Geophysical Logs

Downhole spontaneous-potential (SP) logs from five of the seven coreholes drilled through the Chickahominy Formation (both inside and outside the crater) show that the Chickahominy is notably less permeable (low SP values) compared to the units that bound it (Figs. 7.7A,B; CD-ROM.7). Where both short normal (16") and long normal (64") resistivity curves are available, the two curves track each other closely, indicating no invasion of the mud filtrate, another indication of low permeability (Fig. 7.7B). The Chickahominy sections at NASA Langley and

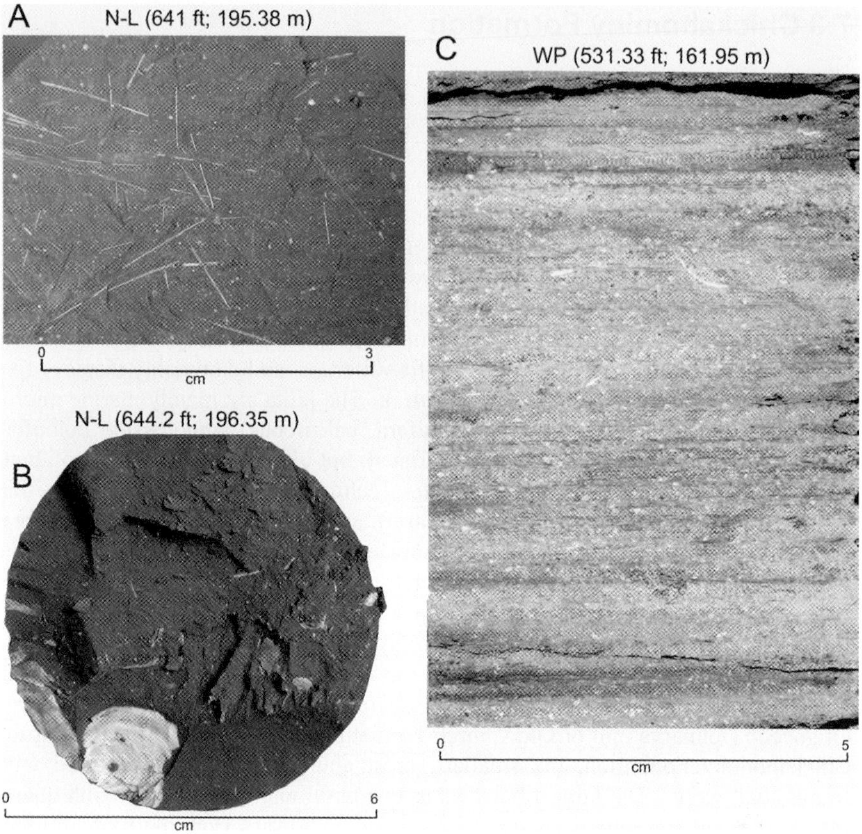

Fig. 7.4. Photographs of Chickahominy Formation illustrating lithic characteristics. *A*, laminar surface of horizontally split fresh core segment showing needle-like echinoid spines distributed parallel to laminae and scattered white spots representing benthic foraminiferal tests (from NASA Langley corehole); *B*, laminar surface of horizontally split fresh core segment showing echinoid spines, thin-shelled clams, and foraminifera (from NASA Langley corehole); *C*, segment of vertically split core (photographed after ten years of dry storage) showing laminated interval within Chickahominy Formation (from Windmill Point corehole). See CD-ROM for color version of this figure.

Exmore are exceptions, however. At NASA Langley, the SP log indicates greater permeability than the underlying Exmore breccia, and at Exmore the upper and lower sections of the Chickahominy have SP values essentially identical to their bounding units.

At all seven of these core sites, the Chickahominy Formation can be partitioned into four principal lithic subunits on the basis of log-defined relative permeability (Figs. 7.7A,B). At each site inside the crater, subunit SP-1 (at the base) is characterized by the lowest permeability. At Exmore, North, Bayside, and Kiptopeke, subunit SP-2 is characterized by gradually upward-increasing permeabilities. At Windmill Point and NASA Langley, on the other hand, the increase is abrupt,

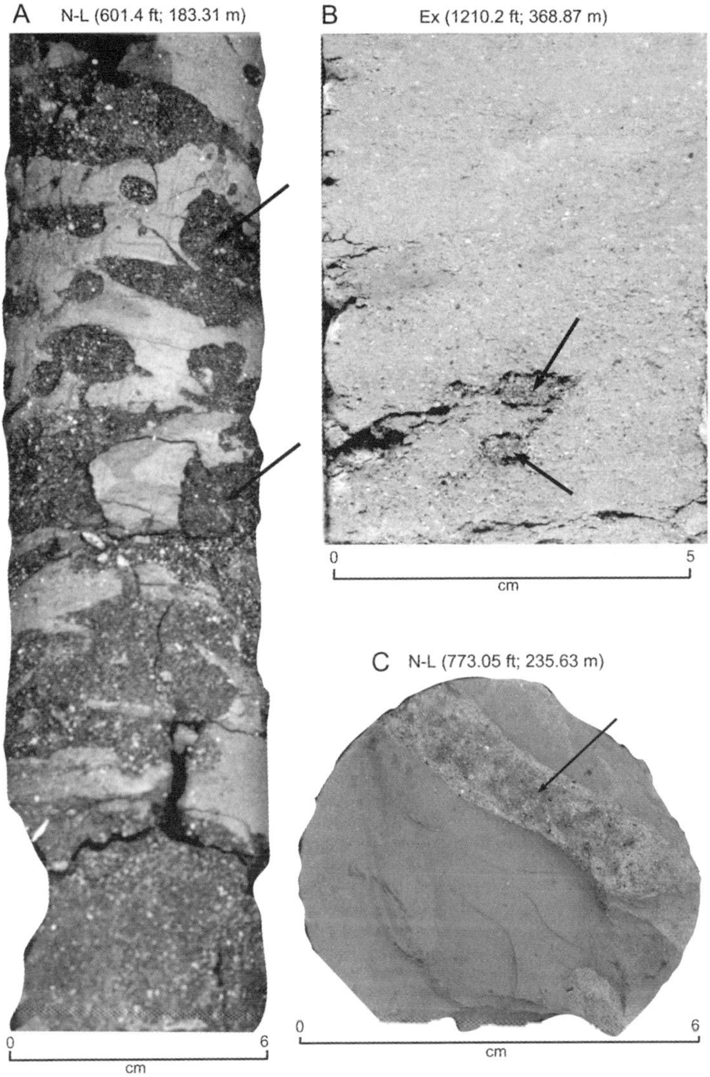

Fig. 7.5. Photographs of core segments showing burrows in upper and lower parts of Chickahominy Formation. *A*, intensely burrowed clay (light-colored) in fresh whole core of upper part of Chickahominy Formation at NASA Langley core site. Burrows filled with greenish-black, glauconitic quartz sand derived from overlying Delmarva beds (lower Oligocene); *B*, Burrows (arrows) in dried out clay of lower part of Chickahominy Formation (vertically split weathered core) at Windmill Point core site. Sand in burrows derived from underlying Exmore breccia; *C*, well-defined sand-filled burrow parallel to bedding plane (horizontal) in lower part of Chickahominy Formation (whole fresh core) at NASA Langley core site; burrow-fill derived from underlying Exmore breccia. See CD-ROM for color version of this figure.

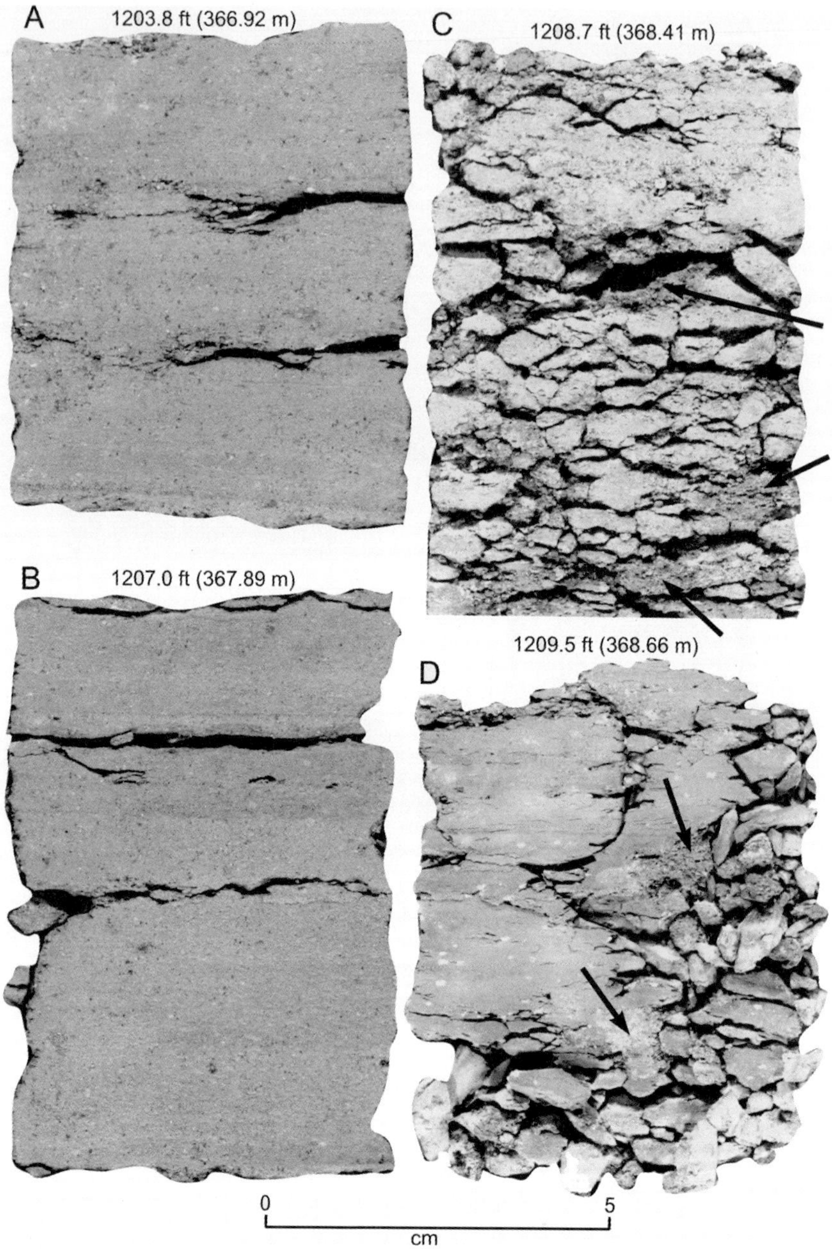

Fig. 7.6. Photographs of weathered split core segments from Chickahominy Formation overlying Exmore breccia at Exmore core site. *A*, *B*, segments typical of hard, finely laminated clay that makes up bulk of Chickahominy at this site; *C*, *D*, segments displaying sand-filled burrows (arrows) near base of Chickahominy at this site; burrows cause weathered cores to fracture and crumble. See CD-ROM for color version of this figure.

reaches highest values for those two coreholes, and then tapers off upward, before declining steeply toward the top (Figs. 7.7A,B). At Newport News, subunit SP-2 is only weakly developed. Subunit SP-3 is more permeable than SP-2 at Exmore, North, Bayside, and Kiptopeke, but less permeable than subunit SP-2 at NASA Langley, Newport News, and Windmill Point. In subunit SP-4, permeabilities decrease relative to SP-3 at Newport News, Windmill Point, North, Kiptopeke, and NASA Langley, but show a relative increase at Exmore (Fig. 7.7B) and Bayside (Fig. 7.7A). A fifth subunit (SP-5) at the top of the formation can be recognized only at Newport News and Kiptopeke. In subunit SP-5, permeability decreases distinctly at its base, then increases appreciably up-section.

At four intracrater sites (NASA Langley, North, Bayside, and Kiptopeke), downhole gamma-ray (GR) logs, which reflect mainly the relative amount of clay and(or) glauconite in the Chickahominy Formation, provide a somewhat stronger definition of downhole lithic changes than do the SP curves (Fig. 7.7A). The GR curves at all four of these sites indicate a fivefold subdivision (A–E) of the Chickahominy. The upward succession of relative GR values, like that of the SP values, is closely similar at North, Bayside, and Kiptopeke. The basal GR subunit (GR-A) displays lowest values at these three sites, but in stark contrast, GR-A gives highest values at NASA Langley. Subunit GR-B shows upward increasing values at North and Kiptopeke, uniformly slightly higher values than GR-A at Bayside, and uniformly much lower values than GR-A at NASA Langley (Fig. 7.7A). In subunit GR-C, the gamma-ray values continue to increase upward at NASA Langley, North, and Bayside, but decrease slightly before increasing again at Kiptopeke. In subunit GR-D, values decrease at all sites relative to GR-C. Maximum GR values are reached at the top of the Chickahominy Formation in subunit GR-E at all four core sites (Fig. 7.7A).

Correlations between SP and GR subunits are complex within individual coreholes, as well as between different coreholes. The most consistent correlation is between SP-3 and GR-C, whose upper and lower boundaries coincide (or nearly coincide) at four sites (Fig. 7.7A). There also is good correlation between SP-1 and GR-A and between SP-2 and GR-B at North, Bayside, and Kiptopeke, but these correlations break down at NASA Langley. At the top of the Chickahominy section, SP-4 is equivalent to GR-D and E, with the exception of Kiptopeke, where SP-5 correlates with the top of GR-E.

Thus the geophysical logs indicate that though the Chickahominy Formation may appear macroscopically to be a lithically uniform deposit, it is far from it. Shifts in sediment sources, sediment composition, accumulation rates, and variable diagenetic processes have created a much more complex lithic unit than we expected to encounter.

For example, the low permeability at the base of the Chickahominy Formation (subunit SP-1) is probably due to extensive diagenesis in this section (pyritization and silicification; see further discussion in Chapter 13), in spite of the prominence of sand-filled burrows, which otherwise should have increased permeability. The unusually low permeability at the top of the two extracrater cores (Windmill Point and Newport News) also may be a response to diagenetic changes.

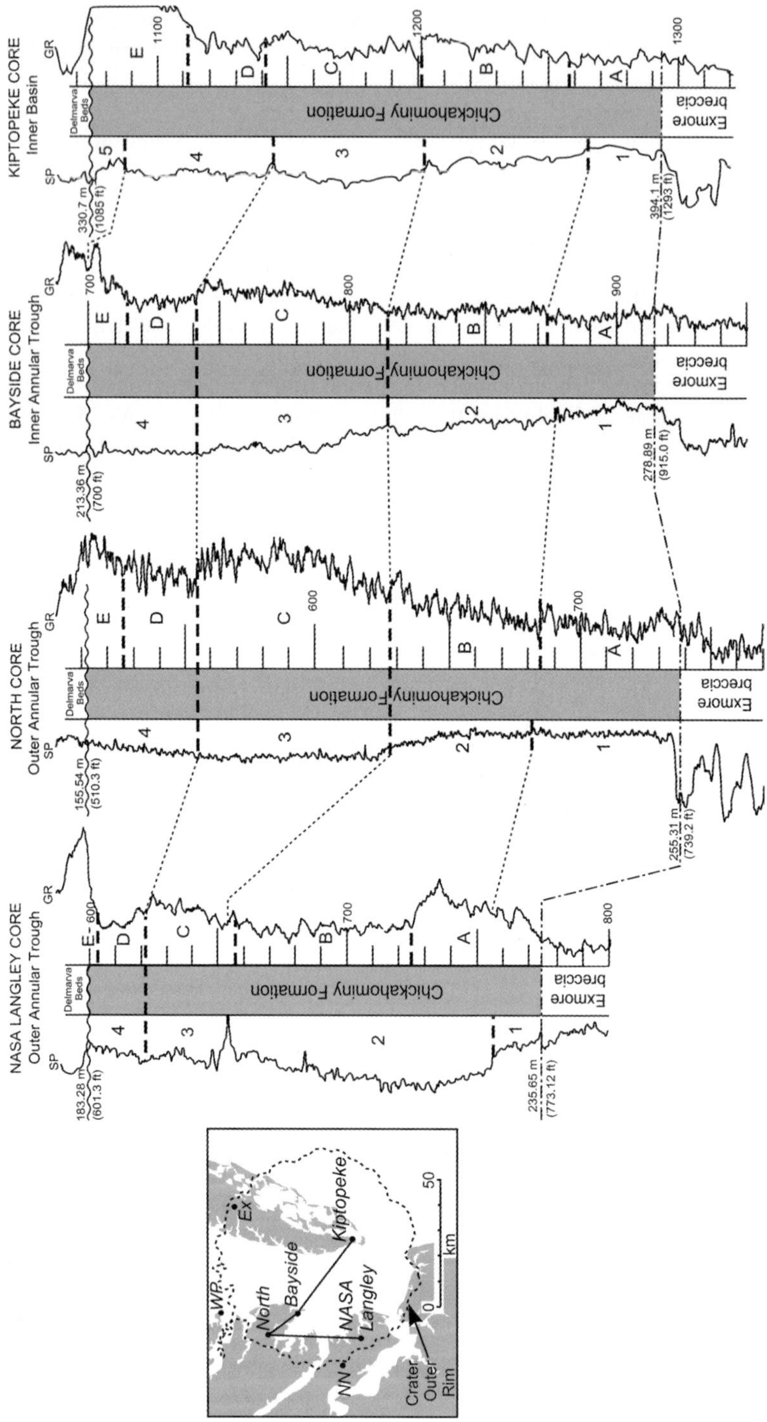

Fig. 7.7A. Expression of Chickahominy Formation on downhole geophysical logs from four core sites inside Chesapeake Bay impact crater. SP = spontaneous potential; GR = gamma-ray. See text for further discussion and CD-ROM.7 for full-scale logs.

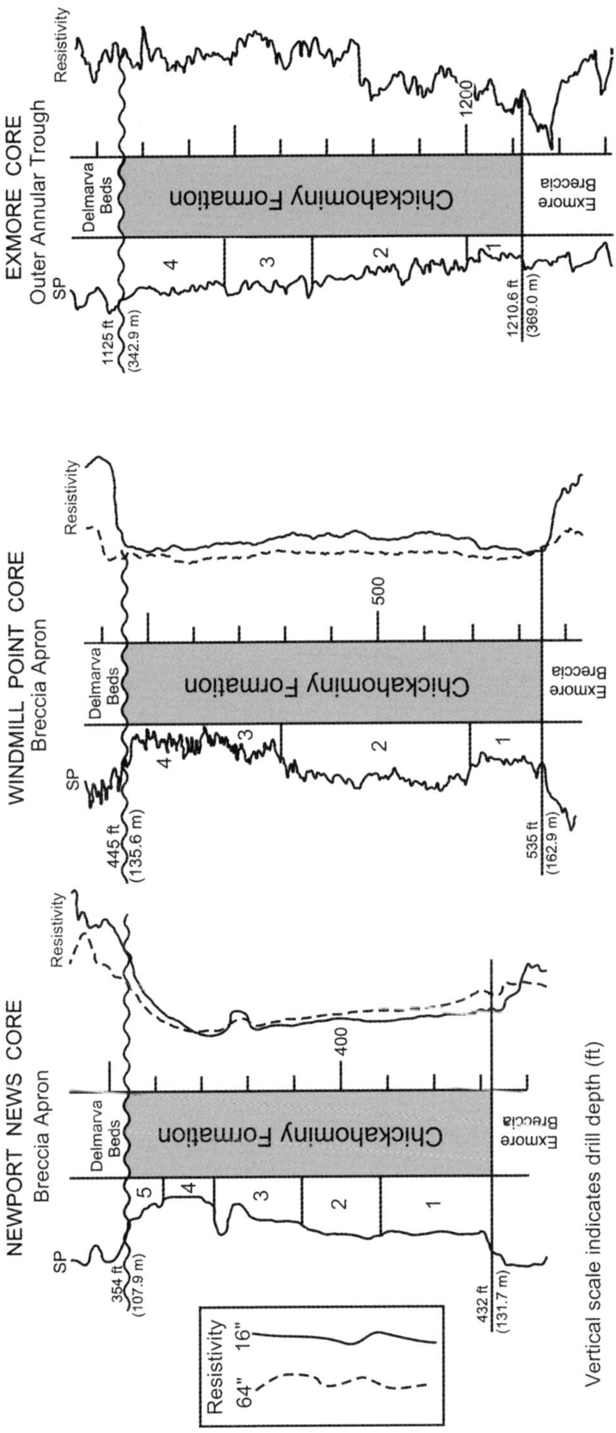

Fig. 7.7B. Expression of Chickahominy Formation on downhole geophysical logs from two core sites outside Chesapeake Bay impact crater and one core site inside crater. *SP* = spontaneous potential; *GR* = gamma-ray. See text for further discussion and CD-ROM.7 for full-scale logs.

During deposition of subunit SP-2, the sedimentary regime at Windmill Point and NASA Langley provided coarser grains (greater permeability) than elsewhere in the crater, and glauconite was probably the primary sand constituent (indicated by the unusually high gamma-ray values in subunit GR-A at NASA Langley). In contrast, nearer the center of the crater (at North, Bayside, and Kiptopeke), glauconite is a more prominent constituent in the upper part of the Chickahominy Formation.

7.3.3 Seismic Signature

Integration of coring and downhole geophysical records allows us to correlate the Chickahominy Formation precisely with its reflection signature on the seismic profiles. Normally, a significant impedance contrast exists between the relatively dense, weakly permeable Chickahominy clay and the loosely consolidated, strongly permeable sands of the overlying Oligocene Delmarva beds (Powars et al. 1992) and(or) the Miocene Calvert Formation (Fig. 2.5). This impedance contrast produces an easily recognized high-amplitude reflection at the upper boundary of the Chickahominy Formation (Figs. 4.3A,B, 4.5A,B, 4.7A,B, 4.9A,B, 4.13, 4.15, 4.18, 4.19, 4.23–4.26, 4.32), which can be traced over the entire crater and for a short distance outside the crater rim. The lower boundary of the Chickahominy also is characterized by a strong impedance contrast and a resultant high-amplitude seismic reflection (PS), where it contacts the underlying unconsolidated sands of the Exmore breccia (Figs. 4.3A,B, 4.9A,B, 4.13, 4.23A,B, 4.25A,B, 4.26A,B, 4.32). Even on profiles where its boundary reflections are weak, the large number of intersections between profiles assures correct stratigraphic placement of both the upper and lower boundaries of the Chickahominy Formation. In the thickest portions of the Chickahominy Formation, as displayed on the seismic profiles, internal reflections indicate the probability of meter-scale bedding. In short, the seismostratigraphic signature of the Chickahominy is easy to recognize and to trace over the crater. Therefore, its present structure (Fig. 7.8) and thickness (Fig. 7.9) can be accurately mapped.

7.3.4 Geometry

The structure, morphology, and distribution of the Chickahominy Formation have been influenced strongly by the original irregular geometry of the upper surface of the Exmore breccia, and by the long-term subsidence differential between the unconsolidated, water-saturated impact breccia inside the crater and the semiconsolidated preimpact sedimentary column outside the crater. Differential subsidence is partly responsible (along with original bathymetric differences between the crater basin and its peripheral lithotopes) for a much thicker section of Chickahominy clay inside the crater than outside the crater (Fig. 7.9). In addition, continued differential subsidence during the roughly 34 million years of Oligocene-Holocene

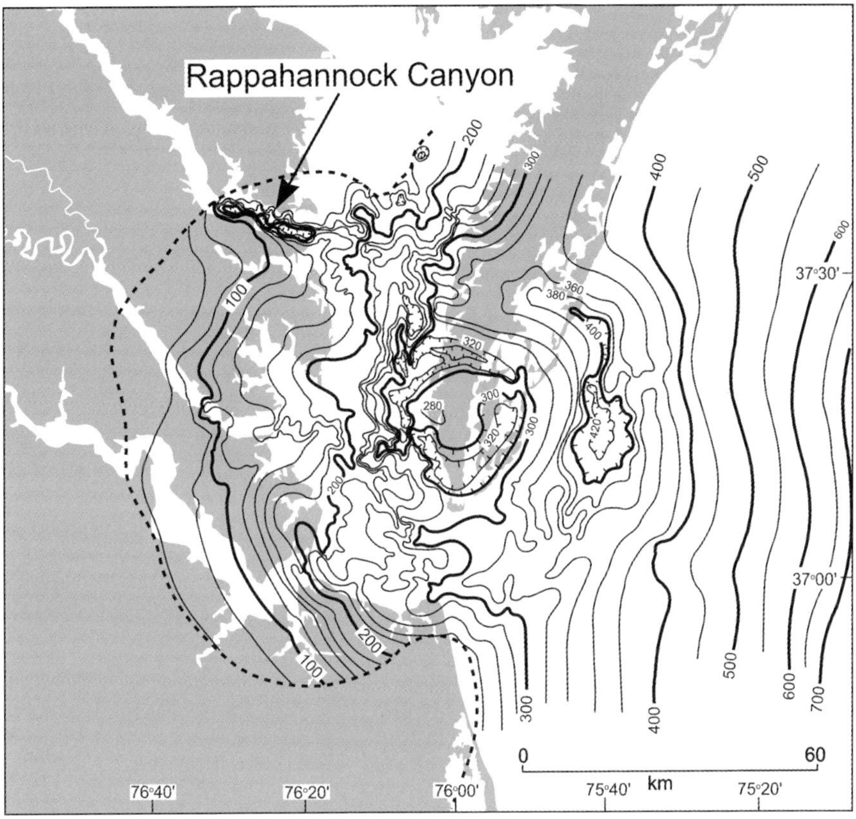

Fig. 7.8. Structure map of upper surface of Chickahominy Formation (contour interval 20 m). Dashed line marks approximate shoreward limit of formation.

time has caused the Chickahominy Formation to sag irregularly over the crater rim in concert with differential compaction of the underlying breccia. That is, the Chickahominy Formation thickens and sags as it crosses into the annular trough and inner basin, just as the Exmore breccia does (Table 7.1; Figs. 4.3A,B, 4.5A,B, 4.9A,B, 4.11, 4.13, 4.14). Likewise, the Chickahominy mimics the geometry of the underlying breccia body by arching up over the peak ring and central peak (Figs. 4.22–4.26, 4.32, 4.34).

Inside the crater, the Chickahominy Formation is ~20–>220 m thick, and averages ~100–120 m (Fig. 7.9). The thickness varies greatly, because the unit fills various pits and troughs in the upper surface of the underlying Exmore breccia, which have been accentuated by the postimpact differential compaction (Figs. 4.7A,B, 4.13, 4.22, 4.26A,B, 4.32). In general, the Chickahominy Formation is thickest where the underlying Exmore breccia is thickest (where the basement surface is deepest) and thins where the Exmore breccia is thinnest (where the basement shallows). The Chickahominy Formation thickens from 20 m to >90 m where it crosses the western part of the outer rim; from 20 m to >150 m across the

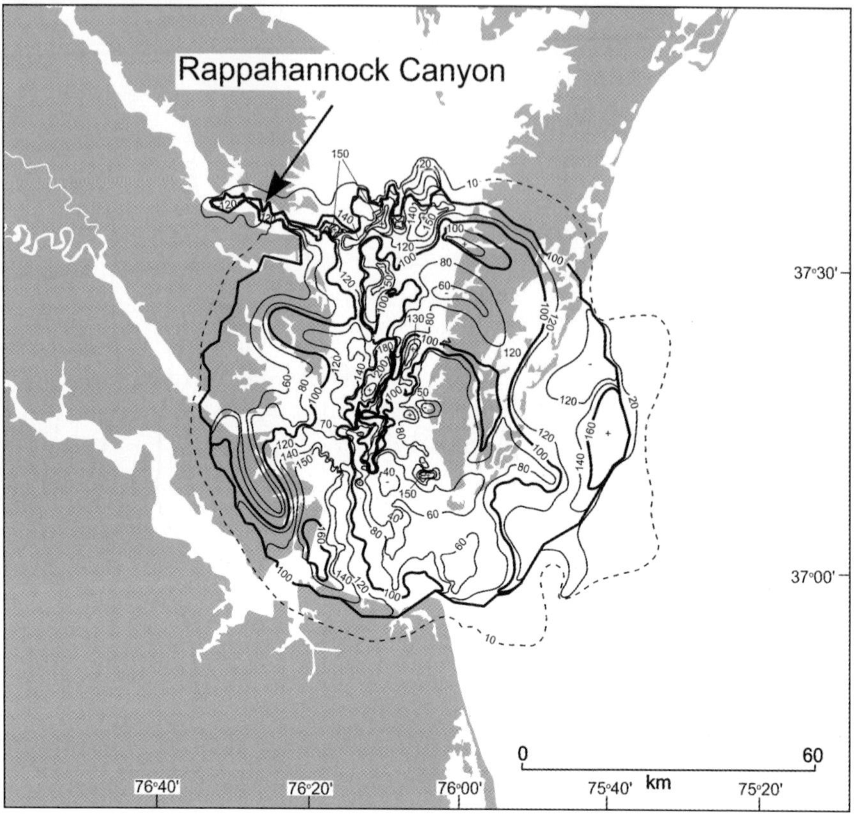

Fig. 7.9. Isopach map of Chickahominy Formation (contour interval 20 m). Note that thickness variations reflect morphology of basement structure under crater (formation thins over basement highs such as peak ring and central peak).

northern part of the outer rim; and from 20 m to >160 m across the eastern and southern parts of the rim (Fig. 7.9; Table 7.1). The thickest measurable part of the formation (>220 m) occupies the western sector of the inner basin. We have no seismic data for the eastern sector of the inner basin. The Chickahominy Formation thins over broad areas of the western, northern, and southern sectors of the annular trough, being thinnest over the southwestern crest of the peak ring and over the central peak. More than 120 m of Chickahominy sediments occupy Rappahannock Canyon (Figs. 4.7A,B, 4.20A, 7.9). The Chickahominy thins rapidly to <10 m within a few kilometers outside the crater outer rim, and is too thin to trace beyond that point on the seismic profiles. The formation is <10 m thick in most of the non-cored boreholes that have penetrated it outside the crater.

Table 7.1. Elevation, sag, and thickness data (from seismic profiles) for Chickahominy Formation where it crosses outer rim of Chesapeake Bay impact crater.

Seismic Profile #	Elevation Outside Rim [1][mbsl]	Elevation Inside Rim [mbsl]	Amount of Sag [m]	Thickness Outside Rim [m]	Thickness Inside Rim [m]	Total Thickness Increase	Percent Thickness Increase
S-2	100	175	75	25	110	85	340
S-3	85	120	35	10	90	80	800
T-13-YR	75	120	45	10	90	80	800
S-16	110	190	80	15	80	65	433
S-17	125	170	45	10	100	90	900
N-3	110	170	60	10	70	60	600
T-11-PR	120	220	100	40	100	60	150
T-9-CB-F	120	210	90	40	100	60	150
S-12	140	180	40	20	140	120	600
S-13	150	200	50	15	110	95	633
T-10-RR	180	230	50	10	140	130	1300
T-1-CB	175	200	25	10	90	80	800
S-4	150	180	30	20	100	80	400
S-10	175	220	45	10	130	120	1200
S-11	160	180	20	10	120	110	1100
S-5	220	240	20	10	120	110	1100
S-6	220	270	50	15	70	55	367
S-8	255	310	55	15	60	45	300
S-9	280	320	40	15	100	85	567
S-19	365	390	25	10	120	110	1100
S-22	370	390	20	10	100	90	900
S-25	350	380	30	30	150	120	400
S-27	315	320	5	20	130	110	550
S-1	365	405	40	20	150	130	650
E-3	310	340	30	25	130	105	420
Average	201	245	44	17	108	91	662

[1]mbsl = meters below sea level

7.3.5 Faults and Fault Systems

In addition to producing the thickening, thinning, and sagging of the Chickahominy Formation, differential compaction of the Exmore breccia also has created a series of normal-offset growth faults and fault systems within the postimpact sedimentary section. These faults break the Chickahominy Formation into discrete fault blocks (Figs. 4.5A,B, 4.9A,B, 4.11, 4.13, 4.20A,B, 4.22–4.26, 4.32). The NASA Langley borehole crossed a postimpact compaction fault, which slices through the Chickahominy Formation at 229.9 m (754.4 ft; Fig. 7.10). The two most prominent systems of compaction faults are expressed on the seismic profiles as complex intervals of disrupted and offset reflections that form distinct grabens. Because these graben structures are present on almost every seismic profile that crosses the outer rim and(or) the wall of the inner basin, we infer that they represent parts of two nearly continuous concentric graben systems that encircle the crater just inside the outer rim and the peak ring (Fig. 7.11).

In addition to the two continuous concentric graben systems, we have documented (on the seismic profiles) more than 700 individual faults and fault clusters (small grabens, horsts, or single normal faults) scattered in mainly concentric orientations throughout the Chickahominy Formation (Fig. 7.11). Most of these compaction faults also extend upward through part or most of the postimpact stratigraphic section (Figs. 4.5A,B, 4.11, 4.13, 4.20A,B, 4.22–4.26, 4.32). Furthermore, the throw on most faults decreases upsection (Fig. 7.12A–C), indicating that they are growth faults, along which long-term continuous or intermittent movement has occurred. Most faults die out before they reach the near-surface Quaternary deposits, but some continue into the Quaternary section as well, to within 10–15 m of the modern bay floor (Figs. 7.13A,B; Colman and Hobbs 1987; Colman and Mixon 1988).

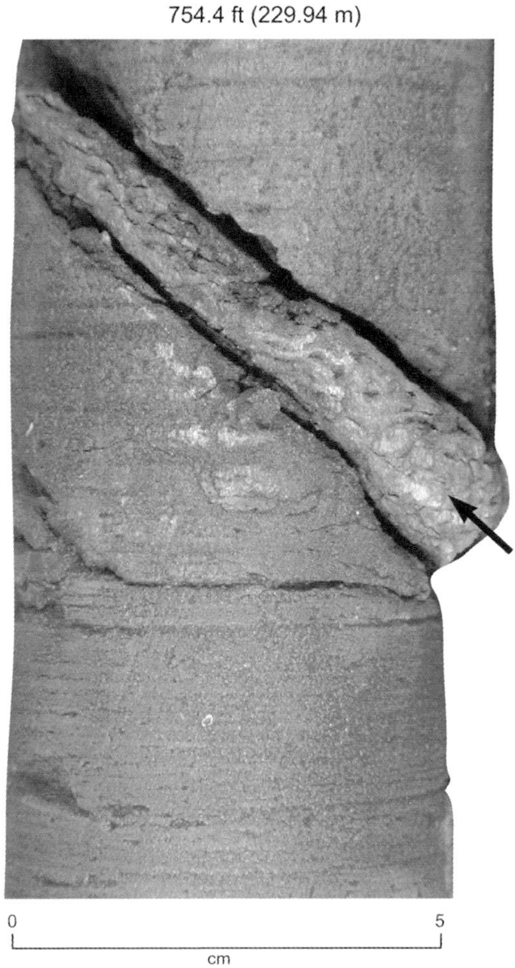

Fig. 7.10. Photograph of core segment, showing minor branch of postimpact compaction fault system, which cuts Chickahominy Formation at 229.94 m (754.4 ft) in NASA Langley corehole. Hanging wall and footwall separated by 1.5-cm-thick layer of pyrite-rich fault gouge. See CD-ROM for color version of this figure.

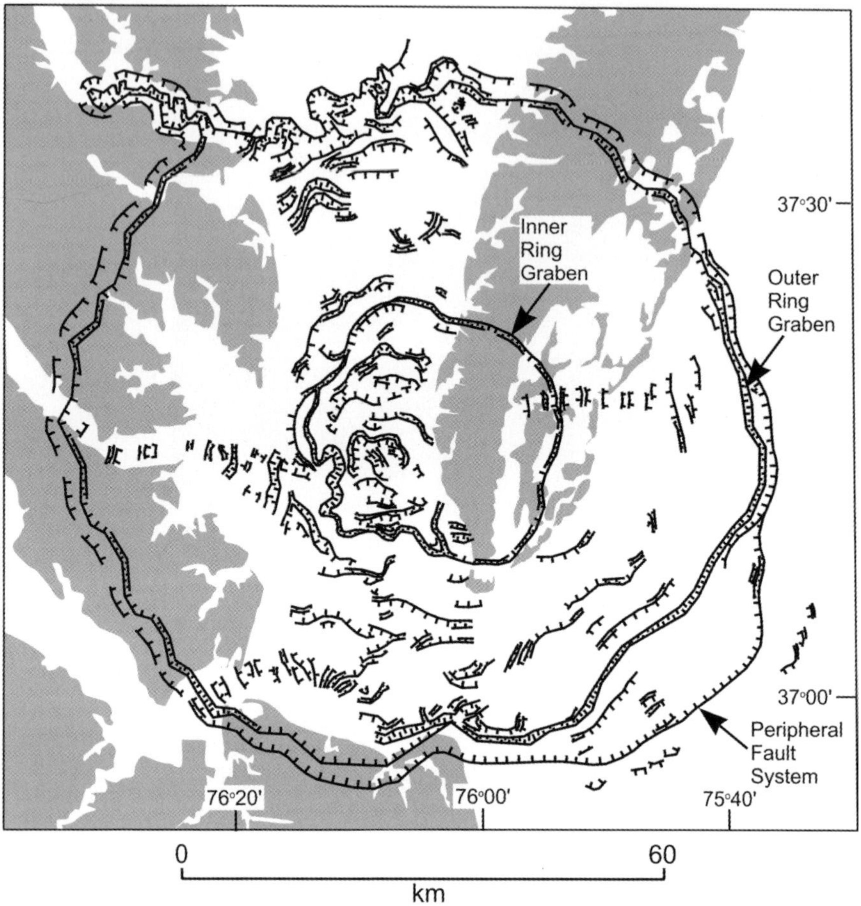

Fig. 7.11. Horizontal distribution of concentric ring-grabens and additional normal faults that offset upper and lower bounding reflections of Chickahominy Formation (mapped on upper surface of Chickahominy Formation). Outer ring-graben marks location of crater outer rim; inner ring-graben marks wall of inner basin. All these faults also extend into postimpact sedimentary section above Chesapeake Bay crater.

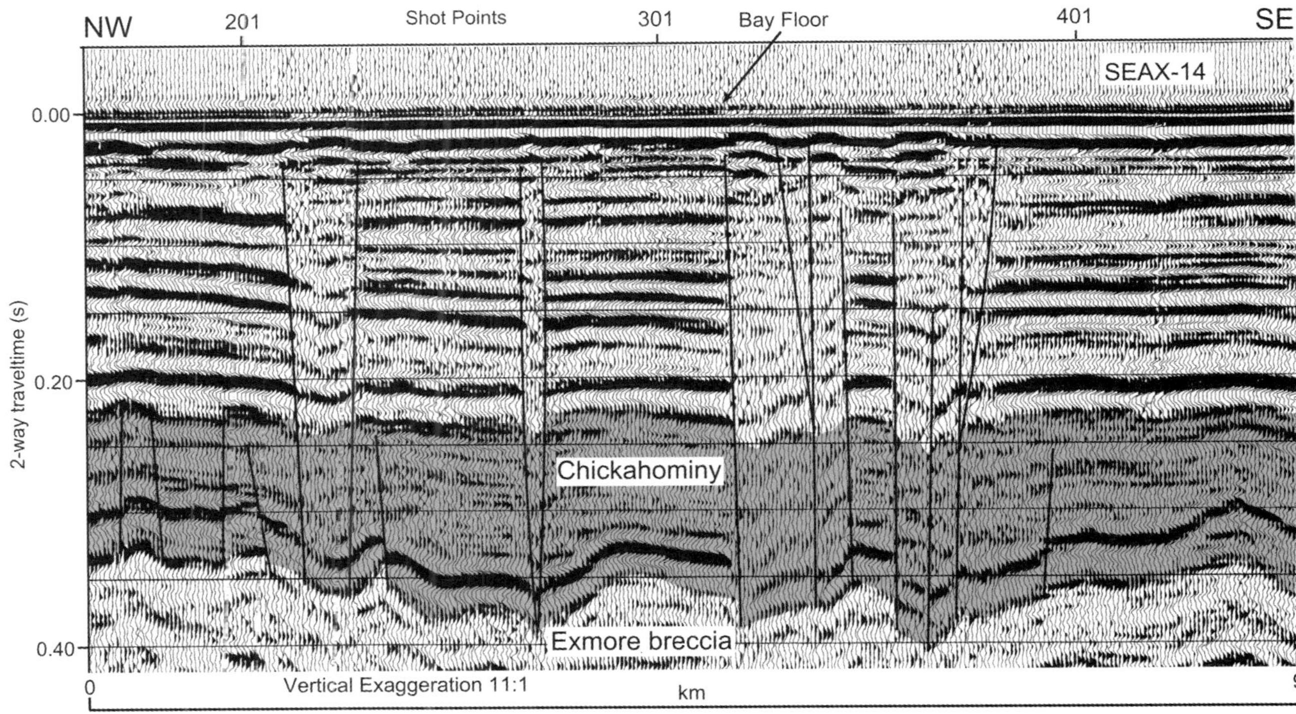

Fig. 7.12A. Interpreted segment of seismic reflection profile SEAX-14 (single channel) across Chesapeake Bay crater, showing incremental displacement along postimpact growth faults. Displacement due to nonuniform differential compaction of underlying Exmore breccia during last 36 myr. Letters at right of each profile segment identify reflections used to demonstrate decreasing upward displacement of strata. Note that some faults nearly reach bay floor.

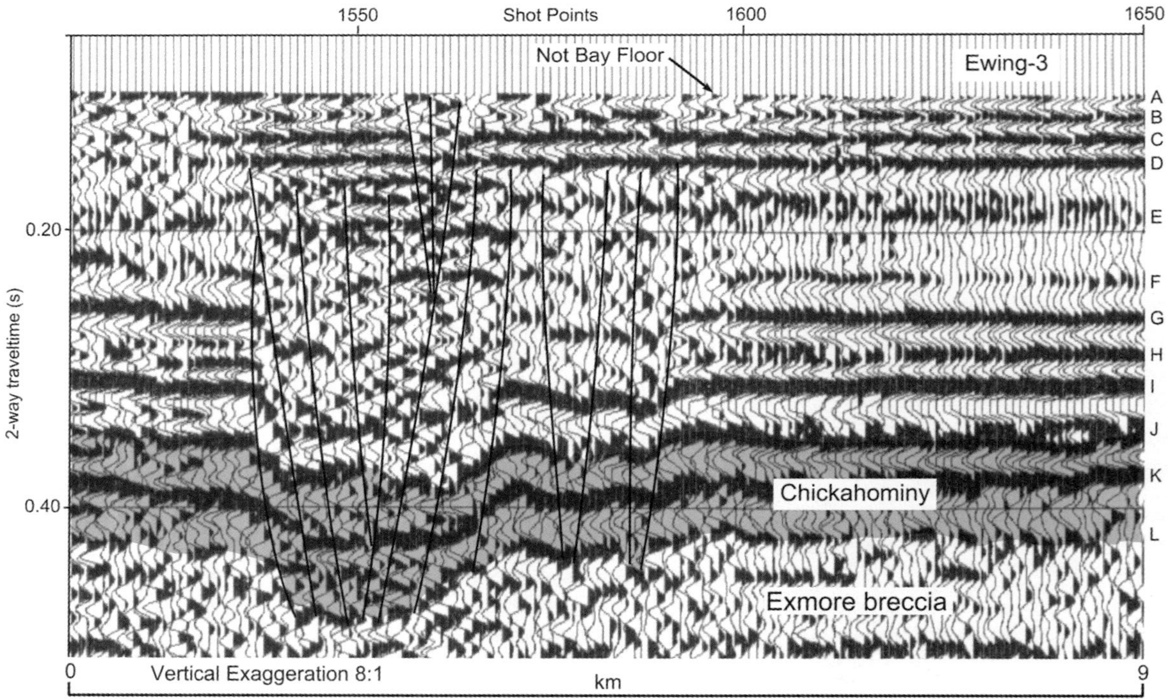

Fig. 7.12B. Interpreted segment of seismic reflection profile *Ewing*-3 (2-channel) across Chesapeake Bay crater, showing incremental displacement along postimpact growth faults. Displacement due to nonuniform differential compaction of underlying Exmore breccia during last 36 myr. Letters at right of each profile segment identify reflections used to demonstrate decreasing upward displacement of strata.

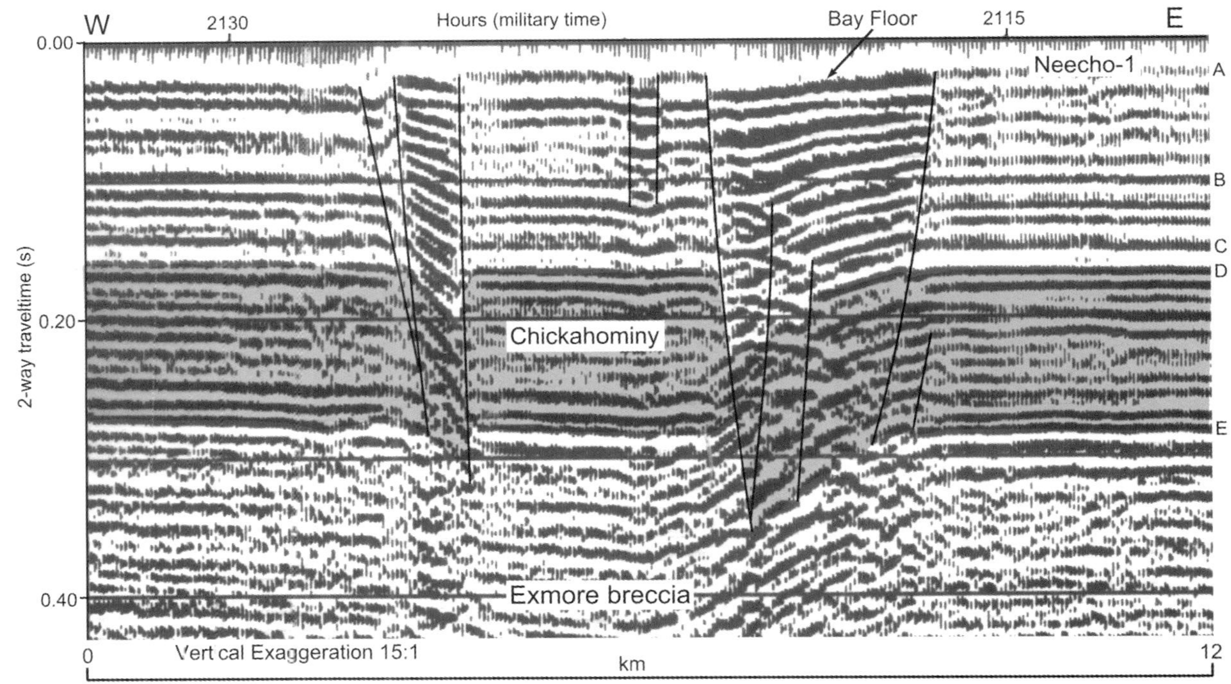

Fig. 7.12C. Interpreted segment of seismic reflection profile *Neecho*-1 (multichannel) across Chesapeake Bay crater, showing incremental displacement along postimpact growth faults. Displacement due to nonuniform differential compaction of underlying Exmore breccia during last 36 myr. Letters at right of each profile segment identify reflections used to demonstrate decreasing upward displacement of strata. Note that some faults appear to reach bay floor.

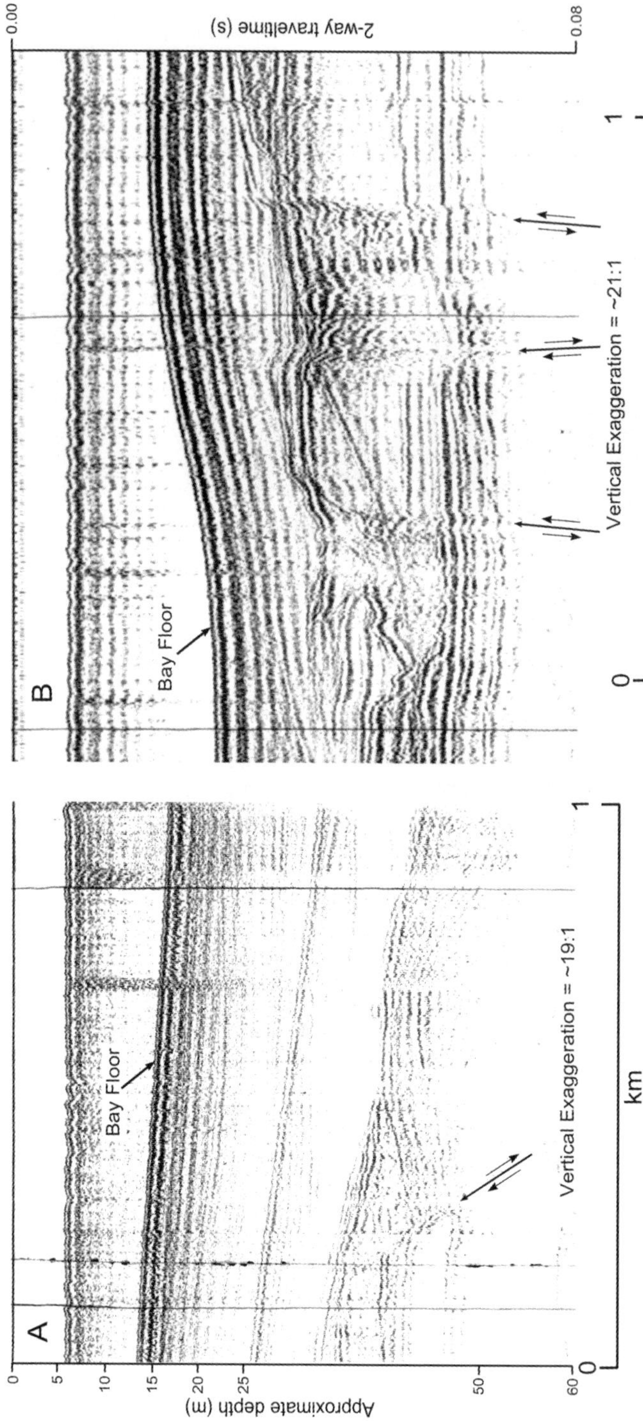

Fig. 7.13A. Interpreted segments of two high-resolution seismic reflection profiles, showing shallow parts of growth faults approaching bay floor above Chesapeake Bay impact crater. Profiles from Colman and Hobbs (1987).

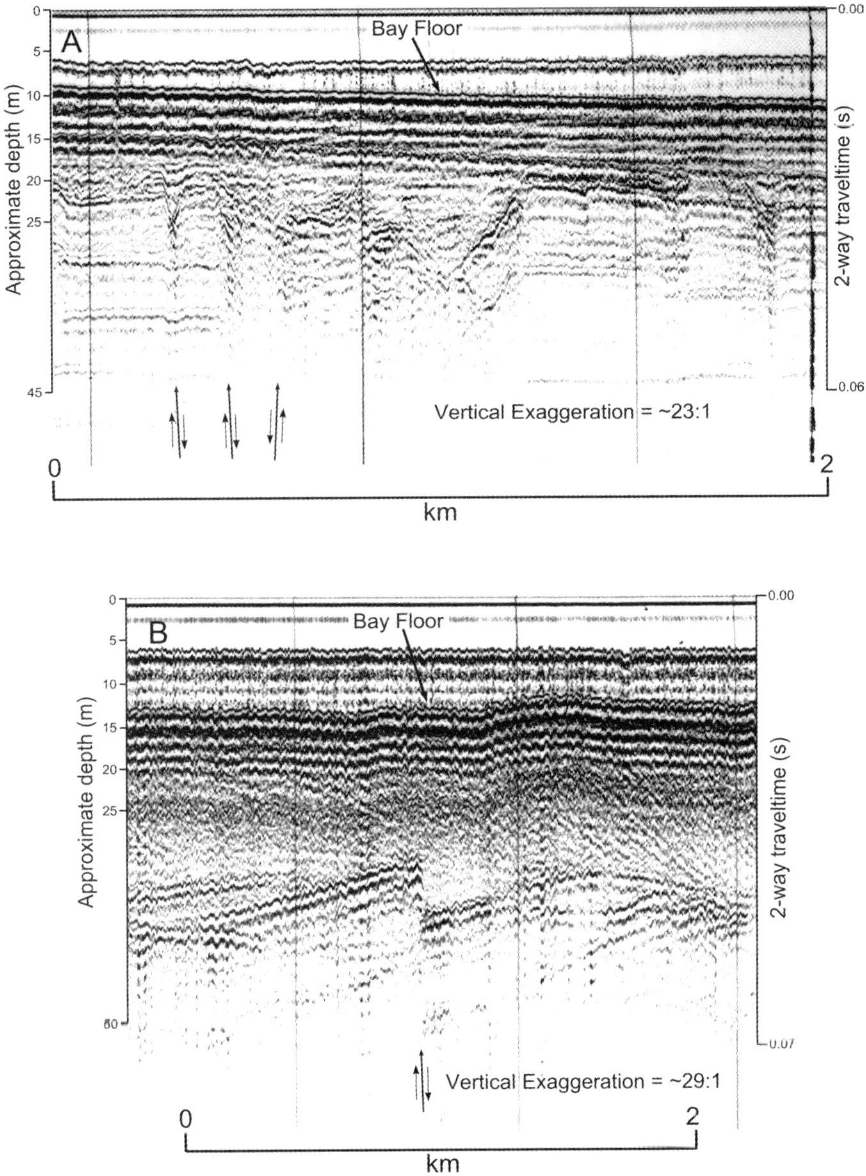

Fig. 7.13B. Interpreted segments of high-resolution seismic reflection profiles, showing shallow parts of growth faults approaching bay floor above Chesapeake Bay impact crater. Profiles from Colman and Hobbs (1987).

8 Age of Chesapeake Bay Impact Crater

8.1 Biochronology

The biochronological age of the Chesapeake Bay impact crater has been determined by documenting the stratigraphic ranges of three groups of microfossils (planktonic foraminifera, bolboformids, and calcareous nannofossils), sampled in two different geographic areas. We compared these ranges to the global geochronostratigraphic framework constructed by Berggren et al. (1995; Fig. 8.1). Sample area one is in and near the crater itself, where Poag and Aubry (1995), Poag and Commeau (1995), and Poag (1997a) first established the detailed planktonic biostratigraphic framework of the Exmore breccia and Chickahominy Formation [Fig. 8.2; see also Poag and Norris (in press) for correlative planktonic foraminiferal biostratigraphy of the Chickahominy Formation in the NASA Langley corehole]. The second sample area for micropaleontological analysis is the New Jersey Continental Slope, where several different investigators have documented microfossil assemblages from impact ejecta cored at DSDP (Deep Sea Drilling Project) Site 612 and ODP (Ocean Drilling Program) Sites 903 and 904 (Fig. 8.3; see Poag and Aubry 1995, for the most recent synthesis and references). The youngest diagnostic foraminifera and nannofossils within the Exmore breccia represent an ~0.8 myr interval (35.2–36.0 Ma) in which the upper part of planktonic foraminiferal Biochron P15 overlaps the lower part of calcareous nannofossil Biochron NP19-20 (Fig. 8.1). This same biochronologic overlap interval is present in the Exmore breccia in all seven coreholes drilled inside and outside the crater, and extends into the lower third of the overlying Chickahominy Formation. The remaining upper two-thirds of the Chickahominy Formation incorporates biochrons P16–P18 (forams) and NP19-20 and NP21 (nannofossils), which represent the final ~1.5 myr of the late Eocene.

Almost identical biostratigraphic relationships are seen at the offshore New Jersey sites. Both the impact ejecta and the immediately overlying pelagic carbonates there belong to the P15/NP19-20 overlap interval (35.2–36.0 Ma), and the remaining offshore upper Eocene section correlates directly with the upper microfossil zones of the Chickahominy Formation (Poag and Aubry 1995; Figs. 8.1–8.3).

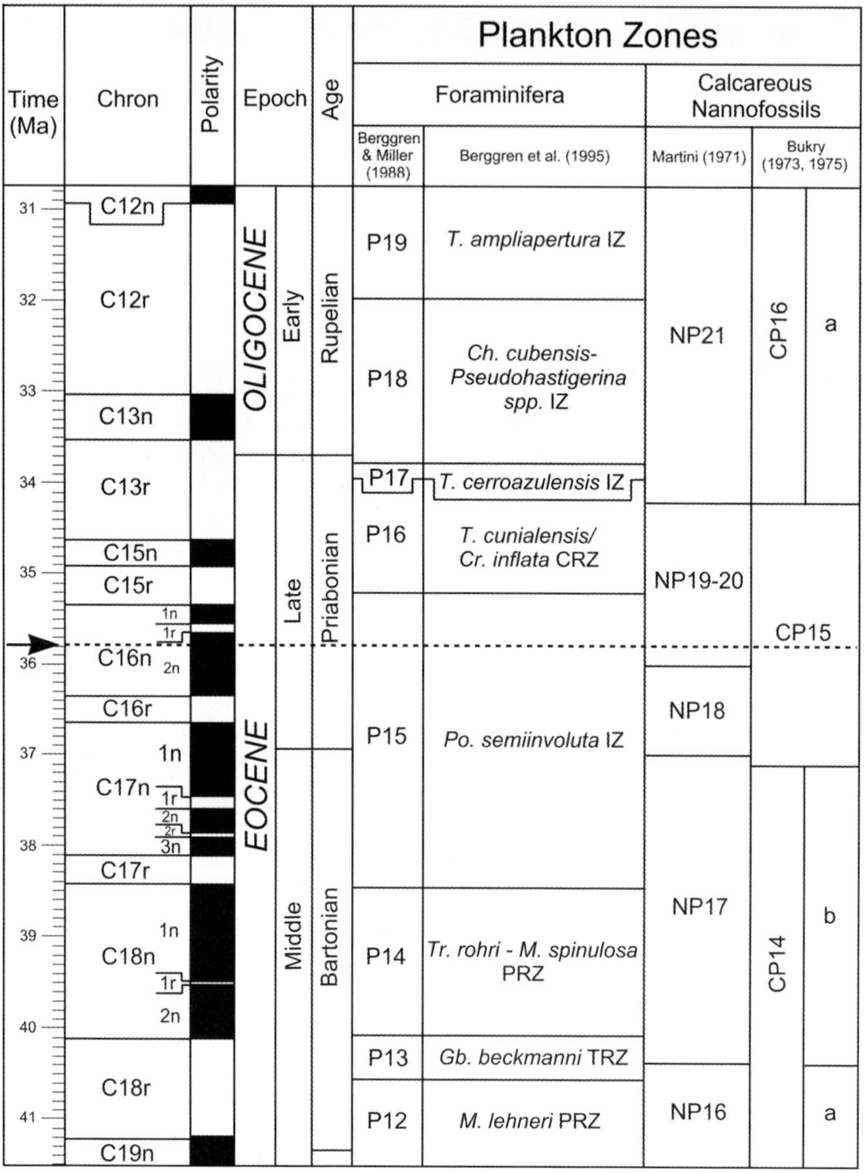

Fig. 8.1. Geochronostratigraphic chart for Eocene and Oligocene epochs used for interpretations in this volume (modified from Berggren et al. 1995). Arrow and dotted horizontal line mark stratigraphic level at which Chesapeake Bay impact took place, as determined by microfossils (planktonic foraminifera, bolboformids, calcareous nannofossils) and magnetostratigraphic analyses of samples from Chesapeake Bay impact crater.

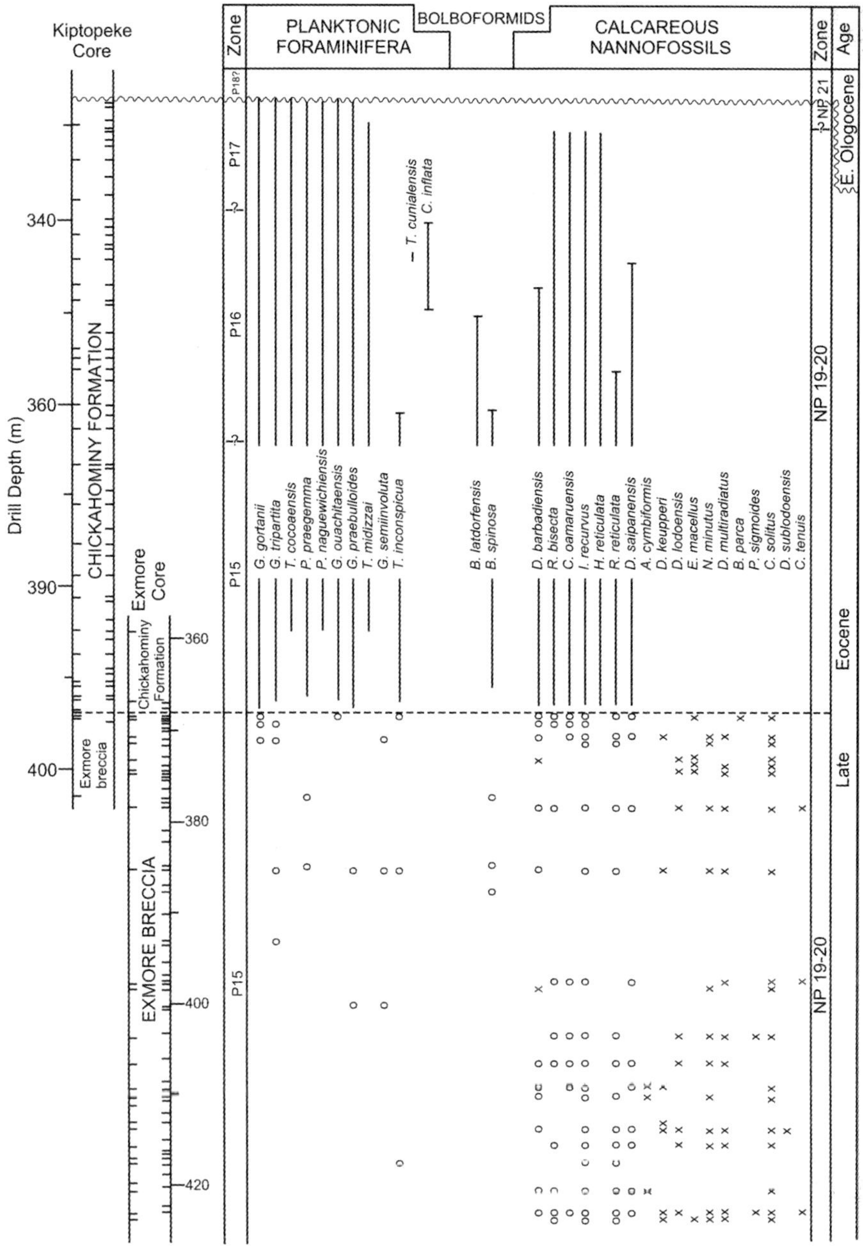

Fig. 8.2. Stratigraphic succession of planktonic foraminifera, calcareous nannofossils, and bolboformids in Chickahominy Formation from Kiptopeke corehole (from Poag and Aubry 1995). o = indigenous late Eocene specimens; x = allogenic specimens (older than late Eocene).

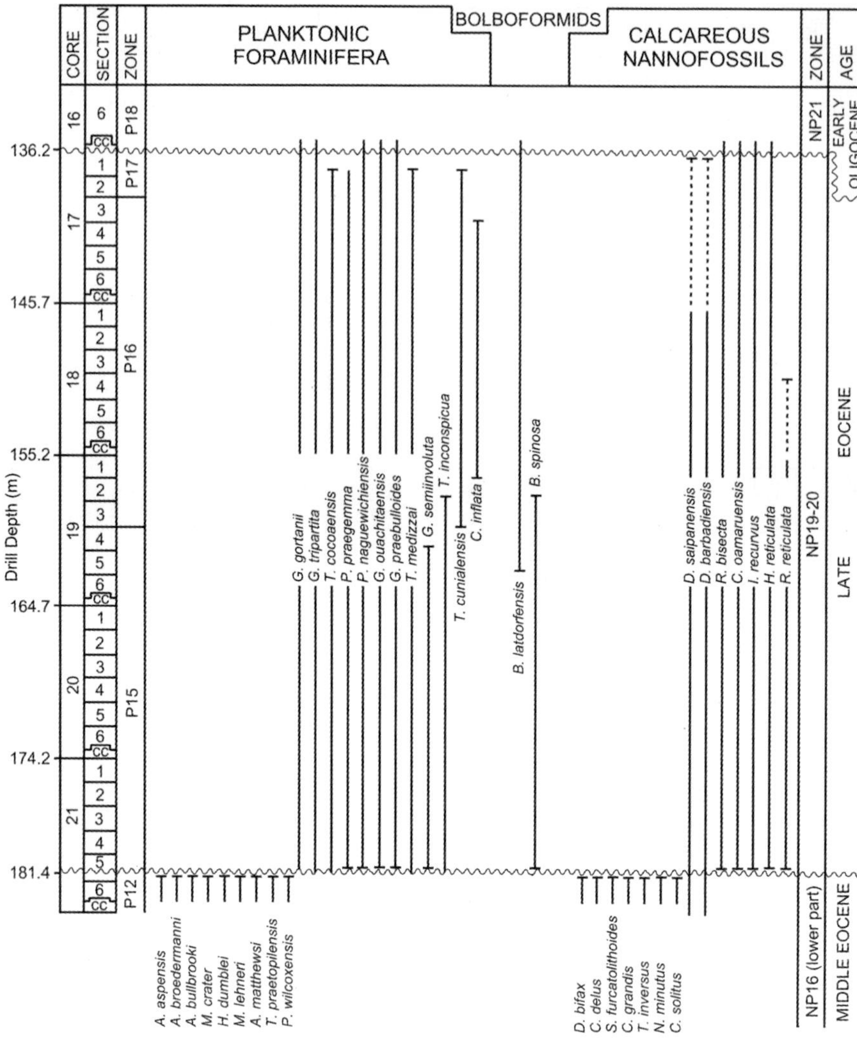

Fig. 8.3. Stratigraphic succession of planktonic foraminifera, calcareous nannofossils, and bolboformids in Chickahominy-equivalent pelagic chalks at DSDP Site 612 (from Poag and Aubry 1995).

8.2 Radiometric Chronology

Radiometric analyses for the Site 612 ejecta (Fig. 2.12), based on $^{40}Ar/^{39}Ar$ plateau values from tektite glass, yield an age range of 35.2 ± 0.3–35.5 ± 0.3 Ma, nearly identical to that obtained from biochronology (Obradovich et al. 1989; Poag and Aubry 1995). We infer that the crater age is equivalent to the ejecta age, but no radioisotopic impact age has yet been obtained from rocks within the Chesapeake Bay crater itself.

8.3 Magnetochronology

Poag et al. (2002) reported the results of a magnetochronological analysis of samples taken at approximately 2-m intervals from the continuously cored, 66-m-thick Chickahominy section in the Kiptopeke borehole (Fig. 8.4). If one assumes a uniform rate of sediment accumulation, the average temporal sampling interval would be approximately 25–38 kyr. The Chickahominy biostratigraphic record can be tied to the magnetostratigraphic record by using Berggren et al.'s (1995) geochronological framework. That framework shows that the boundary between planktonic foraminiferal Zones P15 and P16 lies near the middle of Chron C15r (Figs. 8.1, 8.4). Therefore, by extrapolation, the normally magnetized basal section of the Chickahominy Formation represents the upper part of Chron C16n.2n. Because the ages at the tops of Chrons C16n.1n and C16n.2n have been determined (Fig. 8.4), we can calculate that the average rate of sediment accumulation for the intervening section was 67 m/myr. By extrapolating this rate to the top of the Exmore breccia, we estimate the time of impact to have been ~35.78 Ma (Fig. 8.4; see Chapter 13 for more thorough discussion of sediment accumulation rates).

8.4 Correlation with Other Craters and Impactites

Poag et al. (2002) concluded that the Chesapeake Bay impact, deposition of the Exmore breccia, deposition of the North American tektite strewn field, and deposition of the ejecta-bearing impactite at Massignano, Italy (5.61 m above the base of the outcrop section), all took place within the upper part of Chron C16n.2n (Figs. 8.4, 8.5). This conflicts with initial biostratigraphic assignment at Massignano, in which the impactite at that Italian outcrop was placed within planktonic foraminiferal Biozone P16, rather than P15 (Coccioni et al. 1988; Montanari et al. 1993). In a more recent study of the Massignano section, however, Spezzaferri et al. (2002) indicated that the impactite there lies within the overlap of Zone P15 (forams) and NP19-20 (nannofossils), just as the Chesapeake Bay impact ejecta (North American tektites) and the Exmore breccia do.

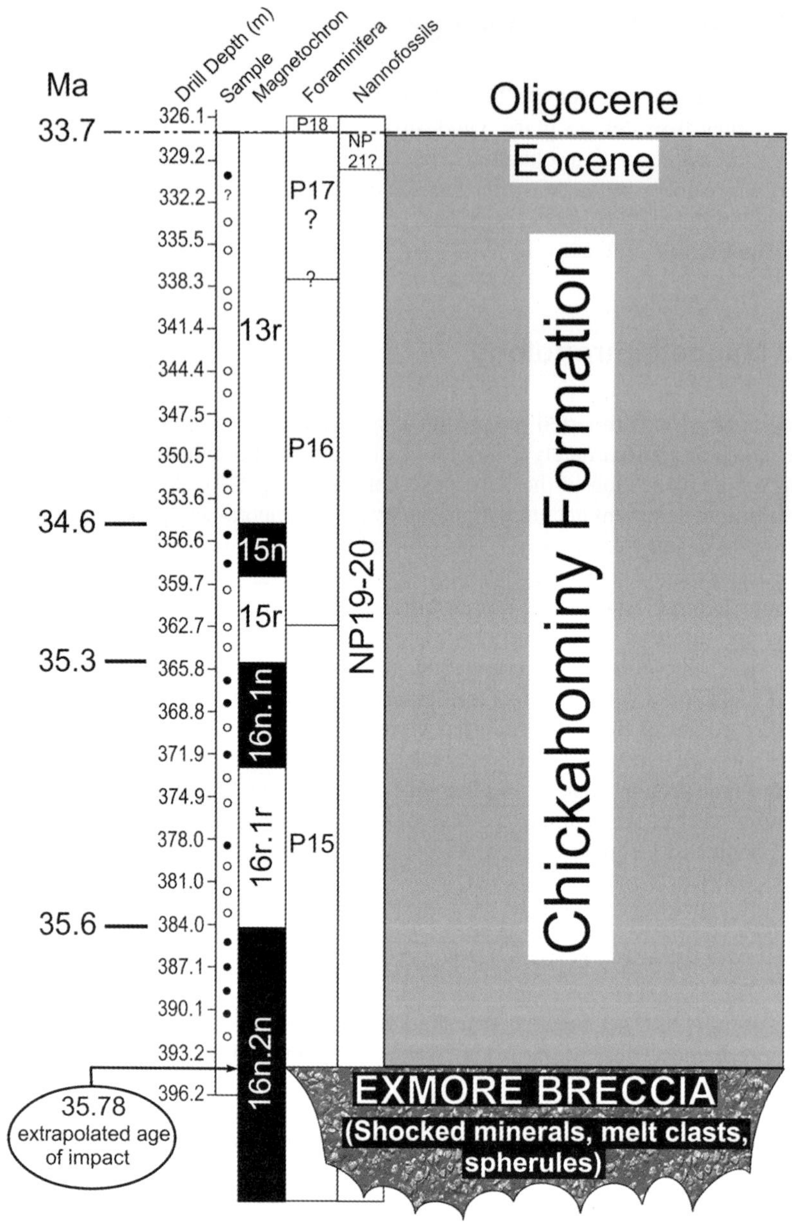

Fig. 8.4. Magnetobiochronology of Chickahominy Formation, which overlies Exmore breccia in Kiptopeke corehole. Age of impact extrapolated from average sediment accumulation rate between magnetochron boundaries at 35.3 and 35.6 Ma. Modified from Poag et al. (2002).

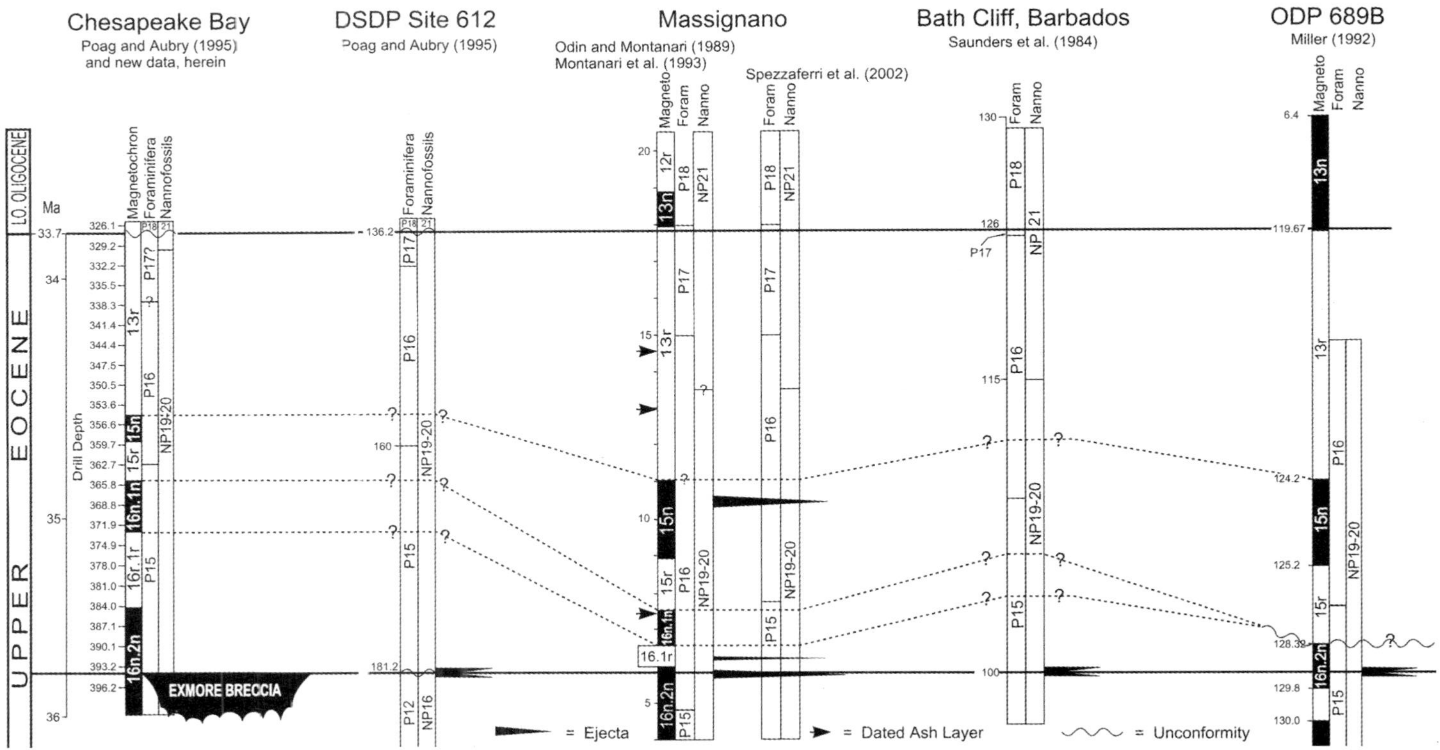

Fig. 8.5. Geochronological correlations between Chesapeake Bay impact crater and late Eocene distal ejecta deposits at DSDP Site 612, Massignano, Italy, Bath Cliff, Barbados, and Maud Rise in Southern Ocean (ODP Site 689B). Depth scales (m) are drill depths (Chesapeake Bay, DSDP 612, ODP 689B) or heights above base of outcrops (Massignano, Bath Cliff). Upper datum is top of Eocene section; lower datum is level of impact debris. Dotted correlation lines based on magnetostratigraphy. Modified from Poag et al. (2002).

Poag et al.'s (2002) paleomagnetic data also have interesting implications regarding late Eocene ejecta cored at Antarctic Site 689B (Fig. 8.5). According to three different authors, Stott and Kennett (1990), Speiss (1990), and Miller (1992), the Antarctic ejecta correlate with either Chron 16n.1n or 15r, both of which are ~0.1–0.9 myr younger than all three known late Eocene impacts. If any of the three published magnetochronologies is correct, then the Site 689B ejecta must represent an unidentified late Eocene impact, which took place in the interval between deposition of the Massignano ejecta layer and a younger iridium-enriched layer (within Chron 13r) in the Contessa Highway section of Italy. However, the stratigraphic relationships and compositions of the ejecta at Site 689B strongly suggest correlation with the Popigai impact (Whitehead et al. 2000; Liu and Glass 2002), and possibly also, in part, with the Chesapeake Bay impact (Glass and Koeberl 1999a). In either case, the ejecta age would cast doubt on the published magnetostratigraphy of Site 689B. Poag et al. (2002) inferred that one or more unconformities have interrupted the stratigraphic succession at Site 689B (Fig. 8.5; see further discussion of the correlations and related features in Chapters 9, 13).

9 Geological Consequences of Chesapeake Bay Impact

9.1 General Nature of Consequences

The most obvious consequences of the Chesapeake Bay impact, for which there is direct evidence, involve immediate (nearly instantaneous) changes in the structure, stratigraphy, and morphology of the sedimentary and crystalline target rocks of southeastern Virginia, and subsequent accumulation of impact-generated deposits (impactites) within and near the primary crater. Consequences of a more regional or hemispheric nature include deposition of distal ejecta and production of far-field seismic effects, such as the possible fracturing of sedimentary beds along the nearby continental shelf edge. Other consequences of a hypothetical nature, for which there is only indirect evidence (or no evidence at all), include possible atmospheric effects (greenhouse warming, climatic cooling), severe stresses imposed upon the global biosphere (extinctions, migrations, faunal or floral turnovers), and, perhaps, even impact-triggered changes in relative sea level (Kendall et al. 1995).

9.2 Reconfigured Basement Structure and Morphology

Seismic reflection profiles gathered during the study of the Chesapeake Bay impact crater (Chapters 4, 5; CD-ROM.8–18) allow us to thoroughly revise earlier interpretations of the configuration of the crystalline basement surface in southeastern Virginia (Figs. 9.1, CD-ROM.3–5). Comparison with three previously published structure maps of the basement surface (Fig. 2.3A,B,C) shows a 20-year evolution in the perception of regional basement morphology.

Basement-imaging marine seismic reflection profiles available from Chesapeake Bay and from the continental shelf east of Delmarva, plus profiles from scattered onshore locations in southeastern Virginia, now total more than 2,000 km of data (Chapter 3; Figs. 3.3, CD-ROM.2; Table 3.2). Several coreholes drilled for studies of geothermal resources of the Virginia Coastal Plain (Costain and Glover 1976–1982) document the basement composition and depth outside the crater (Fig. 2.2; Table 2.1). Two recent coreholes (USGS-NASA Langley, October, 2000; USGS Bayside, August, 2001) provide the first direct assessment of basement composition and possible shock-alteration of *in situ* basement rocks

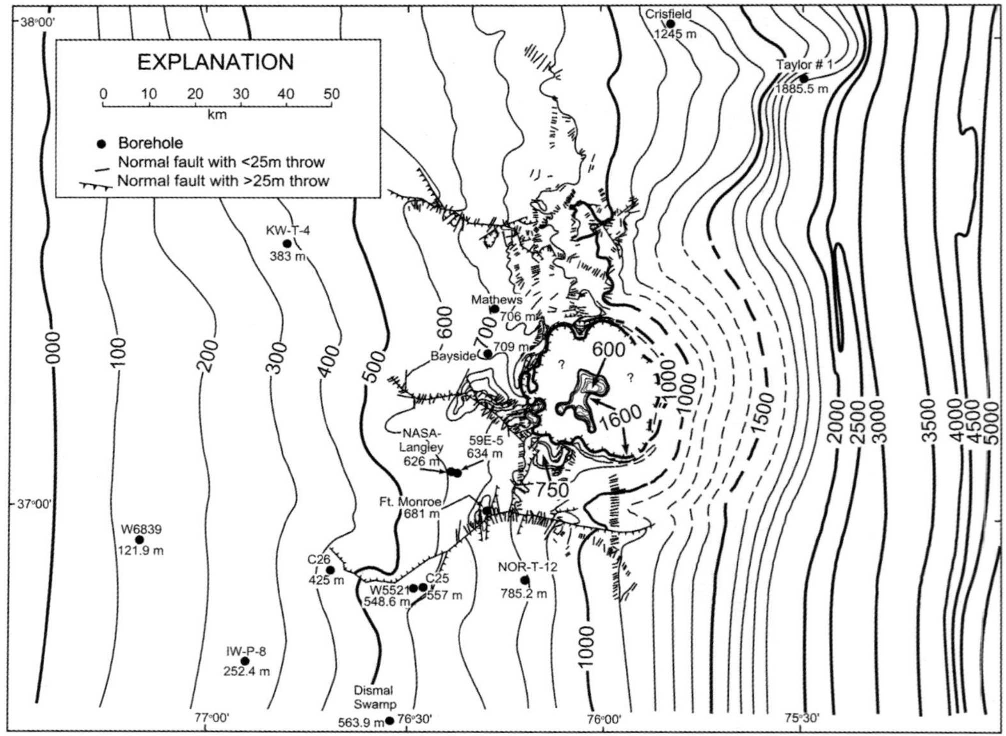

Fig. 9.1. New structure map of crystalline basement surface beneath and surrounding Chesapeake Bay impact crater (contours in m; contour interval variable). Solid dots indicate boreholes that penetrate crystalline basement (see Table 2.1). See Fig. CD-ROM.5 for more structural detail.

inside the crater (Horton et al. 2001, 2002; Powars et al. 2001; Gohn in press). We have integrated the old and new data to construct a new version of the basement structure map, which gives an entirely different picture from the previously published maps cited earlier herein (Figs. 9.1; CD-ROM.5).

On this new structure map, five features stand out (see also Chapter 4): (1) The broad regional sweep of structure contours is dramatically disrupted by a deep, irregularly subcircular depression in the basement surface, centered approximately beneath the town of Cape Charles on the western shore of the Delmarva Peninsula. This depression is the inner basin of the Chesapeake Bay impact crater (Poag 1997a); (2) The inner basin is surrounded by an irregular, knobby, annular ridge, which constitutes the peak ring of the Chesapeake Bay crater (Poag 1997a); (3) An irregular massif of basement rock occupies the center of the inner basin; this is the central peak of the crater (Poag et al. 1999); (4) The basement surface outside the inner basin is marked by extensive systems of normal faults oriented radially and concentrically to the peak ring and inner basin; (5) Interspersed among these faults are numerous small welts and low ridges, many of which also are concentrically oriented around the inner basin and peak ring; and (6) The structural gradient of the eastward-deepening basement surface steepens notably under the continental shelf, beneath the eastern flank of the peak ring, and even more markedly approximately 50 km east of the center of the inner basin (Figs. 9.1, CD-ROM.5).

9.2.1 Central Peak

Poag et al. (1999) pointed out that several of the seismic reflection profiles near the center of the inner basin display faint arcuate reflections believed to represent side echoes from an irregular central peak. Poag and Foster (2000) showed seismic and gravity data from the *Ewing* cruise to firmly establish the presence of an irregular central peak, centered approximately beneath the town of Cape Charles, Virginia (Figs. 4.32, 9.1, CD-ROM.5; see also Chapter 4). We have mapped the outline of this uplift to resemble a misshapen dumbbell (Figs. 9.1, CD-ROM.5); its highest elevation is marked by a relative positive gravity anomaly, centered approximately below the town of Cape Charles (Figs. 4.35, 4.36).

9.2.2 Inner Basin

In its western two-thirds, the irregular periphery of the inner basin is firmly constrained at 20 intersections between seismic reflection profiles (Figs. 4.21, 9.1; CD-ROM.2). In the eastern third, the outline is interpolated to be roughly symmetrical, consistent with the near-symmetry of the negative Bouguer gravity anomaly that underlies the inner basin (Poag 1997a; Figs. 4.35, 4.36) and of the positive gravity anomalies that signify the peak ring (Figs. 4.35, 4.36). For mapping purposes, we assumed that the inner basin walls are nearly vertical (at the map scale) down to an estimated depth of 1.6 km. This assumption is based on the Texaco seismic profiles, the *Ewing* seismic profiles, and Pilkington and Grieve's

(1992) model calculations. For comparison, Sabet (1973) constructed a gravity model for a single profile across the southern Delmarva Peninsula, which showed a deep depression with a maximum depth of ~1.2–2.2 km beneath Cape Charles. Our gravity model (Chapter 4) suggests that the structural floor of the crater could be as much as 2 km deeper than the lip of the crater's outer rim (Fig. 4.37B).

9.2.3 Peak Ring

The diameter of the crest of the Chesapeake Bay peak ring is ~40 km on average (Table 4.5). Seismic profiles show that the ring displays a series of irregular peaks and ridges, the heights of which vary significantly along the western two-thirds of the ring (Figs. 9.1, CD-ROM.5). Detailed physiography of the eastern third of the peak ring is not known due to a lack of appropriate seismic profiles in that region. The gravity signature, however, suggests that the peak ring is roughly symmetrical (Chapter 4).

9.2.4 Normal Faults

The basement surface of southeastern Virginia has heretofore been considered to be relatively fault-free (Poag 1997a). In contrast, we have documented more than 500 individual normal faults, or groups of normal faults, which offset the basement surface along the seismic reflection profiles (Figs. 9.1, CD-ROM.5). We can correlate a few of the fault traces between two or more seismic profiles, but for the majority of faults, there is too little control to determine their spatial orientations. We mapped the faults mainly as subperpendicular to the seismic lines to give a sense of random orientations, but we cannot substantiate this pattern without more closely spaced seismic profiles. Undoubtedly, many faults with smaller offsets are present, but cannot be distinguished at the resolution of our data. For simplicity, we divided the normal faults into two categories: (1) those with vertical throws of 25 m or less; and (2) those with throws of >25 m. By far the majority of the normal faults have relatively small throws (type 1). Such faults can be recognized on all the seismic profiles, regardless of the seismic collection system used.

The type 2 normal faults, on the other hand, appear to have a nonrandom distribution. They occur mainly as concentric faults along the northwest periphery of the inner basin, and as radial fault systems, which originate near the peak ring and extend up the Rappahannock, York, and James Rivers (Figs. 9.1, CD-ROM.5). We expect to find additional radial fault systems on the eastern side of the crater, as soon as better seismic data are available from that sector. Though the positions of the radial fault systems on the western side of the crater appear to be related to the courses of the rivers (or vice versa), their apparent restriction to these locations may be an artifact of our data distribution. Because the river-mouth profiles are more closely spaced than those anywhere else in the bay, they provide plausibility for our choice of connections between individual segments of the radial faults.

9.2.5 Reverse Faults

Reverse faults are relatively rare on the seismic profiles, and some appear to have formed prior to the impact (that is, during the Appalachian orogeny in Paleozoic time). The most prominent set of reverse faults in the basement rocks is present on the Potomac River profile (the Cobb Island fault system of Poag 1997a), where apparently it was reactivated as a series of normal faults by the Chesapeake Bay impact (Fig. 5.12).

9.2.6 Compression Ridges

On a regional scale, the basement surface of southeastern Virginia is gently warped into a series of broad ridges and noses, with long axes that are approximately parallel or slightly diagonal to the dip of the basement. Where we have closer control near the crater, however, numerous smaller welts and ridges can be distinguished (Figs. 4.7A,B, 4.11, 4.20A,B), and their inferred complexity increases with increased structural control. Most of these smaller features are elongated concentrically to the inner basin and its surrounding peak ring (Figs. 9.1, CD-ROM.5). Farther offshore (outside the crater), where the basement gradient steepens, the structural grain changes also. Here, the long axes of the ridges and valleys are perpendicular to the regional dip.

In summary, our collection of more than 2000 km of seismic profiles in the bay and offshore, and the recent coring of basement rocks, reveal that the basement structure beneath Chesapeake Bay is much more complicated than previously suspected. Of the broad embayments and noses presumed to be regional basement features by earlier authors (Fig. 2.3A–C), only the principal axis of the northern embayment is substantiated by the new data (Figs. 9.1, CD-ROM.5). The other features are replaced by a series of smaller-scale ridges, wrinkles, and troughs. These long, narrow features form concentric adjuncts to the 35-km-diameter basement excavation that forms the inner basin of the impact crater. Because of their structure, geometry, and distribution, we interpret the elevated concentric structures as compression ridges, created directly by the compressional shock wave that radiated from ground zero. As a whole, though, the preponderance of normal faults in the crystalline floor of the annular trough indicates that extensional stresses dominated this area during the impact.

The traditional concept of the Norfolk arch (Chapter 2; Owens and Gohn 1985; Gibson 1971; Powars et al. 1992) as a positive basement feature, appears not to be substantiated by the new map. Rather, the local elevation differences cited as evidence for this basement arch appear to be due to the subsidence differential between the crater and surrounding rocks.

9.3 Disruption of Preimpact Sedimentary Column

In addition to such dramatic alteration of the basement configuration, the Chesapeake Bay impact created concentric zones of vaporization, excavation, collapse, and fracturing within the preexisting sedimentary rock column of southeastern Virginia. Our structure maps and cross sections show that the geometry of the preimpact sedimentary units is strikingly different from that published by most previous authors (e.g., Brown et al. 1972; notable exceptions are Powars and Bruce 1999, Powars 2000). Traditionally, the onshore structural and stratigraphic trends of these units have been extrapolated, virtually unchanged, beneath Chesapeake Bay and the Delmarva Peninsula (e.g., Brown et al. 1972; Mixon et al. 1989). The contrast is clearly exemplified by comparing the structure map of the Lower Cretaceous Potomac Formation of Brown et al. (1972; their Cretaceous F unit; our Fig. 9.2) and our new map of the same formation (Fig. 9.3). Brown et al. (1972) mapped the top of Cretaceous Unit F in southeastern Virginia as a broad, gently eastward dipping, slightly flattened, structural nose, with embayments to the north and south of it. The structural contours strike nearly north-south beneath Chesapeake Bay, and the dip of the formation increases markedly at about 300 m depth beneath the eastern margin of Chesapeake Bay (Fig. 9.2). The flattening of the nose gives a hint of the effects of the underlying crater displayed graphically on our new map.

In contrast to the map of Brown et al. (1972), our new map of the top of the Potomac Formation (Fig. 9.3) shows the 300-m contour following the western margin of the bay in the northern half of the map, but in the southern half, the contour arches strongly westward following the rim of the crater, indicating regional uplift during the cratering event. Seismic profiles from the inner continental shelf, combined with the results of the gravity modeling (Chapter 4), allow us to infer a comparable arching of contours around the eastern rim of the crater. North of the crater, the eastward slope of the surface increases at 200 m depth and again at 400 m, between which there is a broad structural shelf. The surface of the Potomac Formation is depressed and relief is quite irregular where the formation has collapsed to form displaced megablocks in the annular trough, though precise elevations of this surface cannot be determined on all seismic profiles. The Potomac Formation is missing altogether (except as breccia clasts) from the zones of excavation and vaporization in the center of the crater (from the peak ring and inner basin; Fig. 9.3).

Structure maps and cross sections of the younger (Paleogene) preimpact sedimentary formations outside the crater show similar regional structural trends to those of the Potomac Formation (Powars and Bruce 1999). Each unit dips eastward outside the crater, and the dip is much steeper east of the crater. However, there is a notable difference in the *intracrater* distribution of Paleogene strata. Seismostratigraphic analysis and extrapolation of corehole results (Poag et al. 1994; Poag 1996; Poag 1997a) suggest that, whereas thick sections of Potomac Formation are present in the displaced megablocks that occupy the floor of the annular trough, Paleogene sediments (as well as youngest Cretaceous strata) have

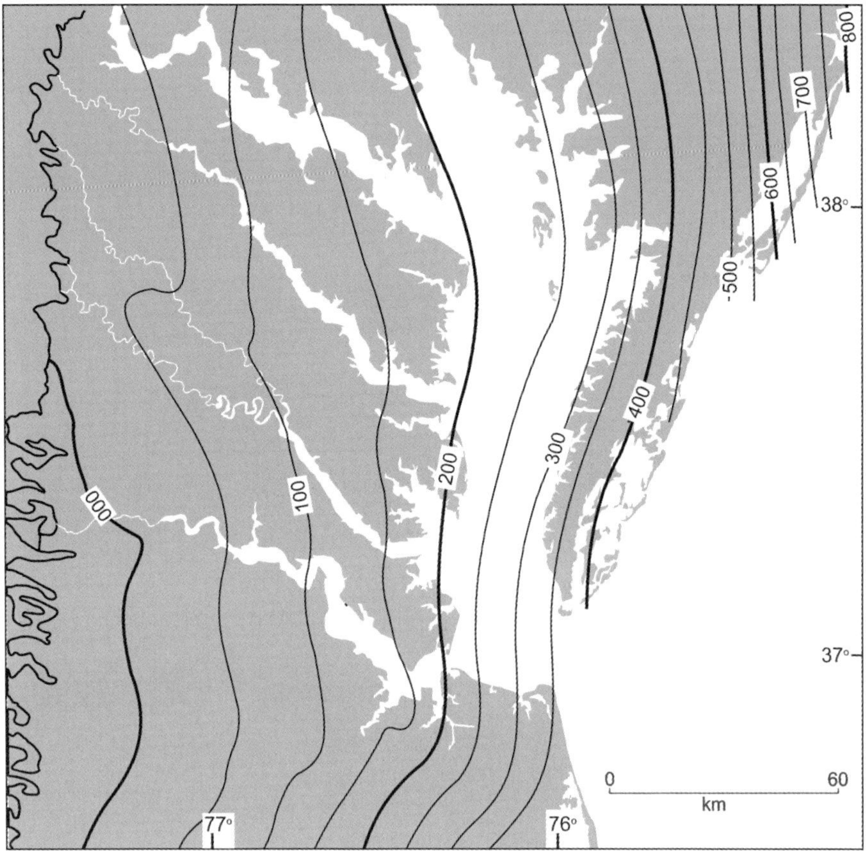

Fig. 9.2. Previously published structure map of upper surface of Lower Cretaceous Potomac Formation (Cretaceous unit F of Brown et al. 1972), modified to show metric contour intervals; contour interval 50 m.

been removed from the tops of the megablocks. This differential removal may be due to the weaker consolidation of Paleogene strata compared to that of the Lower Cretaceous units. Thus it appears that the only preimpact Paleogene strata preserved inside the crater occur as clasts within the Exmore breccia, or as comminuted components of the glauconite-quartz sand matrix.

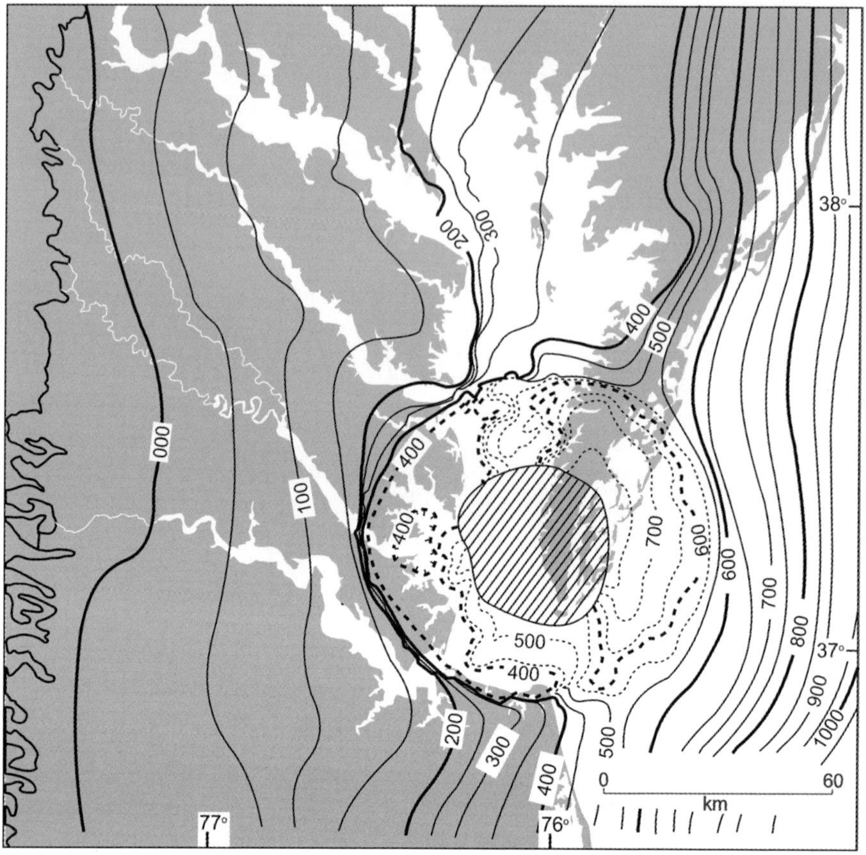

Fig. 9.3. New structure map of upper surface of Lower Cretaceous Potomac Formation based on our seismic reflection profiles and borehole data (contour interval 50 m). Hachured area in center indicates absence of Potomac Formation in inner basin and over peak ring. Potomac Formation is present in annular trough in the form of displaced megablocks and breccia clasts.

9.4 Source of North American Tektite Strewn Field

9.4.1 General Distribution of Distal Ejecta

The most significant documented regional effect of the Chesapeake Bay impact was deposition of distal ejecta. Centimeter-scale layers containing tektites, microtektites, and shocked minerals correlative with the Chesapeake Bay and Toms Canyon craters, are widespread in the western North Atlantic, Caribbean, and Gulf of Mexico. Scattered individual tektites of this age also are present in the coastal plains of Georgia and Texas. Ejecta from 10 sites in this 9 x 10^6 km^2 region represent the bulk of the North American tektite strewn field (NATF; Fig. 2.11; Glass 1989; Koeberl 1989). Vonhof and Smit (1999), Glass and Koeberl (1999a, b), and Liu et al. (2000) recently expanded the possible distribution of the NATF to include ODP Site 689B on the Maud Rise (Leg 113; Barker, Kennett et al. 1990), ODP Site 216 on Ninetyeast Ridge in the eastern Indian Ocean (Leg 22; von der Borch, Sclater et al. 1974), and ODP Site 1090B (Leg 177 Gersonde, Hodell et al. 1999) near the Agulhas Ridge (S. Atlantic) (Fig. 9.4). Latest reconsideration of these distal sites by Liu and Glass (2002), however, casts doubt on their inclusion in the NATF. The radiometric age of the NATF is estimated from $^{40}Ar/^{39}Ar$ step-heating dating of microtektite glass sampled at two locations: (1) A northern location, cored at DSDP Site 612, located 20 km south of the Toms Canyon structure and 330 km northeast of the Chesapeake Bay crater (Figs. 2.12, 9.4), yielded ages of 35.2 ± 0.3 to 35.5 ± 0.3 Ma (Obradovich et al. 1989; Poag and Aubry 1995); (2) A southern location, sampled at Bath Cliff, Barbados, 3,500 km southeast of Chesapeake Bay (Fig. 9.4), yielded an essentially identical age of 35.4 ± 0.6 Ma (Glass et al. 1986). On the basis of its geographical proximity, similar age, and similar geochemistry, the North American tektite strewn field is now believed to have been produced mainly by the Chesapeake Bay impact (Poag et al. 1994; Koeberl et al. 1996; Glass et al. 1998; Glass and Koeberl 1999a,b), though some of the Site 612 ejecta may have come from the Toms Canyon crater (Poag and Aubry 1995; Poag and Poppe 1998; Fig. 2.12).

At deep-sea sites containing ejecta of the North American strewn field, the deposit is dominated by microtektites and shocked quartz, with minor amounts of stishovite and coesite (Fig. 9.4; Glass and Wu 1993; Glass et al. 1998). The microtektite layer is often underlain by a thin concentration of microkrystites (spherules containing clinopyroxene crystals and(or) nickel-rich spinels) (Glass et al. 1985; Glass and Burns 1987; Pierrard et al. 1998, 1999; Fig. 2.13) associated with a moderate iridium anomaly (Sanfilippo et al. 1985; Montanari et al. 1993; Pierrard et al. 1998). These two layers are generally separated by 2–25 cm of sediment or less (equivalent to 10–20 kyr; Glass et al. 1998; Glass and Koeberl 1999a, b), and both layers reside in the P15/NP19-20 overlap interval, which embraces the Chesapeake Bay impact (Poag 1997a,b; see also Chapters 8, 13). Some sites in the North American strewn field, however, contain only one or the other of these layers, or in some cases, the layers are so closely spaced as to be stratigraphically indistinguishable. In contrast to the distribution of North American microtektites, only the microkrystite layer has been compellingly documented at Pacific sites (Whitehead et al. 2000; Fig. 9.4).

Ejecta inferred to have come from the Popigai crater has been reported from the world stratotype of the Eocene-Oligocene boundary at Massignano, Italy (Odin and Montanari 1989; Berggren et al. 1995; Figs. 8.5, 9.4). Geographically, the Massignano section appears to fit within the expanded concept of the NATF (near its northern limit). The composition of the Massignano ejecta, however, resembles a mixture of the microtektite and microkrystite layers. Nickel-rich spinels, shocked quartz, and an Ir anomaly have been documented (Montanari et al. 1993; Clymer et al. 1996; Langenhorst 1996; Pierrard et al. 1998, 1999), and the ^{3}He content is unusually high (Farley et al. 1998). The presence of the nickel-rich

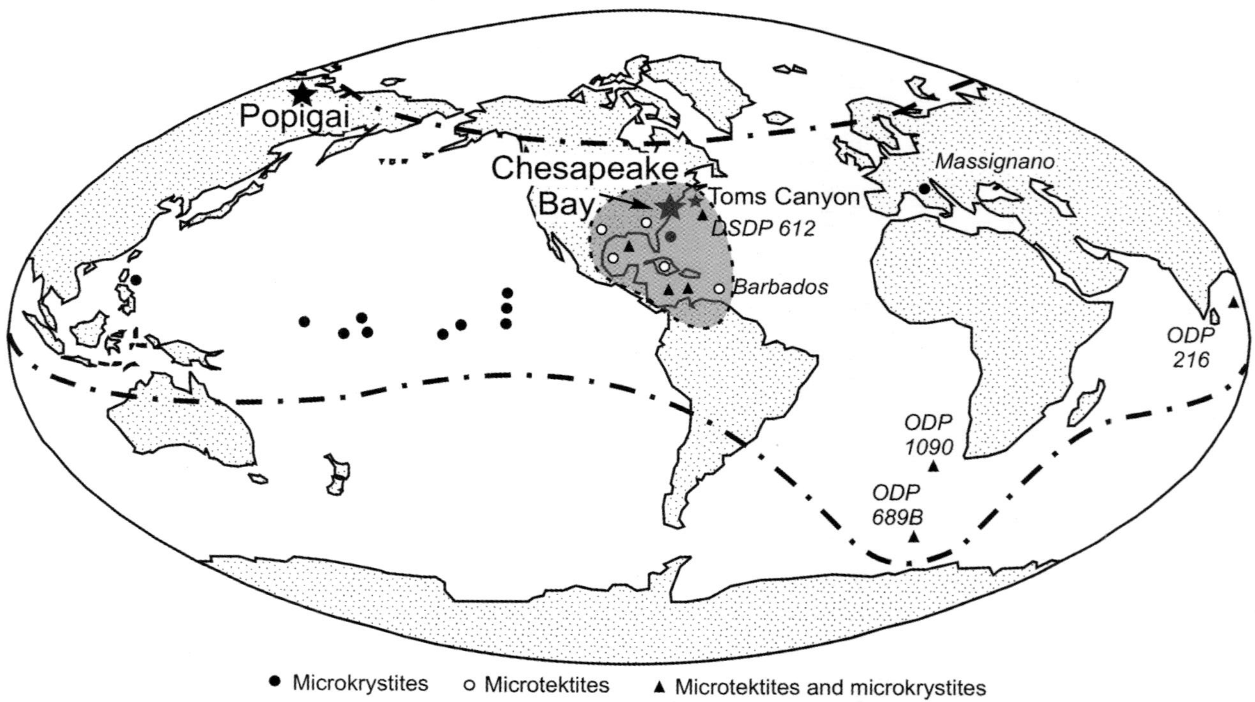

Fig. 9.4. Geographic distribution of North American tektite strewn field (shaded) and microkrystite strewn field (heavy dash-dot line). Circles and triangles indicate boreholes and outcrops from which late Eocene ejecta has been documented. ODP = Ocean Drilling Program; DSDP = Deep Sea Drilling Project.

spinels at Massignano has been taken as evidence of its correlation with the microkrystite layer and the Popigai impact (Vonhof and Smit 1999; Whitehead et al. 2000). Though the Massignano ejecta appears to occur in a single layer, and microtektites have not yet been identified, it has been speculated that this Italian occurrence may represent a condensation of the microtektite and microkrystite layers, where the separate identities of the two layers have been obscured by bioturbation (Vonhof and Smit 1999; Huber et al. 2002) and low deposition rates in this relatively thin (12 m) upper Eocene section. Liu et al. (2001) attempted to determine whether or not both ejecta layers are present at Massignano, but their results were not conclusive.

Magnetostratigraphic studies of the Massignano section reveal that the ejecta deposit resides within the upper part of Chron 16n.2n (Bice and Montanari 1988; Lanci et al. 1996), and is, therefore, >35.6 myr old. A radiometric age of 35.7 ± 0.4 Ma for the Massignano ejecta has been derived indirectly by complicated interpolations between: (1) a weighted mean age estimate from an ash bed at Massignano (35.4 ± 0.3 Ma); and (2) a similarly derived estimate (36.4 ± 0.4 Ma) from an ash bed located ~80 km away in the Contessa Highway section, which was extrapolated by means of magnetochronology to the Massignano section (summarized by Montanari et al. 1993; see also Chapter 8).

9.4.2 Correlation Problems

The biostratigraphic position of the Massignano ejecta was reported by Coccioni et al. (1988) and Montanari et al. (1993) to be within planktonic foraminiferal Biozone P16, whose biochron spans the interval from 34.0 to 35.2 Ma (Berggren et al. 1995). If true, the Massignano ejecta would be biochronologically younger, but radiometrically older, than the Chesapeake Bay and Toms Canyon impacts (and their related impactites). However, Spezzaferri et al. (2002) apparently have resolved this anomaly by demonstrating that the Massignano ejecta layer does belong in Biozone P15, as defined by Berggren et al. (1995). Nevertheless, it has been convincingly demonstrated that the stratigraphic ranges of key species of planktonic foraminifera and calcareous nannofossils are not isochronous between the Massignano section, DSDP Site 612, and the Chesapeake Bay crater (Exmore breccia and Chickahominy Formation; Miller et al. 1991; Montanari et al. 1993; Poag and Aubry 1995).

Biochronological correlations with other pertinent upper Eocene sections are hampered by poor specimen preservation in deep-sea sediments (including Bath Cliff, Barbados) and by significant faunal and floral disparities due to biogeographic differences between tropical micro- and nannofossil assemblages (Caribbean and Gulf of Mexico), mixed tropical-temperate assemblages (DSDP Site 612, Chesapeake Bay), Tethyan assemblages (Massignano), and Antarctic assemblages (ODP Site 689B; Fig. 9.4). Inconsistent biostratigraphic interpretations at two additional coreholes near Site 612 (ODP Sites 903, 904; McHugh et al. 1993; Snyder et al. 1993; Aubry 1993) have further confused correlations with the offshore New Jersey sites (Figs. 2.11, 2.12). For example, at Site 904A, these three groups of authors placed the Zone P15–P14 boundary (planktonic foraminifera), the Zone NP19/20 lower boundary (calcareous nannofossils), and the Zone NP16 lower boundary (calcareous nannofossils) at significantly different core depths.

Although the magnetostratigraphic records at Massignano and Site 689B have helped to resolve some of the biostratigraphic and radiometric correlation problems (Chapter 8), critical inferences, extrapolations, and interpolations are still required, which weakens some of the correlations. The magnetostratigraphic and chemostratigraphic (stable isotopic) data from the Chesapeake Bay impact site further constrain the age of the impact (Chapters 8, 13) and clarify correlations among ejecta sites. The Chesapeake Bay data also provide stratigraphic constraints that will assist in the search for additional occurrences of late Eocene impact ejecta.

9.5 Far-Field Seismic Effects

Hypothetically, powerful seismic waves generated by a large bolide impact would produce significant far-field effects, possibly fracturing and faulting preimpact strata at distances of several hundred kilometers from ground zero. In the case of the Chesapeake Bay impact, there is evidence (gathered by deep-diving submersibles, drilling, and swath bathymetric surveys) of extensive fracturing and brecciation (Fig. 9.5) and unusual steep-walled channels (Fig. 9.6) along a broad outcrop band of lower and middle Eocene limestones on the continental slope 120 km east of Atlantic City, New Jersey (Robb et al. 1983; Farre and Ryan 1985; McHugh et al. 1993, 1995). Though this area of unusual fracturing is 300 km northeast of the Chesapeake Bay crater, Poag et al. (1992) and McHugh et al. (1995) have suggested that the fractures may have been generated by the Chesapeake Bay and(or) Toms Canyon impacts. We propose that the enigmatic, steep-walled channels also may be distal erosional products of the Chesapeake Bay impact.

Fig. 9.5. (Opposite page) Seafloor photographs of outcrops of intensely fractured middle and lower Eocene pelagic limestones on New Jersey Continental Slope adjacent to Toms Canyon crater and ~300 km northeast of Chesapeake Bay crater. *A*, fractured limestone beds form near-vertical cliff in upper half of photograph (arrow). *B*, terrace of highly fractured limestone (arrows mark fractures). *C*, talus of angular limestone clasts at base of cliff. Photographs from submersible dive courtesy of David C. Twichell. See CD-ROM for color version of this figure.

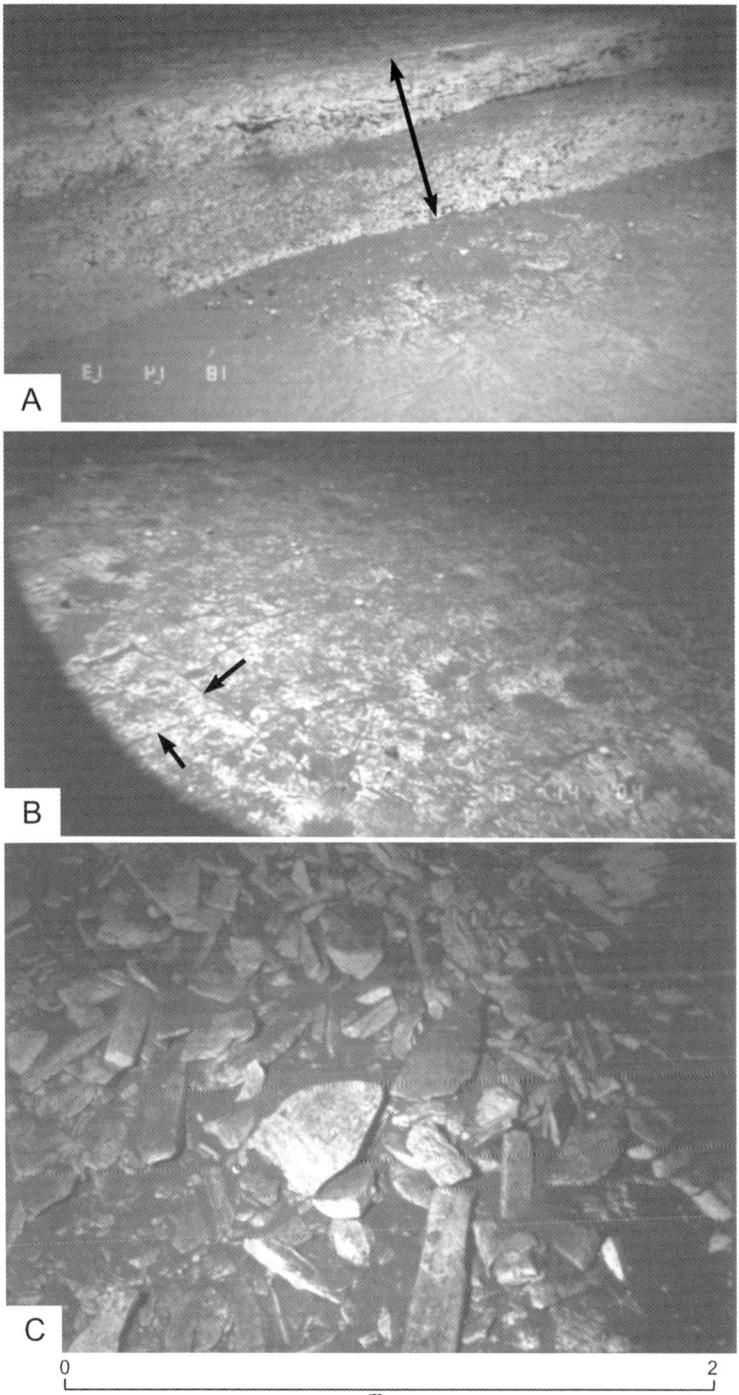
A
B
C
0
2
m

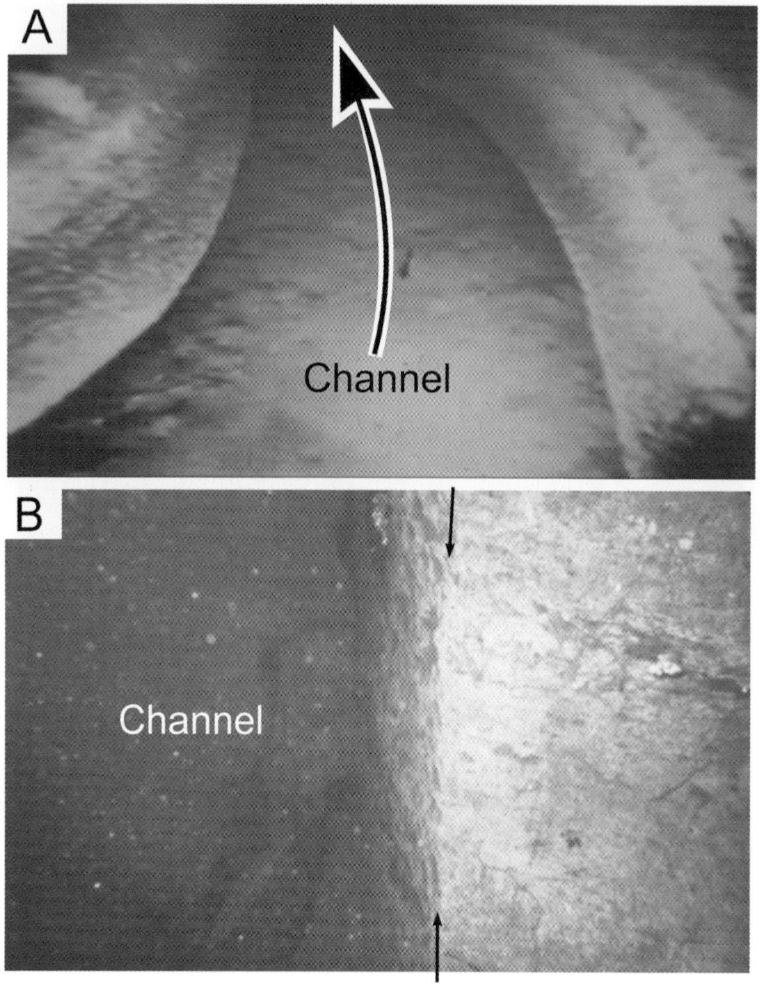

Fig. 9.6. Seafloor photographs showing unusual steep-walled channels in Eocene pelagic limestones exposed at foot of New Jersey Continental Slope. *A*, view down axis of channel (channel ~3 m wide). *B*, vertical view of channel and nearly vertical channel wall; arrows indicate approximately right-angle intersection between channel wall and fractured flat limestone surface that forms top of channel wall. Photographs from submersible dive courtesy of James M. Robb.

10 Comparisons with Other Impact Craters

10.1 Terrestrial Craters

The Chesapeake Bay crater is large and well preserved, and exhibits more impact-generated features (including secondary craters) than most other terrestrial impact structures yet studied, but it appears not to have caused an immediately subsequent mass extinction. These properties make it an important benchmark for improving our understanding of the dynamics of crater formation, ejecta generation and distribution, breccia origin and deposition, and consequent environmental perturbations, or lack thereof. A few known craters in its size class (75–100 km diameter), and several smaller ones, exhibit some or most features of the Chesapeake Bay crater (Table 10.1), but in several key aspects, the Chesapeake Bay crater does not conform to general conceptual models widely applied to explain the formation of complex terrestrial and planetary craters. Some of its unusual features (perhaps all) appear to be related to its original submarine location, which presented a three-layered target composed of: (1) a moderately deep (~300 m) water column; (2) a water-saturated, unconsolidated sediment column (300–500 m thick); and (3) a basement of consolidated crystalline (granitoid and metasedimentary) rocks. Other considerations, such as impactor size, composition, and trajectory, however, also may be pertinent to explaining some of the differences. In order to identify the principal differences, we compare the attributes of the Chesapeake Bay crater with those of other known subaerial and submarine craters on Earth, as well as with some of those on other planetary bodies.

10.1.1 Subaerial Craters

Among the few well-preserved and well-documented impact craters originally formed in subaerial target rocks, the closest analogues for the Chesapeake Bay peak-ring/central-peak structure appear to be the Popigai crater (85-km diameter, late Eocene, Anabar Shield, Northern Siberia, Russia; Masaitis 1994; Masaitis et al. 1999; Whitehead et al. 2002) and the Ries crater (24-km diameter, late Miocene, southern Germany; Pohl et al. 1977; Newsom et al. 1990; Figs. 1.1, 10.1). Both of these craters formed in mixed sedimentary-crystalline target rocks, are clearly marked by steep outer-rim escarpments, and have distinct peak rings of upraised crystalline basement rocks. Each crater also is characterized by a broad annular trough outside the peak ring and a deep inner basin inside the peak ring. Some authors have inferred a crystalline central peak for each, though the deep

Table 10.1. Comparison of Chesapeake Bay crater and its impact-generated deposits with other selected submarine and subaerial craters and impactites.

Crater Diam [km]	Outer Rim	Megablocks (Slumpback Deposits)	Peak Ring	Inner Basin	Central Peak	Breccia on Central Peak	Proximal Breccia	Distal Ejecta	Surgeback Deposits	Tsunami Washback Deposits	Flowin Deposits	Fallout Deposits	Estimated Paleodepth [m]
Chesapeake Bay [85]	sedimentary, faulted	above décollement	crystalline, breached?	crystalline	crystalline	yes	Exmore breccia	N.Am. tektites	yes	yes	yes	yes	~300
Montagnais [45]	sedimentary, eroded	no décollement	crystalline	crystalline	crystalline, depression	yes	yes	none	probably	no	no	no	200-600
Mjølnir [40]	sedimentary, faulted	above décollement	sediments	sediments	sediments	eroded	yes	core 7430 /10-U-01	probably	?	?	?	300-500
Kamensk [38]	sedimentary, faulted	?	sediments	sediments	crystalline	yes	yes	?	yes	?	no	no	100-200
Manson [37]	sedimentary, faulted	above décollement	crystalline	crystalline	crystalline, depression	yes	yes	Crow Creek Member Pierre Sh.	probably	Crow Creek Member Pierre Sh.	no	no	<200
Lockne [24]	sedimentary, faulted	?	crystalline, breached	crystalline	crystalline	yes	Tandsbyn breccia	none	Lockne, Loftarsten breccias	?	?	no	>200
Toms Canyon [21]	sedimentary, faulted	no décollement	none	sediments	none	----	yes	?Site 612, 903, 904?	probably	no	no	no	500-1000
Ames [15]	sedimentary, faulted	?	crystalline	crystalline	none	----	yes	none	?	no?	no	no	10-50
Popigai [85]	sedimentary faulted	décollement ?	crystalline	crystalline	crystalline	yes	yes	Cpx spherules	no	no	no	yes	subaerial
Ries [24]	sedimentary, faulted	no décollement ?	crystalline	crystalline	crystalline	yes	Bunte Breccia	Central European tektites	no	no	no	yes	subaerial

See text for references

interior of neither of these two craters has been fully documented.

In terms of age and general morphology, Popigai is practically a twin of the Chesapeake Bay crater. Their ages are indistinguishable [Chesapeake Bay = 35.78 Ma, new date herein (see Chapters 8, 13); Popigai = 35.7 ± 0.4 Ma; Bottomley et al. 1997]. These two craters also appear to be virtually identical in size. Though the diameter of Popigai is commonly listed as ~100 km (e.g., Grieve et al. 1995; Masaitis et al. 1999; Masaitis personal communication 2002), the geologic maps of Masaitis et al. (1999) and Whitehead et al. (2002) indicate an average diameter of ~85 km. Aside from minor irregularities in the peripheral outlines, the outer rims of the two craters can hardly be distinguished from one another when superposed (Poag et al. 1999; Fig. 10.2). Other morphological features are quite similar, but not identical. For example, there is a slight asymmetry to the Popigai structure, in which the peak ring and central peak are offset to the west relative to the center of the outer rim. The diameters of the peak ring and central peak also may be slightly larger at Popigai. The peak ring and central peak are buried (for the most part) by impact-generated deposits at both of these sites (Fig. 10.1; see also Chapter 11). However, at Popigai, the crystalline rocks of the peak ring crop out in two narrow strips on the northwest side of the crater. Masaitis et al. (1999) estimated a maximum depth for the Popigai annular trough to be ~1 km to the northwest, but twice that (~2 km) to the southeast (Fig. 10.1C). At Chesapeake Bay, a similar deepening of the annular trough takes place from northwest (~0.7 km) to southeast (~1.5 km). There is an enormous difference between the two structures, however, in the maximum structural relief from the deepest part of the annular trough to the crest of the peak ring. This relief at Chesapeake Bay is ~200–600 m, whereas at Popigai it is more than 2 km (Fig. 10.1C).

Significant structural differences in the annular troughs also can be inferred from the drilling and geophysical data at Popigai and Chesapeake Bay. Masaitis et al. (1999) showed (based partly on drilling) that crystalline basement rocks in the Popigai annular trough are deeply faulted, have a high local relief, and incorporate displaced megablocks composed of sedimentary target rocks. At Chesapeake Bay, in contrast, the crystalline surface of the annular trough exhibits only moderate to low structural relief (see Chapters 4, 9). Furthermore, the crystalline rocks of the annular trough at Chesapeake Bay appear not to enclose any sedimentary blocks. Instead, the surface of the trough acts as a zone of detachment, along which the overlying sedimentary megablocks have slid, slumped, and collapsed.

The deeper parts of the inner basin at Popigai, as at Chesapeake Bay, have not yet been drilled, but gravity modeling suggests that the inner basin of each crater may be as deep as 2 km below the lip of the outer rim (Fig. 10.1B,C). As a working hypothesis, however, we use a depth of ~1.6 km for the inner basin at Chesapeake Bay (see Chapter 4). Masaitis et al. (1999) reported a low-relief central peak at Popigai on the basis of three-dimensional computer modeling of gravity data.

Cross sections of the Ries crater based on drilling and geophysical surveys (seismic reflection, gravity, geomagnetics) show comparable structure and morphology to Chesapeake Bay and Popigai, except for its smaller size (with con-

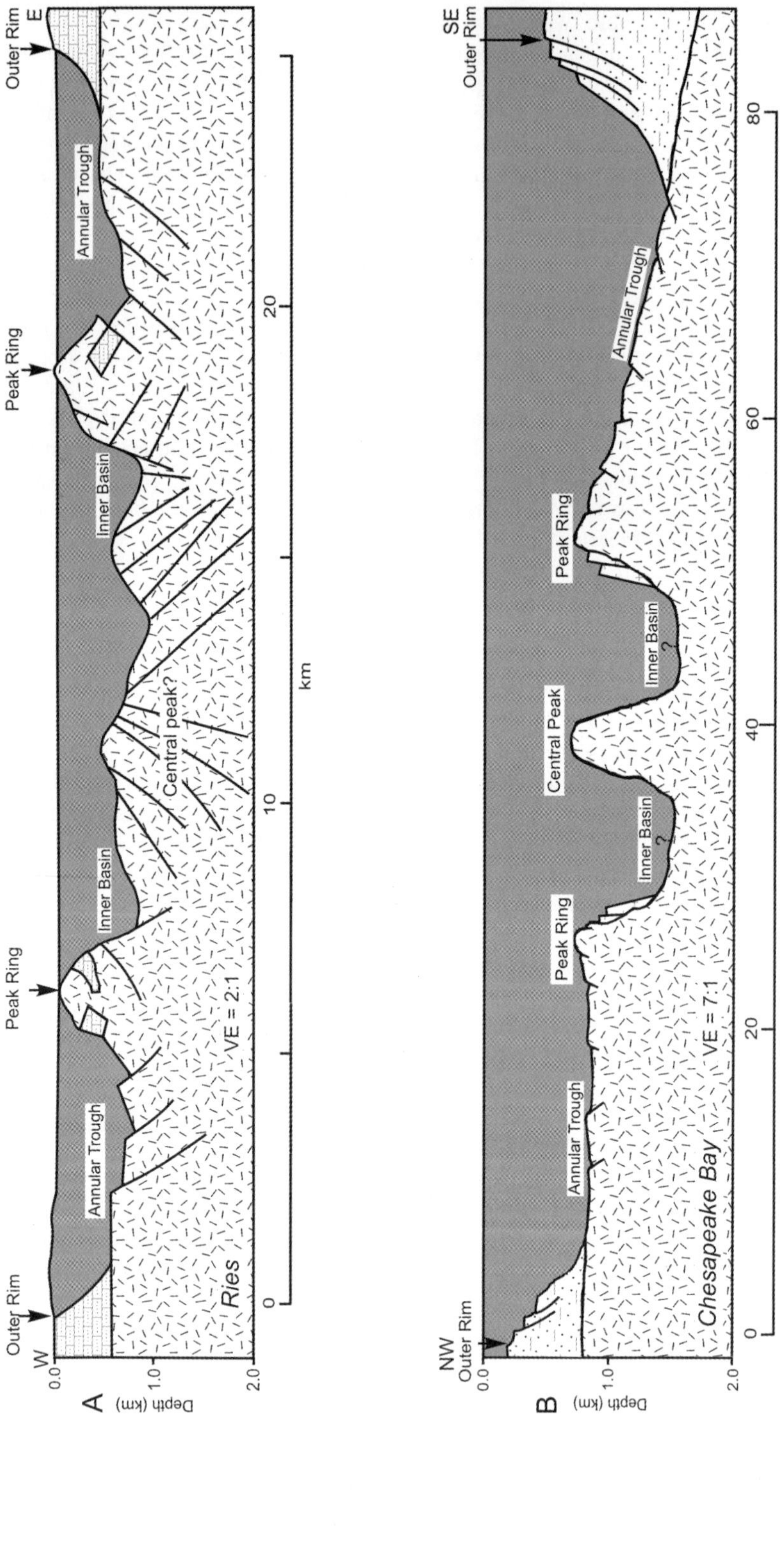
A
Ries
W
E
Outer Rim
Peak Ring
Annular Trough
Inner Basin
Central peak?
VE = 2:1
Depth (km)
0.0
1.0
2.0
0
10
20
km
B
Chesapeake Bay
NW
SE
Outer Rim
Annular Trough
Peak Ring
Inner Basin
Central Peak
VE = 7:1
Depth (km)
0.0
1.0
2.0
0
20
40
60
80
km

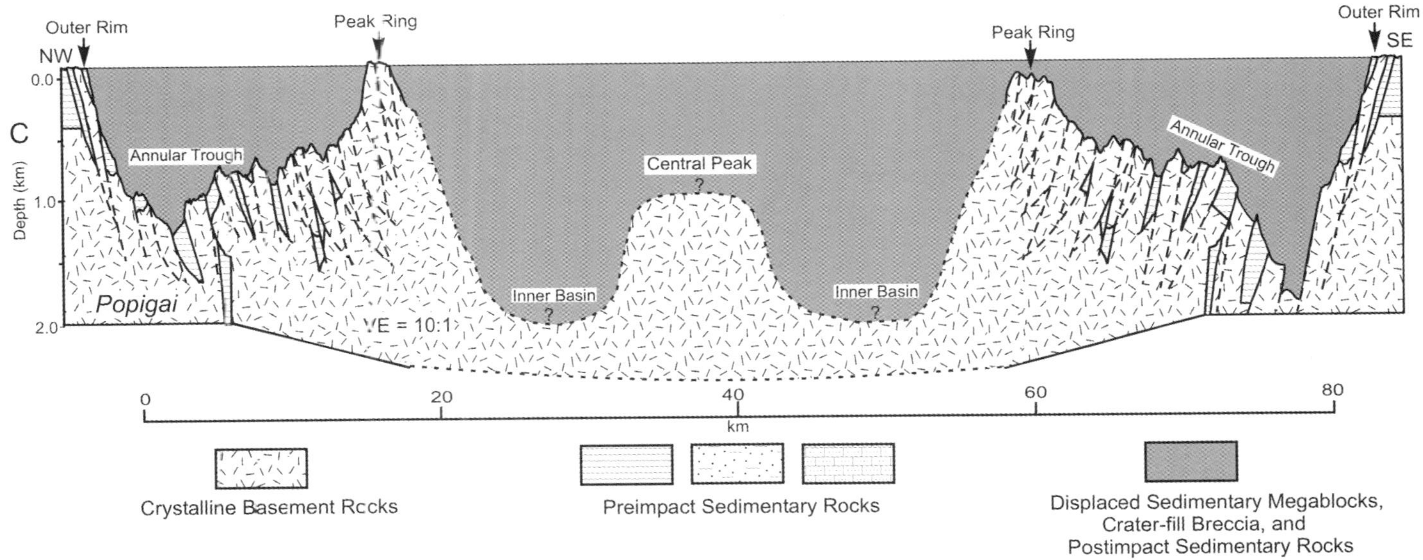

Fig. 10.1. Comparative geologic cross sections of Ries impact crater, southern Germany (*A* opposite page) modified from Pohl et al. (1977); Chesapeake Bay impact crater (*B* opposite page); and Popigai impact crater, Northern Siberia (*C* constructed from two different cross sections of Masaitis et al. 1999; central peak and inner basin interpolated). Note contrasting vertical exaggerations.

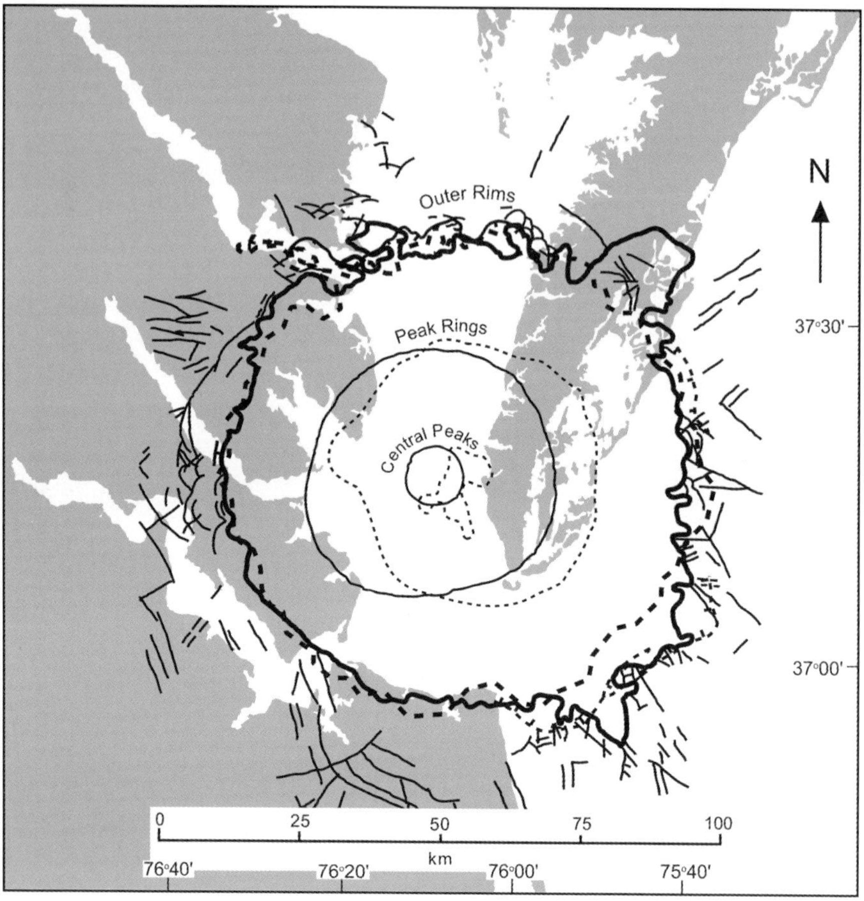

Fig. 10.2. Superposed peripheral outlines of Chesapeake Bay and Popigai impact craters showing similarity in size and geometry of principal features. Solid lines represent Popigai; dashed lines represent Chesapeake Bay. Short concentric and radial fault traces outside craters are from Popigai. Modified from Poag et al. (1999).

sequent smaller values for other properties, such as structural relief, and depth of the annular trough and inner basin) and the undetermined presence of a central peak (Fig. 10.1A). The crest of the crystalline peak ring crops out at Ries, and at least one deep corehole has penetrated the crystalline floor of the inner basin (Stöffler 1977; see also Chapter 11). Like most other intermediate to large terrestrial craters, however, details of the deeper structure at Ries have not been thoroughly investigated, leaving open such questions as whether the sedimentary megablocks are detached from the crystalline basement, or what is the nature of the central peak.

10.1.2 Submarine Craters

Several authors have addressed the hypothetical aspects of how impacts into the ocean might produce craters that differ from those generated by subaerial collisions (Higgins and Butkovitch 1967; Kieffer and Simonds 1980; McKinnon and Goetz 1981; Melosh 1981; Gault and Sonett 1982; McKinnon 1982; Silver 1982; Roddy et al. 1987; Ormö and Lindström 2000; Artemieva and Shuvalov 2002; Ormö and Miyamoto 2002; Shuvalov et al. 2002; Wünnemann and Lange 2002; Ormö et al. 2002). Unfortunately only about a dozen submarine craters have been identified, and fewer have been carefully studied in the field. Four of these are still wholly or partly submerged and also are completely buried by postimpact sedimentary rocks (Chesapeake Bay, Toms Canyon, Montagnais, and Mjølnir; Figs. 1.1, 10.3). Of those remaining, six also are entirely buried by postimpact deposits (Ames, Manson, Granby, Kärdla, Kamensk, and Kaluga). Only two submarine craters (Lockne and Brent) are wholly or partly exposed (Fig. 1.1).

Among submarine craters, Montagnais (45-km-diameter, early Eocene (~51 Ma), Nova Scotian Shelf, Canada; Jansa et al. 1989; Pilkington et al. 1995; Poag et al. 2002) is next in size to Chesapeake Bay, and was the first submarine crater to be discovered (Figs. 10.3–10.7). This structure is known mainly from 1,000 km of seismic reflection surveys, and a single deep (but uncored) borehole that penetrated its crystalline central peak. We have reinterpreted certain aspects of the Montagnais structure after examining a full-scale version of profile 3203-82 (Fig. 2A of Jansa et al. 1989; Figs. 10.5, 10.6, CD-ROM.17).

The outer rim of the Montagnais crater is a steep escarpment cut into sedimentary rocks (Figs. 10.3, 10.5), similar to the outer rim at Chesapeake Bay and Ries. The annular trough, ~7 km wide, has a flat, relatively horizontal floor excavated into lithified Barremian sedimentary rocks known as the Roseway Unit (Fig. 10.5). The excavation surface is parallel to bedding of the Roseway Unit. The crystalline basement surface forms a marked depression beneath the annular trough on the east side of the structure, but descends deeply beneath a thick sedimentary wedge on the western side of the crater (Figs. 10.3, 10.5). A subtle arch on the upper surface of the basement is present on each side of the crater (Figs. 10.3, 10.6), and this may represent a low-relief peak ring. If so, the peak ring's position at 0.5–1.0 km below the surface of the annular trough is quite unusual. A narrow inner basin (3–5 km wide) separates the possible peak ring from a prominent central peak. This central peak is 16 km in diameter at its base and 11 km across at the top. The central peak is bounded by high-angle normal faults, and rises more than 1 km above the floor of the inner basin, and ~200 m above the outer rim of the crater. A distinct structural depression (200 m deep) occupies the center of the peak. The large diameter and great height of the Montagnais central peak are in stark contrast to the relatively narrow and low central peak at Chesapeake Bay and that inferred at Ries (Fig. 10.1). The central peak at Popigai appears to be of similar height above the inner basin to that of Montagnais, but may be narrower in diameter.

Pilkington et al. (1995) found no negative gravity anomaly over the inner basin of the Montagnais crater, and attributed its lack to the unconsolidated nature of the

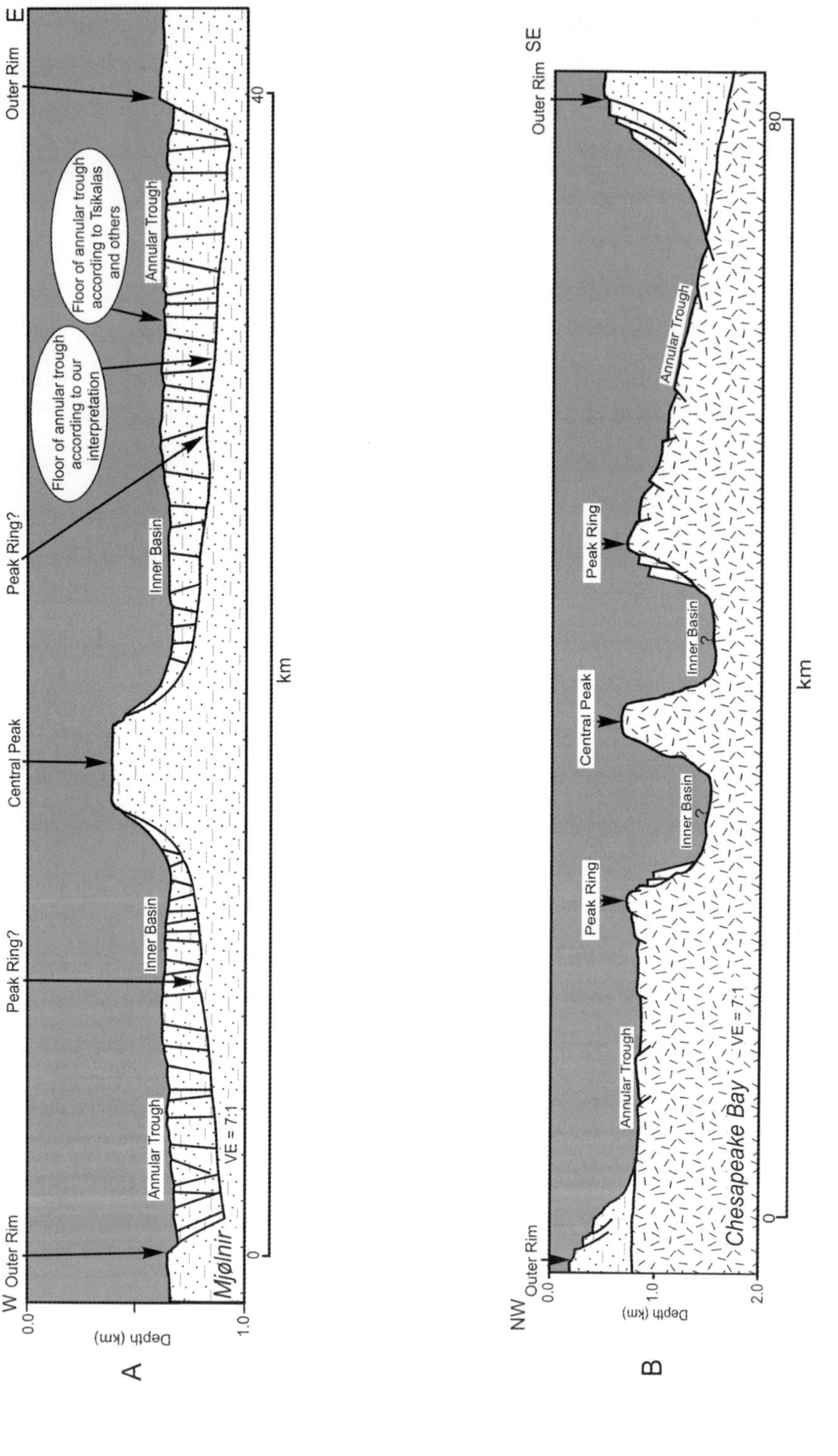
A
W
E
Outer Rim
Peak Ring?
Central Peak
Peak Ring?
Outer Rim
Floor of annular trough according to our interpretation
Floor of annular trough according to Tsikalas and others
Annular Trough
Inner Basin
Inner Basin
Annular Trough
Mjølnir
VE = 7:1
Depth (km)
0.0
1.0
0
40
km
B
NW
SE
Outer Rim
Peak Ring
Central Peak
Peak Ring
Outer Rim
Annular Trough
Inner Basin
Inner Basin
Annular Trough
Chesapeake Bay
VE = 7:1
Depth (km)
0.0
1.0
2.0
0
80
km

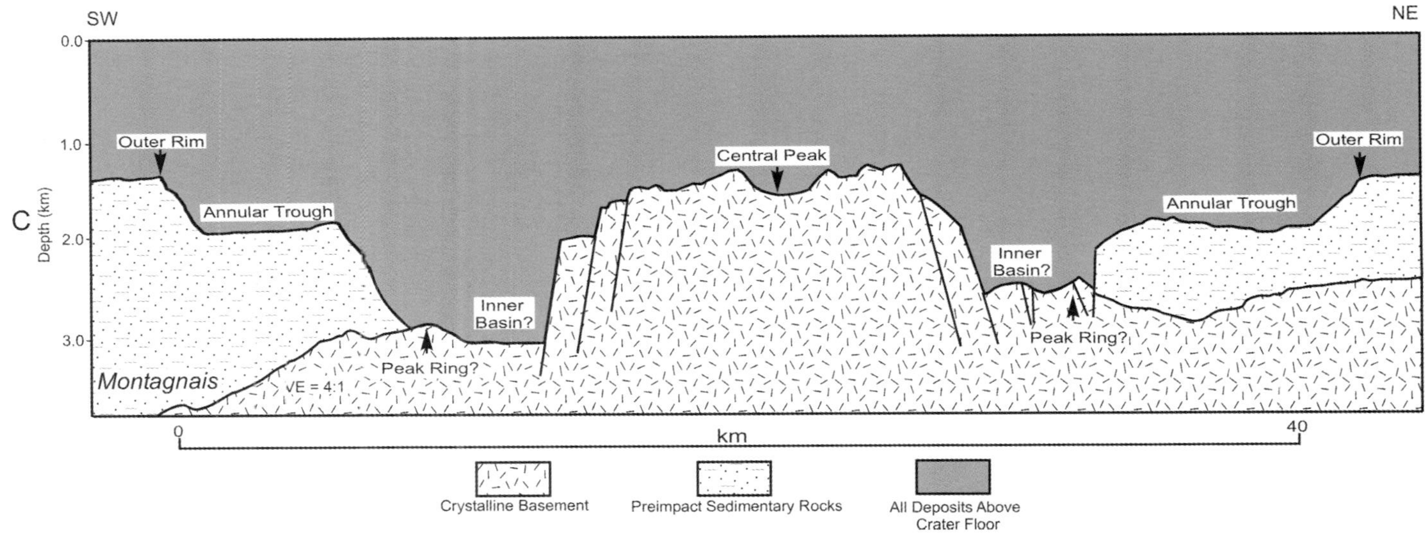

Fig. 10.3. Comparative geologic cross sections of Mjølnir crater, Barents Sea (*A* opposite page) modified from Tsikalas et al. (1998a); Chesapeake Bay crater (*B* opposite page); and Montagnais crater, Scotian Shelf (*C*) modified from Jansa et al. (1989).

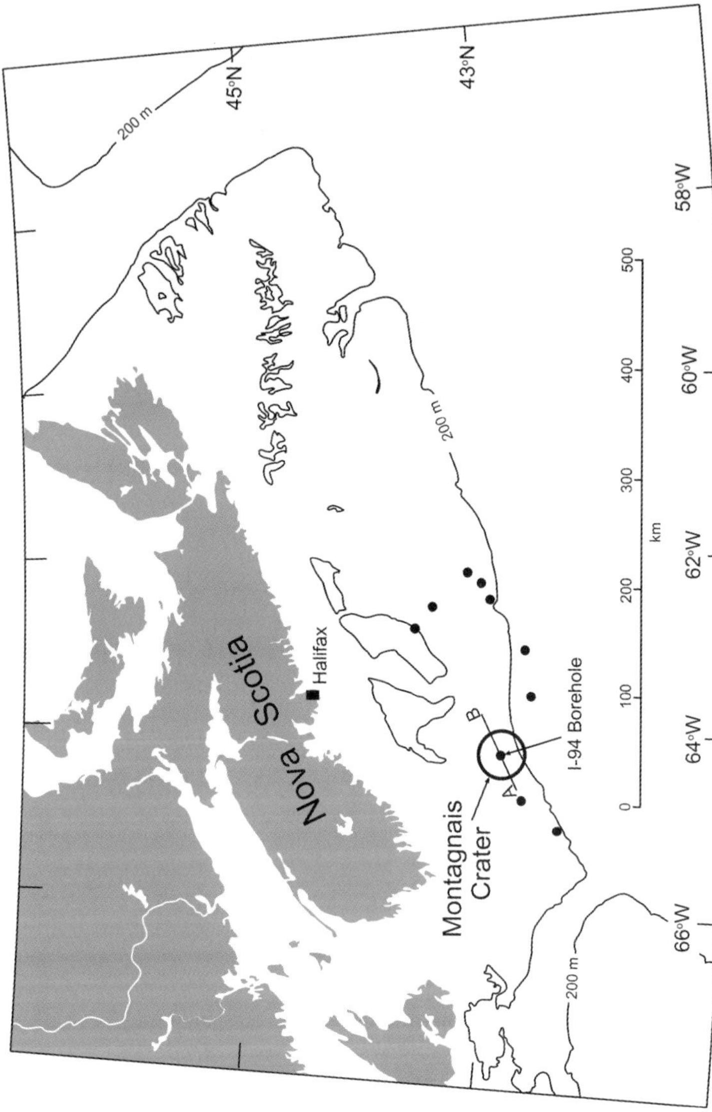

Fig. 10.4. Location map for Montagnais crater, seismic trackline *A-B* (see Figs. 10.5, 10.6, CD-ROM.17), and relevant boreholes (solid black dots). Modified from Jansa et al. (1989). Offshore bathymetry given in meters.

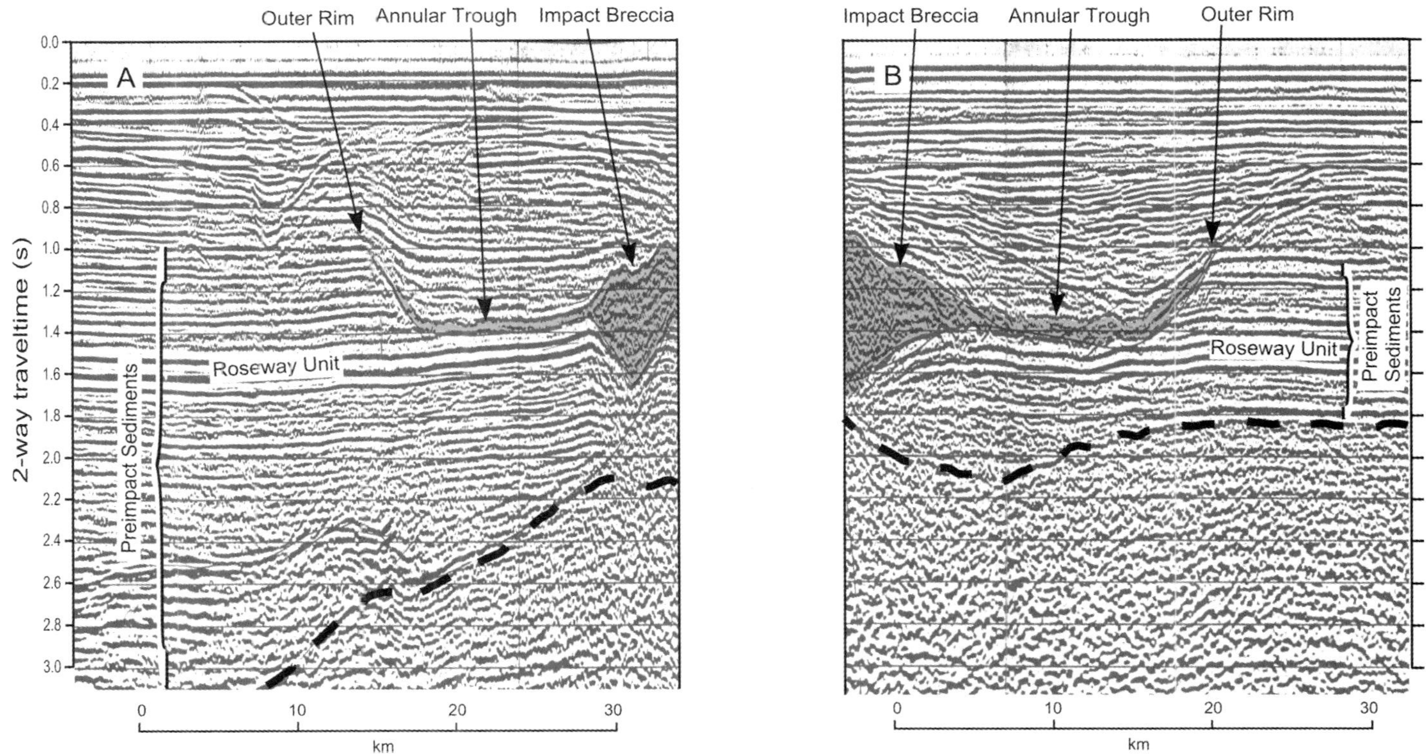

Fig. 10.5. Segments of seismic profile across annular trough of Montagnais crater (Poag et al. 2002; *A*, western side of crater; *B*, eastern side of crater). Impact breccia shaded; heavy dashed line indicates surface of crystalline basement. Prominent seismic signature of Roseway Unit provides stratigraphic datum for correlation. See Fig. 10.4 for location of profile and CD-ROM.17 for full-scale profile. Profile courtesy of LF Jansa.

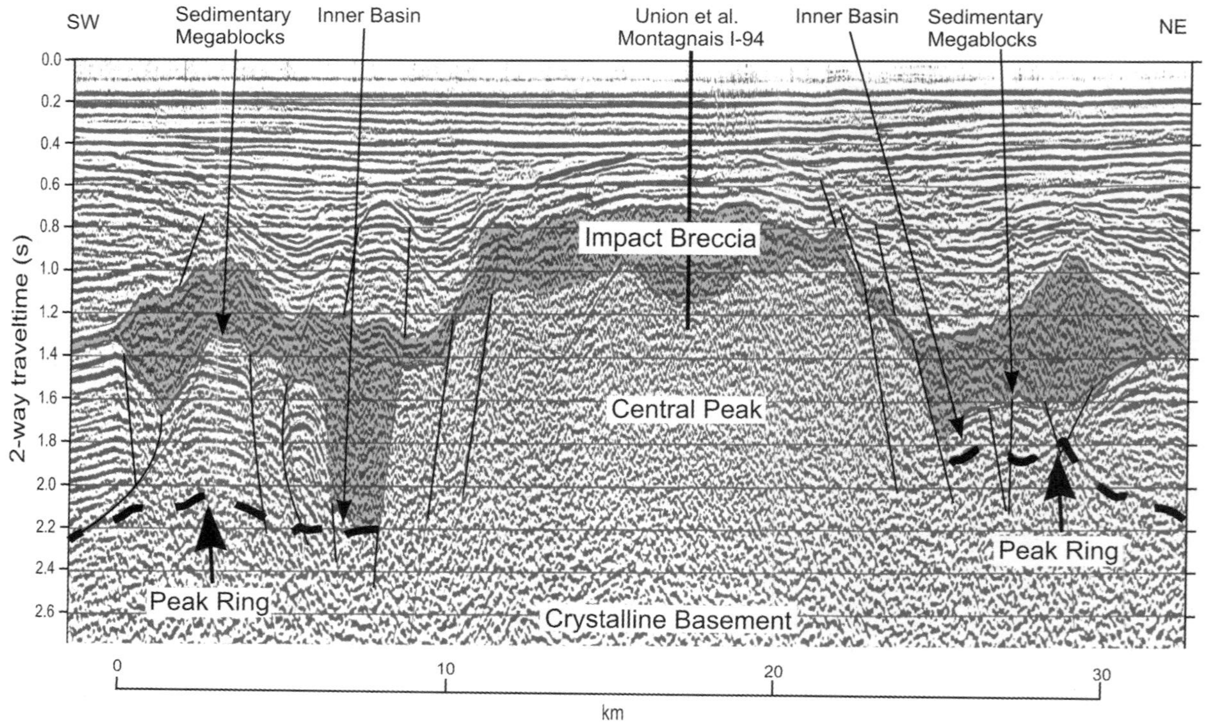

Fig. 10.6. Segment of seismic profile across central peak of Montagnais crater (interpreted by Poag et al. 2002). Impact breccia shaded; heavy dashed line indicates surface of crystalline basement. See Fig. 10.4 for location of seismic profile and CD-ROM.17 for full-scale profile. Profile courtesy of LF Jansa.

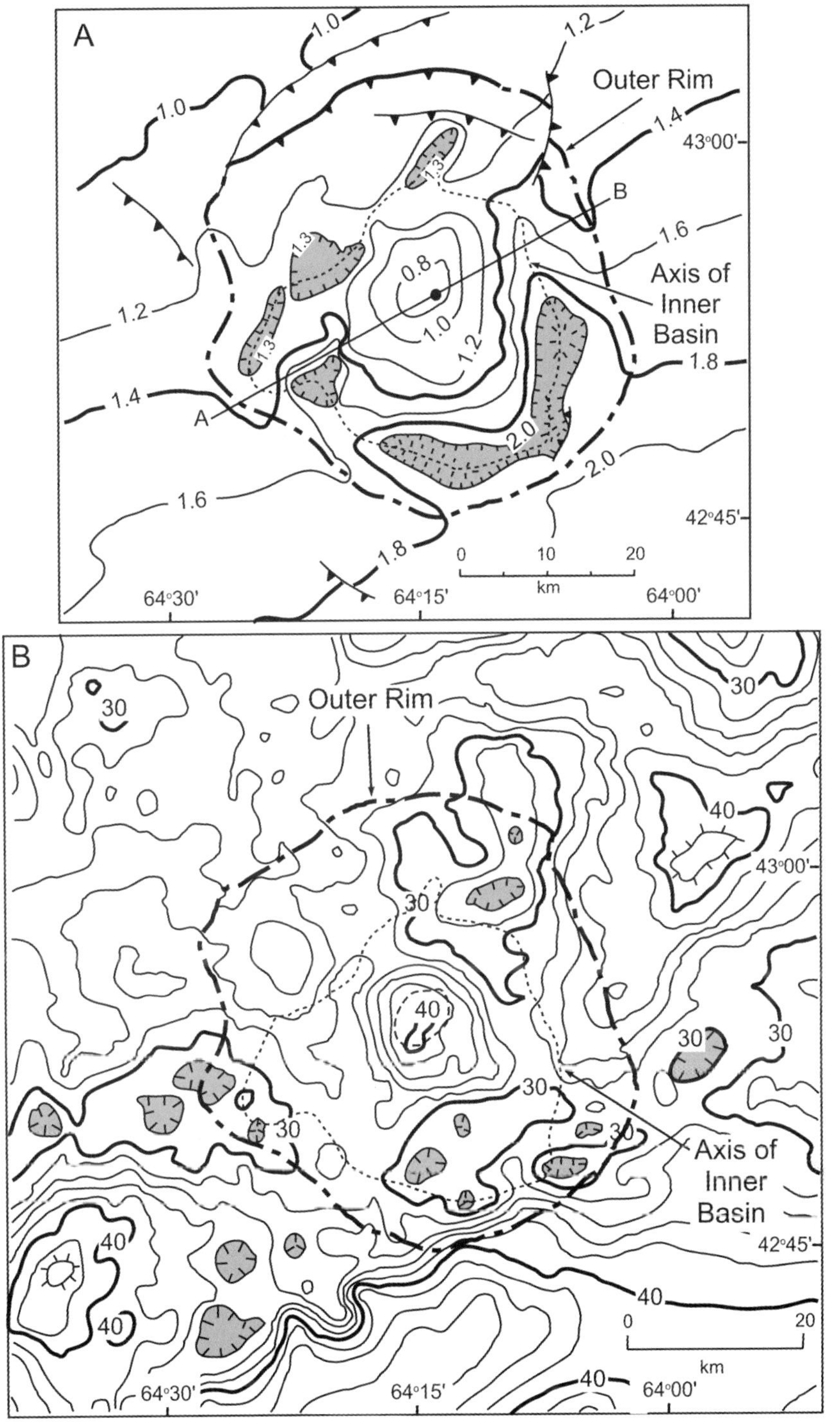
A
1.0
1.2
Outer Rim
1.4
43°00'
B
1.6
1.3
0.8
1.0
1.2
Axis of
Inner
Basin
1.8
A
2.0
1.6
42°45'
0
10
20
km
64°30'
64°15'
64°00'
B
30
Outer Rim
40
43°00'
30
40
30
30
30
30
30
Axis of
Inner
Basin
42°45'
40
40
40
0
20
km
64°30'
64°15'
40
64°00'

sedimentary target rocks. They inferred that the unconsolidated shelf deposits were compacted, but not fractured by the impact, thus preventing the low-density contrast that causes the gravity low characteristic of many subaerial craters. Pilkington et al. (1995) implied that the lack of a central negative gravity anomaly might be characteristic of submarine impact structures. On the other hand, Poag et al. (2002) superimposed the structure and gravity maps of Montagnais and concluded that a subcircular negative gravity anomaly marks the narrow inner basin of Montagnais (Fig. 10.7). Furthermore, the presence of a distinct central anomaly at Chesapeake Bay (Chapter 4) argues against the general lack of such an anomaly in submarine impact craters.

The next smallest of the impact craters still buried by marine waters is the Mjølnir crater (40-km diameter, Late Jurassic, Barents Sea, Norway; Dypvik et al. 1996; Tsikalas 1996 and papers cited therein; Tsikalas et al. 1998a,b, 1999; Shuvalov et al. 2002; Figs. 10.3A, 10.8–10.10). A recently drilled continuous core on the central peak of Mjølnir penetrated 83 m of chaotically deposited sediments (ejecta and slumped debris; Smelror et al. 2001). The nearest other analyzed deep borehole is 30 km outside the Mjølnir crater rim. Extensive seismic reflection surveys have been carried out over the crater, however (872 km of high-resolution, single-channel data; 174 km of shallow, 60-fold, multichannel data; 1081 km of conventional multichannel data; Fig. 10.8).

The most obvious difference between the Mjølnir crater and the Chesapeake Bay, Ries, Popigai, and Montagnais craters is the complete lack of crystalline rocks in the Mjølnir structure. Nevertheless, the general morphology of Mjølnir is similar to that of the other structures (Figs. 10.3A–C, 10.9, 10.10). The outer rim of Mjølnir is irregularly ovate, and is expressed as a steep fault scarp with no raised lip (Figs. 10.3A, 10.8–10.10). A 12-km wide annular trough is shallow, nearly flat, is floored by sedimentary rocks, and forms a décollement (detachment zone) separating the trough floor from the sedimentary megablocks lying above (Figs. 10.3A, 10.9, 10.10). Displaced megablocks present in the outer part of Mjølnir's annular trough are decoupled from the trough's floor by listric normal faults, as at Chesapeake Bay. We should point out that Tsikalas (1996) and Tsikalas et al. (1999) interpreted the top surface of the megablocks to be the floor of the annular trough, whereas we place the trough floor at the décollement beneath the megablocks (as at Chesapeake Bay and Montagnais). Our placement of the trough floor doubles the height of the outer-rim escarpment (to ~300–500 m) compared to Tsikalas's interpretation (Figs. 10.3A, 10.9, 10.10).

On the east side of Mjølnir, a slight arch of the décollement may mark the presence of a subtle, low-relief peak ring. On the west side, the presence of a peak ring cannot be confirmed. A narrow, slightly depressed décollement interval on the east side of the crater could conceivably be considered as an inner basin (Fig.

Fig. 10.7. (Previous page) Comparison of structure map and gravity anomaly map for Montagnais crater. *A*, structure map; contour interval 0.2 km; enclosed depressions of inner basin shaded; *A-B* is trackline for seismic profile (see Figs. 10.5, 10.6). *B*, gravity anomaly map with main structural features superposed to show coincidence between inner basin and ring of gravity lows. Modified from Poag et al. (2002).

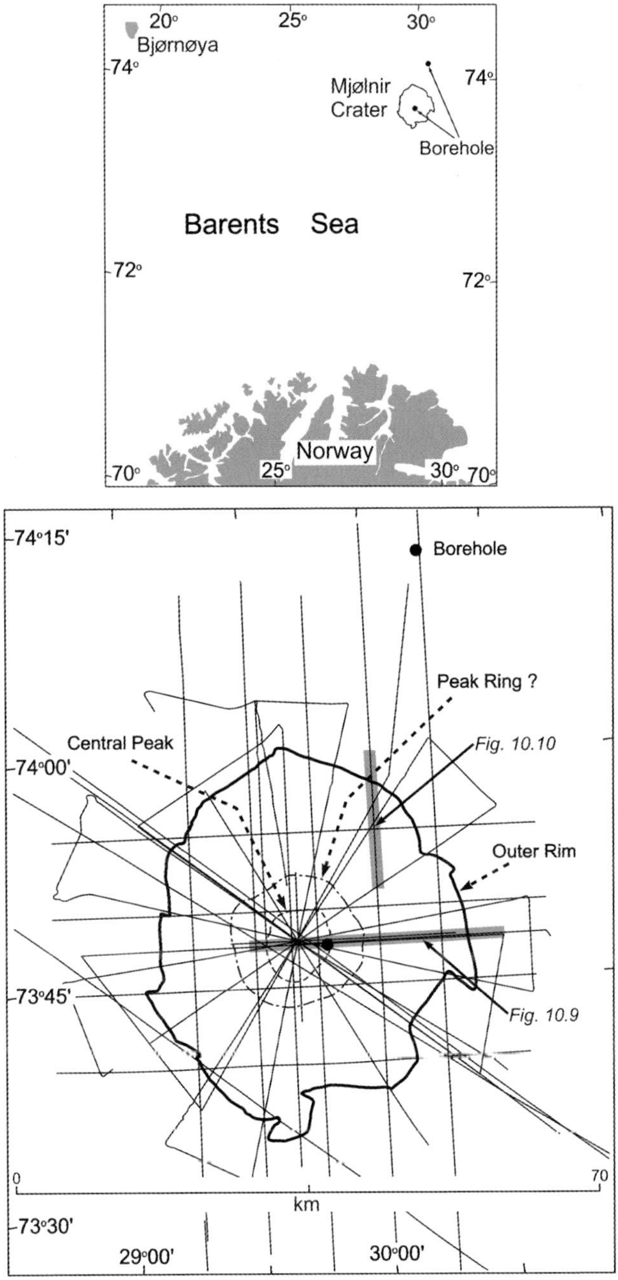

Fig. 10.8. Location maps showing main structural features of Mjølnir crater, seismic tracklines, and boreholes (modified from Tsikalas et al. 1999). Shaded segments of seismic lines are those interpreted by us in Figs. 10.9, 10.10.

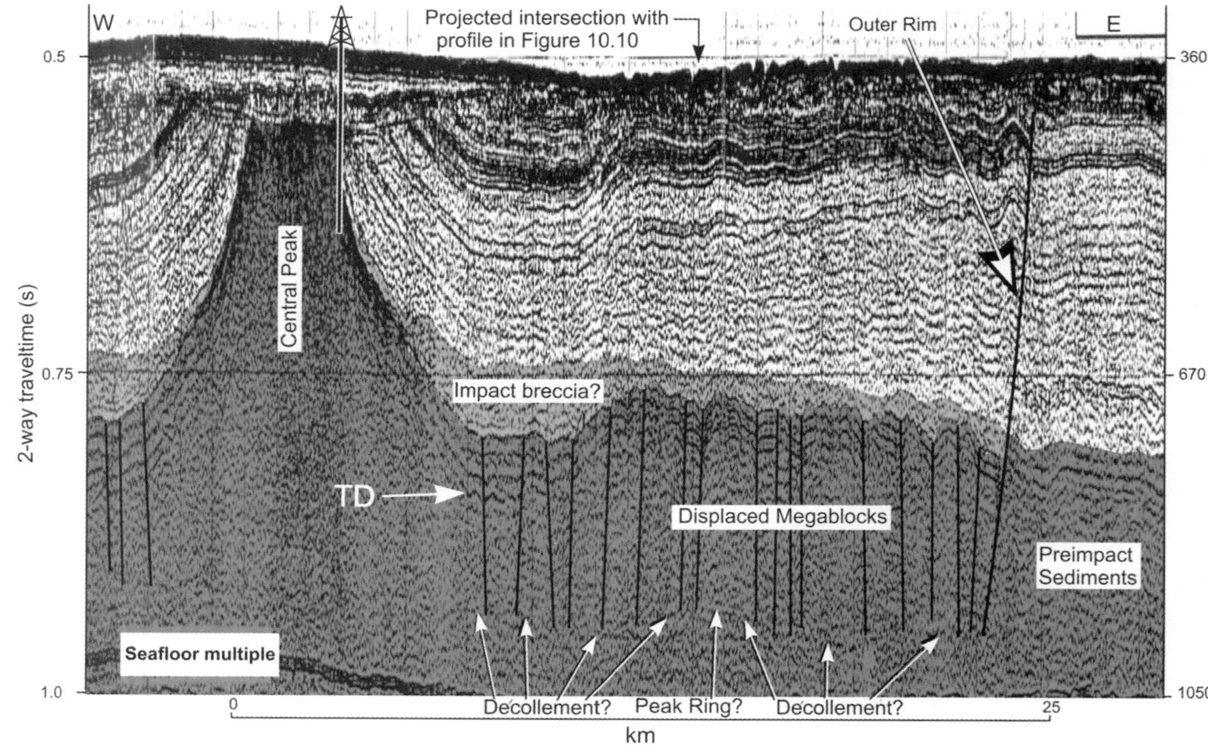

Fig. 10.9. Segment of single-channel seismic profile across central peak and annular trough of Mjølnir crater. Preimpact sediments in darker shading; impact breccia(?) in lighter shading. Interpretation modified from Tsikalas et al. (2002), who interpreted reflection at "TD" as floor of crater, whereas we consider the décollement as structural floor (see section 10.2). See Fig. 10.8 for location of seismic trackline.

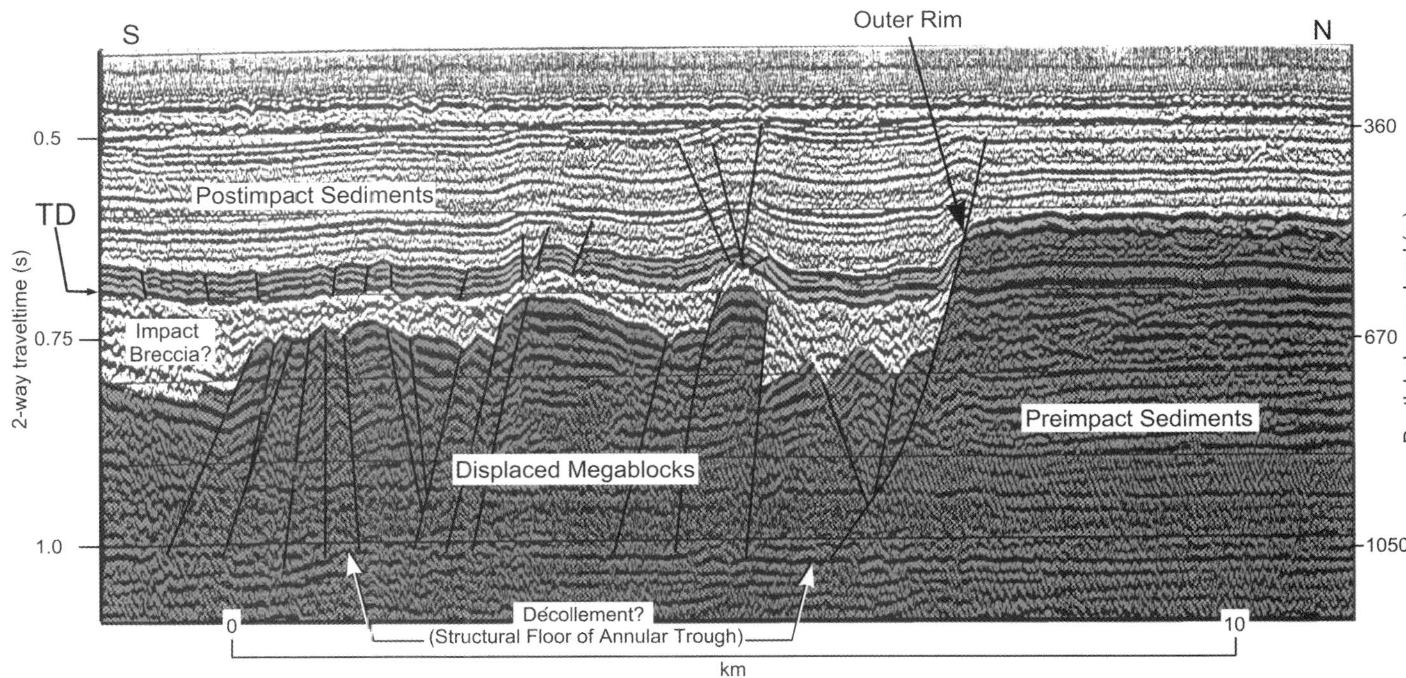

Fig. 10.10. Segment of shallow multichannel seismic profile across annular trough and outer rim of Mjølnir crater. Preimpact sediments in darker shading; initial stratified postimpact sediments in lighter shading. Interpretation modified from Tsikalas et al. (1998a). Tsikalas et al. (1998a) interpreted "TD" as floor of crater (= morphological floor in our terminology; see section 10.2) See Fig. 10.8 for location of seismic trackline

10.9). Mjølnir's prominent central peak (12-km diameter across the base; 8-km diameter at the crest; composed entirely of sedimentary rocks) rises 300 m higher than the lip of the outer rim, and ~500 m above the floor of the inner basin (~250 m higher than the inner basin according to Tsikalas; Figs. 10.03A, 10.9, 10.10). The elevation of the peak was originally even greater, because its crest has been truncated by Late Cenozoic glacial erosion. The central peak is covered directly by postimpact sediments (no impact breccia indicated over the peak).

The Manson impact crater is considered to be a *possible* shallow-water submarine crater (36-km diameter; Late Cretaceous; 74 Ma; Manson, Iowa; Koeberl and Anderson 1996a and papers therein; Figs. 10.11–10.13). Though now located in the middle of the North American craton, Manson has been interpreted by some researchers to have originated from an impact into the Late Cretaceous interior seaway (an epicontinental sea; Izett et al. 1993; Anderson and Witzke 1996; Steiner and Shoemaker 1996). Coring (1,283 m recovered from 12 boreholes), combined with seismic reflection, gravity, and geomagnetic surveys, has revealed the following concentric features at Manson (Fig. 10.12): (1) a fault-bounded outer rim of 36-km diameter, constructed within consolidated Paleozoic and Mesozoic sedimentary rocks (sandstone, siltstone, shale, limestone), surrounding; (2) a zone of terraced, folded, and overturned megablocks (annular trough) ~7 km wide; we have examined a full-scale copy of the north-south seismic profile across Manson crater (Fig. 3 of Schultz and Anderson 1996; our Figs. 10.12, CD-ROM.18) and we interpret the slumped megablocks of the annular trough to have dropped down along listric normal faults, which sole out along a décollement at the top of crystalline basement, just as at Chesapeake Bay; (3) a "moat," (inner basin) ~6 km wide, floored by plutonic, and metasedimentary basement rocks; and (4) a crystalline central peak ~10 km in diameter at the base and ~8 km across the crest. The central peak, which exhibits a notable central depression, is capped by scattered deposits of four different breccia units (Witzke and Anderson 1996). The upper part of the Manson structure, including the central peak, is truncated by an erosional surface, which was subsequently buried by a layer of Quaternary glacial till. Thus, though the Manson crater lacks a peak ring, in many other respects it resembles a scaled-down version of the Chesapeake Bay crater (Fig. 10.13).

Ormö (1998) and Ormö and Lindström (2000) summarized most previous studies of submarine impacts (marine-target impacts), presented new data from detailed field studies (particularly of Sweden's 24-km-diameter Lockne crater; Figs. 1.1, 10.14–10.16), and provided a synthesis of those features that appear to separate submarine craters from subaerial ones on Earth and on Mars. Some of the results published by Ormö and his colleagues help to refine our interpretation of the Chesapeake Bay structure.

The general stratigraphic framework of preimpact rocks at Lockne, the structure, morphology, and impact deposits associated with the crater, and the disposition of postimpact sedimentary rocks closely resemble those of the Chesapeake Bay crater. However, Ormö (1998) used a different descriptive terminology than ours, which he derived from Quaide and Oberbeck's (1968) study of "concentric" lunar craters. Quaide and Oberbeck (1968) hypothesized that when a bolide strikes a two-layered target in which the lower layer is denser than the upper layer,

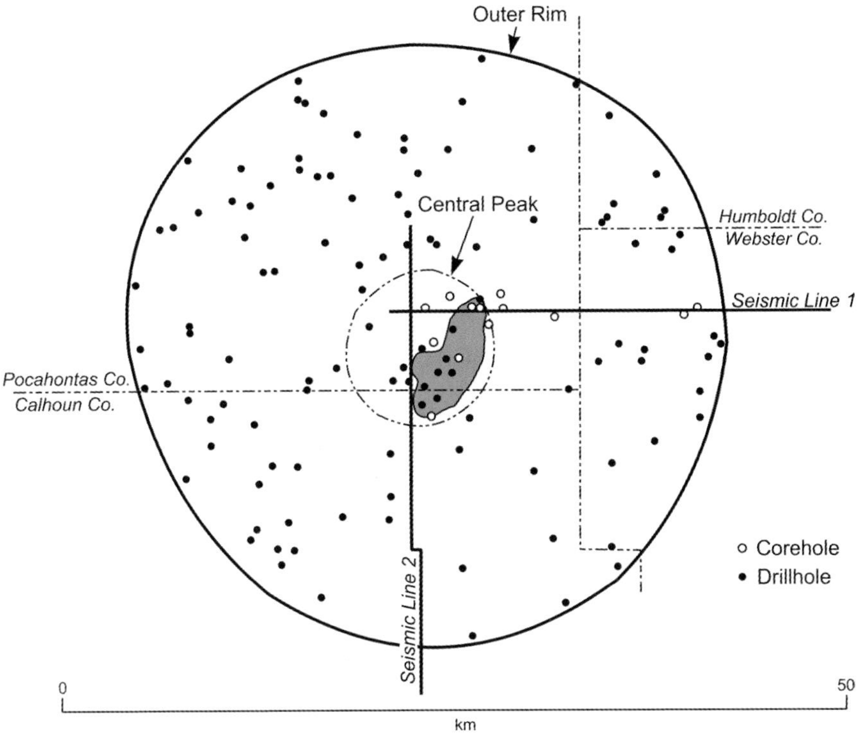

Fig. 10.11. Map view of Manson impact crater showing seismic tracklines and boreholes. Central shaded area indicates structural depression on central peak. Modified from Witzke and Anderson (1996). Part of seismic Line 2 is shown in Figs 10.12, CD-ROM.18.

the impact creates two concentric craters (or several concentric craters in the case of multilayered substrates). Quaide and Oberbeck emphasized, however, that their conclusions applied only to craters with diameters measured in meters, not kilometers. They specified that the morphology of a crater wider than a few hundred meters does not reflect differences in strengths of individual near-surface rock layers. The Quaide-Oberbeck cratering model, therefore, seems inappropriate for submarine craters the size of the Chesapeake Bay crater.

The outer rim of the Lockne crater is an erosional escarpment excavated from sedimentary rocks, though they are much older than preimpact rocks at Chesapeake Bay and are strongly lithified (Lindström et al. 1996). Sixty-five meters of bedded Ordovician limestone overlie 15 m of Cambrian black shale. A gently dipping, flat-floored annular trough ("outer crater" of Lindström et al. 1996; Ormö 1998; von Dalwigk and Ormö 2001), 3 km wide, surrounds a subtle peak ring and an inner basin (7.5 km in diameter; "inner crater" of Ormö 1998), formed in Revsund Granite, which constitutes the Proterozoic crystalline basement rock. Ormö (1998), following Quaide and Oberbeck (1968), referred to the inner basin as a "nested crater." Drilling and gravity measurements indicate that the inner

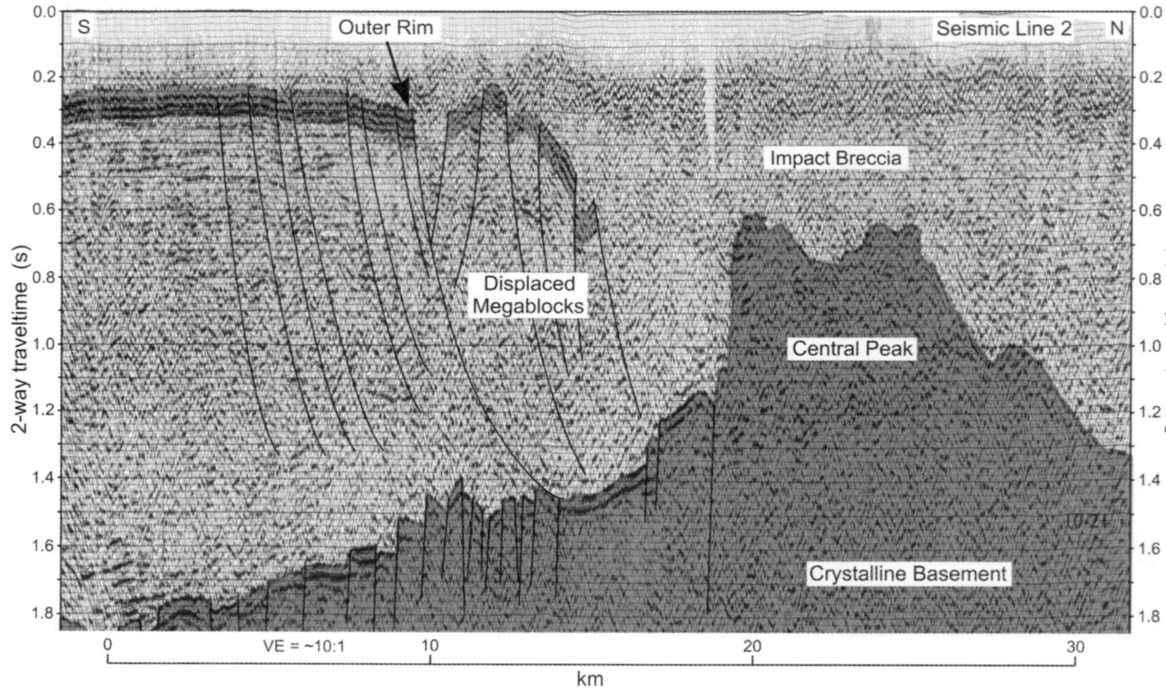

Fig. 10.12. Interpreted segment of seismic reflection profile 2 across Manson impact crater (profile courtesy of Charles Pillmore). Upper shaded interval is series of high-amplitude reflections at top of preimpact sedimentary section. Note that seismic signature of crystalline basement is high-amplitude couplet similar to that of Chesapeake Bay crater. See Fig. 10.11 for location of seismic profile and CD-ROM.18 for full-scale profile. Interpretation modified from Schultz and Anderson (1996).

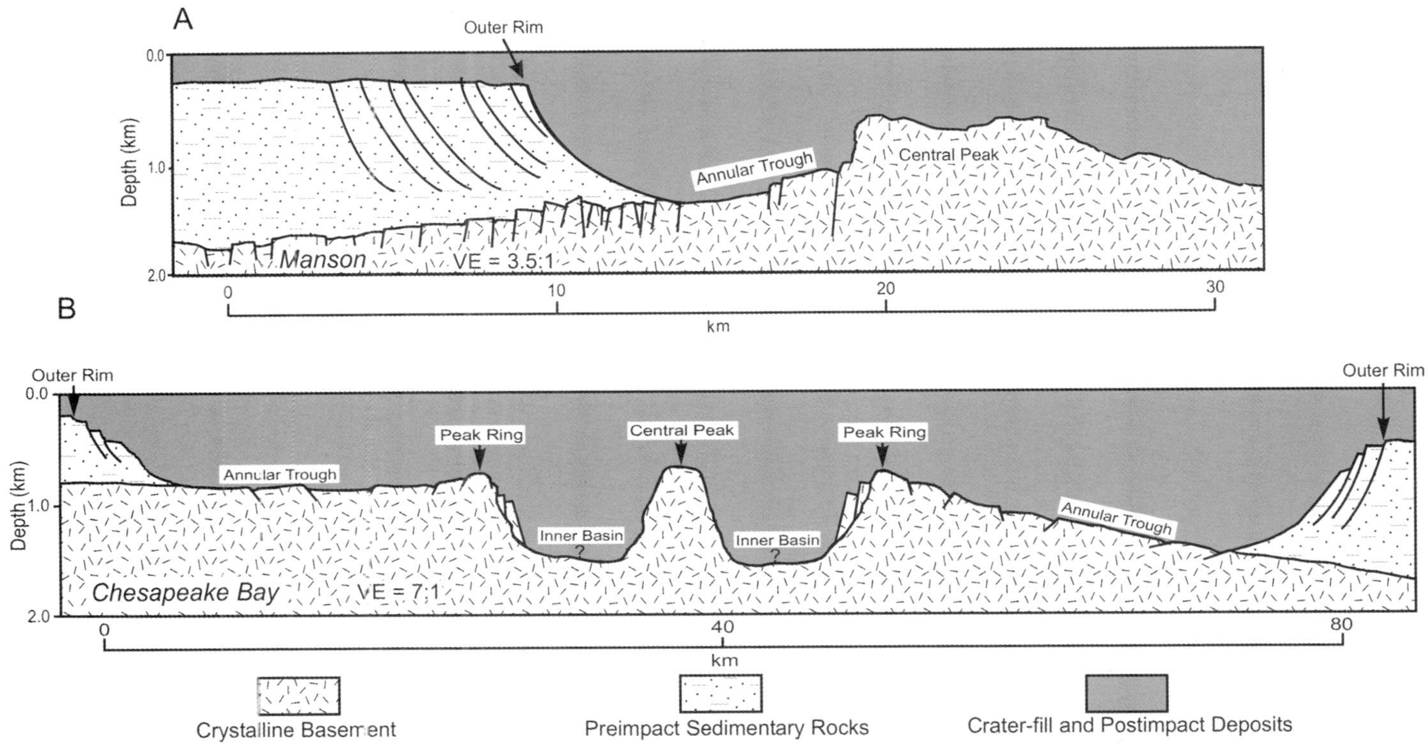

Fig. 10.13. Comparative cross sections for Manson crater (*A*) and Chesapeake Bay crater (*B*). Note difference in vertical exaggeration.

basin at Lockne also contains a low-relief crystalline central peak, though in contrast to Chesapeake Bay, the Lockne peak is not buried by impact breccia (Lindström et al. 1996). Autochthonous, monomictic, Tandsbyn Breccia fills the inner basin of Lockne crater (see Chapter 11). The basal contact between Tandsbyn Breccia and crystalline bedrock varies from sharp to gradual.

The feature that sets Lockne apart from known subaerial craters on Earth is the presence of four broad, breccia-filled gullies (maximum size is 1 km wide, 3 km long, 100 m deep), which slice radially through the annular trough and peak ring and extend into the inner basin (Fig. 10.15). Ormö (1998) and von Dalwigk and Ormö (2001) referred to these gullies as *resurge* features (we prefer *surgeback* as a parallel construction to the terms *fallback*, *slumpback*, and *washback*; see Chapter 12). In Ormö's hypothesis, the surgeback process involves violent collapse of the ejecta-filled water column and turbulent bottom-flow back toward the center of the crater during the modification stage of crater formation. This process is believed to have hydraulically eroded fracture-weakened zones within the peak ring and crater floor, to produce surgeback gullies. The surgeback gullies subsequently filled with surgeback breccia.

Ormö and Miyamoto (2002) applied numerical modeling to estimate the magnitude of surgeback flow at Lockne. They concluded that 1.2×10^{11} m^3 of sea water would be required to fill in the impact excavation at Lockne. Assuming an initial water depth of 200 m at Lockne, Ormö and Miyamoto (2002) calculated an average surgeback velocity of 27.5 m s^{-1}, which took 2200 s (36.7 min) to refill the excavation cavity. Such a high-velocity flow created a maximum erosive force (unit stream power) of 1.9×10^5 W m^{-2}. This is the same order of magnitude of erosive force calculated for the glacial Lake Missoula Flood, which produced the high-relief Channeled Scablands of Washington State (Baker and Milton 1974). At Chesapeake Bay, the combination of greater paleodepth, a much larger impact crater, and the poorly consolidated nature of the sedimentary target, would have produced truly prodigious surgeback erosion on that late Eocene seafloor.

As at Chesapeake Bay, marine sedimentation resumed at Lockne immediately following the impact. But unlike Chesapeake Bay, the Lockne crater and its breccia deposits are covered by a moderately deep-water (>200 m paleodepth) carbonate deposit (the Dalby Limestone). The biochronological age of the Dalby Limestone is the same (middle Ordovician) as the youngest preimpact sediments in the floor of the crater. Postimpact sedimentary deposits thicken over the Lockne crater, just as they do over the Chesapeake Bay crater.

The general structure and morphology of the Chesapeake Bay crater appear to be very similar to those of the Lockne crater, but on a grander scale (Fig. 10.16). Especially notable at Lockne are: (1) the unraised sedimentary outer rim; (2) the flat, shallow, annular trough; (3) the low-relief, crystalline peak ring; (4) the deep inner basin excavated into crystalline basement; and (5) the subtle crystalline central peak buried by postimpact sediments. The two craters differ in detail, however. The Chesapeake Bay crater is, of course, nearly four times larger, the peak ring has greater relief, and the inner basin is much deeper than Lockne. Furthermore, the outer rim at Chesapeake Bay is a fault scarp, rather than an ero-

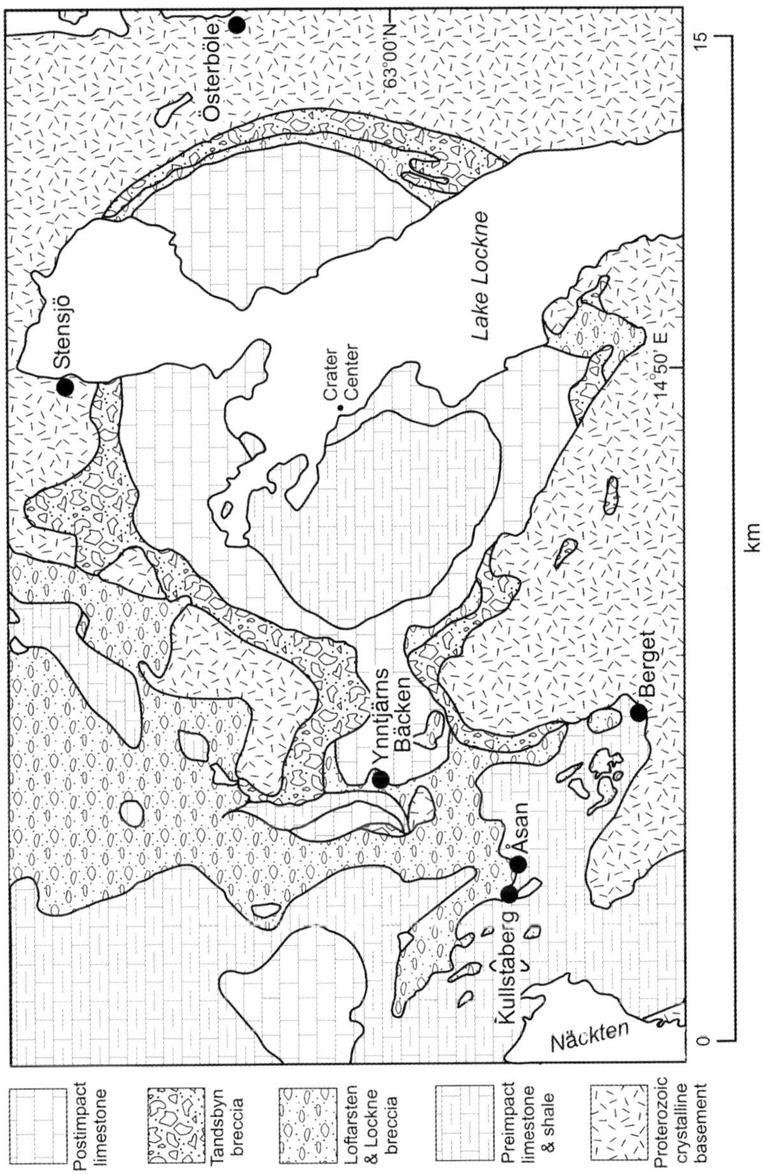

Fig. 10.14. Geological map of Lockne impact crater (modified from Ormö 1998).

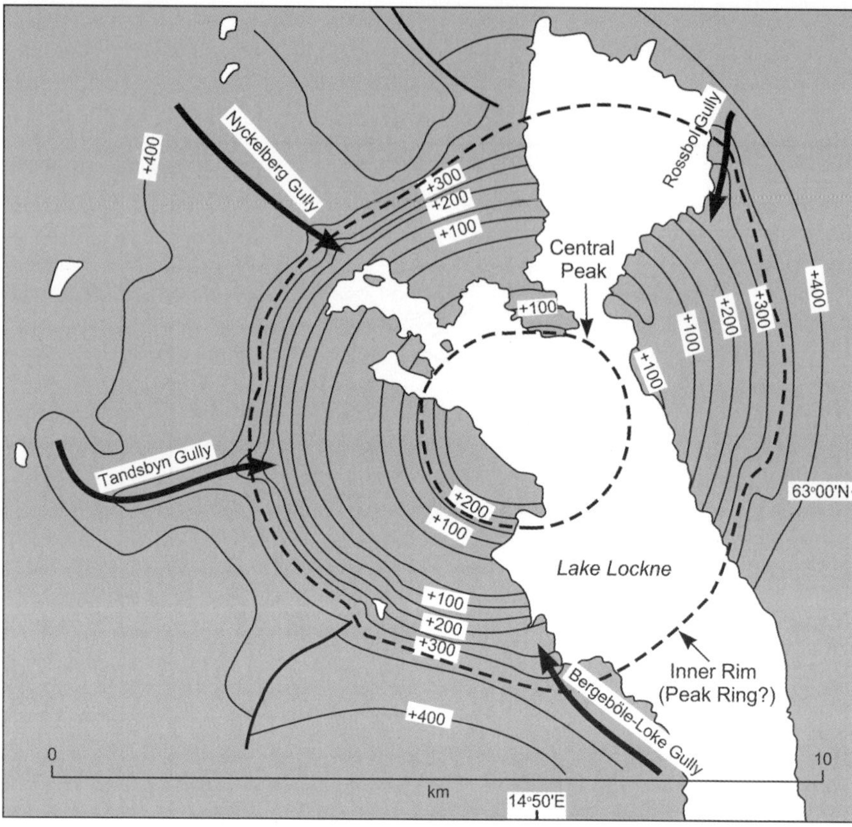

Fig. 10.15. Structure map of Lockne crater showing elevation relative to sea level (contour interval 50 m) and position of four surgeback gullies. Modified from Ormö (1998).

sional one; the floor of the annular trough is crystalline basement rather than sedimentary rock; and in the annular trough, kilometer-scale megablocks lie between the impact breccia and the basement. Also, the crater-fill breccia at Chesapeake Bay is much thicker and completely buries the peak ring and central peak. The gravity signature over Chesapeake Bay suggests that the peak ring may be breached in the southeast quadrant, which could indicate the presence of a surgeback gully, but we have no deep seismic data in that area to confirm a gully-like morphology, nor are there core data to determine the possible presence of surgeback breccia at that location.

Ormö (1998) identified two additional exposed submarine craters with breccia-filled surgeback gullies. The Kamensk crater (Fig, 1.1), a structure buried near the Ukraine-Russia border, has been explored by more than 330 boreholes (Movshovich and Milavsky 1990). Kamensk displays 12 branching gullies, some 100 m deep, filled with allogenic breccia and attributed by Ormö (1998) to surge-

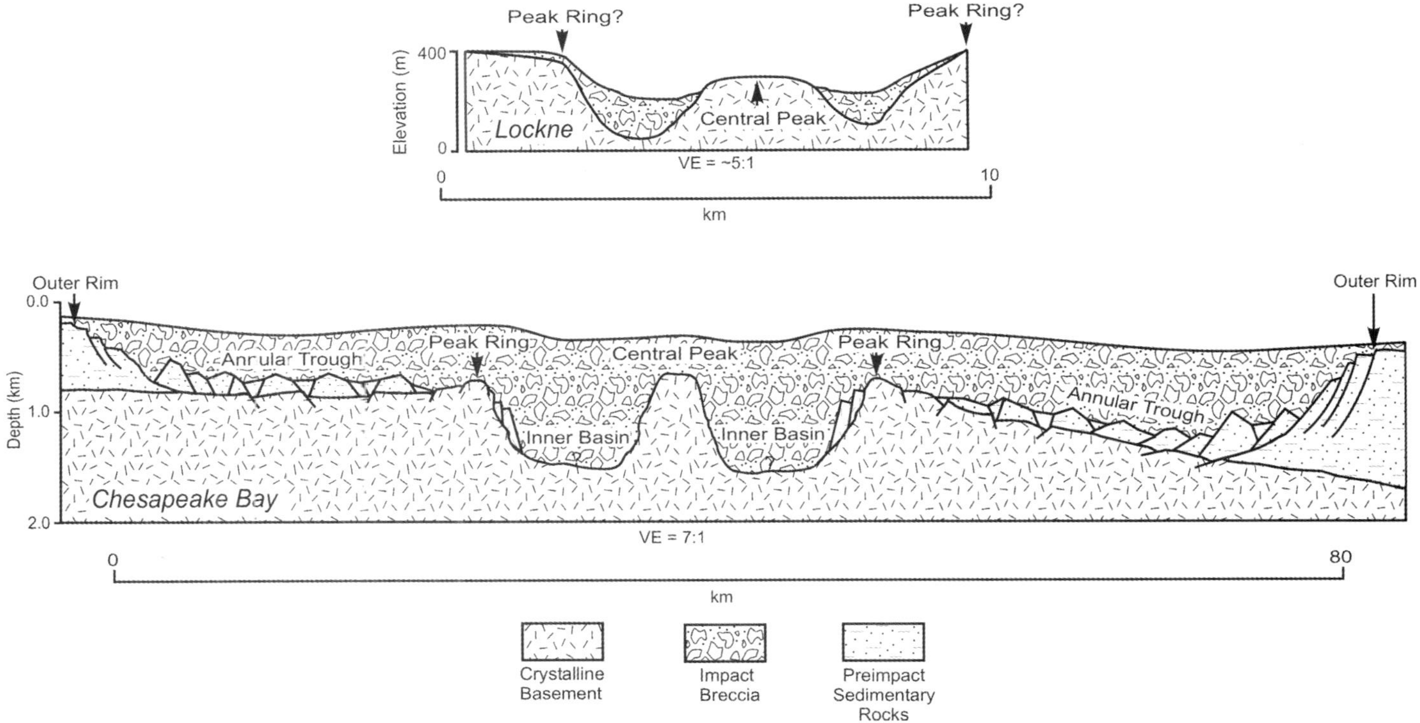

Fig. 10.16. Comparative geologic cross sections for Lockne crater and Chesapeake Bay crater (postimpact deposits removed). Note difference in vertical exaggeration.

back processes. Like Lockne, the Kamensk crater has an annular trough (9 km wide), the inward-sloping floor of which is eroded into preimpact sedimentary rocks. A 20-km-diameter peak ring is excavated more deeply into the sedimentary strata but does not penetrate crystalline basement rocks. A sedimentary central peak also is present, and is buried by a thick lens (~2 km) of authigenic breccia overlain by ~200 m of allogenic, polymict, sedimentary breccia.

Kärdla is a much smaller, buried, Ordovician submarine crater (4-km diameter) excavated into crystalline rocks of the Baltic Shield in Estonia (Suuroja et al. 2002; Fig. 1.1). The structure and stratigraphy of Kärdla are known from 160 boreholes, one single-channel seismic reflection profile, and gravity and magnetic surveys. Ormö (1998) and Suuroja et al. (2002) identified two surgeback gullies with associated surgeback breccia at Kärdla.

From present evidence, then, it appears that the Chesapeake Bay crater is the only unequivocal peak-ring/central-peak crater among the larger complex submarine structures currently known. Montagnais may have a genuine peak ring, but only one seismic profile across Montagnais has been published, and the ring is difficult to ascertain on the west side of the structure. Mjølnir has been described as having a peak ring, but the supposed ring is a subtle feature, formed entirely by sedimentary rocks; it's precise nature is obscure. Definition of Mjølnir's morphology is complicated by the fact that it has undergone postimpact compaction, which may have altered the original morphology of a peak ring. Kamensk has a sedimentary peak ring; Lockne's central peak is poorly developed; the Toms Canyon structure formed entirely within sedimentary strata and, so far, has given no hint of a peak ring or central peak; the Manson structure has no peak ring.

10.2 Extraterrestrial Craters

Geometrical analyses (morphology, structure) of impact craters on other planetary bodies are limited by the inability to image the *structural* floor of the craters. That is because the craters are partly filled with deposits such as impact breccia, basalt flows, and impact melt sheets, whose upper surfaces form the observable *morphological* floor. Peak rings and central peaks cannot be recognized on other planets unless they protrude above the crater-fill deposits. In the case of the Chesapeake Bay crater, on the other hand, we have numerous seismic profiles that image the peak ring, and a few that document the central peak, even though these features are deeply buried by Exmore breccia. A datum comparable to the morphological floor of a typical planetary crater would be at a level within the Exmore breccia, above which the peak ring and central peak would protrude (Fig. 10.17). Such a surface, however, is a conceptual feature, which cannot be traced within the chaotic reflections of the Exmore breccia. The best available proxy for that surface is the upper surface of the Exmore breccia. Because of differential breccia compaction, the morphology of this breccia surface mimics the morphology of the under-

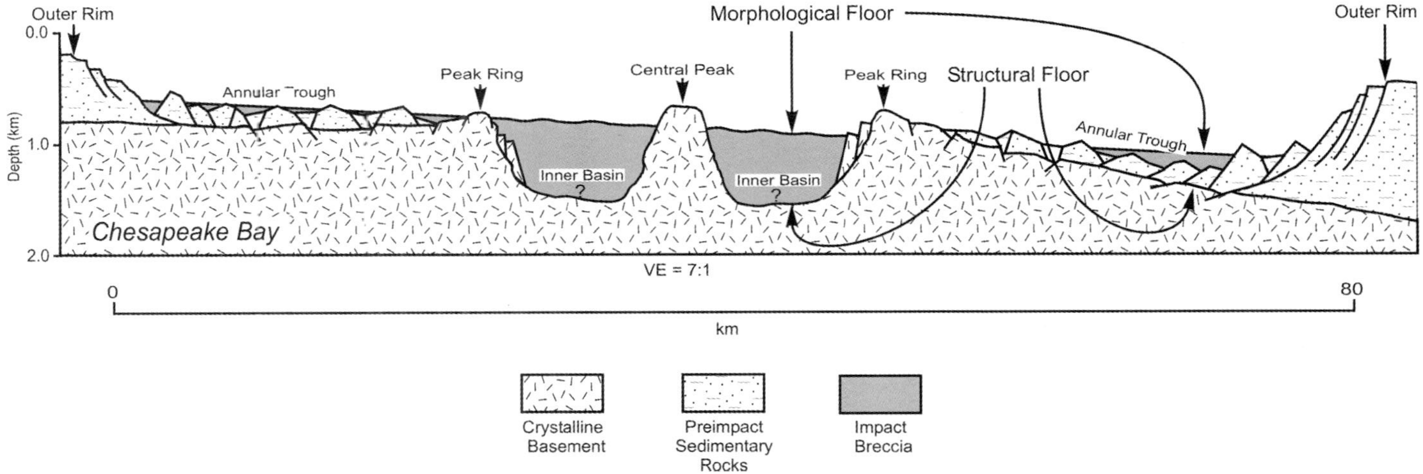

Fig. 10.17. Cross section of Chesapeake Bay crater, showing hypothetical conditions necessary to simulate typical planetary crater as imaged from Earth or spacecraft (thin layer of crater-fill breccia and lack of postimpact deposits would expose crest of peak ring, crest of central peak, and tips of displaced megablocks). Note difference between morphological and structural floor of crater.

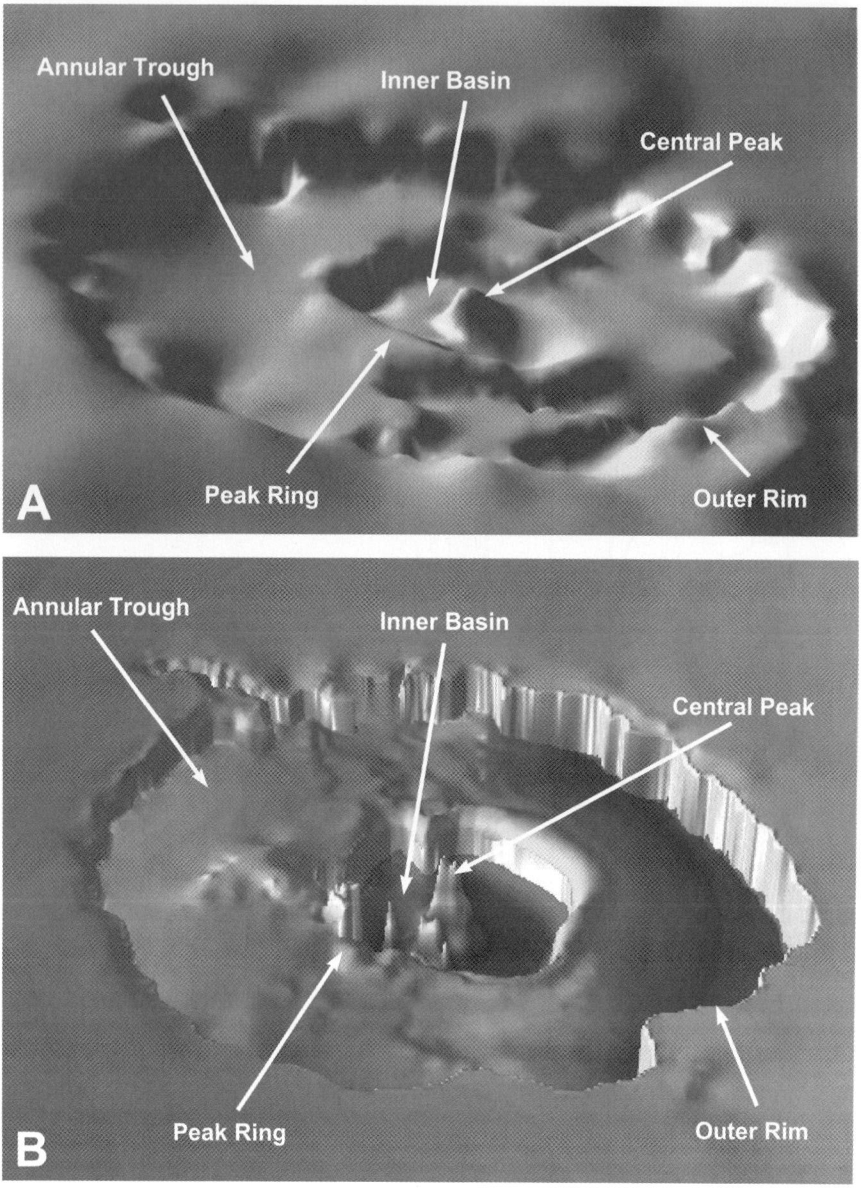

Fig. 10.18. 3-D perspectives comparing morphological features of surface of Exmore breccia (*A*, this is morphological floor of crater) with those of surface of crystalline rocks (*B*, this is structural floor of crater). See CD-ROM.4 for more details of crater morphology.

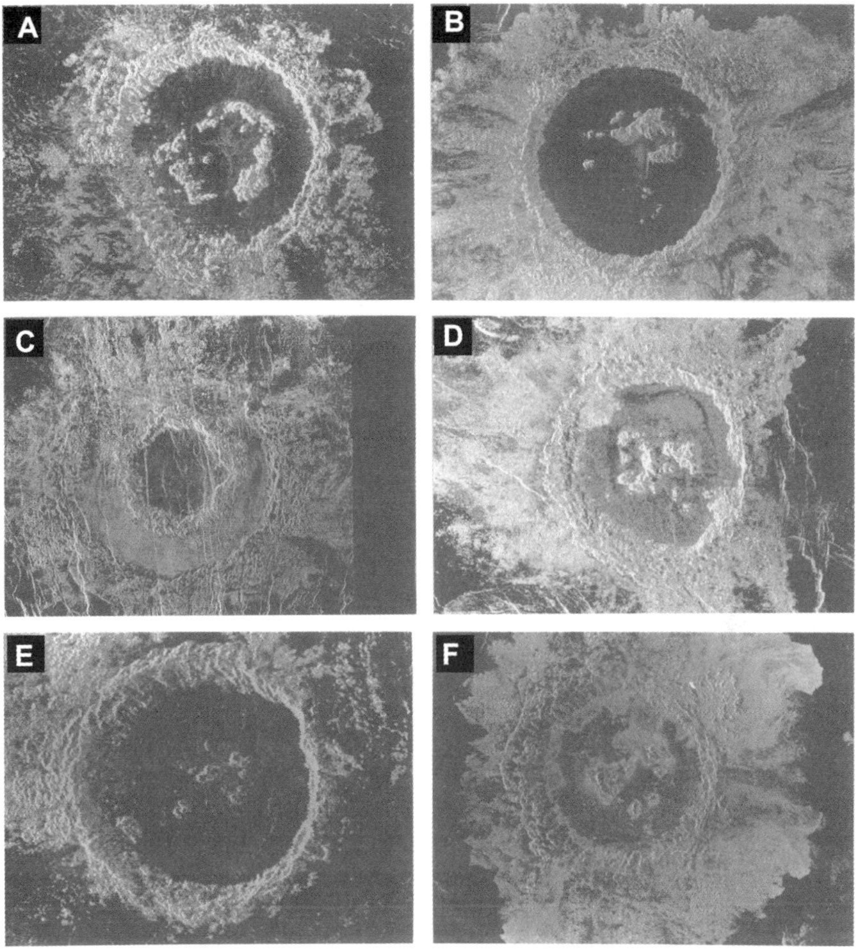

Fig. 10.19. Selected Magellan radar images of Venusian peak-ring/central-peak craters, showing variability in geometry and distribution of peak rings and central peaks. *A*, Bartan crater; 51-km diameter; type A crater with nearly complete ring of highly irregular ridges and peaks, including several isolated knobs; *B*, Aglaonica crater; 62-km diameter; type B crater with partial ring of extremely irregular ridges, peaks, and isolated knobs; *C*, Bonheur crater; 102-km diameter; type C crater displaying continuous knobby ridge with steep scarp facing inner basin; *D*, Charpentier crater; 48-km diameter; type D crater with closely clustered group of isolated irregular peaks and ridges; *E*, Lagerof crater; 57-km diameter; type B crater with highly irregular partial ring of isolated peaks surrounding irregular central peak; *F*, Yablochinka crater; 66-km diameter; type A crater with nearly complete ring of irregular peaks and ridges surrounding small central peak. Images courtesy of WB McKinnon. See Alexopolous and McKinnon (1994) for discussion of craters types A–D.

lying basement surface. Thus, a structure map of the Exmore breccia surface (i.e., the morphological floor of the crater; Figs. 10.18A, CD-ROM.4) displays the same principal features (outer rim, annular trough, peak ring, central uplift) as the buried structural floor (Fig. 10.18B), though the vertical relief on the breccia surface is much reduced.

Morphometric studies of peak-ring and central-peak craters on the Moon (Hale and Head 1979, 1980), Mercury (Hale and Head 1980; Pike 1988; Neukum et al. 2001), Mars (Wood 1980), and Venus (Alexopolous and McKinnon 1992, 1994; Wood and Tam 1993) indicate that craters displaying both peak rings and central peaks (protobasins of Pike 1988; Type D craters of Alexopolous and McKinnon 1994; Fig. 10.19) are comparatively rare in the solar system. Melosh's (1982) hydrodynamic model for ring and central peak formation explains this by treating peak rings as late-stage products of central-peak collapse (see also O'Keefe and Ahrens 1999). Thus peak-ring/central peak craters are envisioned as evolutionary transitions between craters with only central peaks and those with only peak rings. Many field data support this model (Hale and Grieve 1982; Alexopolous and McKinnon 1994), but agreement is not unanimous (Hale and Head 1980). Morphologic and structural variability may depend on such factors as the preimpact geology of target rocks, the velocity, size, and incidence angle of the impactor, and the degree to which crater-fill deposits bury the structural crater floor.

The general morphology of the Chesapeake Bay crater agrees well with that of both Martian protobasins and Venusian Type D craters in most respects. Chesapeake Bay differs, however, in having a larger diameter than most Type D craters. Alexopolous and McKinnon (1994) showed that Type D craters cluster between diameters of 40–50 km, though one example was 70 km wide. In contrast, the diameters of protobasins on Mars cluster at 100–150 km with maximum diameters of 215–285 km (Wood 1980). Alexopolous and McKinnon (1994) suggested that the diameter differences might be related to geological differences. This would fit the increasing body of evidence indicating that the geological history of Mars is more comparable to Earth's than to that of Venus.

The King crater on the Moon's farside appears to be (morphologically and structurally) an approximate lunar analogue to the Chesapeake Bay crater (e.g., Greeley 1994; Schultz and Anderson 1996; Fig. 10.20). The King crater is approximately the same diameter (75 km) as Chesapeake Bay, and displays a distinctly breached crystalline peak ring, from which Schultz and Anderson (1996) inferred a low-angle impactor trajectory. Missing from King is a visible central uplift, though the possible presence of a buried central peak cannot be ruled out. The King crater also differs from the Chesapeake Bay crater in that the peak ring and displaced megablocks at King are not covered by crater-fill breccia; the outer rim is distinctly raised; the surrounding ejecta blanket is well developed; and the ratio of outer-rim diameter to peak-ring diameter is not proportional between the two craters (Fig. 10.21).

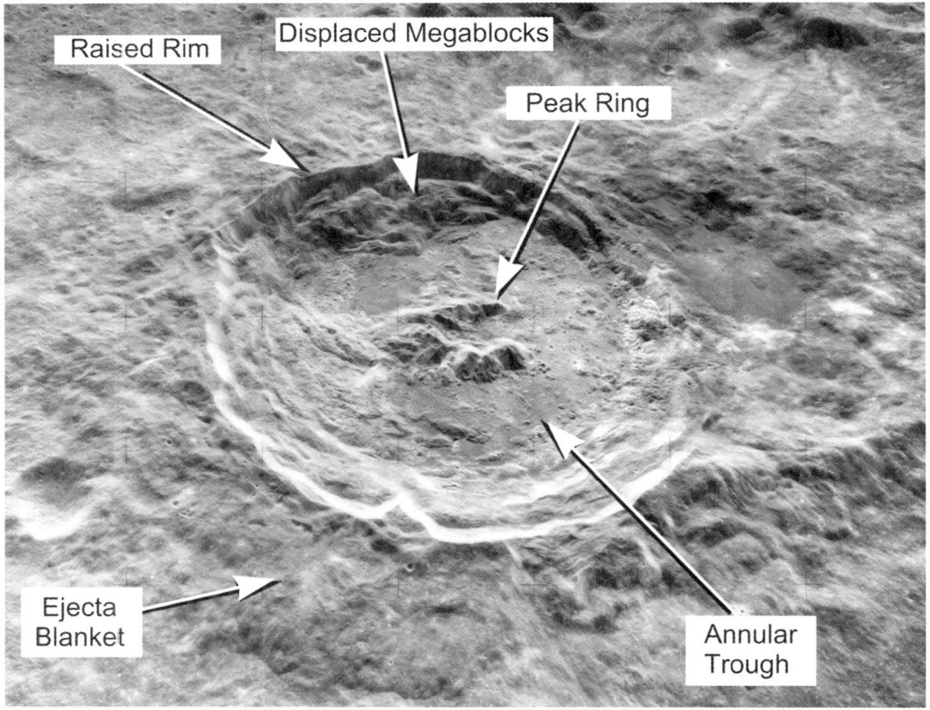

Fig. 10.20. Apollo 16 image of King impact crater (~75-km diameter) on lunar farside, showing principal morphological features. Note general similarity to those of Chesapeake Bay crater, except that here: (1) rim has distinct raised lip; (2) ejecta blanket is strongly developed; and (3) lack of thick crater-fill breccia exposes displaced megablocks, floor of annular trough, and peak ring. Image courtesy of Ronald Greeley.

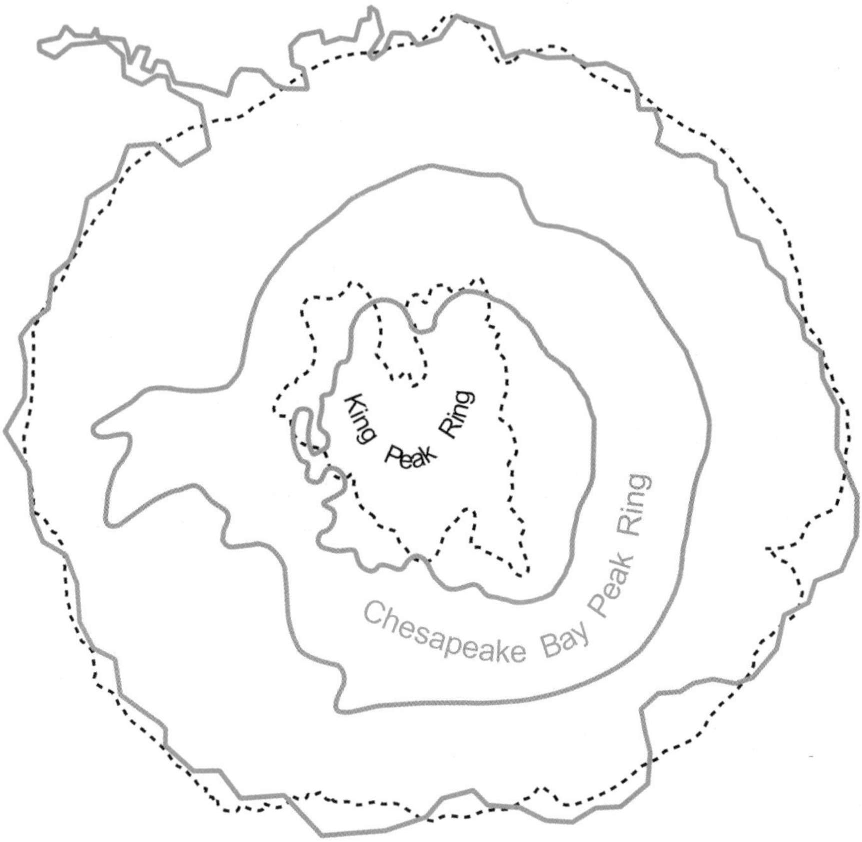

Fig. 10.21. Superposed outlines (using identical diameters) of Chesapeake Bay crater (solid lines) and King crater (dashed lines), showing non-proportional peak rings (outer diameter of King peak ring approximates inner diameter of Chesapeake Bay peak ring).

10.3 Comparison with Chicxulub Multiring Impact Basin

The Chicxulub multiring basin is the largest impact structure to be surveyed with a state-of-the-art multichannel seismic reflection system. The seismic data provide excellent 2-dimensional cross-section profiles of the northern third of Chicxulub's general structure and morphology (Snyder and Hobbs 1999b). Though its much greater size (~200 km diameter) and more complex structure and morphology place Chicxulub in a different category of impact structures (Morgan et al. 1997, 2002; Morgan and Warner 1999a,b; Grieve and Therriault 2000), it has many features analogous to those of the Chesapeake Bay crater. Like the Chesapeake Bay impactor, the Chicxulub bolide struck a broad continental shelf. The resultant crater was subsequently buried by ~1 km of postimpact marine sediments, and has not been perturbed by any significant tectonism. Several onshore boreholes provide stratigraphic evidence of the impact (crater-fill breccia, impact melt rocks). Offshore seismic surveys, especially the four deep-crustal multichannel profiles acquired by the British Institutions Reflection Profiling Syndicate (BIRPS; Morgan et al. 1997, 2002; Brittan et al. 1999; Morgan and Warner 1999a,b; Snyder and Hobbs 1999a; Figs. 10.22–10.27), provide structural and morphological cross sections of the northern third of the basin. Unfortunately, the borehole stratigraphy must be extrapolated long distances (75–230 km) to the seismic profiles (for example, the recently completed Yaxcopoil corehole is 85 km from its projection on the nearest offshore seismic profile; Smit et al 2002). Experience at other craters clearly demonstrates that such long-distance extrapolations can be seriously misleading. Nevertheless, a rough stratigraphic calibration of the seismic profiles has been attempted (Brittan et al. 1999). The published seismostratigraphic interpretations agree roughly with P-wave velocity interpretations, but correlations with gravimetric data are ambiguous (Brittan et al. 1999).

The outer ring of the Chicxulub basin is recognizable at ~80–100 km radial distance from the basin center on the BIRPS profiles (Figs. 10.22–10.26). (We should point out that three different sets of terminology have been applied to the principal topographic and structural features of the Chicxulub basin; we use the terminology of Morgan and Warner 1999b.) The outer ring is manifested by a zone of significant disruption of the preimpact sedimentary section (numerous megablocks bounded by moderate-offset normal faults) located farthest from the center of the basin. Morgan and Warner (1999b) described this feature as a graben linked to a fault or shear zone that dips 30–40 degrees toward the basin center and penetrates to the Moho. Because of the smaller size and different structural style of the Chesapeake Bay structure (it is a complex crater rather than an impact basin), we do not expect to find an analogue of the Chicxulub outer ring at Chesapeake Bay.

The next significant feature toward the center of the Chicxulub basin is the inner ring (Morgan and Warner 1999b). Seismic expression of the Chicxulub inner ring is marked by a distinct structural sag and thickening of the postimpact sedimentary section (Figs. 10.23–10.26). A nearly identical analogue is represented by the outer rim of the Chesapeake Bay crater (Fig. 10.27). A stratigraphically deeper structural change also marks the Chicxulub inner ring. This change is most clearly manifested by a series of prominent high amplitude, continuous, horizontal-to-subhorizontal reflections (inferred from borehole extrapolation to be a succession of carbonate rocks and evaporites), which can be traced laterally from their undisturbed counterparts outside the basin's outer ring (at ~3–3.5 km depth; Figs. 10.23–10.26). Approximately at the Chicxulub outer ring, the section represented by these prominent reflections begins to break up along a series of down-to-the-basin normal faults. Farther basinward, at the inner ring, the high-amplitude section is abruptly displaced downward by about 1–3.5 km, and breaks up into numerous tilted megablocks. This structural change is similar to that observed at the outer rim escarpment of the Chesapeake Bay crater (Fig. 10.27). At Chicxulub, these megablocks can be traced inward across the annular trough to approximately beneath a topographic high that has been called a peak ring; this

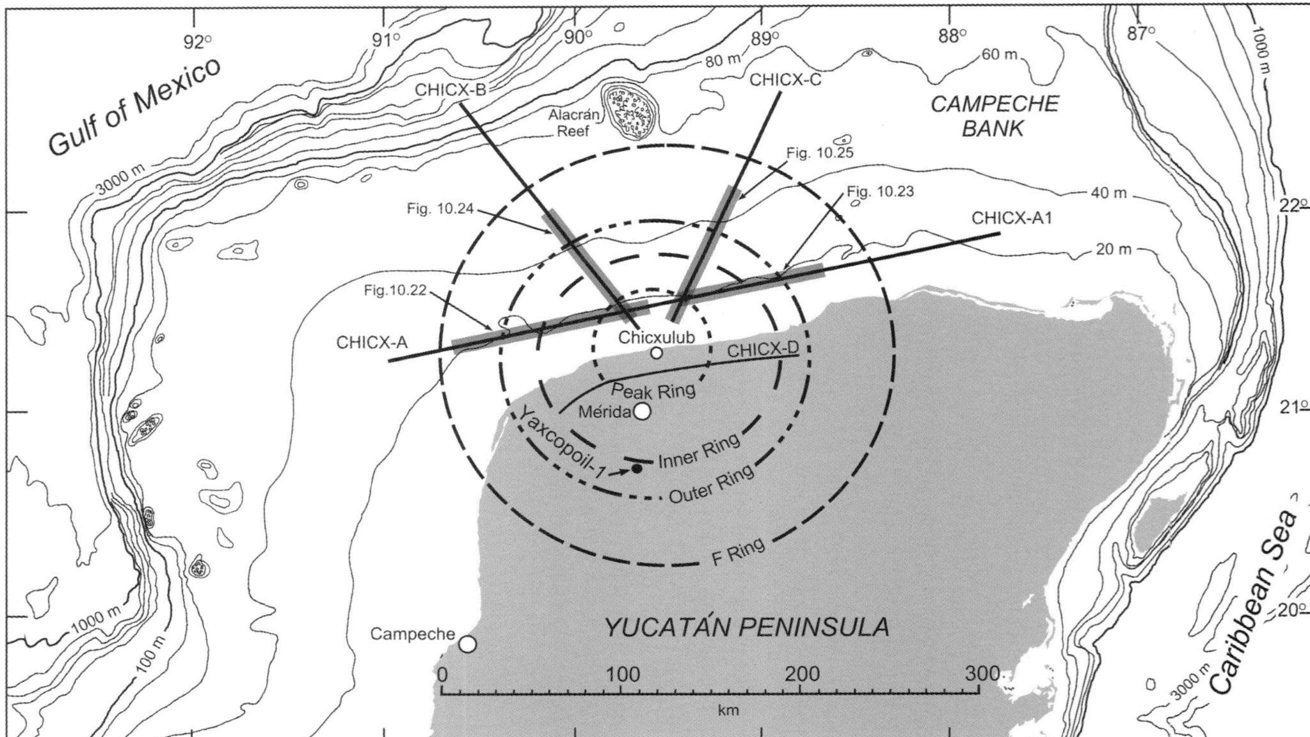

Fig. 10.22. Location map of Chicxulub multiring impact basin (Yucatán, Mexico), showing offshore multichannel seismic reflection tracklines across northern sector of basin. Modified from Morgan et al. (1997).

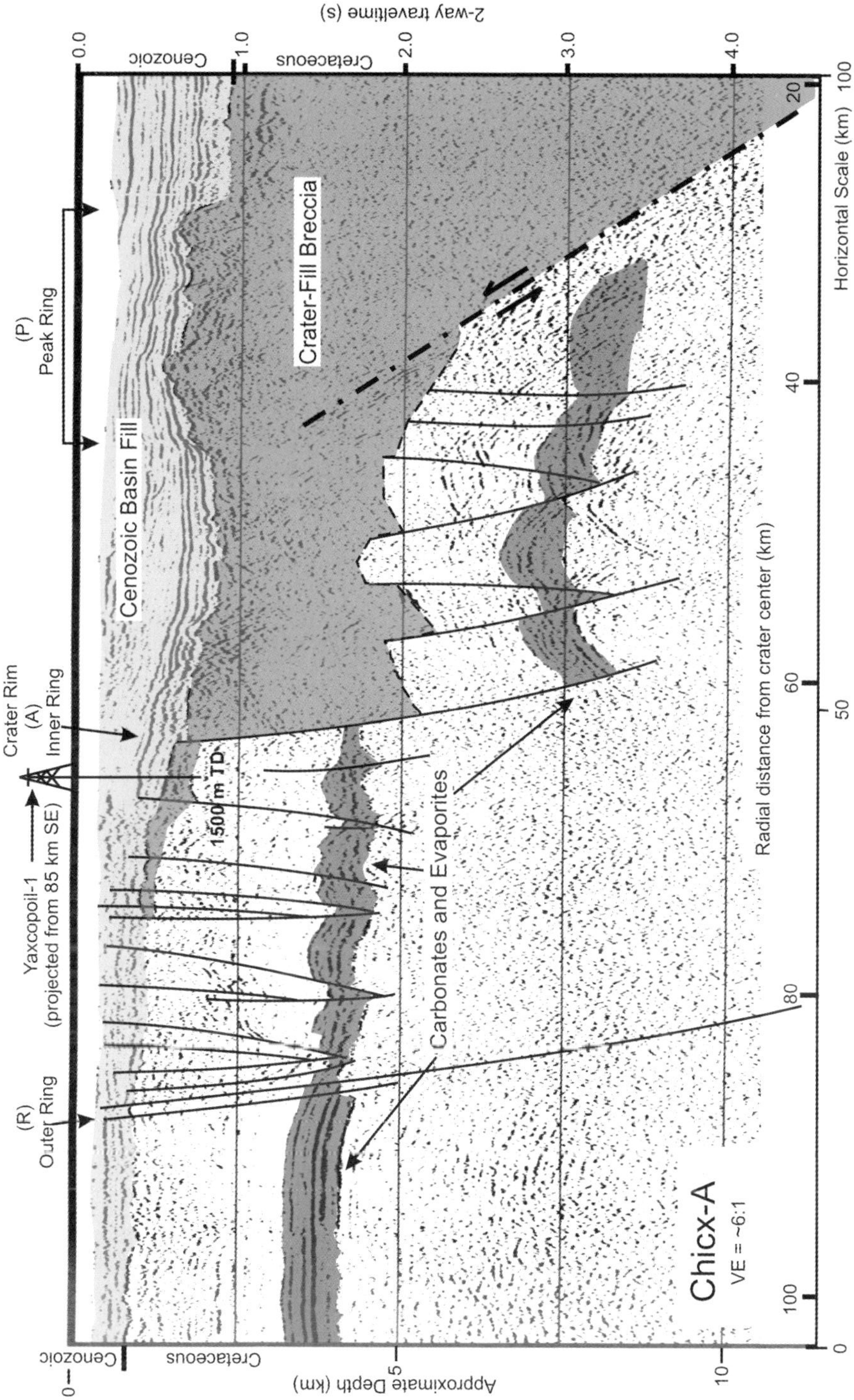
Chicx-A
VE = ~6:1
(R)
Outer Ring
Yaxcopoil-1
(projected from 85 km SE)
Crater Rim
(A)
Inner Ring
(P)
Peak Ring
Cenozoic Basin Fill
Crater-Fill Breccia
Carbonates and Evaporites
1500 m TD
Cenozoic
Cretaceous
Approximate Depth (km)
2-way traveltime (s)
Radial distance from crater center (km)
Horizontal Scale (km)

distribution also is similar to the megablock distribution at Chesapeake Bay (Fig. 10.27).

If one assumes that the entire 1- to 3.5-km-thick preimpact section above the prominent carbonate/evaporite reflections is preserved in each of the displaced megablocks, one can trace the approximate tops of the megablocks across the annular trough (Figs. 10.23–10.27). This exercise shows that there is (at minimum) a 2- to 4-km-thick stratigraphic section between the tops of the megablocks and the base of the postimpact sediments. We infer from the incoherent seismic signature and the moderate P-wave velocities (Brittan et al. 1999) that this 2–4 km section is composed of basin-fill breccia (Morgan and Warner 1999b). The displaced megablocks at Chesapeake Bay are buried in a similar fashion by crater-fill breccia (Fig. 10.27; Chapter 6).

The annular trough at Chicxulub is topographically expressed as a depression in the top surface of the basin-fill breccia. Inward toward the basin center, a distinct-to-subtle topographic elevation (relief of 400–700 m above the topographic floor of the annular trough) has been defined as a peak ring (Morgan et al. 1997; Brittan et al. 1999; Snyder and Hobbs 1999a; Figs. 10.23–10.27). The seismic signature of the topographic peak ring is expressed as a pair of high-amplitude reflections that represent the earliest postimpact (Cenozoic) sedimentary layer, which is draped over the upper surface of the basin-fill breccia. Thus defined, the peak ring has an apparent diameter of ~80 km along composite seismic profile Chicx-A/A1 (this profile does not pass through the center of the basin, however). An irregular circular gravity high appears to only roughly approximate the seismically identified peak ring (Brittan et al. 1999). P-wave velocities indicate that the rocks below the topographic peak ring are comparable in density to sedimentary deposits

Fig. 10.23. (Previous page) Interpretation of segment of seismic reflection profile Chicx-A across northwestern rim of Chicxulub impact basin. Profile from Morgan et al. (1997); interpretations modified from Brittan et al. (1999) and Morgan and Warner (1999b). Shaded interval of high-amplitude reflections in Cretaceous carbonate and evaporite section illustrates impact-generated displacement of preimpact sediments along series of normal down-to-the-basin faults. Heavy dashed diagonal line indicates position of inferred differential motion between outward-thrusted material from central-peak collapse and inward-thrusted material from transient-crater collapse, which may have created topographic peak ring (Brittan et al. 1999; Collins et al. 2002). Deep corehole Yaxcopoil-1 (completed February, 2002) is projected from ~85 km southeast of profile. See Fig. 10.22 for location of profile. Ring terminology: crater rim (after Morgan and Warner 1999a); outer ring, inner ring, peak ring (after Morgan and Warner 1999b); *R* ring, *A* ring, *P* ring (after Snyder and Hobbs 1999a).

Fig. 10.24. (Opposite page) Interpretation of segment of seismic reflection profile Chicx-A1 across northeastern rim of Chicxulub impact basin. Profile from Snyder and Hobbs (1999b); interpretations modified from Brittan et al. (1999) and Morgan and Warner (1999b). Shaded interval of high-amplitude reflections in Cretaceous carbonate and evaporite section illustrates impact-generated displacement of preimpact sediments along series of normal down-to-the-basin faults. See Fig. 10.22 for location of profile. Ring terminology: crater rim (after Morgan and Warner 1999a); outer ring, inner ring, peak ring (after Morgan and Warner 1999b); *R* ring, *A* ring, *P* ring (after Snyder and Hobbs 1999a).

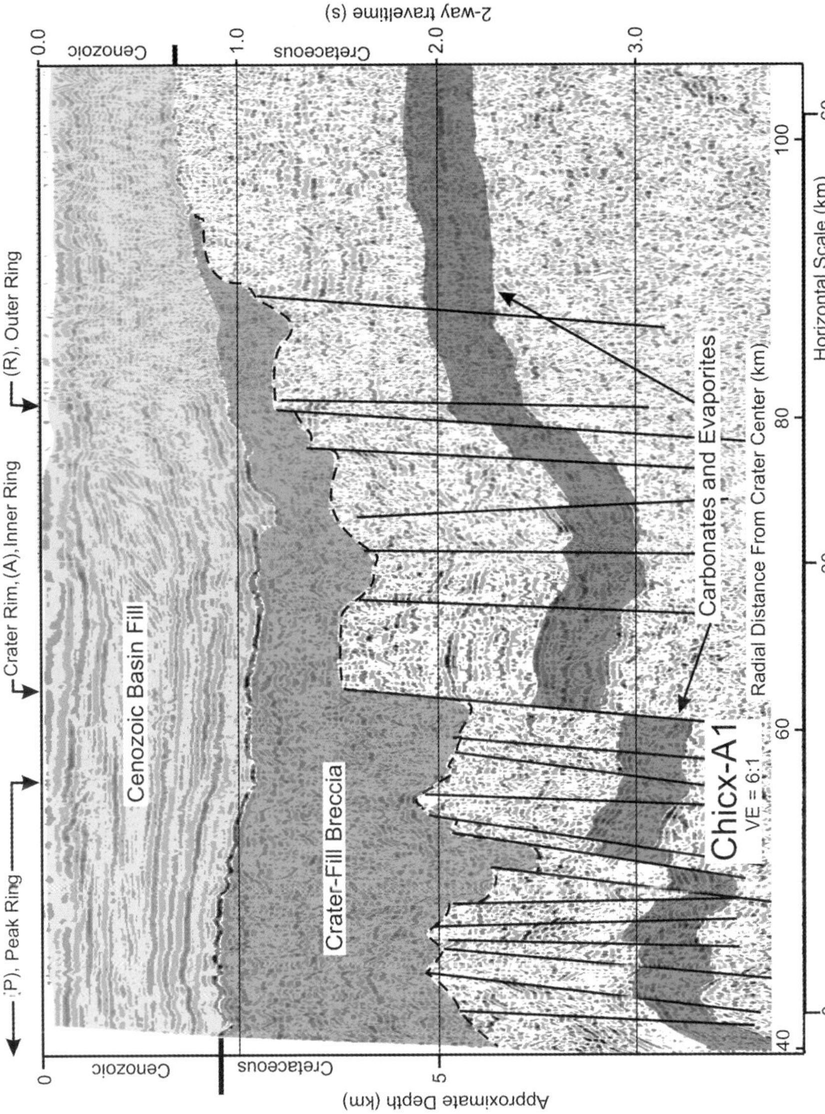
(P), Peak Ring
Crater Rim,(A),Inner Ring
(R), Outer Ring
Cenozoic Basin Fill
Crater-Fill Breccia
Carbonates and Evaporites
Chicx-A1
VE = 6:1
Cenozoic
Cretaceous
0.0
1.0
2.0
3.0
2-way traveltime (s)
0
5
Approximate Depth (km)
40
60
30
80
100
Radial Distance From Crater Center (km)
0
60
Horizontal Scale (km)

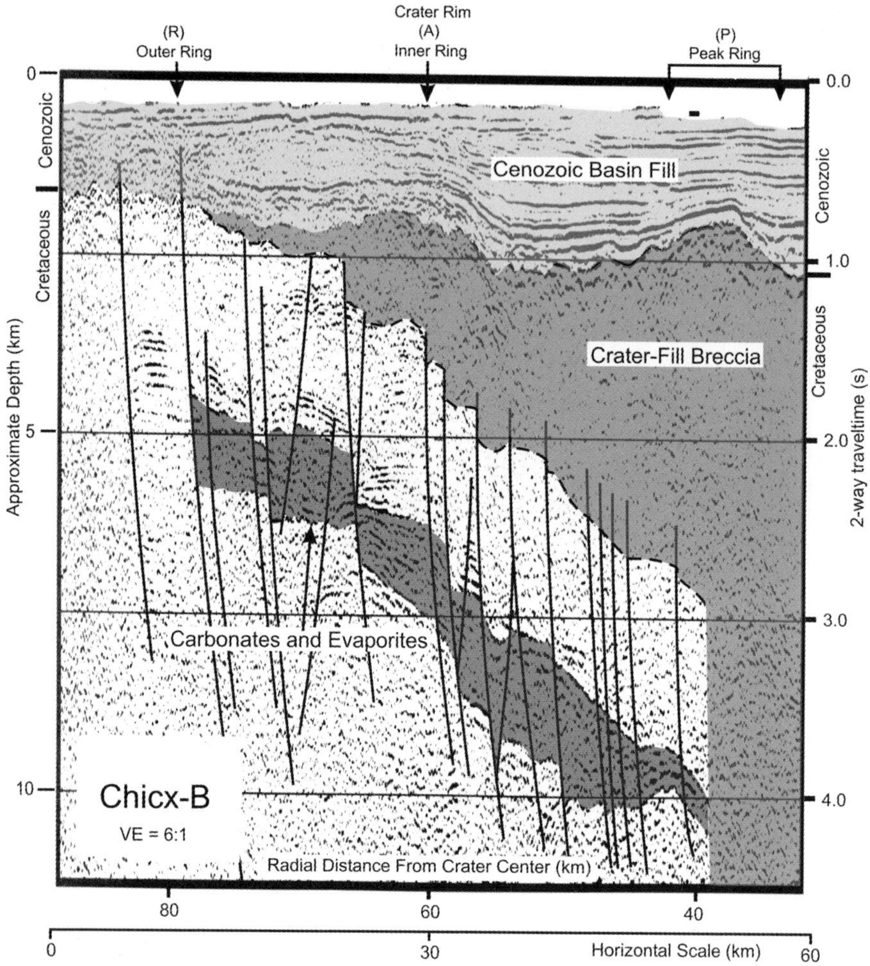

Fig. 10.25. Interpretation of segment of seismic reflection profile Chicx-B across northwestern rim of Chicxulub impact basin. Profile from Morgan et al. (1997); interpretations modified from Brittan et al. (1999) and Morgan and Warner (1999b). Shaded interval of high-amplitude reflections in Cretaceous carbonate and evaporite section illustrates impact-generated displacement of preimpact sediments along series of normal down-to-the-basin faults. See Fig. 10.22 for location of profile. Ring terminology: crater rim (after Morgan and Warner 1999a); outer ring, inner ring, peak ring (after Morgan and Warner 1999b); *R* ring, *A* ring, *P* ring (after Snyder and Hobbs 1999a).

in the lowest part of the basin (Brittan et al. 1999). This interpretation is opposed to the interpretation of Sharpton et al. (1993), who speculated that the peak ring consisted of fractured, uplifted, deep, crystalline basement rocks. The interpretation of Brittan et al. (1999) is closer to that of Pilkington et al. (1994), who concluded that the Chicxulub topographic peak ring consists of low-density breccia.

At Chesapeake Bay, the peak ring also can be recognized as a topographic high on the surface of the crater-fill breccia (Figs. 4.23, 4.25–4.29, 10.18A, 10.27), but there, the topographic peak ring is underlain by a seismically and gravimetrically defined structural high in the crystalline basement. The elevated topographic expression of the peak ring (upper surface of the Exmore breccia) at Chesapeake Bay is the result of differential compaction of the underlying breccia across the underlying structural peak ring in the basement. The seismic profiles at Chicxulub display no prominent high-amplitude reflections at shallow depth (above 4-km depth) that we could unambiguously interpret as a structural peak ring (Brittan et al. 1999; Snyder and Hobbs 1999a). However, on seismic profiles Chicx-A, B, and C, at ~6–7.5 km depth, there are indistinct, arched, or inclined reflections, which could be interpreted as possible manifestations of a structural peak ring in the higher-velocity crystalline basement rocks. Brittan et al. (1999) and Collins et al. (2002) interpreted the topographic peak ring at Chicxulub to be a result of differential motion, in which outwardly-thrusted crystalline breccia from the collapsing central peak overrode inwardly-slumping sedimentary megablocks that were produced by collapse of the transient-crater (Fig. 10.23).

Prior to the BIRPS seismic studies, several investigators of the Chicxulub structure speculated that a central peak composed of uplifted crystalline basement rocks explained the structure's positive central gravity anomaly (Pilkington et al. 1994; Sharpton et al. 1996; Hildebrand 1997). More recently, however, studies of the deep structure of the central Chicxulub basin using wide-angle ocean-bottom seismometers, have failed to document a central peak (Christeson et al. 1999). On the other hand, Snyder and Hobbs (1999) noted an elevated zone of dipping reflections (dips of 15–25 degrees) near the center of composite profile Chicx-A/A1 (reaching from 25 km up to ~15 km depth), which they attributed to shear zones and possible melt intrusions in the crystalline basement rocks. It is not unreasonable to infer that this zone of dipping reflections may represent the fractured flank of a central peak, whose highest prominence is south of profile Chicx-A/A1. Christeson et al. (1999) also concluded that a central peak, if present, must lie south of profile Chicx-A/A1 and north of a parallel onshore refraction profile labeled Chicx-D (Fig. 10.22).

Fig. 10.26. (Next page) Interpretation of segment of seismic reflection profile Chicx-C across northeastern rim of Chicxulub impact basin. Profile from Morgan et al. (1997); interpretations modified from Brittan et al. (1999) and Morgan and Warner (1999b). Shaded interval of high-amplitude reflections in Cretaceous carbonate and evaporite section illustrates impact-generated displacement of preimpact sediments along series of normal down-to-the-basin faults. See Fig. 10.22 for location of profile. Ring terminology: crater rim (after Morgan and Warner 1999a); outer ring, inner ring, peak ring (after Morgan and Warner 1999b); *R* ring, *A* ring, *P* ring (after Snyder and Hobbs 1999a).

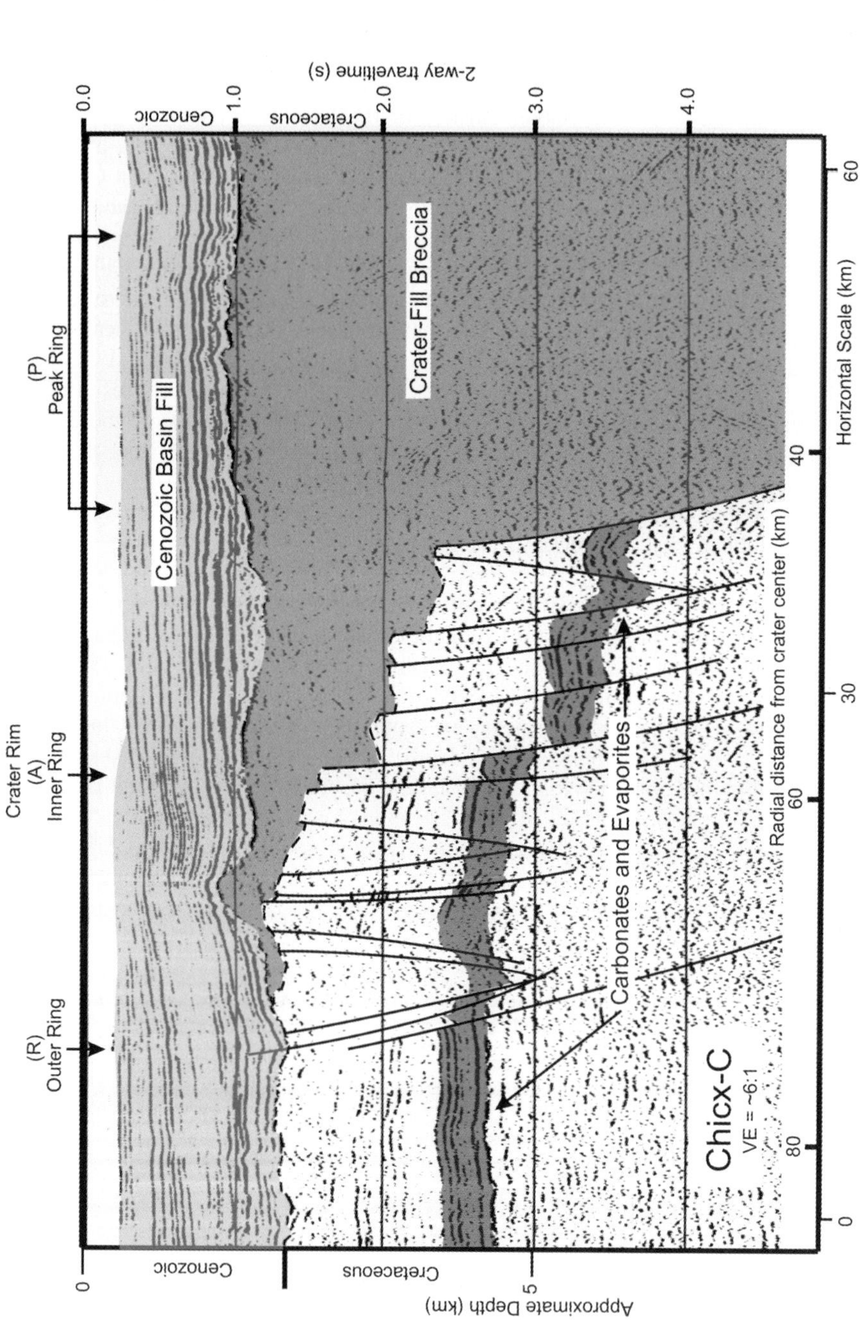
(P) Peak Ring
Crater Rim (A) Inner Ring
(R) Outer Ring
Cenozoic Basin Fill
Crater-Fill Breccia
Carbonates and Evaporites
Chicx-C
VE = ~6:1
Radial distance from crater center (km)
Horizontal Scale (km)
2-way traveltime (s)
Approximate Depth (km)
Cenozoic
Cretaceous

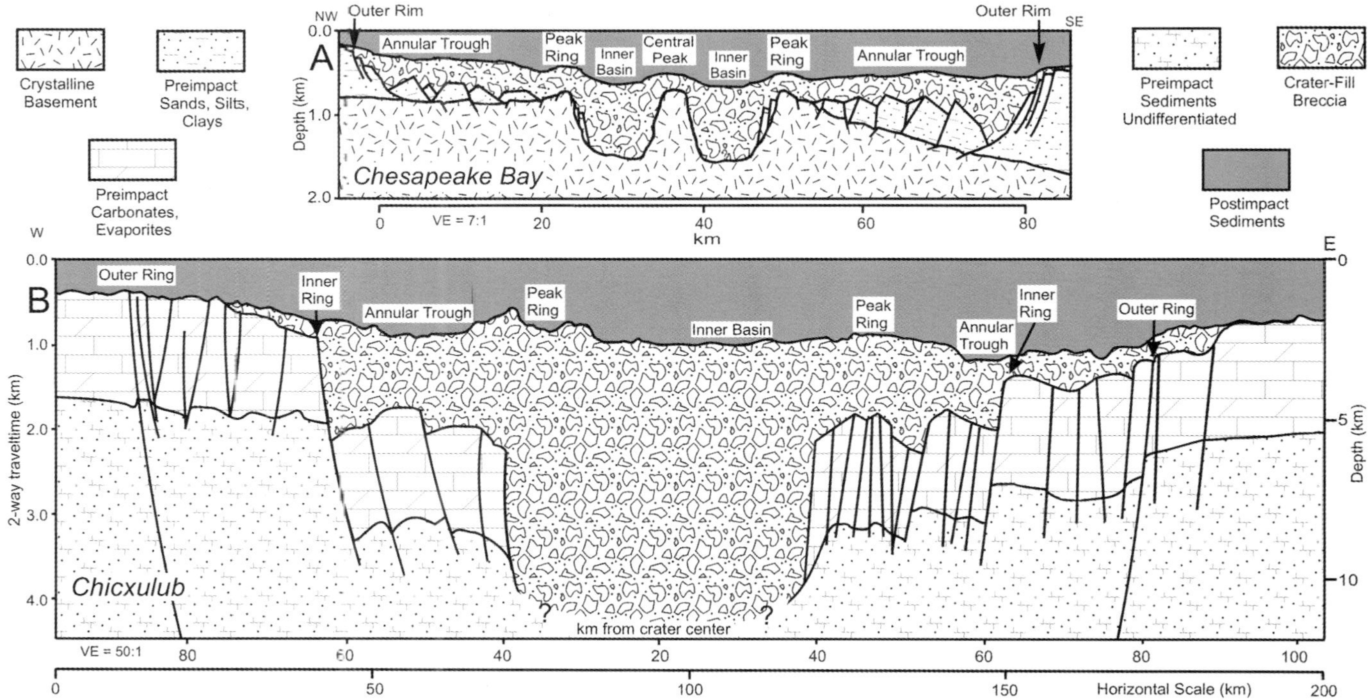

Fig. 10.27. Geologic cross sections comparing principal features of Chesapeake Bay peak-ring/central-peak impact crater (*A*) with Chicxulub multiring impact basin (*B*). Note general morphological and structural similarities, except for much greater size of Chicxulub structure, presence of its outer ring (and the outermost F ring, not shown), sediments beneath displaced megablocks, and lack of structural peak ring (section passes north of the center of Chicxulub basin, therefore its inferred central peak is not shown).

11 Comparisons Between Impactites

11.1 Terrestrial Impactites

The term *impactite* has been used in a variety of narrowly defined senses (Masaitis 1994; Stöffler and Grieve 1994), but usually applies to shock-metamorphosed rock bodies (Jackson 1997). We use impactite in the broader general sense of French (1998) and Koeberl (1998) to embrace any rock body created directly (e.g., impact breccia) or indirectly (e.g., impact tsunamiite) from processes associated with a bolide impact.

The most common and voluminous type of impactite found in and around most terrestrial impact structures is impact breccia. Field studies of these breccia deposits reveal some persistent, broad-scale characteristics, which can serve as general standards for comparing gross composition and stratigraphic relationships, and, thereby, to infer depositional processes. We will examine the breccia records at the Ries crater (southern Germany; Pohl et al. 1977), the Manson crater (Koeberl and Anderson 1996a), the Lockne crater (Ormö 1998), the Popigai crater (Masaitis 1994; Masaitis et al. 1999), the Montagnais crater (Jansa et al. 1989), and the Sudbury multiring basin (Stöffler et al. 1994) for comparison with Chesapeake Bay breccias (Figs. 11.1–11.7). We selected these structures for comparison because their breccias are well studied from outcrops and(or) boreholes.

11.1.1 Ries Breccias

Pohl et al. (1977), Stöffler (1977), Hörz and Banholzer (1980), Hörz et al. (1983), Engelhardt et al. (1995), and Engelhardt (1997), among others, have documented four principal breccia units from the 24-km-diameter Ries crater (Miocene, Figs. 11.1–11.3). These units are separated from each other by stratigraphic position, by general lithic composition, and by degree of shock metamorphism exhibited.

The stratigraphically lowest breccia unit documented at Ries is known as *dike breccia*, and has been sampled in one deep borehole (Figs. 11.2, 11.3). As the name implies, this breccia occurs as narrow dikes or veins, irregularly distributed through the upper 600 m or more of the crystalline basement. Dike breccia contains angular fragments of crystalline and sedimentary rocks and of individual mineral fragments, all supported by a fine-grained matrix. Some of the rock fragments display shock metamorphism, but many have been recrystallized strongly enough to obliterate the shock features. Dike breccias at Ries vary in

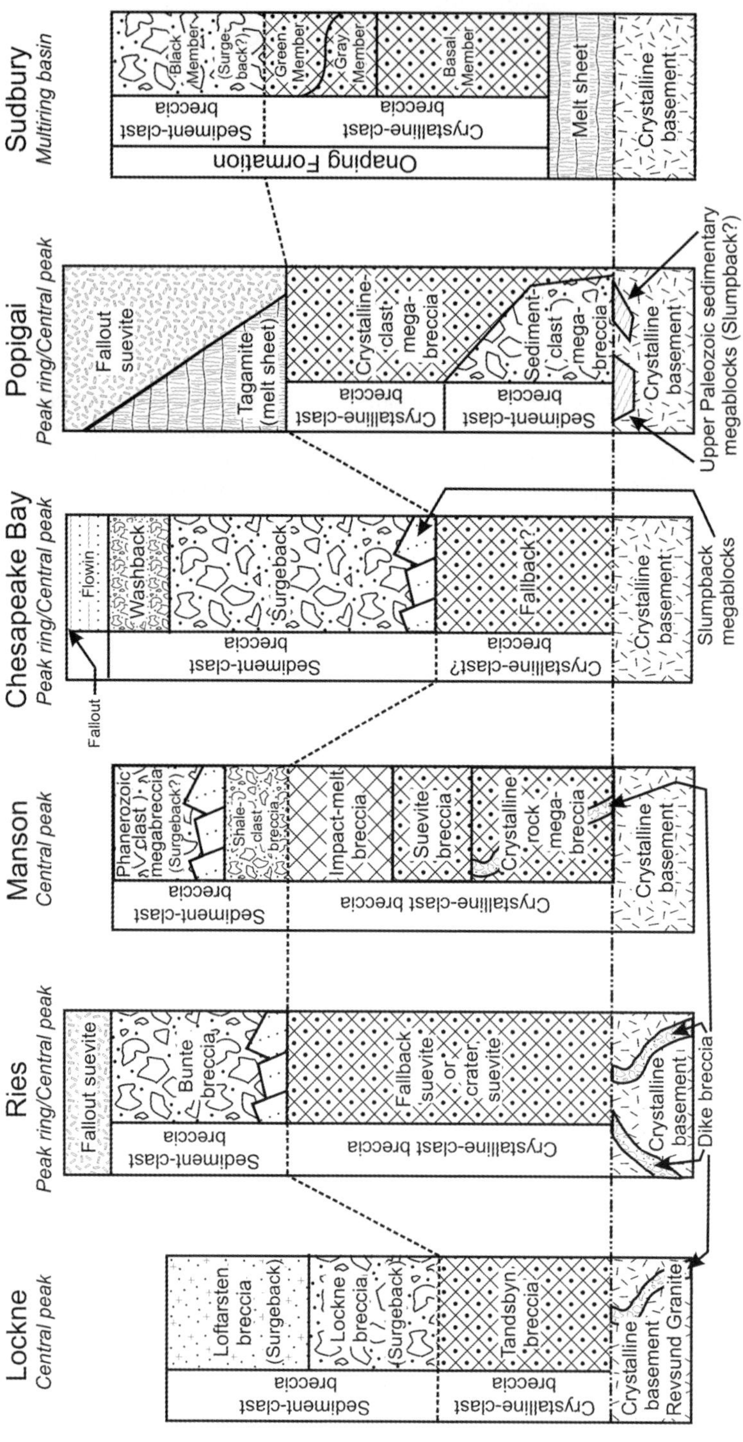

Fig. 11.1. Comparison of breccia stratigraphies, general compositions, and depositional origins at Lockne, Ries, Manson, Chesapeake Bay, and Popigai impact craters, and at Sudbury multiring basin. Note that slumpback megablocks, which underlie the Bunte Breccia at Ries (column 2) and the Phanerozoic-clast megabreccia at Manson (column 3), actually rest on crystalline basement, but this is difficult to illustrate in this figure.

color from grayish to red-brown, and vary in thickness from a few millimeters to a meter. Narrow veinlets of pseudotachylitic breccia also have been documented from the zone of dike breccias (Reimold 1995, 1998).

Stratigraphically above the dike breccias is a glass-bearing, crystalline-clast breccia unit known as *fallback suevite* or *crater suevite* (as much as 400 m thick). Fallback suevite covers the crystalline floor of the inner basin and, presumably, buries the inferred central peak (Fig. 11.3). Rock clasts in the fallback suevite are derived primarily from crystalline basement. Shock-metamorphic features representing shock pressures of 45–60 GPa and postshock temperatures of 900–1300° C are exhibited by some crystalline clasts in the fallback suevite. Glassy melt rock occurs in the suevite as small inclusions or as isolated larger bodies. In the 8–28 mm size range, glass bodies make up 2–39 vol.% of the fallout suevite. The matrix (particles smaller than 1 mm) of the fallback suevite typically consists of 3–13 vol.% glass. As indicated by their chemical compositions, glasses in the fallout suevite are believed to represent shock melt formed from a variety of different igneous and metamorphic rocks that constitute the crystalline basement.

Stratigraphically above the fallback suevite, and cropping out extensively, especially outside the outer rim, is the *Bunte Breccia*. The Bunte Breccia is a dominantly sediment-clast breccia, locally as thick as 100 m, containing clasts derived from consolidated limestones, shales, and sandstones of Jurassic and Triassic age (Figs. 11.1–11.3). Clasts in the Bunte Breccia are poorly sorted and chaotically mixed, ranging from millimeter-sized fragments to blocks tens of meters long, many of which have been strongly distorted by plastic and(or) brittle deformation. These clasts are supported in a fine-grained matrix derived from local unconsolidated sands, silts and clays of Cenozoic age. Only trace quantities of crystalline basement rocks are present in the Bunte Breccia, and they exhibit evidence of only low-level impact shock. Though no obvious vertical sorting has been documented, the Bunte Breccia does exhibit a fining-outward radial grading (Hörz et al. 1983).

The Bunte Breccia is essentially confined to the annular trough and the surrounding ejecta blanket of Ries crater. It is thought to have been deposited by a combination of ballistic ejection and lateral ground surge. These two processes would have moved large volumes of debris radially outward from the transient crater during the excavation stage of crater formation.

The fourth, and stratigraphically highest Ries breccia, termed *fallout suevite*, is present as small patches (as thick as 30 m) on top of the Bunte Breccia, both in the annular trough and on the ejecta blanket (Figs. 11.1–11.3). Clasts in the fallout suevite consist mainly of ejected glass bombs (60–80 vol.% of clasts) and fragments of crystalline basement enveloped by a matrix (80 vol.% of breccia) of dominantly fine-grained glass particles, montmorillonite, and quartz (Engelhardt 1990; Engelhardt et al. 1995). The greater glass content, presence of aerodynamically shaped bombs, and near lack of sedimentary clasts differentiate fallout suevite from fallback suevite. Fallout suevite is believed to have settled out from a turbulent gas cloud as a mixture of ejected basement melt (melt temperatures of 2000° C; shock pressures of >80 GPa) and unmelted particles of crystalline basement (plus minor amounts of sedimentary debris).

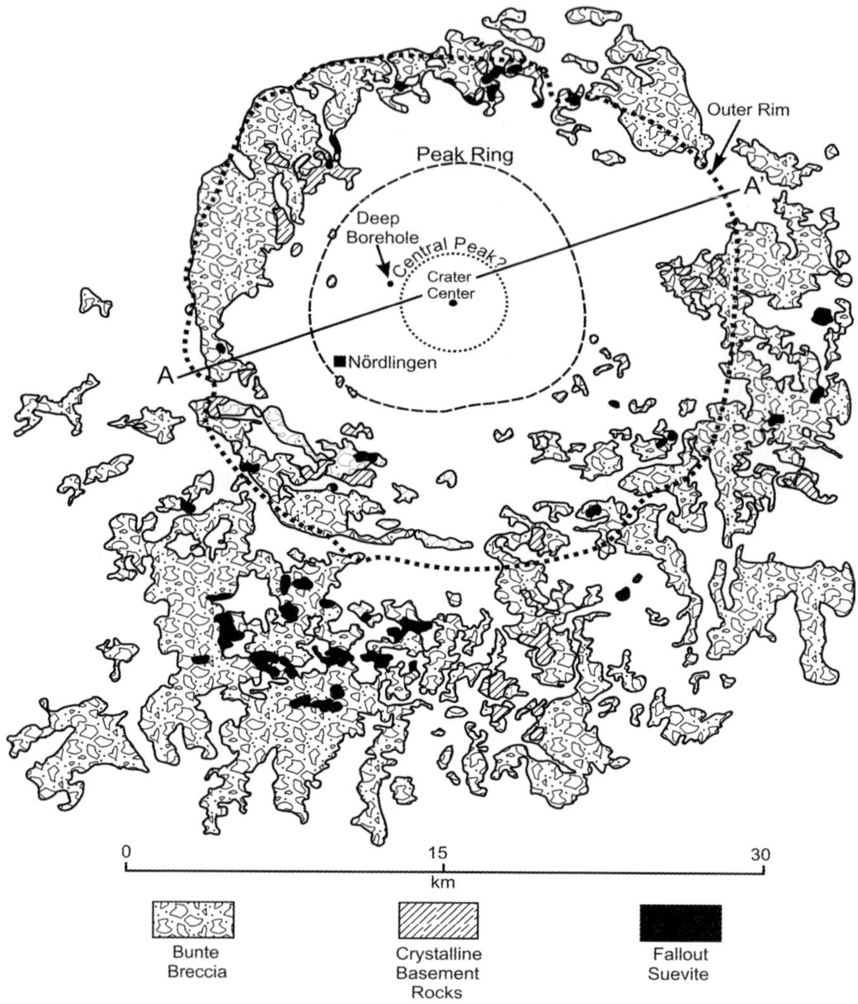

Fig. 11.2. Map of breccia and ejecta distribution on modern topographic surface in and near Ries impact crater (modified from Pohl et al. 1977). Profile *A-A'* is illustrated in Fig. 11.3.

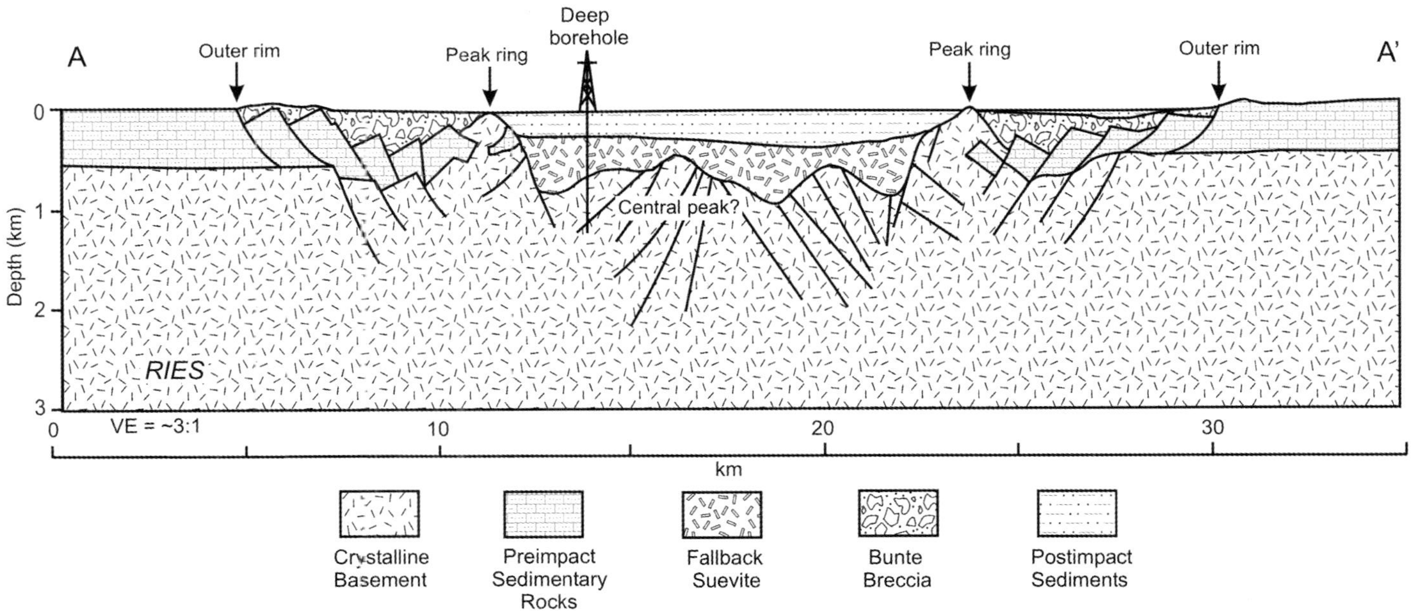

Fig. 11.3. Geologic cross section through outer rim, annular trough, peak ring, and possible central peak of Ries impact crater (modified from Pohl et al. 1977). See Fig. 11.2 for location of cross section.

In the annular trough of the Ries crater, the Bunte Breccia is separated from the crystalline basement by a fifth crater-fill unit. This unit is a slumpback deposit, composed of sedimentary megablocks, which have been displaced from the outer rim of the crater (Figs. 11.1, 11.3).

11.1.2 Manson Breccias

Anderson et al. (1996) and Witzke and Anderson (1996) identified five principal impact breccia units in cores taken from the annular trough and central peak of the 35-km diameter Manson crater (Late Cretaceous; Figs. 11.1, 11.4). The stratigraphically lowest breccia unit in the Manson crater is a crystalline-clast breccia, termed *crystalline-rock megabreccia.* This megabreccia includes shocked clasts of crystalline basement rocks, which commonly are penetrated by veins of impact melt rock and fine-grained breccia dikes, similar to the dike breccias of the Ries crater. The matrix is composed of sand- to silt-sized granitic rock fragments and mineral grains, nearly all of which display evidence of shock metamorphism. The crystalline-rock megabreccia is known only on the central peak, where its maximum drilled thickness is 66 m. Anderson et al. (1996) inferred that the crystalline-rock megabreccia represents impact-brecciated crystalline basement derived from the floor of the transient crater.

The stratigraphically next highest breccia unit on Manson's central peak also is a crystalline-clast breccia, termed *suevite breccia*, which resembles the fallback suevite at Ries (Figs. 11.1, 11.4). Clasts in the suevite breccia are dominantly centimeter- to meter-sized fragments of crystalline basement. Matrix dominates this unit, however, and consists of silt- to sand-sized grains of crystalline basement and minerals derived therefrom, plus minor amounts of sedimentary rocks and impact melt rock. All constituents of the suevite breccia display shock-metamorphic features.

Stratigraphically above the suevite breccia on the Manson central peak is *impact-melt breccia*, in which large clasts are conspicuously sparse. The few large clasts present are dominantly composed of melt rock (some angular, others plastically deformed). The matrix is mainly quartz, which displays a wide variety of shock-related features (multiple sets of PDFs, planar fractures, partial to total melting, isotropization, annealing, and recrystallization).

The next-to-highest breccia unit on the central peak at Manson is a sediment-clast breccia termed the *Keweenawan shale-clast breccia* (Figs. 11.1, 11.4). As its name suggests, the principal clasts of this breccia are centimeter- to meter-sized fragments of dark gray shale and mudstone of Keweenawan age (Proterozoic), mixed with devitrified impact melt-rock clasts, and minor amounts of siltstone and sandstone. The matrix of this breccia unit consists of sand- to clay-sized particles of the same constituents.

The stratigraphically highest breccia unit at Manson is the *Phanerozoic-clast breccia.* This breccia unit covers the entire Manson structure (maximum drilled thickness 191 m, but may reach 2 km; Figs. 11.1, 11.4). Clasts are dominantly

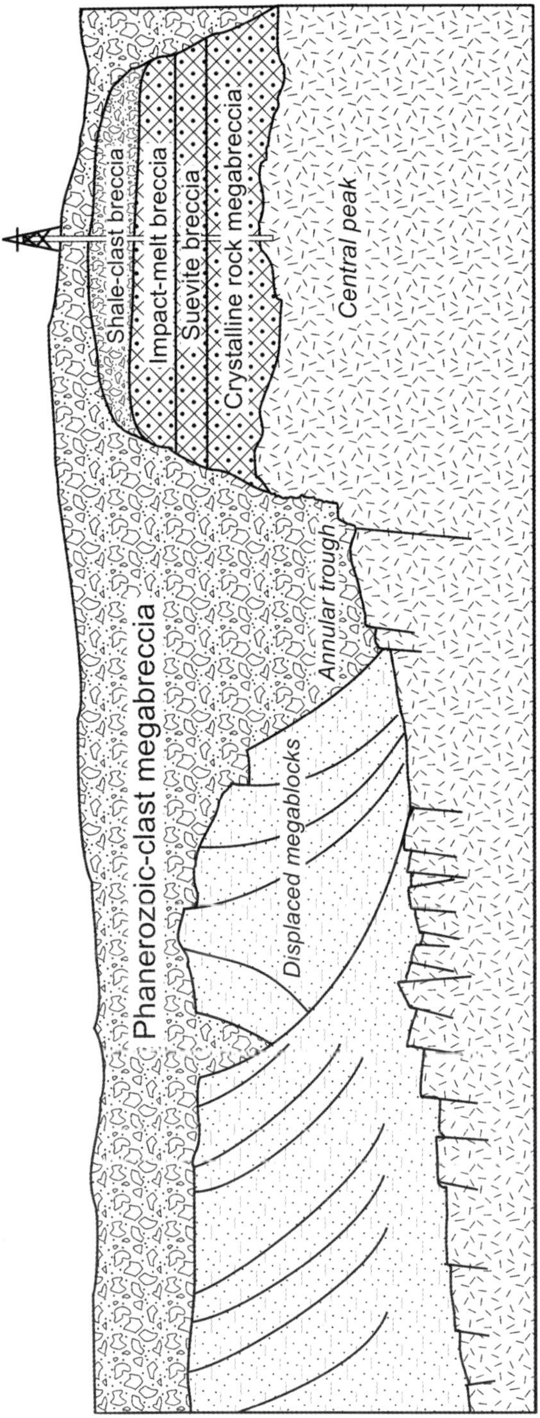

Fig. 11.4. General stratigraphic distribution of impact-generated breccias cored at Manson impact crater. Data from Anderson et al. (1996) and Witzke and Anderson (1996).

Cretaceous shales and mudstones mixed with fewer fragments of Paleozoic carbonates and rare Keweenawan Red Clastics. The matrix is composed of sand- and silt-sized grains of the same constituents, plus rare fragments of crystalline basement and impact melt rock. Shock-metamorphic features are rare. The Phanerozoic-clast breccia resembles the Bunte Breccia at Ries in its abundance of shallow-target rock types and its sparsity of shock features, but differs in its content of rare impact melt-rock clasts. The Phanerozoic-clast breccia is the only breccia unit identified in the annular trough at Manson.

Anderson et al. (1996) interpreted the Phanerozoic-clast breccia to be a landslide or debris-flow deposit, which originated at the outer rim of the crater during the late stages of crater modification. Anderson et al. (1996) speculated that the Phanerozoic-clast breccia may have formed in a shallow epicontinental sea, and subsequently was transported into the inner basin by turbulent flow produced by the collapsing water column (analogous to surgeback breccia). So far, no impact breccias have been found outside the crater rim at Manson.

As at Ries, there is an additional principal crater-fill unit in the annular trough at Manson. This unit is a slumpback deposit composed of displaced sedimentary megablocks that have collapsed from the outer rim of the crater. These megablocks lie between the crystalline basement and the Phanerozoic-clast breccias of Anderson et al. (1966) and Witzke and Anderson (1966; Figs. 11.1, 11.4).

11.1.3 Lockne Breccias

The stratigraphically lowest Lockne breccia unit, which fills the inner basin and covers the central peak, is the *Tandsbyn Breccia* (Lindström et al. 1996; Ormö 1998; von Dalwigk and Ormö 2001; Figs. 10.14, 10.16, 11.1). The Tandsbyn Breccia is a clast-supported deposit of angular fragments (<1cm to several decimeters in diameter) derived from the Revsund Granite (crystalline basement). Many of the clasts are internally brecciated. The Tandsbyn Breccia matrix is composed of finely crushed Revsund Granite. Tandsbyn Breccia also has been injected as sills and dikes into the flanks of the peak ring, a phenomenon which contributes to the ring's elevation above the floor of the annular trough.

Stratigraphically higher than the Tandsbyn Breccia are two different sediment-clast breccias – the *Lockne* and *Loftarsten Breccias* (Lindström et al. 1996; Figs. 10.14, 10.16, 11.1). The Lockne Breccia is widespread around the crater, and is inferred to be the initial surgeback deposit formed by the collapsing marine water column. Lockne Breccia is dominated by sedimentary clasts, principally Ordovician limestone, with minor amounts of Cambrian shale, Tandsbyn Breccia, and Revsund Granite. Lockne Breccia is poorly sorted and unbedded; clasts range in size from granules to meter-sized blocks. A vertical succession of two lithofacies can be recognized, however, within the Lockne Breccia. The basal lithofacies is a monomictic, clast-supported deposit resembling a debriite, which predominates on the floor of the annular trough. Here, huge blocks of bedded Ordovician limestone are separated from underlying undisturbed equivalents by a zone of shattered limestone. The upper Lockne lithofacies also is widespread in the annular trough, but, in addition, fills the surgeback gullies (von Dalwigk and Ormö 2001; see Chapter 10) and forms a thick fill above the Tandsbyn Breccia in the inner basin. The upper Lockne Breccia lithofacies is polymictic and pebble-graded, and contains more crystalline clasts than the basal Lockne lithofacies. Maximum thickness of the Lockne Breccia is 155 m in the inner basin.

Loftarsten Breccia represents the final and stratigraphically highest surgeback deposit of the Lockne crater (Figs. 10.14, 10.16, 11.1); sedimentologically it resembles a turbidite. Loftarsten Breccia is distributed mainly in the inner basin, where it attains maximum thickness of 45 m. The lower, coarse-grained part of the deposit is graded, whereas the upper finer-grained part displays current lineations, cross bedding, and dewatering structures (Simon 1987). Lithologically, the Loftarsten is a graywacke-like sandstone, which resembles the upper lithofacies of the Lockne Breccia. Sand grains of the Loftarsten are derived from both the crystalline basement and preimpact sedimentary beds. Simon (1987) reported as much as 20 vol.% melt lapilli in the Loftarsten, which appear to be fallback (fallout?) particles of impact melt incorporated into the surgeback flow (Sturkell and Ormö 1997).

11.1.4 Popigai Breccias

Masaitis (1994) and Masaitis et al. (1999) have described the impact breccias of the 85-km-diameter Popigai crater (late Eocene subaerial impact; Figs. 11.1, 11.5, 11.6). The stratigraphically lowest breccia unit at Popigai (Figs. 11.1, 11.6) is allogenic, blocky, polymict, *sediment-clast megabreccia* containing mixtures of crystalline and sedimentary clasts, supported by either fine-grained clastic matrix (coptoclastite) or impact melt-rock matrix. This megabreccia reaches ~1 km in thickness and rests mainly above the displaced megablocks and crystalline fault blocks of the annular trough.

Stratigraphically above the sediment-clast megabreccia is *crystalline-clast megabreccia*, supported in part by impact melt-rock matrix. This megabreccia occurs in the annular trough (400–500 m thick), on the flanks of the peak ring, and in the inner basin (presumably burying the crystalline central peak; Figs. 11.1, 11.5, 11.6). The crystalline-clast megabreccia is now missing from the exposed crest of the peak ring (Fig. 11.6), but appears to have originally covered it.

The stratigraphically highest breccia unit at Popigai is *fallout suevite* (maximum thickness of 193 m), which is separated from the crystalline-clast megabreccia by a thick (>600 m) impact melt sheet (*tagamite* of the Russian literature; Figs. 11.1, 11.5, 11.6). Fallout suevite at Popigai is composed of fragments and bombs of impact glass and numerous sedimentary and crystalline clasts enclosed by a finely comminuted and partly altered matrix of the same materials. This composition closely resembles that of the fallout suevite at the Ries crater.

Investigators at Popigai have not identified a crater-fill unit equivalent to the slumpback megablocks we recognize at Ries, Manson, and Chesapeake Bay. However, the cross section of Masaitis et al. (1999) shows megablocks of upper

Fig. 11.5. Distribution map of breccia and other impact deposits on modern topographic surface in and near Popigai impact crater. Modified from Masaitis et al. (1999). Cross section *X-X'* illustrated in Fig. 11.6.

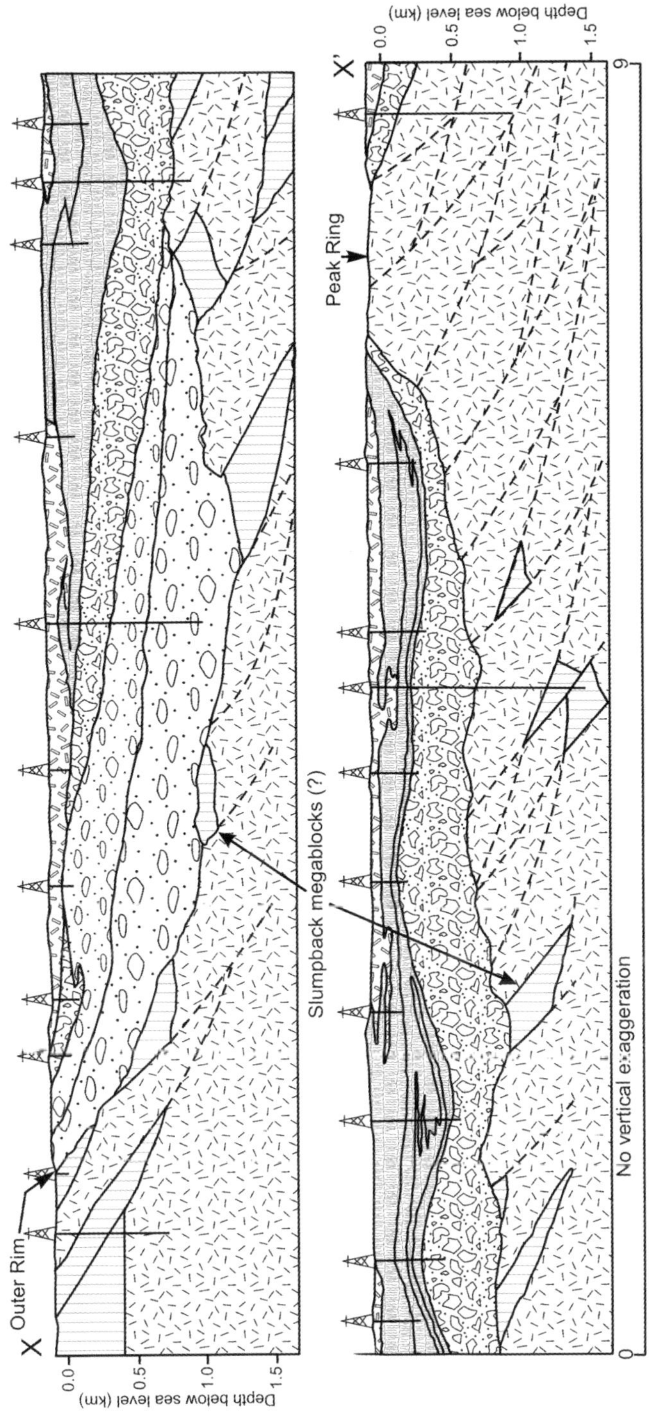

Fig. 11.6. Geologic cross section through outer rim, annular trough, and peak ring in southwest sector of Popigai impact crater (modified from Masaitis et al. 1999). See Figure 11.5 for location of cross section.

Paleozoic sedimentary rocks lying on, or emplaced within, crystalline basement rocks and covered by sediment-clast megabreccias (Fig. 11.6). These megabreccias may be analogous to the displaced sedimentary megablocks of our terminology (Figs. 4.3A,B, 4.7B, 4.9A,B, 6.1, 6.2).

11.1.5 Montagnais Breccias

Impact breccias from the Montagnais impact structure are known only from a single borehole (Montagnais I-94) on the crest of the central peak (Figs. 10.4, 10.6, 11.7). Jansa et al. (1989) identified the following impact deposits in rotary cuttings taken from that borehole, which penetrated to 1,646 m in a structural depression (listed from deepest to shallowest): (1) a basal interval of autochthonous crystalline-clast breccia (391 m thick), which includes two zones of crystalline impact melt rock; (2) a 131-m-thick interval of allochthonous sediment-clast breccia including clasts of limestone, granite, siliciclastic sediments, and metamorphic basement rocks; and (3) a cap of suevite, 38.5 m thick, composed of dominantly glass fragments (Fig. 11.7). Fewer than 10 vol.% of all these different breccia clasts display shock metamorphic features. Moderate- to high-pressure (35–45 GPa) shock metamorphism is indicated by isotropization of quartz and feldspars, as well as the presence of impact glass (Jansa et al. 1989). *In situ* basement rocks at Montagnais are part of the lower Paleozoic Meguma Group – low-grade metasedimentary rocks (metasubgraywacke, phyllite, metaquartzite) that exhibit extensive hairline fractures. Weakly developed shock-deformation features in quartz grains in the basement rocks indicate low shock pressures in the 6–8 GPa range.

Seismic reflection profiles indicate that crater-fill impact breccia of unknown composition covers the entire Montagnais structure, from outer rim to central peak, but the breccia is not identifiable seismically outside the crater (Figs. 10.5, 10.6). Neither has impact breccia been identified in wells outside the crater. By analogy with the other submarine craters that we have discussed in this volume, we infer that most of the Montagnais crater-fill is sediment-clast breccia. The Montagnais crater-fill breccia is thickest (~2 km) in the inner basin, and is also relatively thick (1-2 km) over the displaced megablocks. The crater-fill breccia thins dramatically in the annular trough, however, and pinches out on the upper part of the erosional scarp that marks the outer rim (Fig. 10.5). We infer from the seismic record that this crater-fill breccia originally was much thicker in the annular trough, but has subsequently been drastically thinned by erosion. Postimpact compaction and erosion have also thinned the crater-fill breccia unit over the displaced megablocks on both sides of the Montagnais structure (on seismic profile).

11.1.6 Sudbury Breccias

Though the Sudbury impact structure (Ontario, Canada; 1850 ± 3 Ma; Table 1.1) is classified as a multiring basin (~200-km diameter; Stöffler et al. 1994; Deutsch et al. 1995; Ivanov and Deutsch 1999), its extensively studied succession of crater-

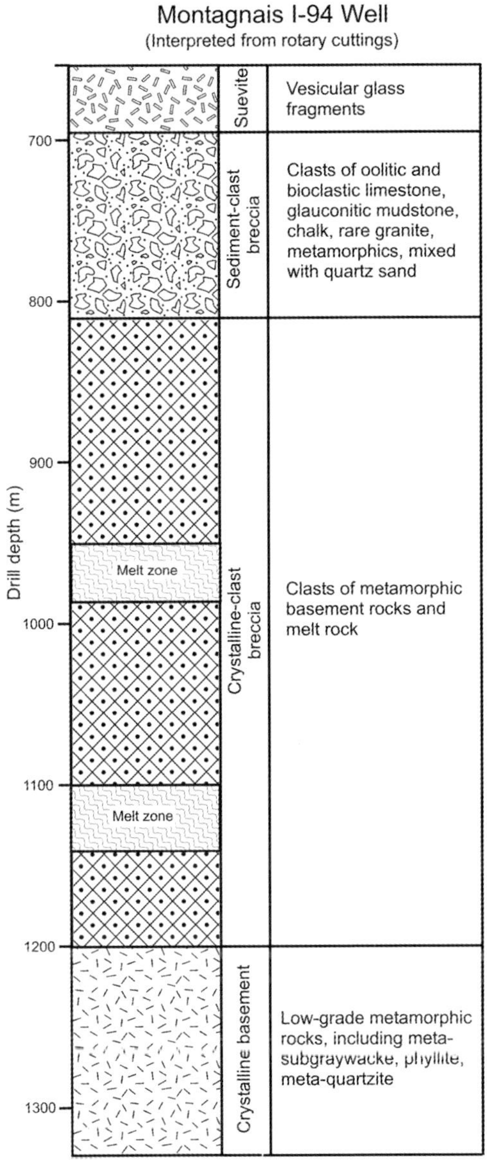

Fig. 11.7. Stratigraphic column of impact-generated breccias inferred from rotary cuttings in I-94 borehole on central peak of Montagnais impact crater. Data from Jansa et al. (1989).

fill breccias is generally comparable to the breccias cited for smaller terrestrial impact structures, though details of origin and emplacement are still vigorously debated. Avermann (1994) and Stöffler at al. (1994), for example, described three principal (thickest) layers of polymict impact breccia, and a fourth, much thinner, breccia layer associated with the Sudbury structure (Fig. 11.1). All four layers are assigned as separate members of the Onaping Formation, which overlies the primary impact melt sheet of the Sudbury Igneous Complex (Brockmeyer 1990; Avermann 1992; Dressler and Reimold 2001).

The stratigraphically lowest breccia, the Basal Member of the Onaping Formation, is 300 m thick (Fig. 11.1). It contains abundant metasedimentary clasts, but only minor amounts of crystalline basement clasts within a crystalline matrix of melt rock. The Basal Member is in sharp to gradational contact with the underlying granophyre of the clast-poor Sudbury impact melt sheet, and is considered by some authors (Brockmeyer 1990; Avermann 1992) to be an integral part of the melt sheet.

Next highest is the Gray Member of the Onaping Formation (Fig. 11.1), a 500-m-thick metasediment-clast breccia, also containing fragments of crystalline basement and irregular melt inclusions in a clastic matrix. Clasts within the Gray Member display shock-metamorphic features. Scattered within the Gray Member are "breccia bodies," which are differentiated by their clast size and different proportions of clasts and impact melt inclusions.

The stratigraphically highest breccia, the Black Member of the Onaping Formation, is also the thickest (~1 km thick), and can be subdivided into an upper unit and lower unit (Fig. 11.1). The Black Member as a whole is a sediment-clast breccia. The Black Member incorporates fragments of the underlying breccia layers and displays evidence of an euxinic aquatic paleoenvironment, possibly representing a surgeback deposit (Bunch et al. 1999). The lower unit of the Black Member differs from the upper unit in containing a few small fragments of shocked crystalline basement, melt inclusions, and chloritized melt particles.

At the contact between the Black and Gray Members of the Onaping Formation, is the irregular, relatively thin (5–70 m) Green Member (Fig. 11.1). The Green Member is unique among documented impact breccias; it comprises fine-grained, often chloritized fragments and small, highly shocked, polymictic mineral clasts in a microcrystalline matrix. The Green Member appears to contain fall-back ejecta, and may represent the collapsed fireball of the Sudbury impact deposit (Avermann 1999).

To summarize, this brief review of well-documented impact breccias reveals a generally uniform stratigraphic succession of broadly defined breccia units. Crystalline-clast breccias or megabreccias usually are at the base of the sequence, where they rest on crystalline basement rocks, and are succeeded upward by sediment-clast breccias of variable compositions. These breccia units are capped, in two of the cited cases, by glass-rich suevite. Impact melt sheets occur at different levels in the crater-fill succession.

11.1.7 Chesapeake Bay Breccias

In broad scope, the seismostratigraphic, lithostratigraphic, and downhole geophysical analyses of the Chesapeake Bay breccia column suggest that here, too, there is a general upward progression from crystalline-clast to sediment-clast breccias. In detail, however, some of the sediment-clast breccias have unique features that have not yet been documented elsewhere. In the inner basin of the Chesapeake Bay crater, the synimpact crater-fill deposits may be divided stratigraphically into six principal subhorizontal units (layers or lithofacies; Figs. 11.1, 11.8; Poag 2000). The three stratigraphically next-to-highest units are encompassed in the general term Exmore breccia (units 3–5 of Figs. 11.1,11.8, 11.9), but each may be recognized as a lithically distinct (but locally variable) unit. Here we briefly describe the physical aspects of these units. See Chapter 12 for further discussion of their origins and depositional processes that formed them.

Crater-fill unit 1 (Cfu-1), at the base of the Chesapeake Bay crater-fill, is known mainly from seismic reflection profiles, and appears to be generally confined to the lower part of the inner basin (Figs. 11.1, 11.8, 11.9). However a 23-m section at the base of the Bayside breccia column might represent part of Cfu-1, because the site is near the inner basin and the core contains an unusual abundance of crystalline and lithified sedimentary clasts that have not been observed at any other core site. This Bayside core section has not yet been analyzed for shock-related features, however.

The relatively high-relief upper surface of Cfu-1 is expressed on seismic profiles by a series of high-amplitude, broad-wavelength, parabolic reflections at 1.0–1.25 s (1250–1500 m below sea level). Maximum estimated thickness of the lowest breccia is 0.65 s (~600 m). Poag (1996) speculated that Cfu-1 is composed mostly of fallback breccia, probably representing a complex, crystalline-clast breccia, similar in composition and origin to the fallback suevite of the Ries crater, the crystalline-rock megabreccias at Manson, the Tandsbyn Breccia at Lockne, the crystalline megabreccias at Popigai, and the Gray and Basal Members of the Onaping Formation at Sudbury. All these cited breccias are typified by large quantities of crystalline clasts, which display a wide range of shock metamorphism, and are enclosed in a highly variable suite of fine-grained matrices.

Crater-fill unit 2 is composed of decimeter-to-kilometer-scale megablocks of mainly Lower Cretaceous stratified sediments, which have slid, slumped, and collapsed from the outer rim of the crater (Figs. 11.1, 11.8, 11.9). At Chesapeake Bay, Cfu-2 appears to be confined to the annular trough of the crater.

Crater-fill unit 3 (third lowest Cfu and basal part of Exmore breccia; Fig. 11.1) at Chesapeake Bay has been sampled by the five continuously cored intracrater boreholes (Fig. 11.8). As we showed earlier (Chapter 6), this layer is largely a clast-supported breccia, confined to the annular trough and inner basin, where it buries the peak ring and central peak (Figs. 11.1, 11.8, 11.9). Most individual clasts documented in this unit are weakly consolidated sediments (clays, silts, sands), with maximum apparent thicknesses of 15–20 m. We interpret Cfu-3 to have formed through hydraulic processes as the marine water column surged back into the crater cavity following impact of the bolide (see Chapter 12).

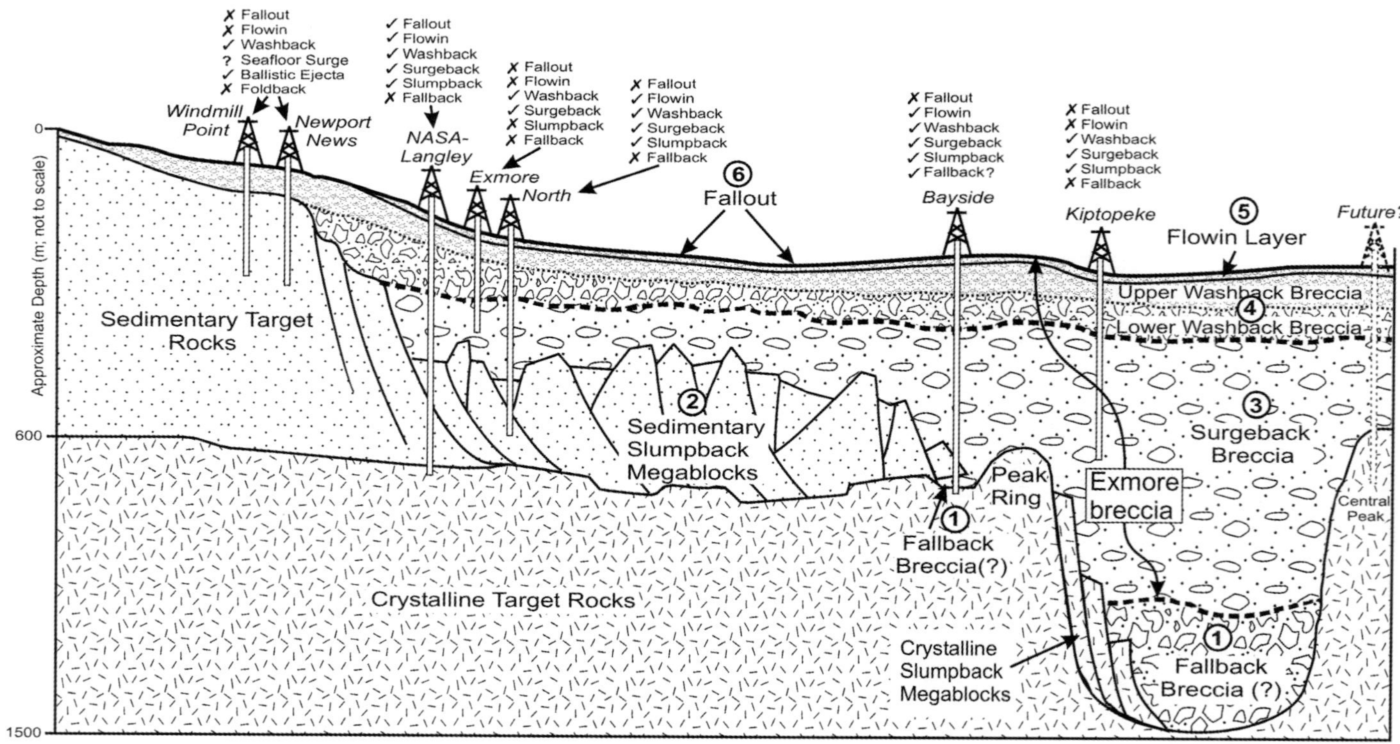

Fig. 11.8. Geologic cross section of western half of Chesapeake Bay impact crater showing relative locations of coreholes and stratigraphic distribution of six principal crater-fill units (Cfu1–Cfu6). Checklist above each core site indicates presence (check) or absence (cross) of inferred lithofacies. See Chapter 12 for further discussion of lithofacies terminology (slumpback, washback, etc.).

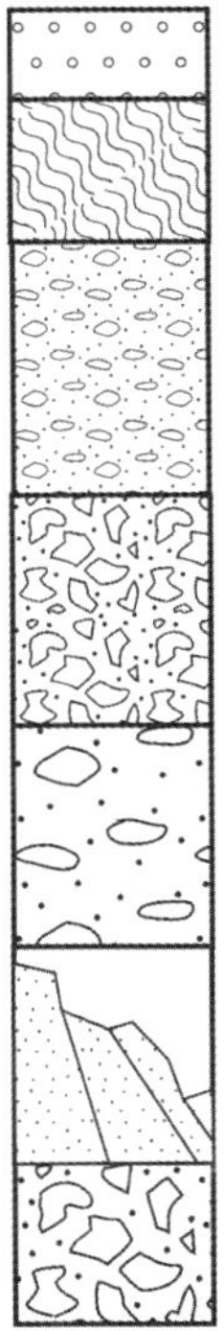

Crater-Fill Unit [Cfu]	Lithic Texture	Genesis	Depositional Regime
Cfu-6	Centimeter-scale silt laminae, pyrite molds of inferred glassy microspherules; cored at 1 site	Fine atmospheric debris and vapor condensates raining back into crater	Fallout
Cfu-5	Thin bedded, slit-rich unit with multidirectional, inclined stratification; cored at 3 sites	Fine-grained, multidirectional gravity flows from impact-generated storms	Flowin
Cfu-4 upper	Dominated by polymictic, meter to centimeter-scale sedimentary clasts supported by glauconite/quartz sand matrix; matrix mainly 50-100% by volume; fining upward; cored at all sites	Debris scoured from seafloor and coastal plain by tsunami waves and washed back out to sea	Washback
Cfu-4 lower	Dominated by polymictic, centimeter to millimeter-scale sedimentary clasts supported by glauconite/quartz sand matrix; matrix ~30% or less by volume; fining upward sequence; cored at all sites	Debris scoured from seafloor and coastal plain by tsunami waves and washed back out to sea	Washback
Cfu-3	Clast-supported breccia; clasts mainly weakly consolidated sediments with 15-20-m maximum apparent thicknesses; fining upward sequence; cored at 5 sites	Debris scoured from seafloor and crater walls and floor by collapsing marine water column	Surgeback
Cfu-2	Decimeter- to kilometer-scale megablocks of mainly stratified Lower Cretaceous sediments; cored at 3 sites	Coarse debris in megaslides and megaslumps derived from collapse of sedimentary outer walls	Slumpback
Cfu-1	Inferred to be composed of dominantly crystalline clasts with abundant evidence of shock metamorphism; not yet cored	Coarse debris derived from crystalline crater floor; ejected short distance before falling back	Fallback

Fig. 11.9. Relationship between lithic texture of crater-fill units, genesis of units, and depositional regime in which each unit formed in Chesapeake Bay impact crater. See Chapter 12 for more details regarding genesis and depositional regimes.

Crater-fill unit 4 at Chesapeake Bay (the middle unit of the Exmore breccia; sampled by five intracrater boreholes) also is dominated by polymictic sedimentary clasts (Figs. 11.1, 11.8, 11.9). We subdivide Cfu-4 into a lower and upper part. In contrast to the underlying Cfu-3, the clasts in the lower part of Cfu-4 are supported by the characteristic glauconitic quartz sand matrix. This matrix also contains a small percentage of centimeter-to-millimeter-size fragments of crystalline basement (granitic and metasedimentary rocks). Though sparse, the basement clasts do contain evidence of shock metamorphism, which ranges from low-pressure fracturing to partial and complete melting (Koeberl et al. 1996; Horton et al. 2001; Chapter 6). Seismic profiles indicate that the lower part of Cfu-4 is present throughout the crater

The upper part of Cfu-4 at Chesapeake Bay (sampled at all seven core sites) is similar to the lower matrix-supported part, but differs in the strong dominance of matrix over clasts (as measured by the relative thickness of matrix intervals versus clasts in the cores; Chapter 6). The upper part of Cfu-4 covers the entire crater and also extends a few kilometers outside the crater rim to form a thin breccia apron (Figs. 11.1, 11.8, 11.9). We interpret Cfu-4 as a two-part tsunami-washback unit (see Chapter 12).

Crater-fill unit 5 at Chesapeake Bay is a thin-bedded clayey silt-rich unit, with distinctly inclined, undulating thin layers and lenses of white, micaceous, fine-to-very fine sand and occasional clasts of medium-to-coarse glauconite-quartz sand derived from the underlying Cfu-4. The direction and angle of bed inclination in Cfu-5 change frequently (commonly reversing) through the section, indicating rapidly changing flow directions during deposition. We interpret this interval to represent the final water-borne synimpact flowin unit, laid down by a series of hypercanes (runaway hurricanes; Emanuel et al. 1995) that developed over the impact region (Poag 2002; see Chapters 6, 12). This flowin unit constitutes the upper part of the Exmore breccia (Figs. 11.1, 11.8, 11.9).

The stratigraphically highest synimpact layer at Chesapeake Bay (Cfu-6) is an ~5-cm-thick fallout unit containing evidence of stacked glass microspherules (Figs. 6.24A,B), which hypothetically might have originally covered all the underlying crater-fill units (Figs. 11.1, 11.8, 11.9). So far, however, this fallout layer has been identified only in the NASA Langley core.

The relative stratigraphic position, extensive spatial distribution, and gross composition of the washback (Cfu-4) and surgeback (Cfu-3) lithofacies of the Exmore breccia at Chesapeake Bay (see Chapter 6) resemble those of the Bunte Breccia at Ries, the Phanerozoic-clast megabreccia at Manson, the Lockne and Loftarsten Breccias at Lockne, and the Black Member of the Onaping Formation at Sudbury. The Exmore differs from the polymict megabreccias at Popigai, however, mainly because the latter are overlain by crystalline-clast megabreccias, and are spatially restricted to the annular trough at Popigai.

Compelling evidence of subaqueous deposition in the Exmore breccia (Chapter 6) also is shared by the Phanerozoic-clast megabreccia at Manson, the Lockne and Loftarsten Breccias at Lockne, and the Black Member at Sudbury. Though the Bunte Breccia displays similar sedimentary evidence of turbulent flow during its deposition, the depositional processes are inferred to have been quite different.

Bunte Breccia presumably was deposited by ballistic ejection and radial ground surge, which were processes of crater excavation. In contrast, the flowin, washback and surgeback facies at Chesapeake Bay, the Phanerozoic-clast megabreccia, the Lockne and Loftarsten Breccias, and the Black Member, formed from turbulent marine processes, which acted to refill the excavated craters.

11.2 Flowin, Fallout, and Dead Zone

The inferred synimpact flowin (Cfu-5) and fallout (Cfu-6) layers and the overlying laminated clay-silt-sand of the initial postimpact deposit (dead zone) at Chesapeake Bay (see Chapters 6,7) have no precise equivalents yet documented at any other terrestrial crater. Though the presence of fallout debris inside a crater is unusual, it is not unique. The Ries and Popigai craters, for example, contain fallout deposits (Fig. 11.1), but in both cases the fallout debris is composed of suevite. The suevite at Ries is overlain, moreover, by 17.2 m of coarse-tail-graded, matrix-rich breccia, which grades further upward to sand and coarse silt. Füchtbauer et al. (1977) interpreted this succession to be either a lacustrine turbidite or a fallout deposit. This graded postsuevite section at Ries most closely resembles a combination of breccia-unit 4 (Cfu-4) and the fallout layer (Cfu-6) at Chesapeake Bay, and is suggestive of a turbidite sequence (at the base) and a fallout layer (at the top). At Sudbury, the inferred fallout unit (Green Member) lies *beneath* an inferred surgeback deposit (Black Member), perhaps indicating that the Black Member is the initial postimpact deposit at Sudbury, equivalent to the dead zone at Chesapeake Bay.

At present, Chesapeake Bay is the only known crater in which proximal fallout materials are part of an unconsolidated, intracrater, marine sediment layer. So far, flowin deposits and a laminated dead zone also are unique to Chesapeake Bay.

11.3 Other Intrabreccia Bodies

In addition to the crystalline-clast breccia (Cfu-1) and the overlying three principal layers of the Exmore breccia (Cfu-3–5), some smaller, distinctive, seismic-reflection features are notable within the inner basin at Chesapeake Bay. For example, a series of high-amplitude, parabolic reflections can be observed within the surgeback breccia between 0.42 and 0.62 s (~525–775 m below the bay floor), just above the inner rim of the peak ring on profile T-8-S-CB-E (Figs. 4.29, CD-ROM.12). These reflections indicate an irregular body of rock with contrasting velocity and(or) density, ~150 m thick and ~7 km long, embedded in the breccia. Other features of similar size and seismic signature are present within the surgeback layer on profiles not illustrated herein. These isolated rock bodies may be: (1) suevitic breccias like those at Ries; (2) "breccia bodies" similar to those of the Gray Member at Sudbury; (3) impact melt rocks (tagamites), such as those at

Popigai; (4) unusually large blocks of sedimentary or crystalline target rocks; or (5) some combination of these deposits.

11.4 Continuous Ejecta Blankets

Grieve (1991) listed only Barringer (a 1-km-wide simple crater of Pleistocene age; also known as Meteor crater), Ries (a 24-km wide complex crater of Miocene age), and Ragozinka (a 9-km-wide complex crater of early Eocene age) as terrestrial craters having *continuous* ejecta blankets, whereas *discontinuous* ejecta blankets have been described around many other craters. Several earlier publications have described a continuous ejecta blanket around the Chesapeake Bay crater as well (Poag et al. 1994; Poag 1997a). As we have pointed out earlier herein, however (Chapter 6), the impact debris encircling the Chesapeake Bay crater cannot be considered a true ejecta blanket. Whatever debris might have composed a possible original ejecta blanket has been remobilized and redistributed as components of the tsunami washback breccia.

With regard to the formation and distribution of impact-generated breccia deposits, then, the location of the Chesapeake Bay impact on a broad continental shelf, not far from the shoreline of a continental landmass, was a critical condition. The presence of a marine water column and a cover of water-saturated, unconsolidated sediments above crystalline basement, resulted in a distinctive suite of breccias and related deposits, whose compositions, stratigraphic succession, and depositional processes are not precisely duplicated at any of the other craters yet studied in relative detail.

11.5 Secondary Breccias

The Chesapeake Bay impact also appears to have produced additional isolated breccia bodies that are deposited outside the primary crater. Each of the inferred Chesapeake Bay secondary craters described herein (Chapter 5) exhibits its own chaotic seismic reflections, which we interpret to represent small bodies of sediment-clast breccia derived from the local sedimentary target rocks. So far, however, none of these secondaries has been cored, though an early report of breccia-like lithologies in a well (borehole 106) near Colonial Beach, Virginia (Cederstrom 1945b; Fig. 5.2), might indicate the presence of secondary breccia. As no other secondary terrestrial craters have been thoroughly documented, we have no basis for comparison with the Chesapeake Bay secondary breccias. There is speculation, however, that some constituents of the Bunte Breccia were derived from isolated local breccias produced by secondary cratering processes (large ejecta blocks might have ploughed through the sedimentary substrate and created an irregular landscape of secondary craters and their respective breccias; Oberbeck 1975; Morrison and Oberbeck 1978).

11.6 Strewn Fields

Aside from the Chicxulub multiring basin, the Chesapeake Bay and Popigai impact structures have produced the best documented, most widespread ejecta strewn fields yet identified (hemispheric to global distribution; see discussion in Chapter 9). Well-known other strewn fields include the Australasian tektites (derived from an as yet unidentified source) and the Central European strewn field (known for its moldavites). The Central European strewn field covers parts of the Czech Republic, Austria and Germany, about 450 km east of its source at the Ries crater. Isotopic evidence indicates that the moldavites represent melt products of the Miocene sands that constituted the surface sediments at the Ries impact site. Koeberl et al. (2001) have analyzed preimpact Cenozoic sediments at Chesapeake Bay for possible isotopic correlation with the tektites of the North American strewn field, but, so far, there is no direct evidence of a match beyond the geochemical data reported by Koeberl et al. (1996). (See Chapter 6 for additional discussion of North American strewn field).

11.7 Impact Melt Rocks

According to Grieve (1987), complex craters the size of the Chesapeake Bay crater usually are characterized by pockets (a few meters thick) or coherent sheets (several hundred meters thick) of impact melt rock, such as the thick tagamite sheets at Popigai (Masaitis 1994; Masaitis et al. 1999). Impact melt sheets often are rather homogeneous in composition, being mixtures of melted and vaporized target rocks that were driven at high velocities down into the expanding transient cavity, as turbulent flows (Dressler and Reimold 2001). Impact melt sheets are, on the other hand, heterogeneous in texture, especially near the upper and lower surfaces where they incorporate mixtures of shocked and unshocked lithic and mineral clasts. The impact melts differ from igneous melts in having been superheated and, therefore, may contain ultra-high-temperature mineral phases, such as baddeleyite, in addition to shock-metamorphosed clasts.

There is no megascopic evidence of massive impact melt rock at Chesapeake Bay. But direct evidence for impact melting is expressed in melt zones and melt-rock particles identified in thin section in some of the smaller crystalline clasts (petrographic studies discussed in Chapter 6). Also, as we have indicated above, there is plausible indirect evidence (in the form of distinctive seismic reflection signatures) that large bodies of impact melt rock could be present in the upper levels of the inner-basin fill at Chesapeake Bay (Figs. 4.29A,B, CD-ROM.12).

12 Implications for Impact Models

12.1 General Conceptual Models and Scaling Relations

12.1.1 Subaerial Cratering

Shoemaker (1963), in his classic study of Meteor (Barringer) crater, divided the impact cratering process into a temporal succession of stages. Shoemaker's pioneering ideas have served as the basis for succeeding conceptual and computer-generated models of the physical processes involved in creating impact structures. Gault et al. (1968) constructed a widely cited model of impact cratering on the basis of extrapolations from nuclear explosions combined with laboratory experiments and computer simulations. Gault et al. (1968) described the impact process in terms of three cratering stages (contact and compression; excavation; modification), while emphasizing that the boundaries between successive stages are not sharply defined. According to the Gault et al. model, the initial few seconds of impact (stage 1) can be viewed as a succession of contact and compressive events. Upon contact, the projectile pushes the target rocks out of its path, thereby compressing and accelerating them, as the target's resistance decelerates the projectile. Resultant shock waves may reach hundreds of GPa, whereupon both the projectile and the target rocks melt or vaporize. Contact and compression constitute the shortest of the three cratering stages (generally <1 s), and are completed as soon as the shock wave and its companion rarefaction wave pass through the impactor.

In the Gault et al. (1968) model, crater excavation (stage 2) lasts for a few seconds to minutes immediately following contact and compression. Excavation flow is promulgated by a hemispherical shock wave and its subsequent rarefaction wave, which set the target rocks in subsonic motion and fragment the near-surface rock layers. Ensuing ejection of this material opens up a transient crater, whose diameter is an order of magnitude greater than that of the impactor. For peak-ring craters, the diameter of the crest of the peak ring approximates the diameter of the transient crater (O'Keefe and Ahrens 1997). The depth of excavation is much shallower than the maximum depth of the crater, being, as a rule of thumb, about one-tenth the diameter of the transient crater (= 0.1 the diameter of the peak ring; Melosh 1989). Applying this estimate to Chesapeake Bay crater, therefore, the predicted maximum depth of excavation would have been about 4 km. Aspects of the excavation stage have recently been reexamined by Melosh and Ivanov (1999), who emphasized the role of acoustic fluidization as a mechanism for temporarily degrading the strength of rocks surrounding the impact site.

Modification of the crater (stage 3) follows completion of excavation (Gault et al. 1968; Melosh 1989), and is characterized by gravity-driven collapse (slumping and sliding) of the nascent crater's outer margin and rebound of target rocks in the crater interior to form a peak ring and(or) central peak. For subaerial craters, the modification stage lasts for tens of seconds. In the case of submarine craters, additional modification of the crater's outer margin is brought about by loading and hydraulic effects as the marine water-column collapses and surges back into the crater (Ormö and Lindström 2000). Furthermore, the final process of tsunami washback might extend the duration of crater modification to several days or weeks, depending upon the water depth and distance to the shoreline.

Kieffer and Simonds (1980) derived a semiquantitative cratering model by applying Shoemaker's (1963) ideas to field data from 32 subaerial terrestrial craters. Kieffer and Simonds (1980) subdivided the cratering process into seven stages. They stressed, in particular, the roles of the composition of target rocks and their contained volatiles in constraining crater development and subsequent impact-generated deposition.

Final structure and morphology of a complex crater can be expressed as, or predicted from, scaled relations between various physical or conceptual features of the crater (such as diameter of outer rim, diameter of central peak, height of central peak, and diameter of transient crater). For example, Melosh (1989) showed that the diameter of the central peak is roughly 0.22 times the outer-rim diameter on all terrestrial planets. For the Chesapeake Bay crater, this relation would predict a central-peak diameter of ~19 km. The average diameter measured by us is ~15 km (~0.18 times the outer-rim diameter).

On the primary basis of field observations, Grieve and Robertson (1979), Grieve (1991), and Pilkington and Grieve (1992) postulated "rule-of-thumb" scaling relations for terrestrial subaerial craters. For example, Grieve (1991) concluded that excavation does not occur all the way out to the outer rim, but is constrained to the central 0.50–0.65 of the outer-rim diameter. Applying this relation to the Chesapeake Bay crater would predict an excavation diameter of 42.5–55.5 km (or ~21–28 km radial distance from the crater center). Our measurements indicate that the diameter of excavation at Chesapeake Bay averages ~50 km (derived from the average diameter of the outer flank of the peak ring). Outside the peak-ring periphery, only the Cenozoic section seems to have been excavated from the tops of the displaced megablocks, which indicates only weak excavation there in the annular trough. We infer that this "excavation" was achieved mainly through surgeback erosion processes during crater modification. Grieve (1991) also determined that stratigraphic uplift of the central peak is 0.09–0.12 times the diameter of excavation. In the case of the Chesapeake Bay crater, this relation would yield a stratigraphic uplift of 4.5–6.0 km, in general agreement with Melosh's (1989) maximum depth of excavation.

Grieve and Robertson (1979) and Pilkington and Grieve (1992) indicated that, for complex craters, the true depth (d_t) of the crater is $0.52D^{0.2}$ (D = diameter of outer rim). If one applies this relation to the Chesapeake Bay crater, the predicted d_t would be 1.2 km. There is no obvious signature of such a structural boundary on our seismic profiles to confirm this prediction. Neither are there any drill data

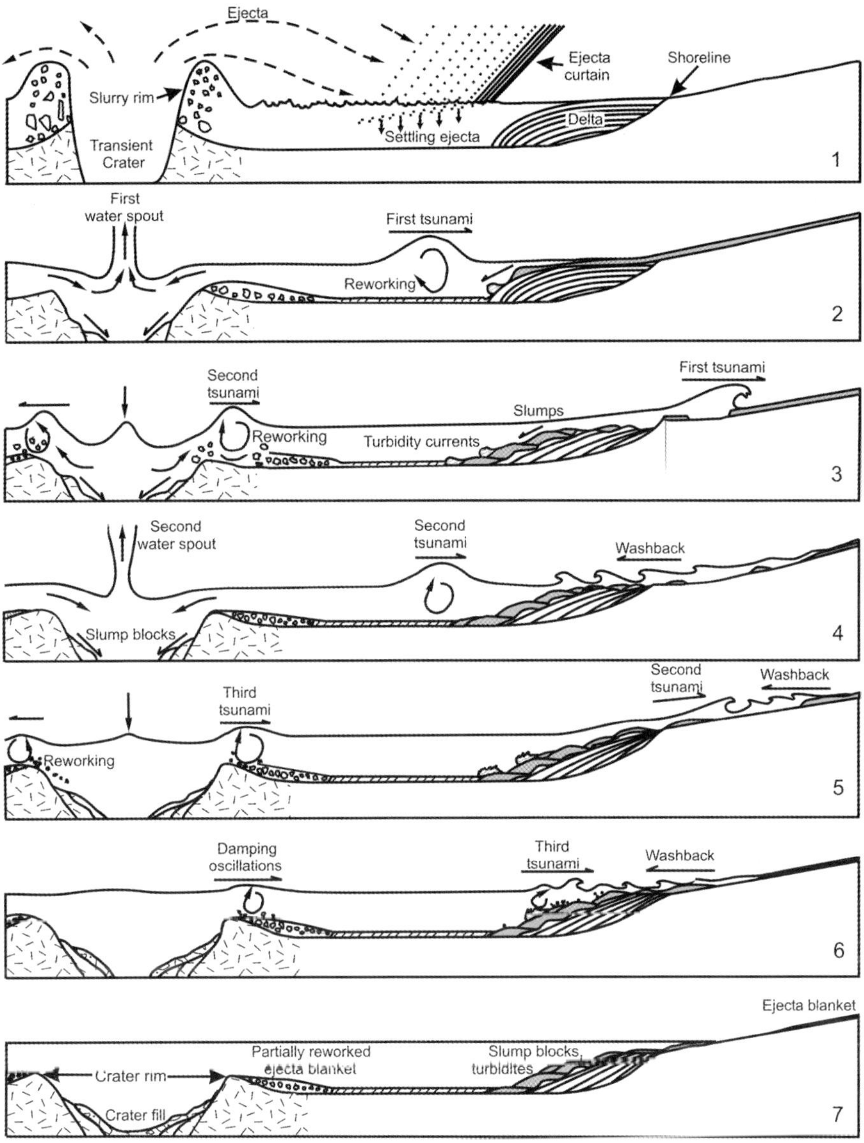

Fig. 12.1. Conceptual model of excavation and deposition associated with a submarine bolide impact (modified from Oberbeck et al. 1993; based on experimental data and hypothetical relations). Seven panels illustrate time slices progressing downward from formation of transient crater to end of crater-fill processes. Compare with Figs. 12.2, 12.3.

yet from this part of the crater.

Overall, where available, the morphometrics of the Chesapeake Bay crater appear to fit rather well the quantitative predictions of Gault et al. (1968), Melosh (1989), Grieve (1991), and Pilkington and Grieve (1992).

12.1.2 Submarine Cratering

It is widely accepted on the basis of field studies and computer modeling, that submarine cratering includes some unique mechanisms or substages due to the presence of a marine water column (Higgins and Butkovitch 1967; Kieffer and Simonds 1980; McKinnon and Goetz 1981; Melosh 1981; Gault and Sonett 1982; McKinnon 1982; Roddy et al. 1987; Ormö and Miyamoto 2002; Shuvalov et al. 2002). Most conceptual and quantitative discussions of submarine cratering, however, have dealt with deep (abyssal) ocean impacts (Ahrens and O'Keefe 1987; Roddy et al. 1987; Nemchinov et al. 1993; Hills et al. 1994; Hills and Goda 1999; Artemieva and Shuvalov 2002; Wünnemann and Lange 2002).

Oberbeck et al. (1993) published one of the earliest and most perceptive conceptual models of an impact into a shallow (neritic) sea (Fig. 12.1). The seven-step Oberbeck model, which is based on field observations and laboratory experiments, focuses only on the excavation and modification stages of crater development. During excavation, a mixture of water and fragmented target rocks would form a *slurry rim,* which extends upward from the sea surface and forms a raised lip around the transient crater (Fig. 12.1, step 1). A cone-shaped curtain of ballistically ejected target fragments proceeds radially away from the crater center. Debris trailing from the ejecta curtain's base settles onto the shallow seafloor.

The slurry rim then moves landward (and seaward) and initiates a giant tsunami wave, which resuspends and erodes seafloor debris (Fig. 12.1, step 2). Subsequently, the wall of the seawater cavity collapses back into the excavation and rushes toward the center of the crater, where it produces a central *slurry spout,* whose collapse initiates a second tsunami wave (Fig. 12.1, step 3), followed by additional slurry spouts and tsunami waves (Fig. 12.1, steps 4, 5), until wave oscillations over the crater are sufficiently damped (Fig. 12.1, step 6).

Oberbeck's model includes a steep-faced delta system at the shoreline (Fig 12.1), which is massively disrupted (slumps, sediment gravity flows) by the impinging train of tsunami waves (Fig. 12.1, steps 3–6). The return flow of the tsunamis produces further nearshore erosion and deposition (Figs. 12.1, steps 4–7). Most effects of the tsunami washback in the Oberbeck model are concentrated in the nearshore region, and have no role in filling the crater, which in this model is a simple crater (no central uplift or peak ring).

Tsikalas (1996) and Tsikalas et al. (1998b, 1999) used the Mjølnir crater to construct a conceptual model for submarine impact into entirely sedimentary target rocks (sandstones, shales, claystones, limestones; Fig. 12.2). A subsequent numerical simulation (Shuvalov et al. (2002) supports and refines the main features of the Tsikalas model. In the Tsikalas model, the transient crater of the excavation stage is shown to be ~4 km deep (from sea level to maximum penetration

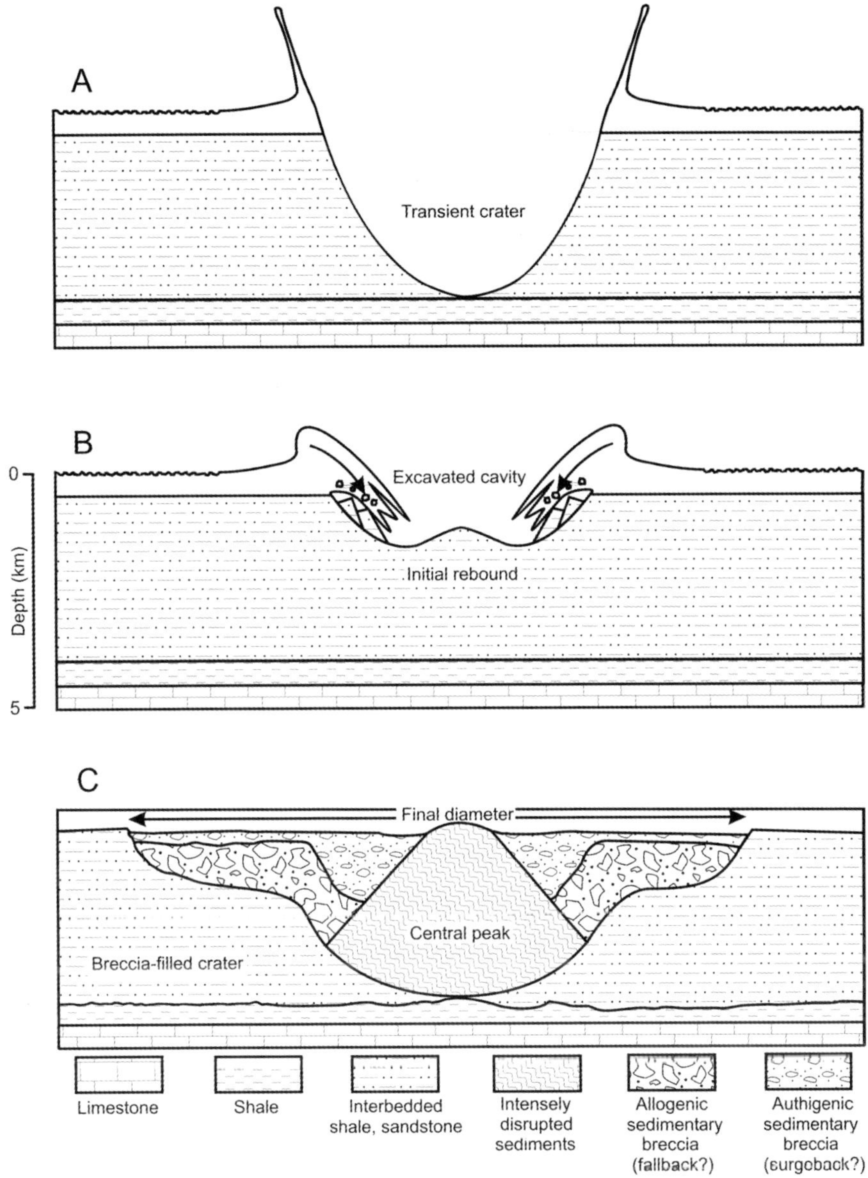

Fig. 12.2. Conceptual model of excavation and deposition associated with a submarine bolide impact (modified from Tsikalas et al. 1998b, 1999; based on Mjølnir crater). Compare with Figs. 12.1, 12.3.

into sediments; ~6 km in the Shuvalov et al. simulation), and is formed of seawater (Fig. 12.2A). As crater modification begins, the sedimentary floor of the crater rebounds (reaching a height of >5 km above the seafloor in the Shuvalov et al. simulation), and the outer margins of the crater begin to slump and slide along a décollement (Fig. 12.2B). This is followed by collapse of the water column, which surges back into the crater, further collapsing the sedimentary walls and eroding the crater floor. Tsunamis were not included in the Tsikalas model, presumably because the Mjølnir impact was too far from a landmass to have produced tsunami washback deposits inside or near the crater. However, the numerical simulation of Shuvalov et al. (2002) created large-amplitude (200 m) tsunamis, as well as surgeback flows with velocities of 50–70 m/s. The final crater at Mjølnir displays a broad, conical central peak, whose relatively low density and distinctive seismic signature are inferred to represent intensely disrupted sediments. Around the periphery of the central peak, the crater floor is inferred to be buried by a relatively thick allogenic breccia (fallback?), which, in turn, is overlain by authigenic breccia (surgeback?). Neither breccia body covers the crest of the central peak.

Tsikalas et al. (1999) calculated a *collapse factor* (ratio of final crater diameter to transient crater diameter) of 2.5 for Mjølnir crater (2.0–2.5 according to Shuvalov et al. 2002). These authors considered this value to be significantly greater than the collapse factor for terrestrial subaerial craters of equivalent transient-crater diameter, which Gault et al. (1968) and Melosh (1989) indicated should be 1.4–2.0. Tsikalas et al. (1999) and Shuvalov et al. (2002), thereby, implied that for a given set of bolide properties (diameter, velocity, density, trajectory), submarine craters would have a greater collapse factor than subaerial craters. The calculated collapse factor for the Chesapeake Bay crater, on the other hand, is 2.1, more in line with Melosh's (1989) estimate for terrestrial subaerial craters.

Ormö (1998) proposed a shallow-water submarine cratering model on the basis of his interpretations of the Lockne crater. Ormö's model was built upon an earlier lunar cratering model of Quaide and Oberbeck (1968), which has no similarities to the more recent model of Oberbeck et al. (1993). Ormö and Lindström (2000) updated the submarine cratering model of Ormö (1998) and emphasized the role of a collapsing and "resurging" marine water-column during the modification stage (Fig. 12.3). The transient crater in the Ormö-Lindström model, like that of Tsikalas et al. (1999), has a raised lip of seawater, but its floor is formed in crystalline rather than sedimentary rocks (Fig. 12.3A). During excavation, an ejecta curtain moves radially away from the impact site (Fig. 12.3B). Crystalline basement is exposed as a flat (perhaps slightly raised) surface around a nested central basin, and the sedimentary cover is further excavated to form the crater's outer rim. At the crater's outer margin, the sedimentary section is folded back and injected with breccia sills and dikes to form a slightly raised lip. Modification is initiated by rebound of the crystalline basement, and the water column collapses and

Fig. 12.3. (Opposite page) Conceptual model of excavation and deposition associated with a submarine bolide impact (modified from Ormö and Lindström 2000; based on Lockne crater). Compare with Figs. 12.1, 12.2.

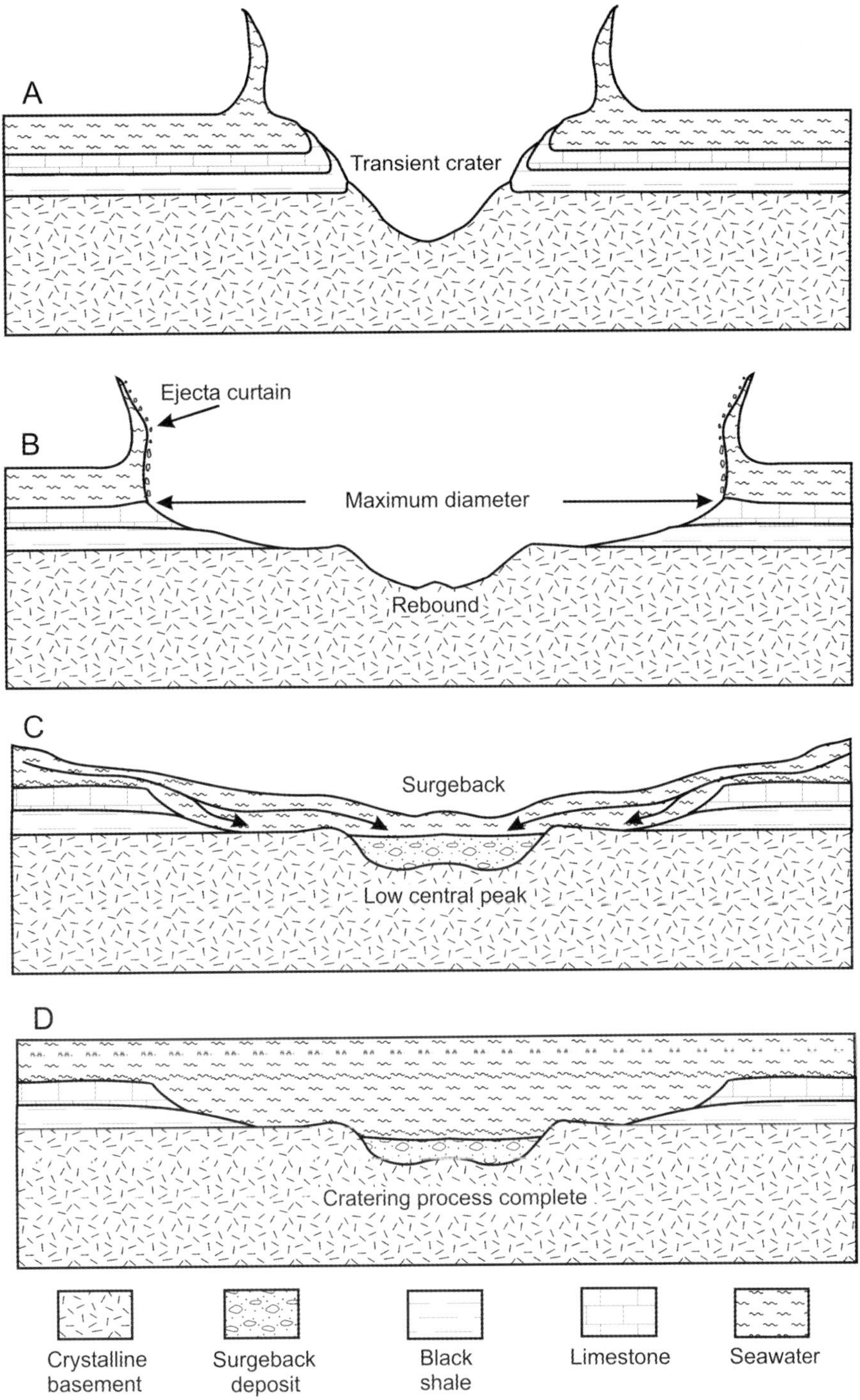
A
Transient crater
B
Ejecta curtain
Maximum diameter
Rebound
C
Surgeback
Low central peak
D
Cratering process complete
Crystalline basement
Surgeback deposit
Black shale
Limestone
Seawater

surges back into the crater (Fig. 12.3C). The inwardly cascading surgeback flow erodes radial channels through the crystalline crater floor (but curiously, not through the sedimentary walls) and fills the channels with surgeback breccia. The crater's central basin and low central peak are also buried by surgeback breccia (any fall-back breccia that may have been present in the central basin is reworked into the surgeback deposit).

Ormö et al. (2002) recently simulated the Lockne impact using the SOVA computer hydrocode (Shuvalov 1999). Using an 800-m-diameter asteroid (density 2.63 g/cm^3) and a paleodepth of 1000 m, the simulation produced impact features similar to those observed or inferred at Lockne. These features included a small, bowl-shaped, central depression, or nested crater, surrounded by a cylindrical water cavity.

As in the Tsikalas model, no tsunami washback is included in the Ormö-Lindström (2000) or Ormö et al. (2002) models. However, Sturkell et al. (2000) described some aqueous sedimentary structures related to the Lockne impact, which crop out 45 km from the Lockne crater. Sturkell et al. (2000) interpreted these features to have formed from Lockne surgeback processes. In our opinion, however, some of the surgeback deposits around the Lockne crater have the same sedimentary features expected of a tsunami or hypercane deposit. In particular, the subaqueously formed sedimentary structures (indicating successive, opposite, flow directions) described from an outcrop 45 km from the Lockne crater (Sturkell et al. 2000) are very close analogues to the silt-banded, multidirectional, flowin facies at the top of the Exmore breccia.

12.2 Conceptual Model for Chesapeake Bay Crater

We present a conceptual model for the Chesapeake Bay crater, which includes a two-dimensional computer simulation of the contact and compression stage, the excavation stage, and the early part of the modification stage. We are especially indebted to David Crawford of Sandia National Laboratories for providing this simulation (Fig. 12.4). The simulation was performed with the CTH shock physics hydrocode (McGlaun and Thompson 1990). Tabular equations-of-state, including accurate renditions of the solid/liquid/gas transitions were used. The materials were treated as strengthless, which is appropriate, since the Hugoniot-elastic-limit is typically exceeded during the early phases of the impact processes depicted here.

We also thank Toshihiro Matsumoto of the Japanese Broadcasting Company (NHK) for providing 3-D snapshots of the impact (Fig. 12.5), derived from a video animation of Crawford's simulations. We have integrated these simulations with our seismic and borehole interpretations (which mainly bear upon the modification stage) to create a holistic conceptual model of the structural, depositional, and morphological aspects of submarine crater formation on the North American Atlantic Continental Shelf ~36 Ma. Nevertheless, we are forced to omit such important aspects as possible suevite and melt-sheet forming processes, because we lack the requisite field data. Moreover, although some of our model concepts may apply to other submarine craters (and to some subaerial ones, as well), each submarine crater identified so far displays enough individual variability to set it apart from all other such craters.

12.2.1 Stage 1 – Contact and Compression

As the impactor for the 2-D simulation, Crawford chose an asteroid that would produce a 40-km-wide transient crater (asteroid diameter 3.3 km; density 3.32 g/cm^3; impact velocity 20 km/s; impact trajectory vertical). This transient-crater diameter matches the average diameter of the Chesapeake Bay peak ring (Chapter 4). The simulated target layers consist of 120 km of air, 300 m of seawater, 200 m of sediments, and 100 km of granite. The composite 500 m of water column plus seafloor sediments are so thin as to be barely discernable at the spatial scale of the simulation, but the general aspects (size, depth, height, temporal succession) of early impact processes and features are clearly expressed (Fig. 12.4).

We have almost no field data bearing on stage-1 processes at Chesapeake Bay, other than the compositions and thicknesses of the target layers, the measured diameter of the peak ring, and an estimate of the preimpact water depth. The diameter of the peak ring probably constrains the diameter of the transient crater, which, in turn, constrains the modeled properties of the impactor. The presence of shock-metamorphosed and melted basement clasts within the Exmore breccia, however, is direct evidence of the high temperatures and pressures produced during stage 1 of the Chesapeake Bay impact, which lasted less than 10 seconds.

12.2.2 Stage 2 – Excavation

At ten seconds after contact (in the Crawford simulation), the Chesapeake Bay transient crater is ~20 km wide, its floor is 15 km below the seabed, and its outer wall (a flask-shaped ejecta curtain composed of a slurry of water and rock debris) rises nearly 40 km above the sea surface. The transient crater depth at this time slice is 18 km below the seafloor (Figs. 12.4, 12.5). The impact fireball is constrained within the crystalline walls of the lower transient crater. A conical shock wave is expanding through the crustal rocks (40 km from ground zero) and the atmosphere, and a hot plume of vaporized bolide, target rocks, and seawater extends ~90 km into the atmosphere.

At 30 seconds after impact, the interior fireball has expanded to 20–25 km height, but is still contained within the ejecta curtain, whose top is at ~40 km (Figs. 12.4, 12.5). The transient crater diameter has expanded to 30 km, and its depth is 20 km. Some ejecta fragments have reached nearly 80 km into the atmosphere, and are beginning to fall back, forming incandescent meteors as friction reheats their outer surfaces. The leading edge of the shock wave is now >60 km

Fig. 12.4. (Next page) 2-D computer simulation of Chesapeake Bay impact, showing five time slices (10 s, 30 s, 60 s, 120 s, 170 s). Simulation generated by D.A. Crawford, Sandia National Laboratories. See text for further explanation and CD-ROM for color version.

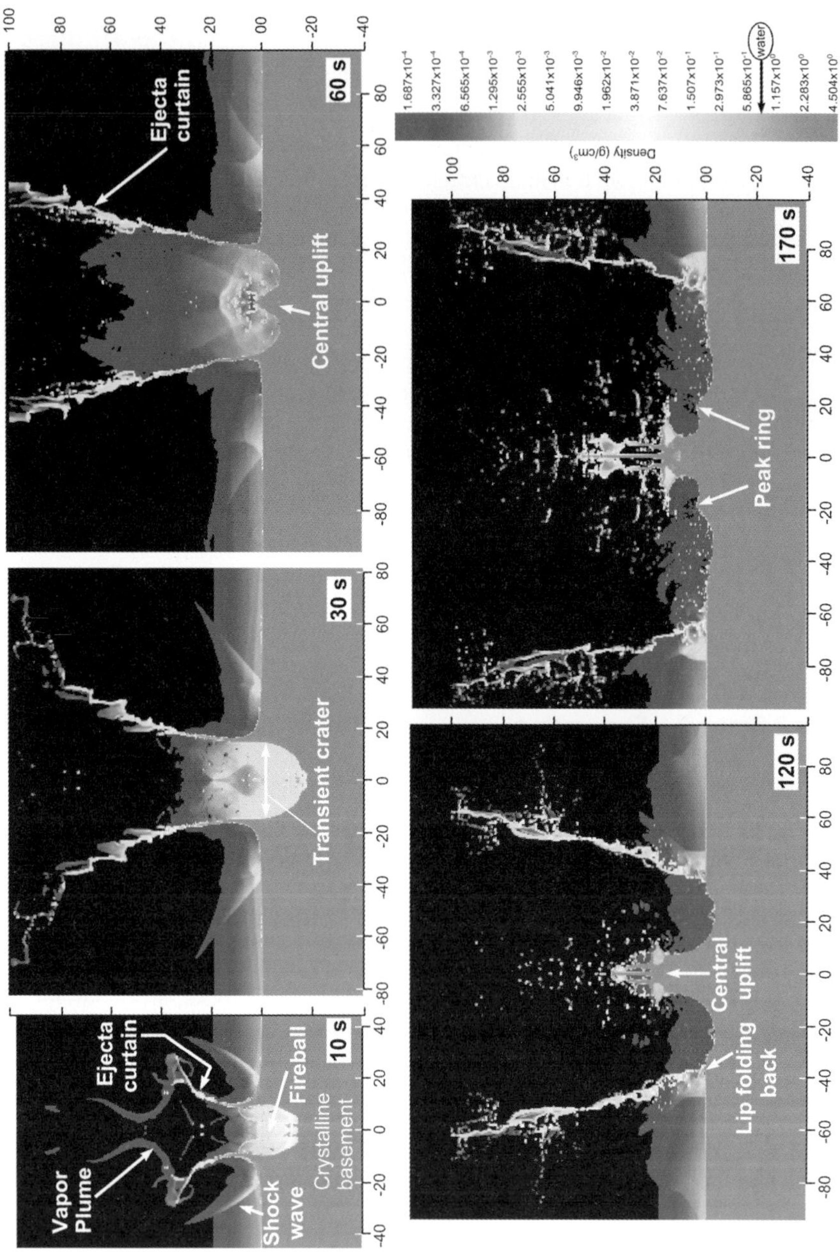
Vapor Plume
Shock wave
Crystalline basement
Fireball
Ejecta curtain
10 s
Transient crater
30 s
Central uplift
Ejecta curtain
60 s
Lip folding back
Central uplift
120 s
Peak ring
170 s
Density (g/cm³)
water

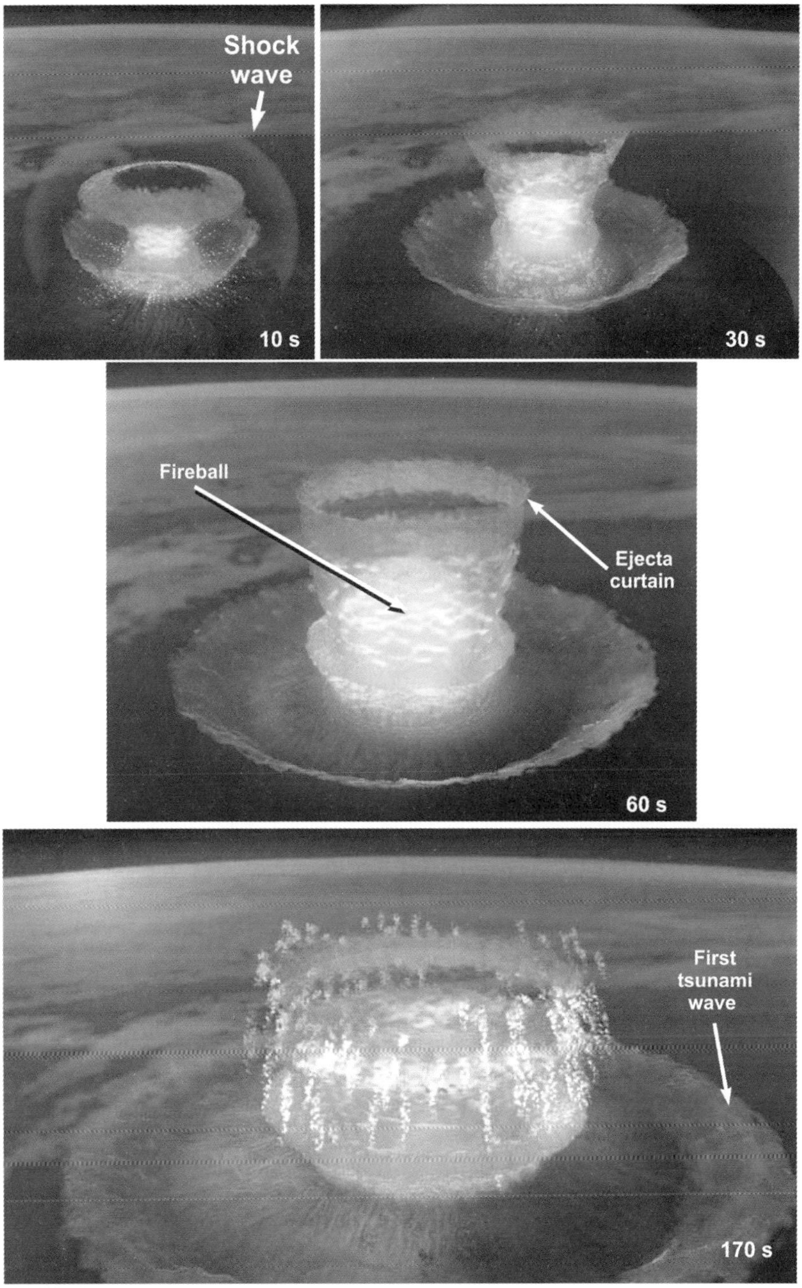

Fig. 12.5. 3-D rendering of four of the time slices shown in Fig. 12.4 (10 s, 30 s, 60 s, 170 s) derived from computer-assisted video animation prepared by NHK (Japanese Broadcasting Company; courtesy of Toshihiro Matsumoto). See text for further explanation and CD-ROM for color version.

from the point of impact.

At 60 seconds after impact, the crater has achieved its maximum transient diameter of 40 km, and the center of the granitic crater floor has rebounded to form a mountainous, 10-km-high, central peak (Figs. 12.4, 12.5). The fireball has begun to collapse and cool within the crater. The ejecta curtain is still well developed up to about 50 km height, but is breaking up above this elevation.

Field evidence of initial excavation-stage processes at Chesapeake Bay can be seen in a combination of properties peculiar to the annular trough. There, the crystalline basement rocks are highly fractured and faulted (seen both on seismic profiles and in cores), and the bases of the overlying sedimentary megablocks have been fluidized into massive, structureless units. We infer that these features were caused by impact-generated compression and rarefaction. The degree of shock metamorphism in the breccia clasts (fragments of crystalline basement) indicates that impact pressures reached >60 GPa. Density differences between the porous, water-saturated, unconsolidated sediments and the denser crystalline basement rocks appear to have been particularly important in their respective reactions to the impact shock. The sediments absorbed significantly more shock energy than the basement, which allowed differential motion between the basement and the sedimentary cover, and enabled huge sedimentary megablocks to detach from the basement surface (zone of décollement), and to slump and slide across its surface. In some places, though, the megablocks show little or no rotation or internal deformation that would indicate lateral motion. Instead, these megablocks appear to have dropped vertically downward as their shock-fluidized bases collapsed.

12.2.3 Stage 3 – Modification

At 120 seconds postimpact, the fireball has dissipated (Fig. 12.4). The central peak has reached maximum height of ~30 km, and is 15–20 km in diameter. Sediment collapse along the crater's outer rim has expanded the crater diameter to ~60 km. The ejecta curtain is breaking up in the atmosphere, as its upper diameter has expanded to 120 km and the sedimentary section of the seafloor is folding back upon itself at the crater's outer rim. A significant volume of weakly ejected debris is raining back into the crater above the central peak.

At 170 seconds after impact, the central peak has collapsed to <20 km height, a 40-km-wide peak ring is forming, and the crater's outer-rim diameter has expanded to its maximum diameter of 85 km (Figs. 12.4, 12.5). The annular trough can now be differentiated, and its floor is ~1–1.5 km below the lip of the outer rim. The disassociating ejecta curtain is now ~180 km in diameter across the top. Figure 12.5 (170s) shows the initial tsunami wave forming and beginning a radial expansion that will take it to the eastern flank of the Appalachian Mountains.

The Crawford simulation did not run long enough to show final collapse of the central peak, subsequent gravity collapse of the oceanic water column, or impact-generated tsunami activity. Other published numerical simulations using different hydrocodes (e.g., Wünnemann and Lange 2002, using a 2-D SALE code;

Shuvalov et al. 2002, using the SOVA multimaterial code) indicate that most of these remaining processes of crater modification, except for tsunami washback and hypercane (flowin) deposition, would be completed by 800–1300 s (13–20 min) after impact. Tsunami and hypercane activity might last for days or weeks, depending on many variables, such as bolide diameter, water depth, physiography of the seafloor, and distance to the nearest shoreline.

12.3 General Conceptual Model of Crater-Fill Deposition

The modification stage of impact cratering embraces a variety of depositional regimes and their resultant deposits or lithofacies, which in themselves, are sufficiently independent from actual crater-forming processes to deserve separate study. The superb preservation of the Chesapeake Bay crater and the abundance of high quality seismic and borehole data have yielded more evidence of these depositional aspects than any previous study of submarine craters. We propose a two-part general depositional model for Chesapeake Bay, which contrasts regimes and lithofacies inside the crater (intracrater) with those outside the crater (extracrater).

12.3.1 Intracrater Regimes and Lithofacies

Previous field studies (including those at Chesapeake Bay; see Chapters 6, 11) and conceptual models indicate that there is a general temporal succession of six crater-filling depositional regimes inside submarine craters (Fig. 12.6A). This succession of regimes is determined from the spatial distribution of the lithofacies they produce (Fig. 12.6B). The basal lithofacies (on the floor of the inner basin and covering the outer flank of the peak ring) are bodies of fallback breccia. Fallback breccia consists primarily of weakly ejected fragments that fall directly back into the crater, mixed with allochthonous debris that never left the crater. For most large craters, individual clasts in the fallback breccia are mainly remnants of the deepest target rocks, which constitute crystalline basement (Fig. 12.7).

Stratigraphically above the fallback breccia is the slumpback lithofacies (slumped megablocks), derived mainly from gravitational collapse of the outer crater walls and of the margins of peak rings and central peaks (Figs. 12.6, 12.7). In cases where crystalline rocks are overlain by sedimentary rocks, the bulk of the slumpback lithofacies is expected to consist of sedimentary megablocks, many of which can be kilometer-sized blocks. Mixing of crystalline and sedimentary clasts is expected to occur at the fallback/slumpback boundary.

The next highest intracrater stratigraphic unit in submarine craters consists of the surgeback lithofacies, derived from loading pressures on, and hydraulic erosion of, the crater walls and surrounding seafloor caused by collapse of the oceanic water column (Fig. 12.6, 12.7). In targets with surficial sedimentary rocks, surgeback lithofacies should be dominated by sedimentary clasts, with a significant component of finer grained matrix.

A

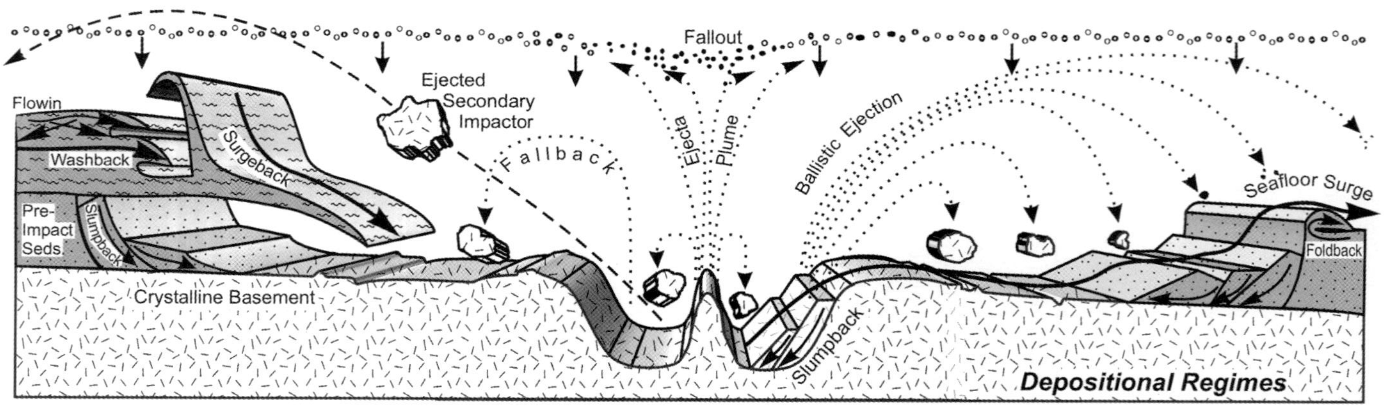

B

EXPECTED LITHOFACIES

Intracrater

Fallout - Finest debris and vapor ejected into or above atmosphere before raining back to Earth's surface; includes microtektites, shocked minerals, etc.

Flowin - Fine debris scoured from seafloor by impact-generated hypercanes and washed back into crater; deposited as multidirectional sediment flows.

Washback - Debris scoured from seafloor and coastal plain by tsunami and then washed back to sea; dominantly sedimentary matrix; upward-fining sequence

Surgeback - Debris scoured from seafloor, crater walls, and crater floor by collapsing water column; complex mixture of mainly sedimentary clasts and matrix; upward-fining sequence

Slumpback - Coarse debris in megaslides and megaslumps derived from sedimentary outer walls and flanks of crystalline peak ring; dominantly sedimentary and crystalline clasts

Fallback - Coarse debris derived from crater floor, ejected short distance before falling back; dominantly crystalline clasts

Extracrater

Fallout - Finest debris and vapor ejected into or above atmosphere before raining back to Earth's surface; includes microtektites, shocked minerals, etc.

Washback - Debris scoured from seafloor and coastal plain by tsunami and then washed back to sea; incorporates previously deposited ejecta; dominantly sedimentary matrix; upward-fining sequence

Ballistic ejecta - Debris ejected into ballistic trajectories, which is deposited from radially expanding debris curtain; laterally graded deposit

Seafloor surge - Debris ejected radially outward from crater as ground-hugging, scouring, flows; laterally graded deposit

Foldback - Beds at outer rim fold back to produce reversed stratigraphy

Fig. 12.6. (Opposite page) *A*, general model of depositional regimes expected to be represented by impact deposits at Chesapeake Bay impact crater, based on previous field studies and conceptual models; those regimes presumably would have been distributed in a symmetrical manner around the center of the crater, but are shown asymmetrically here for simplicity. *B*, general description of depositional lithofacies expected at Chesapeake Bay crater based on the regime model in Fig. 12.6A. Compare with Fig. 12.7.

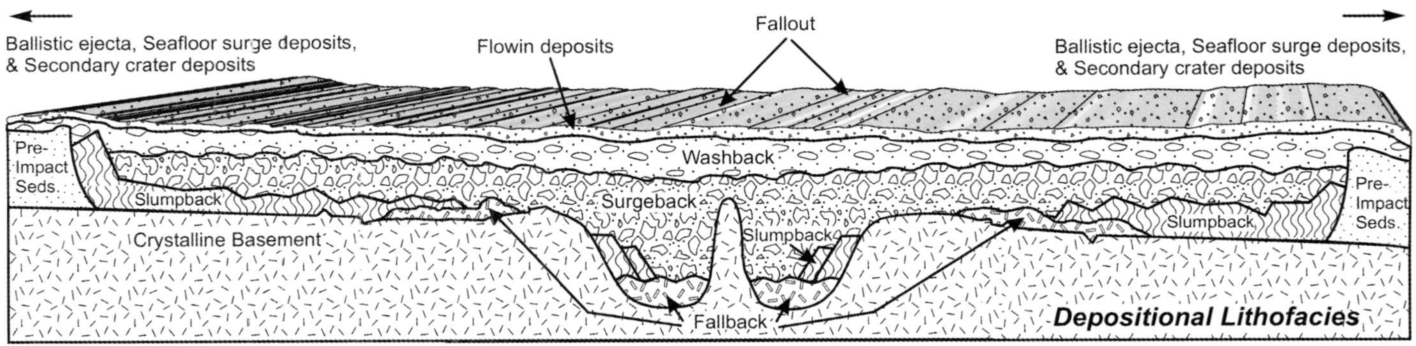

Fig. 12.7. General impact depositional lithofacies model for Chesapeake Bay impact crater, based on actual data from seismic reflection profiles and coreholes. Compare with Fig. 12.6.

Above the surgeback lithofacies, and mixed with it near their mutual boundary, is the washback lithofacies (Figs. 12.6, 12.7). This lithofacies contains principally sedimentary debris scoured from the surrounding seafloor and adjacent coastal plain by the runup and washback of tsunami wavetrains. A major component of fine-grained matrix is expected to be incorporated in washback lithofacies. All coreholes within Chesapeake Bay crater have documented an upward-fining sequence of breccia in the washback lithofacies (Figs. 12.6, 12.7; see also Chapter 6). So far, Chesapeake Bay is the only crater in which a washback lithofacies has been specifically delineated.

On the basis of the Chesapeake Bay cores (Chapter 6), we propose a sixth depositional regime for this general conceptual model – the flowin regime. In the uppermost few meters of the Exmore breccia at Chesapeake Bay, the sediments differ from the preceding lithofacies in three important ways. *First*, this upper interval contains finer sediment – mainly fine sand, silt, and clay; *second*, the upper section loses the massive character of the underlying units and becomes bedded, or laminated on a centimeter-scale; *third*, the bedding is dominantly inclined, and the dip direction changes frequently over short vertical distances (centimeter-scale). These features indicate a complex regime in which flow direction and velocity changed relatively rapidly through time, in contrast to the generally unidirectional washback inferred for the tsunamis. This depositional style suggests a succession of postimpact storms, which stirred the seafloor around the crater with sufficient energy to send small debris flows or turbidity currents into the crater. The story is complicated, however, by the fact that benthic foraminifera in the initial marine sediments overlying the dead zone at Chesapeake Bay suggest paleodepths of around 300 m (Chapter 13). At such depths, the seafloor is not likely to be stirred by normal storm action, so we speculate that the stirring agent may have been a succession of runaway hurricanes, or *hypercanes*.

The concept of hypercanes was introduced by Emanuel et al. (1995), who described them as runaway hurricanes. These authors proposed that hypercanes could develop when the sea-surface temperature exceeds a critical threshold, such as might occur in the vicinity of a marine bolide impact. Emanuel et al. (1995) focused on the ability of hypercanes to inject massive volumes of water and aerosols into the stratosphere, but here we are more interested in the potential for moving seafloor sediment in deep water.

A computer simulation by Emanuel et al. (1995) showed that a bolide that could produce sea-surface temperatures of 50°C over an area of 50 km or more could generate a hypercane with maximum wind velocities of 300 m/s (650 mph), or about three times the maximum horizontal velocity of known hurricanes. Crawford's computer simulations (Fig. 12.4), combined with the shock-metamorphic features documented in the Chesapeake Bay cores (Chapter 6), show that the Chesapeake Bay bolide would have produced even higher temperatures over a larger area, thus we consider the presence of hypercanes to be a plausible inference. Hypercane flowin deposits also might be expected in other well-preserved submarine craters excavated into shallow continental shelves.

The final synimpact, intracrater, depositional regime is that of fallout. Fallout debris is composed of millimeter- to submicrometer-sized particles, some of which could remain entrained in or above the Earth's atmosphere for weeks-to-months following the impact (Figs. 12.6, 12.7). Microtektites, microkrystites, and shocked minerals are among the components of the fallout lithofacies.

12.3.2 Extracrater Regimes and Lithofacies

Field studies of submarine impact deposits immediately outside the crater rims have documented two distinct depositional regimes (Fig. 12.6). These regimes are expected to produce two of the three stratigraphically highest synimpact lithofacies found inside the crater – washback lithofacies overlain by fallout lithofacies.

In subaerial craters, two additional synimpact extracrater regimes have been recognized. The stratigraphically lowest subaerial extracrater depositional regime is that of ground surge, in which a viscous mixture of fragmental target material is ejected radially along the ground surface (Fig. 12.6). Hypothetically, the equivalent regime in submarine craters would be seafloor surge. This regime operates close to the seabed, flushing debris outward from the crater, scouring radial channels or striations, and also leaving a laterally graded deposit (coarser debris closest to the crater). No seafloor-surge lithofacies has been recognized at Chesapeake Bay, presumably, in part, because its constituents would have been incorporated into the washback lithofacies, and thus would be extremely difficult to differentiate.

The remaining extracrater depositional regime of subaerial craters is that of ballistic ejection, which produces a conical curtain of debris moving outward in ballistic trajectories, and lays down a laterally graded deposit (Fig. 12.6). When the larger clasts within this ballistic ejecta curtain eventually impact the Earth's surface, they may create additional concussive debris, or even excavate secondary craters. Obviously, in many cases it may be difficult, or impossible, to confidently separate products of ballistic sedimentation from those of the seafloor-surge and washback regimes. At Chesapeake Bay, any ballistic ejecta would have been stirred into the washback lithofacies. The numerous sand-sized and occasional cobble-sized crystalline clasts within Chesapeake Bay extracrater washback deposits probably represent reworked ballistic ejecta.

Some modelers also expect a separate extracrater deposit to be formed from high-velocity jetting during the early stages of impact (Melosh, 1989), but, so far, we have found no field evidence for such a regime related to the Chesapeake Bay impact.

12.4 Differentiating Crater-Fill Lithofacies at Chesapeake Bay

On the basis of cores, geophysical logs, and seismic reflection profiles, the broadly defined Exmore breccia can be divided into six stratigraphically distinct synimpact lithofacies (Fig. 12.7), which can be compared to the general model of crater-fill lithofacies (Fig. 12.6). From the nature of these lithofacies, their respective depositional regimes can be inferred (Table 12.1).

In order to apply the crater-fill model to Chesapeake Bay impact lithofacies, we must know: (1) the gross lithologies of the deposits, which can be derived from cores and downhole geophysical logs; (2) the stratigraphic succession (stacking order) of different lithic units within and near the crater, which can be derived from cores, downhole geophysical logs, and seismic reflection profiles; (3) the geometry of the lithic units, which can be determined from seismic profiles; and (4) the location of each core relative to the morphological and structural features of the crater, as determined from seismic profiles.

The stratigraphically deepest lithofacies at Chesapeake Bay, assumed to contain fallback breccia, appears to be restricted mainly to the deepest parts of the inner basin (Figs. 11.8, 12.7). However, the deep inner basin has not yet been cored, and none of the currently available cores contain breccia dominated by crystalline clasts. In fact, basement clasts larger than sand size are extremely rare (Chapter 6). Moreover, there is no clear signature of fallback breccia on the seismic profiles, though Poag (1996a) identified a possible upper boundary for a presumed fallback unit. Near the bottom of the Bayside corehole, however, is a 20.33-m-thick, crystalline-clast breccia (688.54–708.87 m) unlike any other so far encountered at Chesapeake Bay (see Chapter 6). This breccia might represent fallback debris. This core section is composed of a matrix-supported breccia with abundant, mainly cobble-sized clasts (Fig. 6.4). The matrix in this unit is a white-to-light gray sand, quite distinct from the typical greenish gray, glauconite-rich Exmore matrix; it appears to be a finely comminuted immature granitic sand. Most cobbles in this section are either hard, indurated sandstones or weathered granite, as opposed to the poorly lithified sand, silt, and clay clasts of the typical Exmore breccia. Also, many of the cobbles display thin white rinds, similar to fusion rinds.

In terms of volume and lateral distribution, deposits from the washback, surgeback, and slumpback regimes are the dominant lithofacies encompassed by the Exmore breccia (Figs. 11.8, 12.7). Poag (1999b, 2000) attempted to quantify the differences between these lithofacies by calculating the relative thickness of matrix versus boulders (M/B ratio) in the cored sections. Where appropriate, we use the M/B ratio in conjunction with several qualitative lithic properties of the cores to differentiate these three breccia types. Poag (1999b, 2000) also attempted to correlate the spontaneous potential (SP) curve from the downhole geophysical logs with individual lithologies from each corehole, in order to determine whether the SP curve could be used to calculate an M/B ratio in the uncored sections of the Exmore and Kiptopeke boreholes. Our further analyses of the NASA Langley, North, and Bayside logs suggest that the lithic variability of the breccias is too great to yield reliable M/B ratios from the logs alone.

Inside the Chesapeake Bay crater, the three cored sections of slumpback lithofacies are characterized by the nearly complete absence of glauconitic quartz sand typical of the Exmore matrix. Neither is there other lithic or microfossil evidence of Cenozoic strata in the displaced megablocks. Instead, the megablocks contain thick sections (as thick as 20 m) of relatively undisturbed (mainly tilted), stratified (bedded or laminated), sedimentary beds (mainly nonmarine sands, silts, and clays, including thick paleosol intervals) of Early Cretaceous age (Figs. 6.7, 6.8). These normally stratified Cretaceous intervals are occasionally interrupted by zones of brecciated sediments, which we interpret as evidence of internal deformation within the displaced megablocks. In the deepest cored megablocks, near the basement surface at NASA Langley and at Bayside, 22–32 m of sedimentary section near the basement surface displays intervals of massive stratal disruption (possibly caused by acoustic fluidization), moderate cementation, and splotches of possible hydrothermal mineralization (Fig. 6.6).

The stratigraphically next highest depositional lithofacies in the Chesapeake Bay crater is composed of surgeback breccia, which was documented at all five intracrater core sites, but not outside the crater (Figs. 12.6, 12.7). Surgeback breccia in most coreholes is clast-supported, having too little matrix to calculate a meaningful M/B ratio. The exception is Bayside, where a greater abundance of matrix allows calculation of an M/B ratio of 1:9. At North and NASA Langley, we calculated instead, a cobble-to-boulder ratio (C/B ratio), which is 2:1 at both sites. Clasts in the surgeback breccia are notable for their extreme plastic deformation, inclined interclast contacts, and high-angle to near-vertical stratal orientations (Figs. 6.9, 6.10). Boulder-sized clasts are commonly 12–20 m thick in this lithofacies.

The stratigraphically next highest crater-fill lithofacies is notable for the dominance of typical Exmore matrix (medium-to-coarse, glauconitic, quartz sand; Fig. 6.14H,I). We calculated M/B ratios in this lithofacies ranging from 2:1 (North corehole), to 3:1 (Bayside corehole), to 10:1 (NASA Langley corehole). Largest clasts within this lithofacies are notably smaller (2–4 m) than those in the surgeback breccia. The unit displays a distinct upward-fining sequence, especially notable at the NASA Langley and Bayside sites, where the upper ~30 m of section contains no clasts larger than a few centimeters. Larger clasts in this lithofacies display the same intense plastic deformation features, inclined contacts, and high-angle stratal orientations, characteristic of the surgeback breccia. We interpret this lithofacies to be composed of tsunami washback breccia.

The washback lithofacies is also present outside the crater at the Windmill Point and Newport News core sites. In fact, it is the only crater-fill lithofacies encountered at sites outside the crater. A small component (<1 vol.%) of impact ejecta (microtektites, sand-sized crystalline clasts, shocked and melted mineral grains) is scattered throughout the washback breccia at all sites. Some of these constituents might represent ballistic ejecta and(or) seafloor-surge deposits, which were remobilized during the washback process.

Inside the crater, the washback lithofacies can be further divided into upper and lower sublithofacies. The lower sublithofacies (Fig.6.4G) is coarser (M/B ratios of 0.01:1 to 2:1) and contains larger clasts (5–7 m) than the upper sublithofacies. The upper sublithofacies approaches nearly 100 percent matrix. Clasts larger than 1–2 cm are rare, which gives M/B ratios of essentially 1:0 (Fig. 6.14I).

Overlying the washback lithofacies at Chesapeake Bay is a flowin deposit, a lithofacies not accounted for in previous models. In sharp contact (at most sites) above the washback lithofacies is a thin layer (~0.6 m) of clayey silt with generally inclined thin laminae and lenses of white, fine-to-very fine sand (mainly a mixed suite of benthic foraminifera and other microfossils redeposited from washback breccia deposits outside the crater rim, plus abundant muscovite flakes and framboidal pyrite). The azimuth and steepness of inclination among these sandy laminae change repetitiously up the core, indicating that their depositional geometry was controlled by multiple flow directions (Fig. 6.23). We interpret this silt-rich flowin layer to represent a succession of localized turbidites or debris-flow deposits produced by hypercanes (Emanuel et al. 1995), which stirred up the shallow seafloor regimes outside the crater basin.

Hypothetically, a fallout lithofacies should be present at the top of the flowin lithofacies. We have found sparse evidence of the fallout regime at Chesapeake Bay, however, because constituents, such as shocked grains and microtektites, are exceedingly rare (<1 vol.%). Presence at the NASA Langley site of *in situ* pyrite lattices containing millimeter-sized spherical cavities (implying a layer of glass microspherules; Poag 2002; Fig. 6.24A,B) is the only evidence that fallout debris accumulated inside the Chesapeake Bay crater (Figs. 12.6, 12.7).

12.5 Comparison of Models

Our field evidence supports many aspects of the Oberbeck impact model (Oberbeck et al. 1993; Table 12.1; Fig. 12.1), which emphasized that submarine impacts generate lithofacies that are laterally extensive, with highly complex structure and extreme spatial variability. The Oberbeck model, however, focused mainly on extracrater deposits, characterized by upward- and distally-fining sequences, which contain sorted and unsorted, massive, graded, finely layered, and crossbedded units. Near the impact site (outside the crater), the Oberbeck model predicts that ejecta deposits would be crudely graded. The lower layers would contain megaclasts, the middle layers would be water-sorted, and the upper layers would be argillaceous and finely laminated. In nearshore areas distal to the crater, Oberbeck et al. (1993) inferred chaotic mixtures of ejecta, reworked seafloor sediment, massive slumps, and turbidites. Landward of the shoreline, high-velocity ejecta would plough and brecciate the subaerial substrate, loosening it up for the impinging tsunami waves to entrain sedimentary clasts and wash them back to sea. The Oberbeck model, however, gives no role to the washback or flowin processes in filling the primary crater.

The Ormö model (Ormö 1998; Ormö and Lindström 2000; Table 12.1; Fig. 12.3) differs considerably from the Chesapeake Bay impact model. In particular, the Ormö model lacks a peak ring, displaced megablocks, fallback breccia, washback breccia, flowin deposits, and a fallout layer. Almost all the impact-generated deposits in the Ormö model are derived from surgeback processes (Ormö and Miyamoto 2002).

The thick and widespread nature of breccia deposits in the Tsikalas model (Tsikalas et al. 1999; Table 12.1; Fig. 12.2) is similar to that of the Chesapeake Bay model, except that the crest of the central peak is not covered by breccia. Our inference of similarity is weakened, however, by the sparsity of core documentation

Table 12.1. Comparison of principal structural, depositional, and erosional features among four conceptual models of submarine bolide impacts

Model	Target Rocks	Central Peak	Peak Ring	Fallback breccia	Slumped megablocks	Surgeback breccia	Washback breccia	Flowin deposits	Fallout layer	Surgeback channels	Washback channels
Oberbeck (General-hypothetical)	sediments over crystallines	none	none	none	at crater margins and near shoreline	none	distal to crater	not indicated	not indicated	none	not indicated
Tsikalas (Mjølnir; 40 km diameter)	entirely sedimentary (consolidated)	distinct; crest not buried by breccia	subtle; crest buried by breccia	inferred from density values	inferred from seismics	inferred from density values; possibly cored	none	none	none	none	none
Ormö (Lockne; 20 km diameter)	consolidated sediments over crystallines	subtle; crystalline	none	none indicated	none	outcrops; dominant crater-fill deposit	none	none	none	several documented at outcrops	none
This volume (Chesapeake Bay; 85 km diameter)	unconsolidated sediments over crystallines	prominent; crystalline; buried by breccia	prominent; crystalline; buried by breccia	possibly cored; inferred from seismics; mainly in inner basin and outer annular trough	extensive in annular trough; cored and inferred from seismics; buried by surgeback and washback breccia	extensive throughout crater; cored and inferred from seismics; buries central peak and peak ring	extensive throughout crater; limited distribution outside crater; cored and inferred from seismics; buries central peak and peak ring	cored inside crater; too thin to image on seismics; covers breccia inside crater	cored inside crater; too thin to image on seismics	none documented	possibly present west of crater

at Mjølnir. The presence of a peak ring at Mjølnir also is not strongly documented, due perhaps to its sedimentary composition and unusually low relief. The Tsikalas model also lacks flowin and fallout deposits.

Our Chesapeake Bay model differs from all other models in emphasizing a strong crater-filling role for tsunami washback processes, and highlights the intracrater breccias and megabreccias that are so widespread and voluminous at Chesapeake Bay. Complete burial of the peak ring and central peak by surgeback, washback, and flowin deposits appear to set the Chesapeake Bay crater apart from all other known impact craters of its size range. On the other hand, it shares this attribute with the much larger Chicxulub impact basin (Chapter 10), which also is the result of a submarine bolide strike into mixed target rocks on a shallow continental shelf.

13 Biospheric Effects of Chesapeake Bay Impact

13.1 Local Paleoenvironmental Effects

The Virginia cores contain an excellent, detailed sedimentological and micropaleontological record of successive postimpact depositional and biotic changes at the Chesapeake Bay impact site. This record is contained mainly in the Chickahominy Formation (Chapter 7), which represents one of the most complete sections of immediately postimpact marine strata yet discovered (apart from deep-sea cores) for the last ~2.1 myr of late Eocene time. The exceptional Chickahominy record is a product of relatively deep-water, fine-grained, microfossiliferous deposition within a closed basin. Deposition took place on a slowly subsiding, passive continental shelf that underwent no syndepositional tectonism or synchronous major eustatic sea-level changes. Subsequent Cenozoic sea-level falls and marine transgressions have eroded the top of the Chickahominy Formation, but it has never been subjected to significant tectonic activity.

13.1.1 Sediment Accumulation Rates

By plotting magnetochronology (Poag et al. 2003) against drill depth in the Kiptopeke corehole, three distinct depositional episodes have been recognized within the late Eocene Chickahominy Formation (Fig. 13.1; Table 13.1). The stratigraphically lowest depositional episode is defined by the 11 samples between paleomagnetic boundaries at 384.0 m and 365.0 m (1259.8–1197.6 ft; 35.60–35.30 Ma), in which sediments accumulated at an average rate of 67 m/myr. This is the most rapid rate of sediment accumulation during Chickahominy time at this site. If one assumes (on the basis of similar lithology) that the 11 Chickahominy samples below 384.0 m (1259.8 ft) represent accumulation at the same rate of 67 m/myr, then the extrapolated age of the first postimpact sample (and, therefore, the approximate age of the impact) is 35.78 Ma (Fig. 13.1; Table 13.1; Chapter 8).

The second Chickahominy depositional episode is represented by sediments between paleomagnetic boundaries at 365.0 m and 354.8 m (1197.6–1166.7 ft; 35.30–34.60 Ma), which accumulated at an average rate of 10 m/myr (Fig. 13.1; Table 13.1). The third Chickahominy depositional episode is represented by the 20 stratigraphically highest samples (354.8–327.7 m; 1166.7–1077.6 ft; 34.6–~33.7 Ma), which accumulated at an average rate of 30 m/myr (Fig. 13.1; Table 13.1).

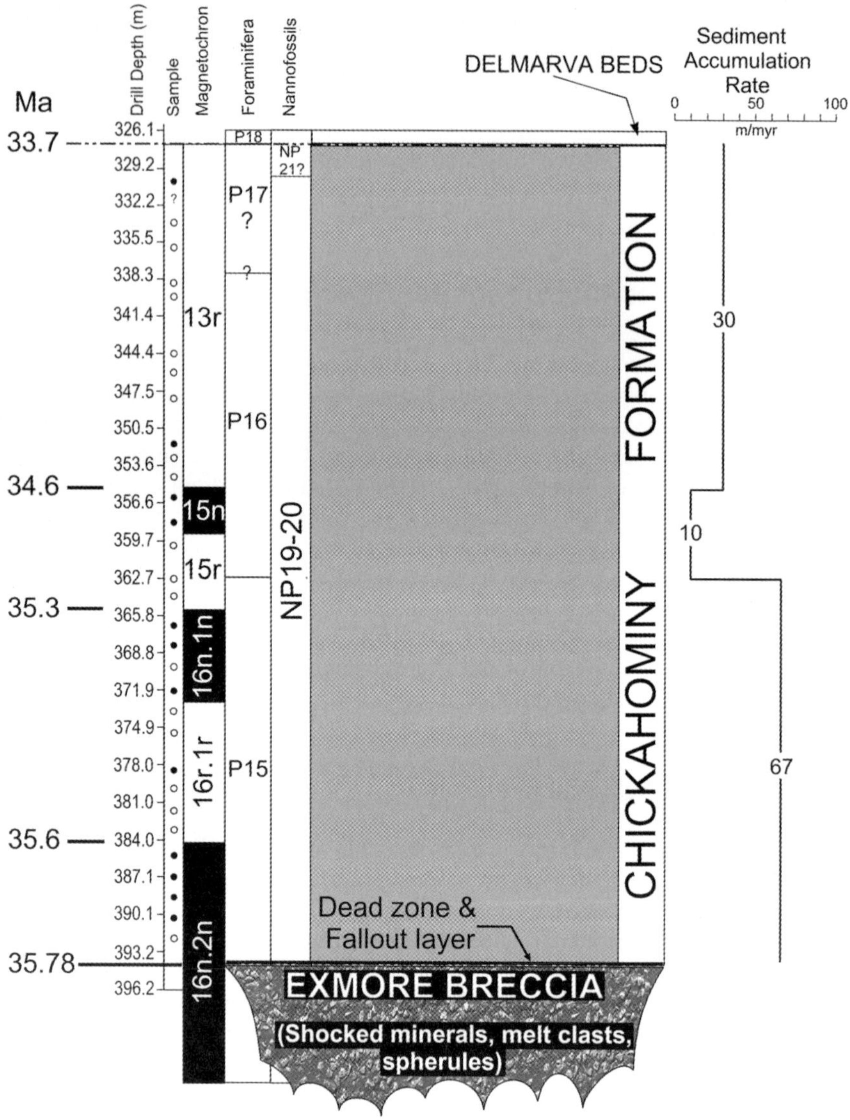

Fig. 13.1. Stratigraphic column for Chickahominy Formation at Kiptopeke core site, showing magnetochronology, biochronology, and sediment accumulation rate curve. Average rate = 32 m/myr. Modified from Poag et al. (2003).

Thus, during the first ~480 kyr following impact, the average sediment accumulation rate was maximum for the late Eocene at Kiptopeke. For the next 700 kyr, the rate dropped to approximately 15% of the initial rate, but then tripled during the final 900 kyr. We have compared this chronodepositional framework with

Table 13.1. Sediment accumulation rates calculated from ages and depths of magnetochron boundaries in the Chickahominy Formation at Kiptopeke.

Upper Chron Boundary Age [Ma]	Lower Chron Boundary Age [Ma]	Upper Chron Boundary Depth [m]	Lower Chron Boundary Depth [m]	Thickness Between Chron Boundaries [m]	Time Between Chron Boundaries [myr]	Sediment Accumulation Rate [m/myr]
33.7	34.6	327.7	354.8	27.1	0.9	30.1
34.6	35.3	354.8	365.0	7.27	0.7	10.4
35.3	35.6	365.0	382.3	20.23	0.3	67.4
35.6	35.78 (extrapolated)	382.3	384.0	11.8	0.18	67.4 (assumed)

Average sediment accumulation rate = 31.92 m/myr
@30.1 m/myr, 1 m = 33 kyr
@10.39 m/myr, 1 m = 92 kyr
@67.43 m/myr, 1 m = 15 kyr

specific biotic characteristics of the late Eocene benthic foraminiferal community at the Kiptopeke site to interpret the paleoenvironmental succession following the impact.

13.1.2 Stratigraphic Attributes of Benthic Foraminiferal Community

13.1.2.1 Preimpact Benthic Foraminiferal Community

To assess the general effects of the impact on the local biota, we have examined the excellent record of fossilized late Eocene benthic foraminifera at and near the impact site. Evidence of the nature of the preimpact foraminiferal community in the target zone was completely destroyed upon impact, through vaporization and excavation. Subsequently, erosion during several successive Cenozoic marine regressions and transgressions has virtually erased the record of late Eocene deposition elsewhere in southeastern Virginia. We know, however, that well-developed marine foraminiferal assemblages (planktonic as well as benthic) were present in the target area prior to impact, because representatives of them are mixed into assemblages of older microfossils within the Exmore breccia (Poag and Aubry 1995; Fig. 8.2). To get a more quantitative estimate of the composition of these preimpact assemblages, however, we have to rely on records from outside the impact area. The closest documentation comes from boreholes in Maryland and New Jersey. Poag and Commeau (1995) reported a diverse late Eocene planktonic foraminiferal assemblage from a corehole in southern Maryland (Ohio Oil-Larry G Hammond #1 well, near Salisbury, Maryland). The planktonic foraminiferal species represented there are typical of mixed tropical-temperate planktonic assemblages found at many other late Eocene sites around the globe. We have studied the benthic foraminifera from the same Hammond corehole, and found a rich, diverse assemblage (Table 13.2), which includes most of the species represented in the Chickahominy Formation in Virginia (Fig. 13.2A,B).

Table 13.2. Species of benthic foraminifera in early late Eocene sediments (preimpact) from Hammond corehole, near Salisbury, Maryland.

Alabamina midwayensis	*Guttulina irregularis*
Anomalinoides bilateralis	*Gyroidinoides aequilateralis*
Anomalinoides "kiptopekensis"	*Gyroidinoides byramensis*
Bolivina gracilis	*Gyroidinoides planatus*
Bolivina "preavirginiana"	*Hanzawaia blanpiedi*
Bolivina virginiana	*Kaleshnikovella* sp.
Bulimina jacksonensis	*Lenticulina virginiana*
Buliminellita curta	*Loxostomina vicksburgensis* f. spinosa
Caucasina marylandica	*Marginulina cocoaensis*
Charltonina madrugaensis	*Massillina decorata*
Cibicidina sp.	*Melonis planatus*
Cibicidoides "chickahominyanus"	*Oridorsalis umbonatus*
Cibicidoides cocoaensis	*Planularia* sp.
Cibicidoides pippeni f. speciosus	*Pullenia quinqueloba*
Cibicidoides renzi	*Pyramidina* sp.
Cibicidoides "rugoumbonatus"	*Spiroplectinella mississippiensis*
Cibicidoides sculpturatus	*Stilostomella* "aduncocostata"
Epistominella minuta	*Stilostomella cocoaensis*
Gaudryina alazanensis	*Stilostomella* "exilispinata"
Globobulimina ovata	*Textularia virginiana*
Globocassidulina subglobosa	*Turrilina robertsi*
Globulina gibba	*Uvigerina gardnerae*
Grigelis annulospinosa	*Uvigerina jacksonensis*
Grigelis "tubulosa"	*Vaginulinopsis jacksonensis*
Guttulina hantkeni	

Total number of species = 49

13.1.2.2 Postimpact Benthic Foraminiferal Community

The initial postimpact response of the foraminiferal community at the impact site is best documented in the NASA Langley core. This corehole, completed in October, 2000, was the first to be drilled after the impact crater's presence had been convincingly demonstrated (Poag et al. 1994). As a result, the cores from NASA Langley were described, photographed, and sampled more carefully and systematically than those from the previous four core sites (Newport News, Windmill Point, Exmore, Kiptopeke). As we showed in Chapter 7, the Chickahominy Formation at the NASA Langley site is overlain by a distinctive 19-cm-thick unit of laminated sand, silt, and clay (top at 235.65 m; 773.12 ft), which is a dead zone, barren of indigenous biota. The lower boundary of this dead zone is interpreted to be a fallout layer, which contains evidence (pyrite lattices) of the prior presence of microspherules (Poag, 2002b; see Chapter 6). This dead zone does contain white laminae, lenses, and burrow casts, however, (Figs. 6.22, 7.3), composed of fine to very-fine-grained quartz sand, glauconite, mica, pyrite, and mixtures of reworked specimens of older (preimpact) foraminifera (middle Eocene, early Eocene, Paleocene, Late Cretaceous, and Early Cretaceous microbiota). Thus, it appears that the dead zone represents an interval of immediately postimpact abiotic deposition. We can roughly estimate its maximum duration to have been ~3 kyr by extrapolating the sediment accumulation rate from the Kiptopeke magnetochronology (Fig. 13.1; Table 13.1). However, the distinctly coarser grain size of the dead zone sediments may indicate considerably more rapid deposition than that of the lower Chickahominy section, possibly representing only ~1 kyr or less of accumulation.

An equivalent dead zone is present in the North and Bayside coreholes, and is represented by the first sample below the Chickahominy Formation (sample 0 at ~394.05–394.23 m; 1292.8–1293.4 ft) in the Kiptopeke corehole. This Kiptopeke sample contains an unusually sparse assemblage made up of single specimens of only 17 different species, all of which appear to have been reworked. As yet, however, no microspherule-bearing interval has been identified at any corehole other than at NASA Langley, so we cannot pinpoint the precise base of the dead zone at the other six sites.

The succession of foraminiferal assemblages in the remaining late Eocene section (the Chickahominy Formation) above the dead zone is best documented from the Kiptopeke core, where the formation is 66.4 m thick (217.8 ft). We have identified 150 species of benthic foraminifera among 47 stratigraphically successive Chickahominy samples taken from Kiptopeke. We have displayed the stratigraphic ranges of the benthic species as a consecutive succession of samples in order of sample depth (Fig. 13.2A,B; CD-ROM.13.2).

The first sample above the dead zone at Kiptopeke (sample 1; 393.9 m; 1292.3 ft) contains representatives of 33 species of benthic foraminifera (Fig. 13.2A). Of these, six represent agglutinated species and 27 are calcareous species. Seven of these species are known from the middle Eocene Piney Point Formation in Virginia and Maryland (Jones 1990), and 14 occupied the site prior to impact (by inference from the Hammond core). The most abundant species in this assemblage are typical of the rest of the Chickahominy samples at this site, and 20 of the species (61%) are represented in nearly every sample to the top of the late Eocene section. We call this the *Cibicidoides pippeni* Assemblage, after one of the most abundant and persistent species in the assemblage. The sediments that contain the assemblage comprise the *Cibicidoides pippeni* Zone (Partial-Range Zone; Table

Fig. 13.2. (Next two pages) Stratigraphic range chart for benthic foraminifera in Chickahominy Formation at Kiptopeke core site. Broad shading indicates nominate taxa for principal benthic foraminiferal subassemblages that constitute *Cibicidoides pippeni* Assemblage. Heavy black lines indicate agglutinated species. Small open circles indicate reworked specimens found in dead zone. Complete, full-size version of this figure can be found on CD-ROM.

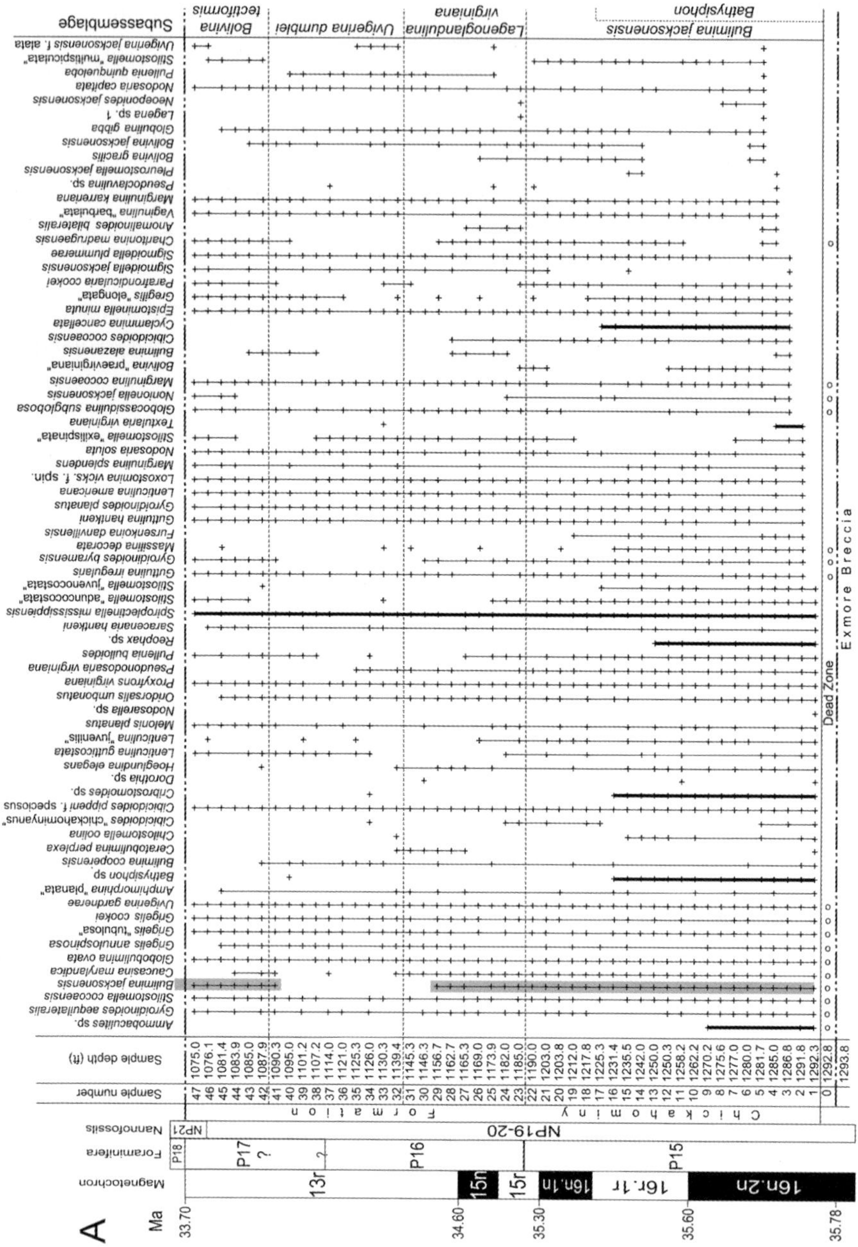

A
Ma
Magnetochron
Foraminifera
Nannofossils
Sample number
Sample depth (ft)
Subassemblage
Bulimina jacksonensis
Bathysiphon
Lagenoglandulina virginiana
Uvigerina dumblei
Bolivina tectiformis
Dead Zone
Exmore Breccia
Chickahominy Formation
NP19-20
NP21
P15
P16
P17
P18

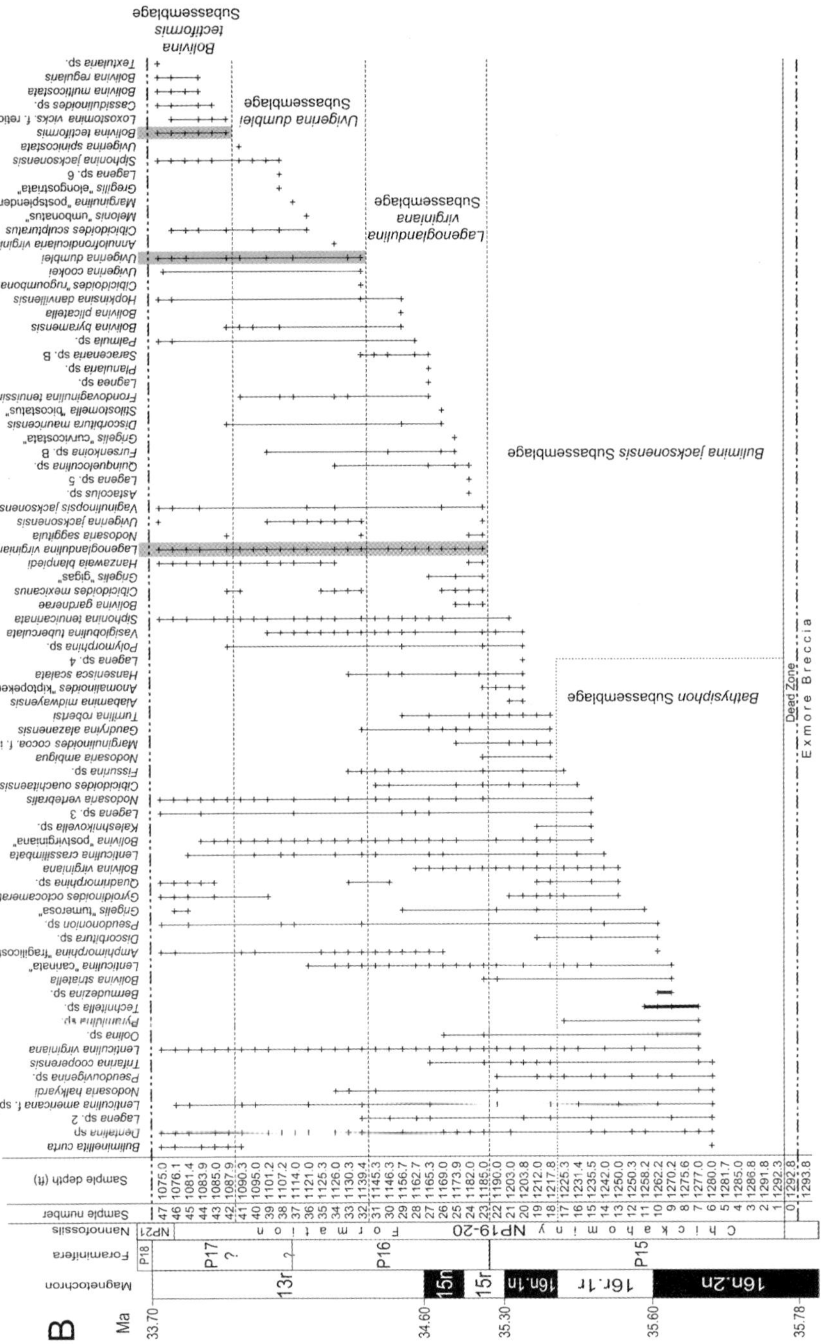
B
Ma
Magnetochron
Foraminifera
Nannofossils
Sample number
Sample depth (ft)
Bolivina tectiformis Subassemblage
Uvigerina dumblei Subassemblage
Lagenoglandulina virginiana Subassemblage
Bulimina jacksonensis Subassemblage
Bathysiphon Subassemblage
Dead Zone
Exmore Breccia
Chickahominy Formation
13r
15n
15r
16n.1n
16r.1r
16n.2n
P18
P17
P16
P15
NP21
NP19-20
33.70
34.60
35.30
35.60
35.78

13.3; Fig. 13.3). The *Cibicidoides pippeni* Assemblage also typically includes several species of unusually large lenticulinids (2–4 mm in diameter), and an association of large marginulinids, stilostomellids, and nodosariids, whose representative specimens reach lengths of 3–7 mm.

Table 13.3. Prominent benthic foraminiferal species of *Cibicidoides pippeni* Assemblage. Species listed are those whose specimens are persistently present and(or) abundant.

Bolivina jacksonensis	*Lenticulina virginiana* *
Bulimina jacksonensis *	*Loxostomina vicksburgensis* f. spinosa *
Caucasina marylandica *	*Marginulina cocoaensis* *
*Charltonina madrugaensis**	*Marginulina karreriana*
Cibicidoides pippeni f. speciosus *	*Marginulina splendens*
Epistominella minuta *	*Melonis planatus* *
Globobulimina ovata *	*Nodosaria capitata*
Globocassidulina subglobosa *	*Nodosaria soluta*
Globulina gibba *	*Oridorsalis umbonatus* *
Grigelis "elongata"	*Parafrondicularia cookei*
Grigelis annulospinosa *	*Proxyfrons virginiana*
Grigelis "tubulosa" *	*Pseudonodosaria virginiana*
Guttulina hantkeni *	*Pullenia bulloides*
Guttulina irregularis *	*Sigmoidella plummerae*
Gyroidinoides aequilateralis *	*Spiroplectinella mississippiensis* *
Gyroidinoides byramensis *	*Stilostomella* "aduncocostata" *
Gyroidinoides planatus *	*Stilostomella cocoaensis* *
Lenticulina americana f. typica	*Stilostomella cookei*
Lenticulina americana f. spinosa	*Stilostomella* "exilispinata" *
Lenticulina "juvenilis"	*Uvigerina gardnerae* *
Lenticulina gutticostata	*Vaginulina* "barbulata"

*indicates species also present in earliest late Eocene (preimpact) sample in Hammond corehole near Salisbury, Maryland

Moving upsection, samples 2–4 (393.7–391.6 m; 1291.7–1285.0 ft) contain 25 additional species (Fig. 13.2A), which, along with the 33 species in sample 1, constitute most of the long-ranging Chickahominy benthic foraminiferal species. In terms of species composition, then, the typical *Cibicidoides pippeni* Assemblage of benthic foraminifera, rich both in species and specimens, repopulated the target site at nearly full strength by approximately 36 kyr pti (by extrapolation of the sediment accumulation rate; pti = postimpact; Fig. 13.4).

On the basis of the stratigraphic ranges of several other distinctive species, the *C. pippeni* Zone (or Assemblage) can be subdivided into five Subzones (or Subassemblages. The *C. pippeni* Zone and its five Subzones are defined as follows (Figs. 13.2A,B, 13.3, 13.4):

Cibicidoides pippeni Taxon-range Biozone – That part of the Chickahominy Formation embracing the stratigraphic range of the nominate species. *Cibicidoides pippeni* appears to have a more extensive stratigraphic range in other localities, however, such as the US Gulf Coast and Caribbean (van Morkhoven et al. 1986).

Fig. 13.3. Nominate species for benthic foraminiferal zone and subzones recognized in Chickahominy Formation. Scale bar = 100 μm. *1 Cibicidoides pippeni* (Cushman and Garrett), 1938. Exmore corehole; umbilical view. *2 Bulimina jacksonensis* Cushman, 1925b. Exmore corehole; lateral view. *3 Lagenoglandulina virginiana* (Cushman and Cederstrom), 1949. Newport News corehole; lateral view. *4 Uvigerina dumblei* Cushman and Applin, 1926. Exmore corehole; lateral view. *5 Bolivina tectiformis* Cushman, 1926. Exmore corehole; lateral view. *6 Bathysiphon* sp. Kiptopeke corehole; lateral view.

Bulimina jacksonensis Interval Subbiozone – That part of the Chickahominy Formation embracing the partial stratigraphic range of the nominate species (Fig. 13.3) between its lowest occurrence and the lowest occurrence of *Lagenoglandulina virginiana.*

Lagenoglandulina virginiana Interval Subbiozone – That part of the Chickahominy Formation embracing the partial stratigraphic range of the nominate species (Fig. 13.3) between its lowest occurrence and the lowest occurrence of *Uvigerina dumblei.*

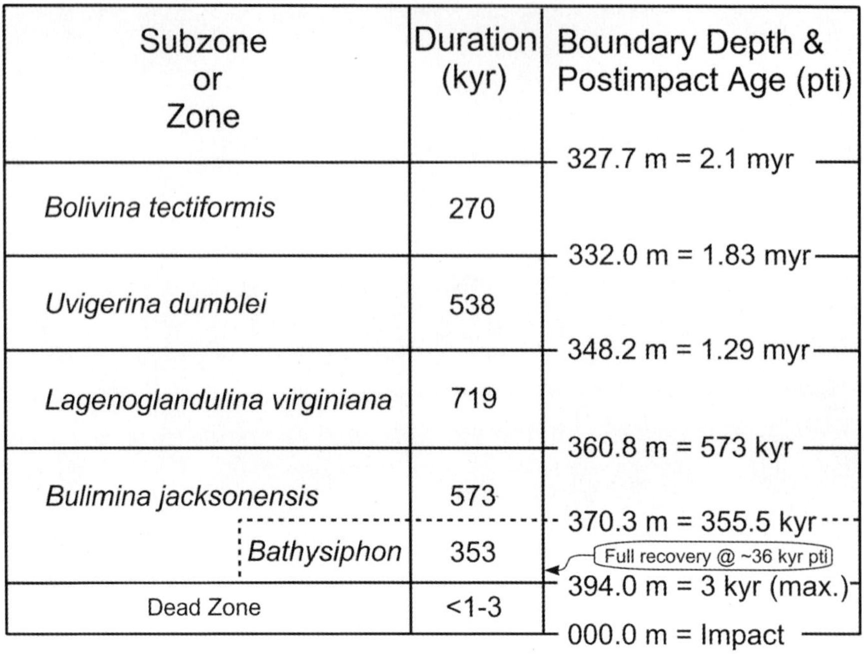

Fig. 13.4. Benthic foraminiferal subzones in Chickahominy Formation at Kiptopeke site, their duration in thousands of years (kyr) and the depths and postimpact (pti) ages of their boundaries.

Uvigerina dumblei Interval Subbiozone – That part of the Chickahominy Formation embracing the partial range of the nominate species (Fig. 13.3) between its lowest occurrence and the lowest occurrence of *Bolivina tectiformis*.

Bolivina tectiformis Taxon-range Subbiozone – That part of the Chickahominy Formation embracing the total range of the nominate species (Fig. 13.3).

Bathysiphon Abundance Subbiozone – That part of the Chickahominy Formation at the base of the *Bulimina jacksonensis* Subzone, which contains the peak development of an agglutinated suite of benthic foraminifera (Fig. 13.5), in which *Bathysiphon* sp. is a notable (persistent and relatively abundant) constituent.

The oldest subassemblage of calcareous taxa is the *Bulimina jacksonensis* Subassemblage, which is present in samples 1–22 at Kiptopeke (Figs. 13.2A,B, 13.4; Table 13.4). Representatives of 42 species in this subassemblage are persistently present and(or) abundant in the *Bulimina jacksonensis* Subzone (Interval Subzone). The *Bulimina jacksonensis* Subassemblage is present throughout planktonic foraminiferal Zone P15, and precedes the appearance of *Lagenoglandulina virginiana* and *Hanzawaia blanpiedi* at Kiptopeke.

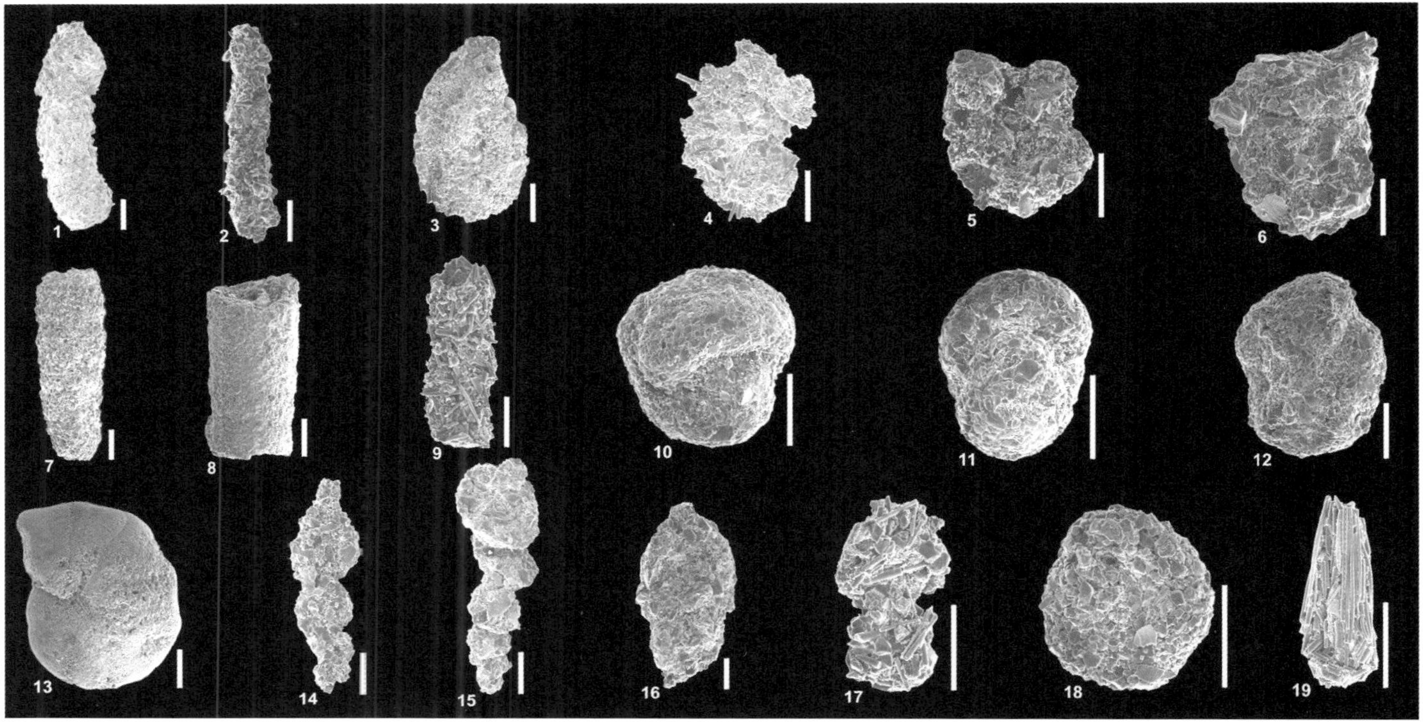

Fig. 13.5. Agglutinated benthic foraminiferal species characteristic of the *Bathysiphon* Subassemblage of the Chickahominy Formation at the Kiptopeke core site. Scale bar = 200 μm. Note prominent sponge spicules in specimens 4, 9, 17, 18. *1–6 Ammobaculites* spp. *7–9 Bathysiphon* spp. *10–12 Cribrostomoides* spp. *13 Cyclammina cancellata* Brady, 1879. *14–17 Reophax* spp. *18* Genus and species indeterminate. *19 Technitella* sp.

Table 13.4. Prominent calcareous benthic foraminiferal species of *Bulimina jacksonensis* Subassemblage. Species listed (64) are those whose specimens are persistently present and(or) abundant or are restricted (or nearly so) to this subassemblage.

Alabamina midwayensis	*Lenticulina americana* f. typica
Amphimorphina "planata"	*Lenticulina americana* f. spinosa
Anomalinoides "kiptopekensis"	*Lenticulina* "carinata"
Bolivina gracilis	*Lenticulina crassilimbata*
Bolivina jacksonensis	*Lenticulina virginiana*
Bolivina "praevirginiana"	*Loxostomina vicksburgensis* f. spinosa
Bolivina striatella	*Marginulina cocoaensis*
Bolivina virginiana	*Marginulina karreriana*
Bulimina jacksonensis	*Marginulina splendens*
Caucasina marylandica	*Massilina decorata*
Charltonina madrugaensis	*Melonis planatus*
Chilostomella oolina	*Nodosaria capitata*
Cibicidoides "chickahominyanus"	*Nodosaria cooperensis*
Cibicidoides cocoaensis	*Oridorsalis umbonatus*
Cibicidoides pippeni f. speciosus	*Parafrondicularia cookei*
Discorbitura sp.	*Proxyfrons virginiana*
Dentalina sp.	*Pseudonodosaria virginiana*
Epistominella minuta	*Pseudouvigerina* sp.
Fursenkoina danvillensis	*Pullenia bulloides*
Globobulimina ovata	*Pyramidina* sp.
Globocassidulina subglobosa	*Saracenaria hantkeni*
Globulina gibba	*Sigmoidella plummerae*
Grigelis "elongata"	*Spiroplectinella mississippiensis*
Grigelis annulospinosa	*Stilostomella* "aduncocostata"
Grigelis "tubulosa"	*Stilostomella cocoaensis*
Guttulina hantkeni	*Stilostomella cookei*
Guttulina irregularis	*Stilostomella* "exilispinata"
Gyroidinoides aequilateralis	*Stilostomella* "juvenocostata"
Gyroidinoides byramensis	*Stilostomella* "multispiculata"
Gyroidinoides planatus	*Turrilina robertsi*
Hoeglundina elegans	*Uvigerina gardnerae*
Kaleshnikovella sp.	*Vaginulina* "barbulata"

The stratigraphically next highest significant species event is a faunal turnover documented in sample 23 at 361.2 m (1185.0 ft; Figs. 13.2A,B, 13.4). At this level, eight new calcareous species first appear, and ~10 species disappear from the section (some exit only temporarily, however). The most distinctive new taxa of this group are *Lagenoglandulina virginiana* (Fig. 13.3) and *Hanzawaia blanpiedi*, both of which persist in most samples from this stratigraphic level to the top of the Chickahominy Formation (Fig. 13.2 A,B). We call this the *Lagenoglandulina virginiana* Subassemblage (Table 13.5), because this taxon is represented in more samples than any other species in the subassemblage (Figs. 13.2A,B, 13.4). This faunal turnover took place at ~35.3 Ma, approximately 573 kyr pti (Table 13.1), and coincided with: (1) a turnover in planktonic foraminifera, which defines the P16/P15 boundary (Berggren et al. 1995); and (2) the boundary between the *Bolboforma latdorfensis* and *B. spinosa* zones (Poag and Aubry 1995). The stratigraphic position of this turnover approximates the level at which the sediment accumulation rate drops from 67 to 10 m/myr (Fig. 13.1).

Table 13.5. Prominent benthic foraminiferal species of *Lagenoglandulina virginiana* Subassemblage. Species listed (55) are those whose specimens are persistently present and(or) abundant or are restricted (or nearly so) to this subassemblage.

Amphimorphina "fragilicostata"	*Hoeglundina elegans*
Anomalinoides bilateralis	*Lagenoglandulina virginiana*
Bolivina gardnerae	*Lenticulina americana* f. typica
Bolivina jacksonensis	*Lenticulina* "carinata"
Bolivina "postvirginiana"	*Lenticulina virginiana*
Bolivina virginiana	*Loxostomina vicksburgensis* f. spinosa
Bulimina alazanensis	*Marginulina cocoaensis*
Bulimina jacksonensis	*Marginulina karreriana*
Caucasina marylandica	*Melonis planatus*
Ceratobulimina perplexa	*Nodosaria capitata*
Charltonina madrugaensis	*Nodosaria cooperensis*
Cibicidoides cocoaensis	*Nodosaria vertebralis*
Cibicidoides mexicanus	*Oridorsalis umbonatus*
Cibicidoides ouachitaensis	*Proxyfrons virginiana*
Cibicidoides pippeni f. speciosus	*Pseudonodosaria virginiana*
Dentalina sp.	*Pullenia bulloides*
Epistominella minuta	*Saracenaria* sp. B
Gaudryina alazanensis	*Sigmoidella jacksonensis*
Globobulimina ovata	*Sigmoidella plummerae*
Globocassidulina subglobosa	*Siphonina tenuicarinata*
Globulina gibba	*Spiroplectinella mississippiensis*
Grigelis annulospinosa	*Stilostomella cocoaensis*
Grigelis "tubulosa"	*Stilostomella cookei*
Guttulina hantkeni	*Stilostomella* "exilispinata"
Guttulina irregularis	*Turrilina robertsi*
Gyroidinoides aequilateralis	*Uvigerina gardnerae*
Gyroidinoides byramensis	*Vasiglobulina tuberculata*
Gyroidinoides planatus	

The stratigraphically next highest significant species event takes place in sample 32 (347.3 m; 1139.4 ft). A faunal turnover at this level features the lowest appearances of *Uvigerina cookei, Uvigerina dumblei*, and *Cibicidoides* "rugoumbonatus" (Figs. 13.2A,B, 13.3, 13.4) and the near absence of *Bulimina jacksonensis, Caucasina marylandica*, and *Gyroidinoides byramensis*, which otherwise, are present through most of the rest of the Chickahominy section. *Uvigerina dumblei* is the most persistent species, and gives its name to this Subassemblage (Table 13.6). The *U. dumblei* Subassemblage appeared at ~1.29 myr pti (Fig. 13.4).

The stratigraphically highest significant species event takes place in sample 42 (331.6 m; 1087.9 ft), ~1.83 myr pti (Figs. 13.2A,B, 13.4). *Bolivina tectiformis* is

Table 13.6. Prominent benthic foraminiferal species of *Uvigerina dumblei* Subassemblage. Species listed (52) are those whose specimens are persistently present and(or) abundant or are restricted (or nearly so) to this subassemblage.

Amphimorphina "fragilicostata"	*Lenticulina crassilimbata*
Bolivina byramensis	*Lenticulina virginiana*
Bolivina jacksonensis	*Loxostomina vicksburgensis* f. spinosa
Bolivina "postvirginiana"	*Marginulina cocoaensis*
Bulimina cooperensis	*Marginulina karreriana*
Cibicidoides mexicanus	*Marginulina splendens*
Cibicidoides pippeni f. speciosus	*Melonis planatus*
Cibicidoides sculpturatus	*Nodosaria capitata*
Dentalina sp.	*Nodosaria cooperensis*
Epistominella minuta	*Nodosaria vertebralis*
Frondovaginulina tenuissima	*Oridorsalis umbonatus*
Globobulimina ovata	*Proxyfrons virginiana*
Globocassidulina subglobosa	*Pullenia bulloides*
Globulina gibba	*Sigmoidella jacksonensis*
Grigelis "elongata"	*Sigmoidella plummerae*
Grigelis annulospinosa	*Siphonina jacksonensis*
Grigelis "tubulosa"	*Siphonina tenuicarinata*
Guttulina hantkeni	*Spiroplectinella mississippiensis*
Guttulina irregularis	*Stilostomella cocoaensis*
Gyroidinoides aequilateralis	*Stilostomella cookei*
Gyroidinoides planatus	*Stilostomella* "exilispinata"
Hanzawaia blanpiedi	*Uvigerina dumblei*
Lagenoglandulina virginiana	*Uvigerina gardnerae*
Lenticulina americana f. typica	*Uvigerina jacksonensis*
Lenticulina americana f. spinosa	*Vaginulina* "barbulata"
Lenticulina "carinata"	*Vasiglobulina tuberculata*

the most prominent new taxon to appear, and gives its name to this Subassemblage (Fig. 13.3; Table 13.7). The appearance of *B. tectiformis* is accompanied by the first appearance of *Loxostomina vicksburgensis* f. reticulata, the persistence of *Buliminellita curta*, and the reappearance of *Bulimina jacksonensis*, *Caucasina marylandica*, and *Gyroidinoides byramensis*.

In summary, a species-rich, but slightly underdeveloped *Cibicidoides pippeni* Assemblage was reestablished at Kiptopeke, <1–3 kyr after the impact (Fig. 13.4). This Assemblage reached full postimpact recovery ~36 kyr pti, and subsequently, three faunal turnovers took place among the Chickahominy calcareous benthic foraminifera (spaced ~0.6 myr apart): at 573 kyr, 1.29 myr, and 1.83 myr pti. Nearly all species that constitute the *Cibicidoides pippeni* Assemblage disappear from the Virginia Coastal Plain record at the unconformable Eocene/Oligocene boundary, to be replaced at the crater core sites by a relatively species-poor assemblage characterized by an abundance of large specimens of *Uvigerina vicksburgensis*.

Table 13.7. Prominent benthic foraminiferal species of *Bolivina tectiformis* Subassemblage. Species listed (61) are those whose specimens are persistently present and(or) abundant or are restricted (or nearly so) to this subassemblage.

Bolivina multicostata	*Loxostomina vicksburgensis* f. reticulata
Bolivina "postvirginiana"	*Marginulina cocoaensis*
Bolivina regularis	*Marginulina karreriana*
Bolivina tectiformis	*Marginulina splendens*
Bulimina jacksonensis	*Melonis planatus*
Buliminellita curta	*Nodosaria capitata*
Cassidulinoides sp.	*Nodosaria cooperensis*
Caucasina marylandica	*Nodosaria vertebralis*
Charltonina madrugaensis	*Nonionella jacksonensis*
Cibicidoides pippeni f. speciosus	*Oridorsalis umbonatus*
Cibicidoides sculpturatus	*Palmula* sp.
Dentalina sp.	*Parafrondicularia cookei*
Epistominella minuta	*Proxyfrons virginiana*
Globobulimina ovata	*Pullenia bulloides*
Globocassidulina subglobosa	*Quadrimorphina* sp.
Globulina gibba	*Saracenaria hantkeni*
Grigelis "elongata"	*Sigmoidella jacksonensis*
Grigelis annulospinosa	*Sigmoidella plummerae*
Grigelis "tubulosa"	*Siphonina jacksonensis*
Guttulina hantkeni	*Siphonina tenuicarinata*
Guttulina irregularis	*Spiroplectinella mississippiensis*
Gyroidinoides aequilateralis	*Stilostomella* "aduncocostata"
Gyroidinoides byramensis	*Stilostomella cocoaensis*
Gyroidinoides octocameratus	*Stilostomella cookei*
Gyroidinoides planatus	*Stilostomella* "exilispinata"
Hanzawaia blanpiedi	*Stilostomella* "multispiculata"
Hopkinsina danvillensis	*Uvigerina gardnerae*
Lagenoglandulina virginiana	*Uvigerina dumblei*
Lenticulina americana f. typica	*Vaginulina* "barbulata"
Lenticulina virginiana	*Vaginulinopsis jacksonensis*
Loxostomina vicksburgensis f. spinosa	

13.1.2.3 Bathysiphon Subassemblage

Poag (1997a) noted that another important stratigraphic feature of the Chickahominy foraminiferal association in the Kiptopeke core is the presence of a distinctive subassemblage of agglutinated benthic species (tests built of sediment grains derived from the seafloor) superimposed on the lower two-thirds of the *B. jacksonensis* Subassemblage in samples 1–17 (393.9–373.5 m; 1292.3–1225.3 ft). This agglutinated subassemblage characterized the first ~355.5 kyr of postimpact deposition at the Kiptopeke site (Figs. 13.2A,B, 13.4, 13.5). The agglutinated subassemblage consists of representatives of 11 species, the stratigraphically most persistent of which are *Bathysiphon* sp., *Cribrostomoides* sp., and *Cyclammina cancellata*; we call this the *Bathysiphon* Subassemblage (Fig. 13.2A,B, 13.4, 13.5;

Table 13.8. Agglutinated benthic foraminiferal species (11) of *Bathysiphon* Subassemblage.

Ammobaculites sp.	*Pseudoclavulina* sp.
Bathysiphon sp.	*Reophax* sp.
Bermudezina sp.	*Sprioplectinella mississippiensis*
Cribrostomoides sp.	*Technitella* sp.
Cyclammina cancellata	*Textularia virginiana*
Dorothia sp.	

Table 13.8). Of these agglutinated species, only one, *Spiroplectinella mississippiensis*, is represented consistently above sample 17 (373.5 m; 1225.3 ft). *Spiroplectinella mississippiensis*, in fact, persists to the top of the Chickahominy Formation. At Kiptopeke, *Gaudryina alazanensis* (Fig. 13.2A,B) is a distinctive agglutinated form in the middle of the *Lagenoglandulina virginiana* Subzone, but is absent in stratigraphically higher subassemblages.

13.1.3 Community Structure of Benthic Foraminiferal Associations

We have used three attributes of community structure to help interpret the paleoenvironmental implications of the Chickahominy benthic foraminiferal associations: (1) generic predominance; (2) generic equitability; and (3) species richness.

13.1.3.1 Predominance and Equitability

Species predominance (the species whose members constitute the greatest relative abundance in an assemblage) and species equitability (a measure of the uniformity of predominance) are properties widely used to evaluate the paleoecology of benthic organisms, including foraminifera (Dodd and Stanton 1981; Hayek and Buzas 1997). Poag (1981) demonstrated that *generic* predominance tends to parallel or mimic *species* predominance among benthic foraminifera and, thus, can be an important foraminiferal community property for assessing the paleoenvironments represented in Cenozoic marine strata. Likewise, generic equitability can be substituted for species equitability. We have, therefore, evaluated generic predominance and equitability among the Chickahominy benthic foraminifera at Kiptopeke. (We use equitability to mean simply the number of predominant or co-predominant genera in a given sample.) We designate sole predominance if there is a spread of 4% or more between a given genus and other genera in the sample. A spread of <4% would constitute co-predominance (Fig. 13.6). In samples in which no single species was solely predominant or clearly clustered, we designate co-predominance to those species whose relative abundances were within 2% of each other (e.g., samples 4, 8, 41, 46; Fig. 13.6).

The predominant benthic genus in sample 1 (393.9 m; 1292.3 ft; 3–4.5 kyr pti) is *Bolivina* (34%; Figs. 3.6–13.8). This strong predominance produces a generic

Magneto-chronology (myr)	Sample Number	Drill Depth (ft) (m)	*Gyroidinoides*	*Grigelis*	*Globobulimina*	*Uvigerina*	*Bulimina*	*Caucasina*	*Bolivina*	*Epistominella*	*Stilostomella*	*Globocassidulina*	Generic Equitability	
	47	1075.0 (327.7)	3	20	7	9	1	1	7	9	11		1	
	46	1076.1 (328.0)	3	10	1	10	+		10	10	10	5	5	
	45	1081.4 (329.6)	5	9	+	8	+	+	15	10	3	23	1	
	44	1083.9 (330.4)	8	7	1	15	+	+	16	22	2	4	1	
	43	1085.0 (330.7)	6	10	1	35	+	+	9	+	+	4	1	
	42	1087.9 (331.6)	6	8	+	12	1	+	23	26	1	1	2	
	41	1090.3 (332.3)	9	6	+	3	1	+	11	20	12	7	1	
	41	1095.0 (333.8)	8	15	+	18	+	+	13	16	1	7	3	
	39	1101.2 (335.6)	4	6		14	+		38	17	+	7	1	
	38	1107.2 (337.5)	6	8	+	18	+	+	6	15	5	2	2	
	37	1114.0 (339.5)	9	3	+	11		1	18	8	22	4	2	P16
	36	1121.0 (341.7)	8	4	+	24	+		17	12	3	11	1	
	35	1125.3 (343.0)	7	3	+	27	1		20	12	1	4	1	
	34	1126.0 (343.2)	8	4	3	17	+	+	21	8	5		1	
	33	1130.3 (344.5)	Fine fraction lost											
	32	1139.8 (347.4)	7	5	2	11		2	31	8	15	2	1	
	31	1145.3 (349.1)	12	9	2	17		5	14	7	11	3	2	
	30	1146.3 (349.4)	8	9	1	20	+	+	15	14	+	7	1	
	29	1156.7 (352.6)	8	8	+	12	+	2	22	15	10	3	1	
	28	1162.7 (354.4)	8	11	1	7	+	2	4	6	10	7	2	
34.60	27	1165.3 (355.2)	9	12	+	23	1	4	4	14	4	8	1	
	26	1169.0 (356.3)	8	10	+	23	+	3	16	12	5	7	1	
	25	1173.9 (357.8)	19	8	+	6	+	7	34	11	2	4	1	
	24	1182.0 (360.3)	8	4	1	21	8	4	18	11	5	3	2	
	23	1185.0 (361.2)	7	2	1	7	2	2	12	39	4	7	1	
	22	1190.8 (363.0)	6	5	1	16	7	4	24	5	9	1	1	
35.30	21	1203.0 (366.7)	10	6	+	19	8	2	11	14	1	6	1	
	20	1203.8 (366.9)	Fine fraction lost											
	19	1212.0 (369.4)	3	11	1	6	4	4	23	12	3	1	1	
	18	1217.8 (371.2)	3	13	1	6	4	6	20	19	1	1	2	
	17	1225.3 (373.5)	6	3	+	10	4	10	24	29	1	1	1	
	16	1231.4 (375.3)	5	8	1	13	4	9	16	20	+	2	1	
	15	1235.5 (376.6)	8	10	3	7	5	17	15	7	2		2	
	14	1242.0 (378.5)	4	5	4	4	3	29	10	13	1	3	1	
	13	1250.0 (381.0)	3	4	4	4	2	23	11	11	2	3	1	
	12	1250.3 (381.1)	5	2	2	13	5	12	20	14	+	6	1	P15
35.60	11	1258.2 (383.5)	6	7	8	5	5	17	9	4	1	3	1	
	10	1262.2 (384.7)	4	3	4	5	3	25	11	4	4	6	1	*Bathysiphon* Subassemblage
	9	1270.2 (387.1)	5	4	11	10	10		7				3	
	8	1275.6 (388.8)	13	4	7	11	10	14	15	1	7	7	3	
	7	1277.0 (389.2)	10	10	11	8	12	+	12	+	4	1	5	
	6	1280.0 (390.1)	+	10	7	8	9	18	+	+	+	+	1	
	5	1281.7 (390.6)	6	10	10	14	10	11	1	1	3		1	
	4	1285.0 (391.6)	8	9	6	10	10		3	+	3		4	
	3	1286.8 (392.2)	11	12	9	8	4	3	3	2	3	+	1	
	2	1291.7 (393.7)	20	7	3	10	14		2	3	1		1	
	1	1292.3 (393.9)	19	8	+	6		7	34	11	2	4	1	
35.78	0	1292.8 (394.1)	Dead Zone											
		1293.8 (394.3)	Exmore Breccia											

Fig. 13.6. Stratigraphic distribution of generic predominance and equitability among benthic foraminifera in Chickahominy Formation in Kiptopeke core. Shading highlights predominant or co-predominant species represented in each sample.

equitability value of 1 for this sample. Many modern and fossil species of *Bolivina* are well known as opportunists – species whose populations have the capacity to rapidly increase in absolute and relative abundance whenever appropriate environmental conditions arise. This opportunism is expressed at the base of the Chickahominy Formation by the fact that *Bolivina* is so strongly predominant in sample 1 (the next most abundant genus represents only 19% of the sample), yet is quite sparse (<1–3%) in samples 2–6 (Figs. 13.6–13.8). The community structure changes dramatically in sample 2 (393.7 m; 1291.7 ft), in which *Gyroidinoides* assumes sole predominance (20%; Figs. 13.6–13.8). Generic equitability remains low at a value of 1. Samples 3–9 (392.2–387.1 m; 1286.8–1270.2 ft) represent an interval of marked co-predominance in which *Gyroidinoides*, *Grigelis*, *Globobulimina*, *Uvigerina*, *Bulimina*, *Caucasina*, and *Bolivina* generally share similar relative abundances (8–18%), and generic equitability ranges from 1 to 5 (Figs. 13.6–13.8).

In samples 10–25 (384.7–357.8 m; 1262.2–1173.9 ft), generic predominance shifts back and forth between principally *Caucasina*, *Bolivina*, and *Epistominella*, and generic equitability is mainly 1 (Figs. 13.6–13.8). *Epistominella* expresses an opportunistic tendency in sample 23 (361.2 m; 1185.0 ft), where it is unusually abundant at 39% (the next most abundant genus represents only 12%). Samples 26–28 (356.3–354.4 m; 1169.0–1162.7) represent a transitional interval: samples 26 and 27 are dominated by *Uvigerina* (23% each; equitability = 1); sample 28 is co-predominated by *Grigelis* (11%), *Stilostomella* (10%), and *Gyroidinoides* (8%) (equitability = 3; Figs. 13.6, 13.8).

Samples 29–46 (352.6–328.0 m; 1156.7–1076.1 ft) represent a thick interval in which predominance shifts frequently between mainly *Bolivina*, *Uvigerina*, and *Epistominella* (10–38%) and equitability is mainly 1 or 2 (Figs. 13.6–13.8). *Grigelis* returns to co-predominance in sample 46 (10%; equitability = 4), and is singularly predominant (20%) in sample 47.

Another way to view the stratigraphic changes in generic predominance is to examine the vertical abundance patterns of individual genera (Fig. 13.8). We see clearly the great relative abundance of *Globobulimina*, *Bulimina*, and *Caucasina* in the *Bulimina jacksonensis* Subzone versus their scarcity higher in the section. *Stilostomella* and *Globocassidulina* have an inverse distribution pattern, being least abundant in the *Bulimina jacksonensis* Subzone. *Gyroidinoides* and *Grigelis* are moderately abundant throughout the Chickahominy section. *Uvigerina*, *Bolivina*, and *Epistominella* are generally relatively abundant, but all display one or more isolated peaks of >30% abundance, a reflection of their opportunistic life strategy.

Fig. 13.7. (Opposite page) Important species from Chickahominy Formation used for paleoenvironmental interpretations. Scale bar = 100 μm. *1 Epistominella minuta* Olsson, 1960. Exmore corehole; umbilical view. *2 Charltonina madrugaensis* (Cushman and Bermúdez), 1948. Exmore corehole, umbilical view. *3 Gyroidinoides aequilateralis* (Plummer), 1927. Exmore corehole; umbilical view. *4 Gyroidinoides byramensis* (Cushman and Todd), 1946. Exmore corehole; umbilical view. *5 Uvigerina gardnerae* Cushman, 1926. Exmore corehole; lateral view. *6 Caucasina marylandica* (Nogan), 1964. Exmore corehole; lateral view. *7 Bolivina gracilis* Cushman and Applin, 1926. Exmore corehole; lateral view. *8 Bolivina virginiana* Cushman and Cederstrom, 1949. Exmore corehole; lateral view. *9 Bolivina* "praevirginiana." Exmore corehole; lateral view. *10 Globobulimina ovata* (D'Orbigny), 1846. Exmore corehole; lateral view. *11 Grigelis cookei* (Cushman), 1933. Exmore corehole; lateral view, final two chambers. *12 Grigelis annulospinosa* (Bandy), 1949. Exmore corehole, lateral view. *13 Stilostomella cocoaensis* (Cushman), 1925b. Exmore corehole; lateral view. *14 Stilostomella* "exilispinata." Exmore corehole; lateral view.

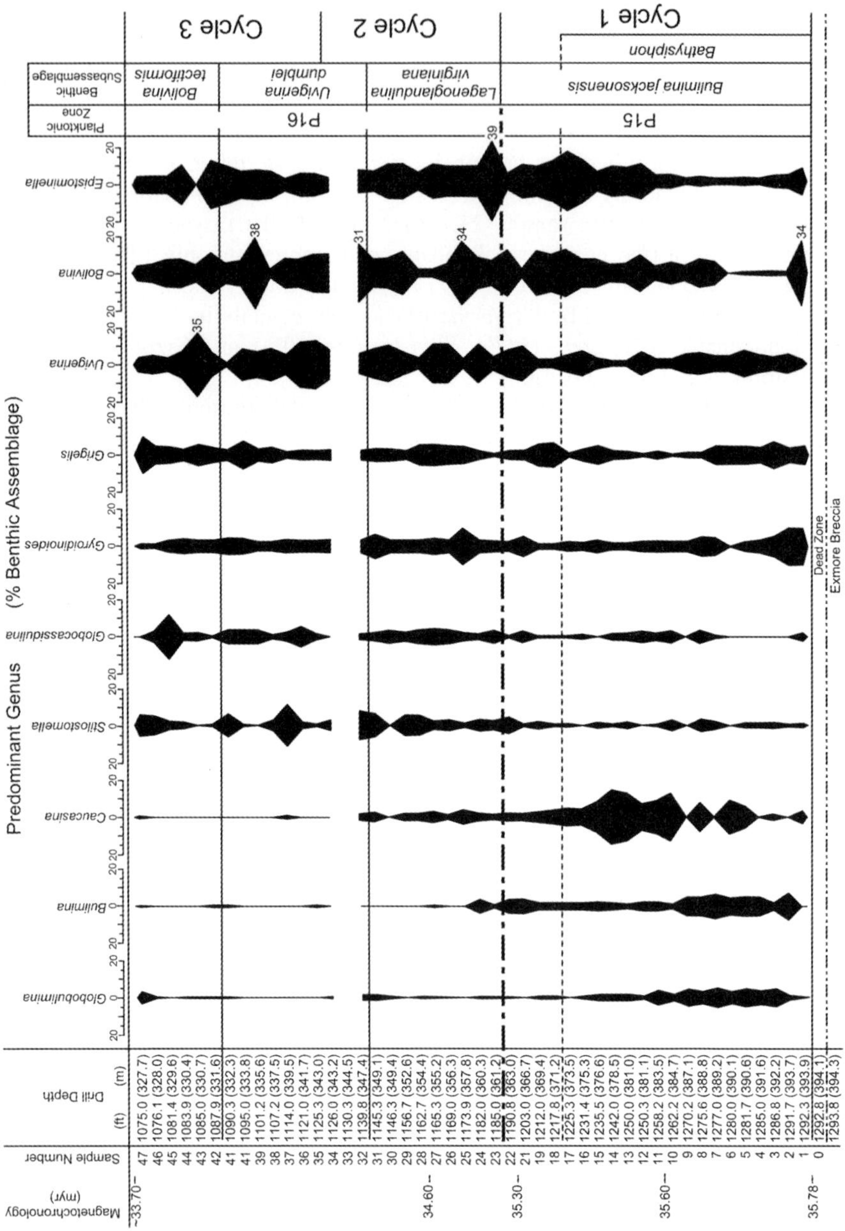
Predominant Genus (% Benthic Assemblage)
Globobulimina
Bulimina
Caucasina
Stilostomella
Globocassidulina
Gyroidinoides
Grigelis
Uvigerina
Bolivina
Epistominella
Planktonic Zone
P15
P16
Benthic Subassemblage
Bulimina jacksonensis
Lagenoglandulina virginiana
Uvigerina dumblei
Bolivina tectiformis
Bathysiphon
Cycle 1
Cycle 2
Cycle 3
Magnetochronology (myr)
Sample Number
Drill Depth (ft) (m)
Dead Zone
Exmore Breccia

13.1.3.2 Species Richness

Species richness (the number of species represented in a sample; a measure of diversity; Dodd and Stanton 1981) is another attribute of community structure that helps to interpret the Chickahominy paleoenvironments. Species richness among the Chickahominy benthic foraminifera at Kiptopeke varies from a minimum of 25 in sample 35 (343 m; 1125.3 ft) to a maximum of 60 in sample 23 (base of Zone P16) and in sample 47 (Fig. 13.9). When we plot species richness on a time scale, three major cycles of low-to-high species richness can be recognized at Kiptopeke (Fig. 13.10). Cycle 1 lasted ~600 kyr, and is nearly coincident with Zone P15 and the *Bulimina jacksonensis* Subzone. Species richness Cycle 2 lasted ~800 kyr, and encompasses the *Lagenoglandulina virginiana* Subzone and the lowest three samples in the *Uvigerina dumblei* Subzone (Fig. 13.10). Cycle 3 embraces the final ~600 kyr of the late Eocene at Kiptopeke, and includes most of the *Uvigerina dumblei* Subzone and all of the *Bolivina tectiformis* Subzone. The boundaries of each of these three richness cycles correlate approximately, but not precisely, with the major shifts in sediment accumulation rate at Kiptopeke (Fig. 13.10). The imprecision of the correlations is most notable for the two major reductions in species richness at the base of richness Cycles 2 and 3. The species richness shifts lagged the changes in sediment accumulation rate by 100–300 kyr. Further study of additional Chickahominy sections in the other Chesapeake Bay coreholes will help to clarify the reasons for this lag effect.

13.1.3.3 Paleoenvironmental Interpretations

Samples from outside the Chesapeake Bay crater show that rich, deep-water benthic foraminiferal assemblages occupied the target site and surrounding seafloor biotopes prior to the bolide impact. As we showed in Chapters 6 and 7, the transition from synimpact to postimpact deposition began with a 1.5–5 cm fallout layer followed by a dead zone, the latter of which represents no more than ~3 kyr of postimpact deposition (Fig. 13.4). Subsequent samples show that initial repopulation of the impact site rapidly renewed the rich preimpact benthic assemblage (Tables 13.1–13.8; Figs. 13.2, 13.4, 13.9). This distinctive assemblage (*Cibicidoides pippeni* Assemblage) recovered fully by 36 kyr pti, vigorously persisted for the remaining ~2.1 myr of the late Eocene, and underwent three moderate faunal turnovers before experiencing a major turnover at the unconformable Eocene-Oligocene contact. The persistence of the basal *Bathysiphon* Subassemblage to 355 kyr pti may reflect lingering effects of the impact on paleoenvironments inside the crater.

Fig. 13.8. (Opposite page) Polygon plot of stratigraphic variability in relative abundance for individual predominant genera in benthic foraminiferal communities in Chickahominy Formation at Kiptopeke site. Scale should be read as the sum of percent shown on each side of zero (e.g., in sample 1, the percentage of *Bolivina* is 17 plus 17 = 34). Numbered polygons (31–39) indicate all relative abundances greater than 30, which are values associated with blooms of opportunistic species. Heavy dashed line is P15/P16 boundary. Cycles 1–3 refer to cycles of species richness displayed in Fig. 13.10.

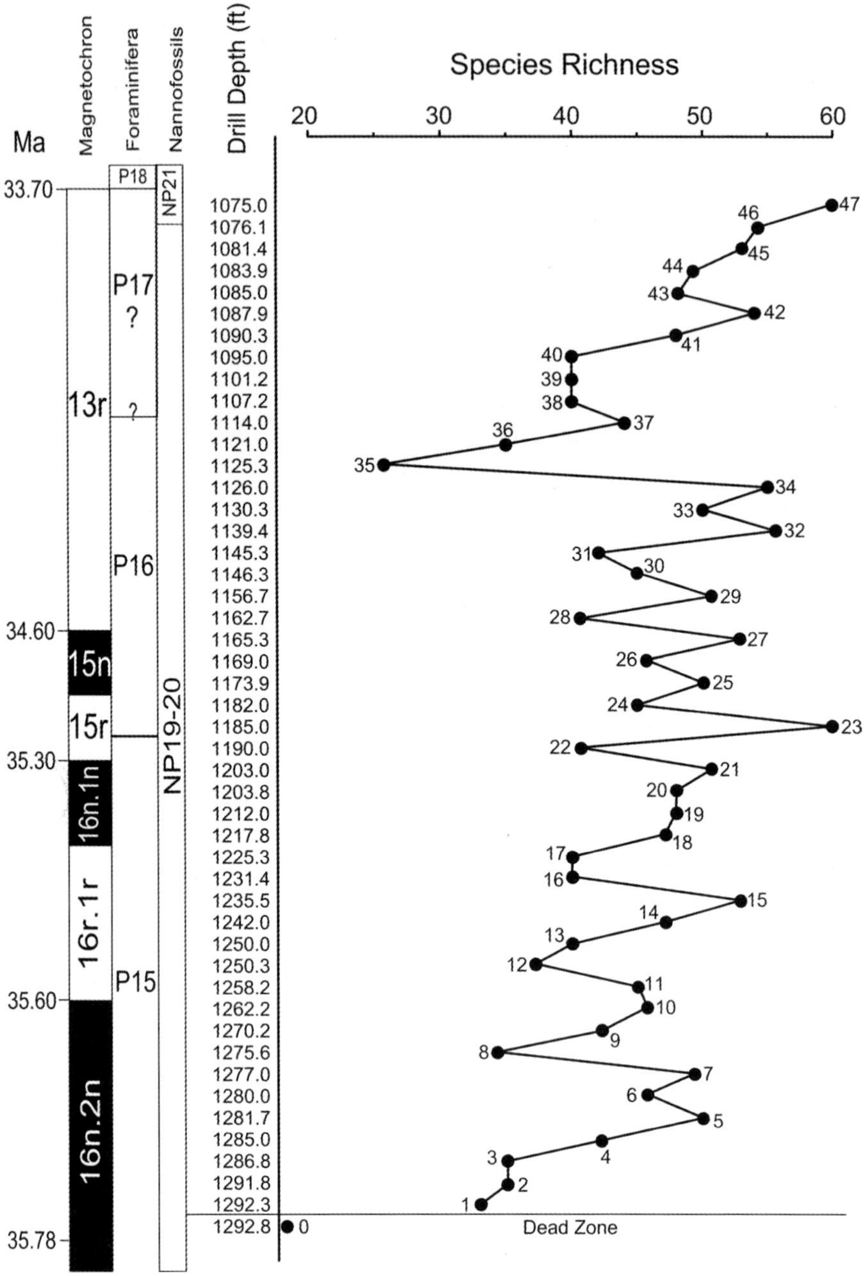

Fig. 13.9. Stratigraphic distribution of species richness among benthic foraminifera in Chickahominy Formation in Kiptopeke core. Numerals beside solid circles indicate sample numbers.

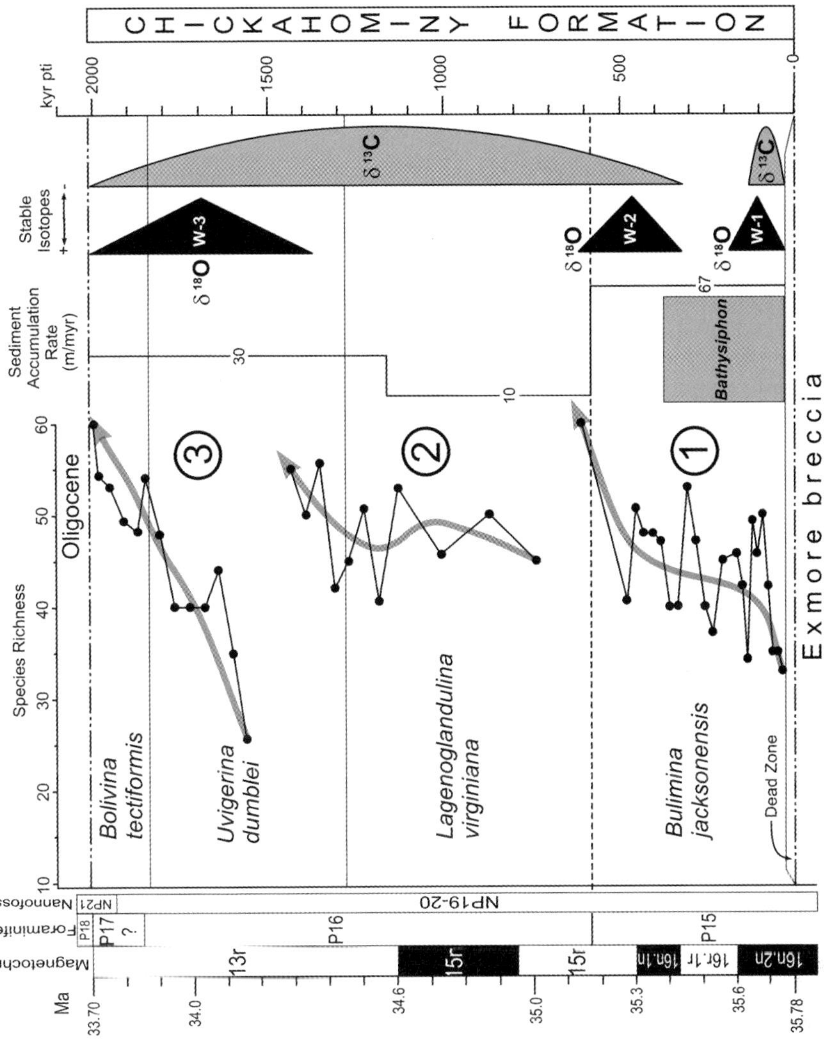

Fig. 13.10. Chronostratigraphic summary, showing foraminiferal, sedimentological, and stable isotopic attributes measured within Chickahominy Formation at Kiptopeke core site. Vertical scale on left side measures mega-annums (millions of years ago), vertical scale on right side measures postimpact years (pti). Shaded arrows and circled numerals designate three different cycles of low-to-high species richness.

The reliability of our assessment of the late Eocene paleoenvironments rests in large part on current knowledge of the environmental limits and preferences of counterpart species and genera in the modern oceans. Several syntheses and reviews of modern foraminiferal ecology provide guidance in this assessment (e.g., Poag 1981; Culver and Buzas 1980, 1981, 1982; Sen Gupta 1999). Our paleoenvironmental interpretations focus mainly on five factors that strongly influence the distribution of modern foraminiferal populations: (1) seafloor physiography; (2) substrate sedimentology; (3) microhabitat; (4) bottom-water chemistry; and (5) nutrient supply. By inference, these factors also were significant for their late Eocene analogues, and, therefore, are reflected in the composition of the Chickahominy assemblages.

13.1.3.3.1 Seafloor Physiography

The NASA Langley and Kiptopeke core sites, prior to impact, occupied the middle part of a broad, gently sloping continental shelf (see Chapter 1). But after impact, of course, the two sites were inside the crater, a partly filled, subcircular excavation, whose upper surface formed a depression or closed basin in the seafloor. Presumably, the depression was somewhat deeper in the center than along the periphery, but the precise relative relief remains to be determined. The Chickahominy Formation is also present at sites outside the crater rim, such as Windmill Point and Newport News. Preliminary examination of the extracrater Chickahominy assemblages indicates slightly shallower paleodepths than inside the crater, but quantitative analyses of the extracrater assemblages have not been completed.

Nearly all the Chickahominy species at Kiptopeke have modern counterparts (in fact, some are still extant), which are most abundant (individually and in similar species associations) in outer neritic to upper bathyal marine biotopes (150–500 m water depths; Table 13.9; Charletta 1980; Poag 1981; van Morkhoven et al. 1986). Many of these species (such as *Bulimina jacksonensis*, *Cassidulina tenuicarinata*, *Hoeglundina elegans*, *Turrilina robertsi*, *Bolivina byramensis*, *Stilostomella* spp.) also occur in other Paleogene outer neritic-bathyal deposits (Beckman 1954; Tjalsma and Lohmann 1983; van Morkhoven et al. 1986). We infer a paleodepth of ~300 m for the Chickahominy assemblage at Kiptopeke. Preliminary semiquantitative analyses of the Chickahominy benthic suites at the NASA Langley site have recently been completed (Poag and Norris in press). The benthic foraminiferal suites at NASA Langley closely resemble those at Kiptopeke (both in species composition and relative species abundance), and, thereby, indicate similar paleoenvironments to those documented herein at Kiptopeke.

13.1.3.3.2 Substrate Sedimentology

The sediments occupied by the Chickahominy benthic foraminiferal communities were soft, fine-grained muds (mainly micaceous, silty to sandy clay). Chickahominy fossil suites (mainly microfossils, echinoid spines, thin-shelled clams, and burrow casts of invertebrate organisms) indicate that these substrates and overlying marine watermasses supported abundant populations of benthos (foraminifera, ostracodes, echinoids, ophiuroids, solitary corals, bivalves, scaphopods, and, occasionally, siliceous sponges), plankton (foraminifera, calcareous nannofossils, radiolarians, diatoms, dinoflagellates, bolboformids), and nekton (fish). The common presence of filamentous organic detritus and pollen grains still in the sediments indicates an abundant supply of terrigenous organic carbon during Chickahominy deposition. The dark color of the Chickahominy clays, and the abundance of pyrite (as framboidal aggregates, burrow casts, thin wafer-like crusts, irregular nodules, and frequent replacements of shell material in many of the fossil groups) indicate that sulfate-reducing conditions commonly existed below the sediment-water interface.

13.1.3.3.3 Microhabitats

Modern benthic foraminifera have been assigned to different microhabitats, mainly according to the depth at which they are most abundant in the substrate (Corliss 1985, 1991; Gooday 1986; Rathburn and Corliss 1994; Jorissen et al. 1995, 1998; Jorissen 1999). Such microhabitats are present consistently in outer neritic, bathyal, and abyssal marine settings, but are not well developed in coarser-grained middle neritic and inner neritic settings (Murosky and Snyder 1994; Lueck and Snyder 1997). These vertically separated deep-water microhabitats are further characterized by their ambient physical, chemical, and biological properties, such as oxygen content, food supply, toxic substances, and potential for interactions with other organisms. The shallowest microhabitat is occupied by epifauna (forms at or protruding above the sediment-water interface). Next deepest is the shallow infauna, which constitutes the uppermost 2 cm of the substrate (Lutze and Thiel 1989; Corliss 1991; Buzas et al. 1993; Gooday 1994; Jorissen 1999). Generally, the epifaunal and shallow infaunal microhabitats are relatively well oxygenated and receive a relatively rich supply of labile, easily metabolizable organic detritus. In the intermediate (2–4 cm depth) and deep (4–10 cm depth) foraminiferal microhabitats, oxygen values generally decrease downward as the organic detritus becomes progressively more refractory and difficult to metabolize.

13.1.3.3.4 Bottomwater Chemistry

Most of the predominant genera and species in the Chickahominy benthic foraminiferal assemblages have modern counterparts notable for their opportunistic life strategies, and their tolerance of, or preference for, oxygen-depleted (disoxic, microxic, anoxic) muds rich in organic detritus. Among the best documented of these modern taxa are the calcareous genera that predominate in the *Cibicidoides pippeni* Assemblage: *Epistominella*, *Bolivina*, *Bulimina*, *Globobulimina*, *Globocassidulina*, *Uvigerina*, and *Buliminella* (counterpart to *Caucasina*) (Phleger and Soutar 1973; Douglas and Heitman 1979; Mackensen and Douglas 1989; Kaminski et al. 1995; Jorissen et al. 1992; Sen Gupta et al. 1996; Bernhard and Sen Gupta 1999; Loubere and Fariduddin 1999; Table 13.9). Most of the members of the Chickahominy *Bathysiphon* Subassemblage also are typical inhabitants of oxygen-depleted, nutrient-rich substrates (Gooday 1994; Kaminski et al. 1995).

Table 13.9. Benthic foraminiferal species used for interpretation of Chickahominy paleoenvironments at Kiptopeke and NASA Langley core sites.

Species	Test Construction	Microhabitat	Oxygen/ Nutrient Tolerance	Preferred Paleodepth	Opportunist
Ammobaculites sp.	agglutinated	infaunal	low/high	outer neritic–upper bathyal	
Amphimorphina "fragilicostata"	calcite	infaunal	low/high	outer neritic–upper bathyal	
Amphimorphina "planata"	calcite	infaunal	low/high	outer neritic–upper bathyal	
Bathysiphon sp.	agglutinated	epifaunal	low/high	Bathyal–abyssal	
Bolivina byramensis	calcite	[1]s infaunal	low/high	outer neritic–upper bathyal	
Bolivina gardnerae	calcite	s infaunal	low/high	outer neritic–upper bathyal	yes
Bolivina gracilis	calcite	[2]i–[3]d infaunal	low/high	outer neritic–upper bathyal	
Bolivina jacksonensis	calcite	i–d infaunal	low/high	outer neritic–upper bathyal	yes
Bolivina multicostata	calcite	s infaunal	low/high	outer neritic–upper bathyal	
Bolivina plicatella	calcite	s infaunal	low/high	outer neritic–upper bathyal	
Bolivina "postvirginiana"	calcite	s infaunal	low/high	outer neritic–upper bathyal	yes
Bolivina "preavirginiana"	calcite	s infaunal	low/high	outer neritic–upper bathyal	
Bolivina regularis	calcite	i–d infaunal	low/high	outer neritic–upper bathyal	
Bolivina striatella	calcite	s infaunal	low/high	outer neritic–upper bathyal	
Bolivina tectiformis	calcite	s infaunal	low/high	outer neritic–upper bathyal	
Bolivina virginiana	calcite	s infaunal	low/high	outer neritic–upper bathyal	yes
Bulimina alazanensis	calcite	i–d infaunal	low/high	outer neritic–upper bathyal	
Bulimina cooperensis	calcite	i–d infaunal	low/high	outer neritic–upper bathyal	
Bulimina jacksonensis	calcite	i–d infaunal	low/high	outer neritic–upper bathyal	yes
Caucasina marylandica	calcite	phytodetrital	low/high	outer neritic–upper bathyal	yes
Charltonina madrugaensis	calcite	d infaunal	low/high	outer neritic–upper bathyal	yes
Cibicidoides pippeni	calcite	epifaunal	high/low	outer neritic–upper bathyal	

Table 13.9. (cont.)

Species	Test Construction	Microhabitat	Oxygen/ Nutrient Tolerance	Preferred Paleodepth	Opportunist
Cribrostomoides sp.	agglutinated	s infaunal	low/high	outer neritic–upper bathyal	
Cyclammina cancellata	agglutinated	s infaunal	low/high	outer neritic–upper bathyal	
Dorothia sp.	agglutinated	d infaunal	low/high	outer neritic–upper bathyal	
Epistominella minuta	calcite	epifaunal	low/high	outer neritic–upper bathyal	yes
Gaudryina alazanensis	agglutinated	d infaunal	low/high	outer neritic–upper bathyal	
Globobulimina ovata	aragonite?	i–d infaunal	low/high	outer neritic–upper bathyal	
Globocassidulina subglobosa	calcite	phytodetrital	low/high	outer neritic–upper bathyal	
Grigelis cookei	calcite	infaunal	low/high	outer neritic–upper bathyal	
Grigelis "curvicostata"	calcite	infaunal	low/high	outer neritic–upper bathyal	
Grigelis "elongata"	calcite	infaunal	low/high	outer neritic–upper bathyal	
Grigelis "elongostriata"	calcite	infaunal	low/high	outer neritic–upper bathyal	
Grigelis "gigas"	calcite	infaunal	low/high	outer neritic–upper bathyal	
Grigelis "tubulosa"	calcite	infaunal	low/high	outer neritic–upper bathyal	
Grigelis "tumerosa"	calcite	infaunal	low/high	outer neritic-upper bathyal	
Gyroidinoides aequilateralis	calcite	s infaunal	low/high	outer neritic–upper bathyal	
Gyroidinoides byramensis	calcite	s infaunal	low/high	outer neritic–upper bathyal	
Gyroidinoides octocameratus	calcite	s infaunal	low/high	outer neritic–upper bathyal	
Gyroidinoides planatus	calcite	s infaunal	low/high	outer neritic–upper bathyal	
Hoeglundina elegans	aragonite	epifaunal	low/high	outer neritic–upper bathyal	
Marginulina cocoaensis	calcite	infaunal	low/high	outer neritic–upper bathyal	
Marginulina karreriana	calcite	infaunal	?/?	outer neritic–upper bathyal	
Melonis planatus	calcite	i–d infaunal	low/high	outer neritic–upper bathyal	
Nodosaria capitata	calcite	infaunal	low/high	outer neritic–upper bathyal	

Table 13.9. (cont.)

Species	Test Construction	Microhabitat	Oxygen/ Nutrient Tolerance	Preferred Paleodepth	Opportunist
Nodosaria pustulosa	calcite	infaunal	low/high	outer neritic–upper bathyal	
Nodosaria saggitula	calcite	infaunal	low/high	outer neritic–upper bathyal	
Nodosaria soluta	calcite	infaunal	low/high	outer neritic–upper bathyal	
Nodosaria vertebralis	calcite	infaunal	low/high	outer neritic–upper bathyal	
Oridorsalis umbonatus	calcite	epifaunal	low/high	outer neritic–upper bathyal	
Reophax sp.	agglutinated	i–d infaunal	low/high	inner neritic–upper bathyal	
Spiroplectinella mississippiensis	agglutinated	d infaunal	low-high	outer neritic–upper bathyal	
Stilostomella "aduncocostata"	calcite	infaunal	low/high	outer neritic–upper bathyal	
Stilostomella annulospinosa	calcite	infaunal	low/high	outer neritic–upper bathyal	
Stilostomella "bicostatus"	calcite	infaunal	low/high	outer neritic–upper bathyal	
Stilostomella cocoaensis	calcite	infaunal	low/high	outer neritic–upper bathyal	
Stilostomella "exilispinata"	calcite	infaunal	low/high	outer neritic–upper bathyal	
Stilostomella "juvenocostata"	calcite	infaunal	low/high	outer neritic–upper bathyal	
Stilostomella "multispiculata"	calcite	infaunal	low/high	outer neritic–upper bathyal	
Technitella sp.	agglutinated	infaunal	low/high	outer neritic–upper bathyal	
Turrilina robertsi	calcite	infaunal	low/high	outer neritic–upper bathyal	
Uvigerina cookei	calcite	s infaunal	low/high	outer neritic–upper bathyal	
Uvigerina dumblei	calcite	s infaunal	low/high	outer neritic–upper bathyal	
Uvigerina gardnerae	calcite	s infaunal	low/high	outer neritic–upper bathyal	yes
Uvigerina jacksonensis	calcite	s infaunal	low/high	outer neritic–upper bathyal	
Uvigerina spinicostata	calcite	s infaunal	low/high	outer neritic–upper bathyal	

[1]s = shallow depth (0–2 cm) below sediment-water interface
[2]i = intermediate depth (2–4 cm) below sediment-water interface
[3]d = deep (4–10 cm) below sediment-water interface

13.1.3.3.5 Nutrient Supply

Outer neritic, bathyal, and abyssal benthic foraminifera, as a whole, depend upon the flux of labile organic carbon for their food source (Gooday 1994; Loubere and Fariduddin 1999). There is considerable evidence from modern oceans that the geographic distribution, test size, and abundance (absolute and relative) of certain benthic foraminiferal species and genera are strongly correlative with the flux of organic detritus to the seafloor (Caralp 1989; Corliss and Fois 1990; Corliss and Silva 1993; Pfannkuche 1993; Linke et al. 1995; Gooday 1996; Morigi et al. 2001). In particular, most of the predominant Chickahominy calcareous genera (and those of the *Bathysiphon* Subassemblage) have modern counterparts that are most abundant, and often have largest test sizes, in organic-rich muds, which are often also oxygen-depleted.

Of special note in the Chickahominy assemblage is an association of small, smooth, thin-walled, hyaline, opportunistic genera, such as *Epistominella*, which in modern oceans live epifaunally on the seafloor within aggregates of phytodetritus (a gelatinous matrix containing the remains of phyto- and zooplankton; Gooday 1993, 1994). These species have opportunistic feeding strategies, and grow explosively into large concentrations during peak development of phytodetritus. Among the predominant Chickahominy taxa, species of *Epistominella*, *Caucasina*, and *Globocassidulina*, are probably representative of this lifestyle (Table 13.9).

13.1.3.3.6 Paleoenvironmental Summary

Overall, then, the Chickahominy benthic foraminiferal associations documented in the Kiptopeke corehole represent consistently diverse, species-rich communities living within the upper 10 cm of fine-grained substrates, in paleodepths of ~300 m, generally typified by high flux rates of organic carbon and by oxygen deficiency. The common presence of burrow casts, however, indicates that the bottom waters and surficial sediments were not anoxic. The development of five stratigraphically successive subassemblages and several shifts in generic predominance, equitability, and species richness point to marked temporal changes in environmental properties other than paleodepth. These changing properties we infer to include sediment delivery rates, prominent-to-subtle changes in substrate chemistry and nutrient flux, and broad-scale shifts in climate (from the $\delta^{18}O$ record) and the global carbon budget (from the $\delta^{13}C$ record; Fig. 13.10).

At the Kiptopeke site, the initial postimpact benthic foraminiferal association in Chickahominy sample 1 contains 33 species, is solely dominated numerically by opportunistic *Bolivina* (34%), and is associated with pyrite crusts and burrow casts, silica diagenesis, and the agglutinant suite that comprises the *Bathysiphon* Subassemblage. These characteristics distinguish this sample from all subsequent samples (*Bolivina* reaches >30 % in three other samples, but in none of these is it associated with the *Bathysiphon* Subassemblage). Thus, the initial postimpact benthic foraminiferal community at Kiptopeke occupied the floor of an enclosed basin rich in organic detritus, which depleted the dissolved oxygen content both at the sediment-water interface (indicated by *Bathysiphon*) and within the shallow infaunal microhabitat (indicated by *Bolivina*). In addition, the living Chickahominy assemblages occupied a basin below which shallowly buried impact breccia contained elevated interstitial salinities. Groundwater salinity as high as 25,000 mg/L characterizes the Exmore breccia at the Kiptopeke site (see Chapter 14 for further discussion). Moreover, this brine probably was warmer than surrounding bottom waters, because of its proximity to superheated crystalline basement rocks, whose elevated temperatures may have endured for 1 myr pti (Chapter 12).

These early postimpact conditions lasted only for ~2 kyr, because in sample 2, the benthic foraminiferal association changed significantly. *Bolivina* was reduced to 2%, and generic predominance shifted to *Gyroidinoides*, probably reflecting a reduction in the flux rate of organic carbon. *Gyroidinoides* remained one of several co-predominant genera during most of the next ~120 kyr, sharing its abundance with *Grigelis*, *Globobulimina*, *Uvigerina*, *Bulimina*, and eventually joined again by *Bolivina* (Fig. 13.6). This association represents full recovery of the Chickahominy benthic foraminiferal community following the impact, which was reached by the time the sediments of sample 4 had accumulated. These microfaunal relations were briefly interrupted by the sole predominance of *Caucasina* (in sample 6), which presumably indicates a short period of increased phytodetritus accumulation.

At ~122 kyr pti, co-predominance shifted to *Bolivina* and *Caucasina* for 131 kyr, and then to *Bolivina* and *Epistominella* for 702 kyr. Both associations indicate continued oxygen depletion and abundance of organic carbon, the latter often available through phytodetrital flux. At 955 kyr pti, *Uvigerina* became the sole predominant form for 196 kyr, followed by one sample (~121 kyr duration) with co-predominant *Gyroidinoides*, *Grigelis*, and *Stilostomella*. These changes probably reflect a shift away from phytodetritus as a major source of nutrients, which may be associated with initiation of the negative shift in $\delta^{13}C$, which took place nearly concomitantly (Fig. 13.10; see section 13.2.4). These changes also may be partly attributable to the low sediment accumulation rate that characterized this time interval.

At 1.27 myr pti, coincident with a shift to higher sediment accumulation rates, *Bolivina* became predominant again, and continued as predominant or co-predominant genus for most of the next 800 kyr. In the early part of this interval, *Bolivina* shared predominance with, or was replaced by, *Uvigerina*; in the late part of the interval, it shared predominance with *Epistominella*. *Grigelis*, *Stilostomella*, and *Globocassidulina* joined or replaced these associations in three single-sample exceptions. We interpret the *Bolivina-Uvigerina* association to indicate continued oxygen depletion, but with a relatively sparsity of phytodetritus. The subsequent increase in *Epistominella* indicates a return to abundant phytodetritus, which may be related to the beginning of the latest Eocene subpulse of warm climate (indicated by a negative shift in $\delta^{18}O$; Fig. 13.10; see section 13.2.3).

At the top of the Kiptopeke Eocene section (sample 47), another faunal shift raised *Grigelis* to sole predominance for the only time during the entire postimpact interval. The significance of this change is not obvious to us, but it may be related to the beginning of a positive shift in both $\delta^{18}O$ and $\delta^{13}C$ isotopic values, along with a major sea-level fall, accompanied by a change in sediment composition (to coarse, glauconitic, quartz sand) and a shift in accumulation rate, all of which took place at the Eocene-Oligocene contact at this site (Fig. 13.10).

Poag and Norris (in press) have shown that a nearly identical biostratigraphic and paleoenvironmental record is represented in the Chickahominy foraminiferal assemblages at the NASA Langley site (borehole 233), 39 km southwest of Kiptopeke (borehole 2), near the outer rim of the Chesapeake Bay crater (Fig. 1.3).

13.1.3.3.7 Comparison with Biotic Changes at K-T Boundary

Several studies of benthic foraminiferal assemblages have been carried out across the impact-produced Cretaceous-Tertiary boundary (Dailey 1983; Keller 1988, 1992; Thomas 1990; Nomura 1991; Kaiho 1992; Schmitz et al. 1992; Widmark and Malmgren 1992; Kuhnt and Kaminski 1993; Coccioni and Galeotti 1994; Speijer and van der Zwaan 1996). Most of the K-T study sections, however, are distal ejecta deposits, thousands of kilometers away from their source, the Chicxulub impact site. In contrast, the Kiptopeke site is inside Chesapeake Bay crater. Three differences stand out between the K-T and Chesapeake Bay scenarios: (1) At neritic to mid-bathyal sites for the K-T boundary, major long-term lineage changes took place between the preimpact and postimpact benthic foraminiferal communities, whereas at Chesapeake Bay, the long-term postimpact community (once it was reestablished) was essentially the same as the preimpact benthic community; (2) The postimpact benthic foraminiferal association at several neritic and upper bathyal K-T boundary sites represents shallower depositional facies than do the preimpact assemblages, whereas at Chesapeake Bay, the paleodepth increased following impact, because the partly filled crater formed a seafloor depression; (3) Benthic foraminiferal changes across the K-T boundary are associated with major extinctions and turnovers in calcareous plankton and a profound productivity crisis in the pelagic ecosystem. At some K-T sites, the extinction event produced massive fluxes of organic carbon to the seafloor, whereas at other sites, reduced productivity significantly diminished the flux of organic carbon (Coccioni and Galeotti 1994; Brinkhuis and Zachariasse 1988; Pospichal 1994; Speijer and van der Zwaan 1996). No analogous ocean-wide pelagic crisis arose from the Chesapeake Bay impact, though there is evidence that plankton communities along the U.S. Atlantic margin may have been stressed (MacLeod 1990; MacLeod et al. 1990). On the other hand, it is fair to say that the presence of a dead zone subjacent to the Chickahominy Formation indicates that local bottom and water-column conditions were hostile to benthic as well as planktonic organisms for ~1–3 kyr or less.

Otherwise, there are some striking similarities in the immediately postimpact K-T and Chesapeake Bay benthic foraminiferal associations and in the sediments that encompass them (Fig. 13.11). A good comparative example is the outcrop at El Kef, Tunisia (the type section for the K-T boundary, which represents outer neritic to upper bathyal paleodepths; Speijer and van der Zwaan 1996; Fig. 13.11). The postimpact boundary clay at El Kef is 50 cm thick, is laminated, and contains pyritized and hematitic burrow casts, much like the 19-cm thick, laminated, silty, dead zone at Chesapeake Bay. The lowest postimpact assemblage at El Kef is

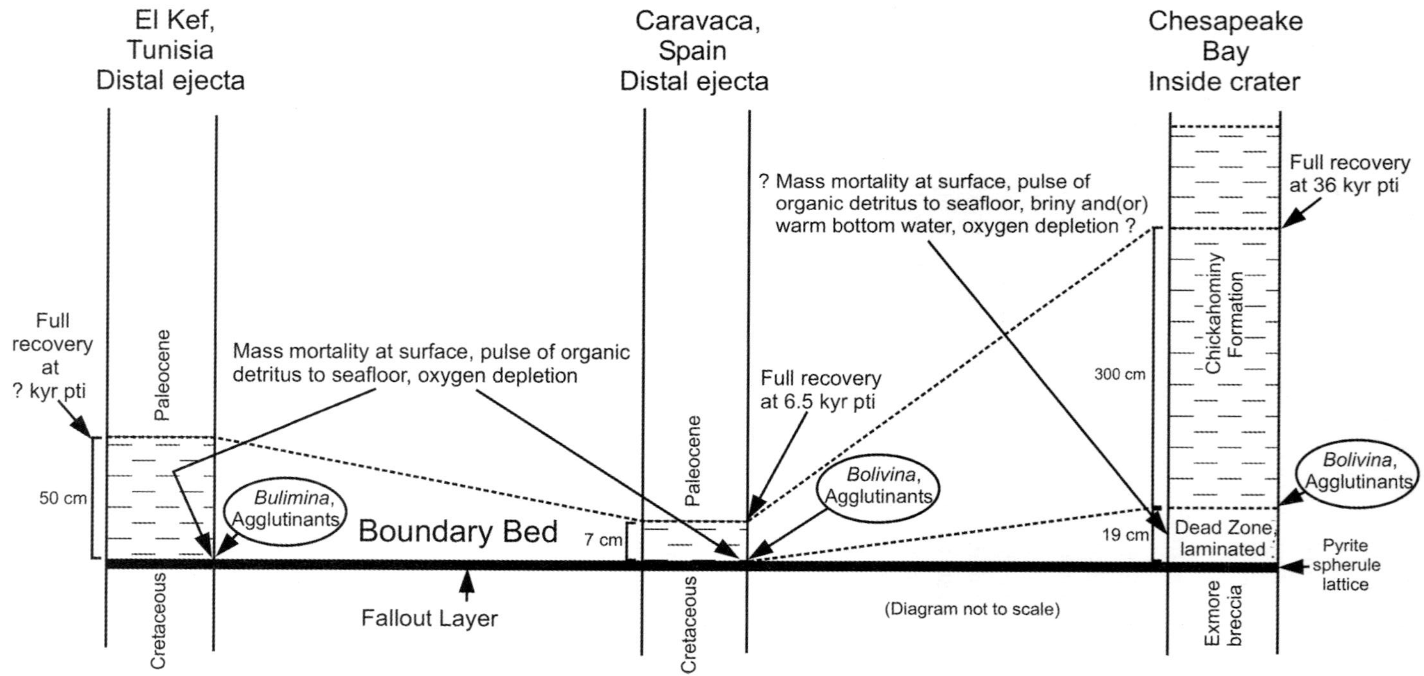

Fig. 13.11. Comparison of postimpact recovery among benthic foraminiferal community at Kiptopeke site inside Chesapeake Bay crater with distal ejecta outcrops correlative to K-T impact at Chicxulub, Yucatán, Mexico.

notable for relatively low diversity (species and genera) accompanied by a unique spike (sole predominance; 22–32%) in the abundance of *Bulimina ovata* (synonymous with *Globobulimina ovata* in the Chickahominy assemblages). This is reminiscent of the low diversity and unusual *Bolivina* abundance spike (34%) in the first postimpact assemblage at Chesapeake Bay. There also is a distinct increase in the abundance of several agglutinated species in the lower few centimeters of the postimpact section at El Kef, just as at Chesapeake Bay. It should be noted, however, that the El Kef study was based on analysis of the >250 μm size fraction, whereas we analyzed the >63 μm size fraction, which encompasses a more complete representation of the Chickahominy benthic association. Indeed, many of the predominant Chickahominy taxa, such as *Grigelis, Caucasina, Bolivina, Epistominella, and Stilostomella*, would have been acutely underrepresented if we had analyzed a coarser size fraction.

Another good example, though representing somewhat deeper paleodepths (middle bathyal) than Kiptopeke, is the K-T boundary section at Caravaca, Spain (also based on analysis of the >250-μm size fraction; Coccioni and Galeotti 1994; Fig. 13.11). At Caravaca, the initial postimpact benthic foraminiferal assemblage (within a 7-cm-thick laminated clay) contains representatives of only two genera; *Bolivina* (calcareous) and *Spiroplectammina* (agglutinated). This low-oxygen, high-nutrient association lasted for an estimated 0.5–0.6 kyr after impact. Following development of this initial low-diversity assemblage, numerous preimpact taxa reappeared progressively upward through the section (commonly known as Lazarus species), until normal polytaxic assemblages regained prominence at ~6.0–6.5 kyr pti. This compares with the ~3-kyr duration of the initial *Bolivina*-dominated association at Kiptopeke and full recovery of the *Cibicidoides pippeni* Assemblage at ~36 kyr pti (Fig. 13.11).

13.2 Possible Global Paleoenvironmental Effects

According to some authors, a bolide impact the size of the Chesapeake Bay event, accompanied by atmospheric perturbations such as those cited in Chapter 12, should have produced a mass extinction severe enough to eliminate ~50% of marine species (Raup 1991a,b). To date, however, only sparse evidence of an immediate, widespread, impact-derived perturbation of the late Eocene biosphere has been proffered (Sanfilippo et al. 1985; Keller 1986; MacLeod et al. 1990; Brinkhuis and Coccioni 1995; Vonhof et al. 2000; Spezzaferri et al. 2002). No major extinction event greater than the normal 5% background value is known to have taken place (Poag 1997b). No evidence of mass mortality, pulsed extinctions, or mass extinctions has been found distal to the known late Eocene craters or associated with late Eocene ejecta deposits.

This dearth of extinctions has major implications for the kill curve proposed by Raup (1991a,b; Fig. 13.12) to relate impact crater size to the resultant percent of marine species loss due to mass extinction (Jansa et al. 1990; Jansa 1993; Poag 1997b; Fig. 13.12). Jansa et al. (1990), Jansa (1993), and Poag (1997b) showed that the Chesapeake Bay data (along with data from the Montagnais crater) invali-

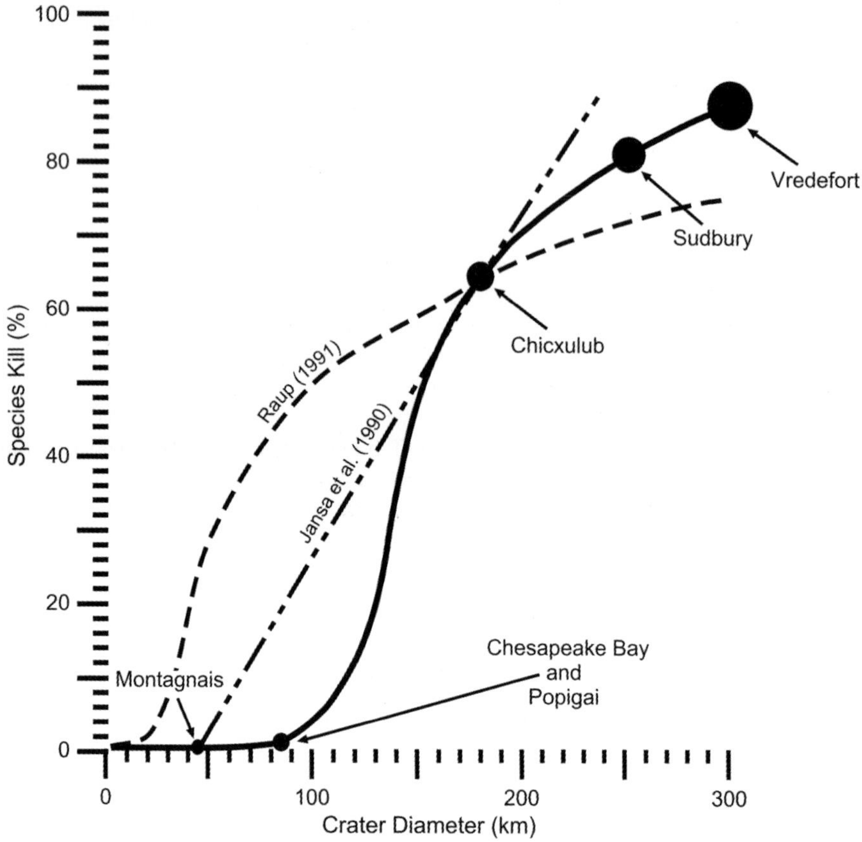

Fig. 13.12. Kill curve designed by Raup (1991a) to relate impact crater size to extinction of marine species. See text for further explanation. Modified from Poag (1997b).

date Raup's kill curve for impact craters smaller than ~100 km in diameter (Fig. 13.12).

It is well established that the severity of impact effects depends on natural variability in such things as: (1) the size, composition, trajectory, and speed of the impactor; (2) the relative consolidation and composition of the target rocks; (3) the latitudinal and topographic location of the impact site; (4) the nature of and ambient stress regime of the preimpact biota; and (5) the general nature of the existent terrestrial, oceanic, and atmospheric environments. The impact at either Chesapeake Bay or Popigai alone, however, would have produced enough energy (~10^7 Mt equivalents of TNT) to significantly alter atmospheric conditions regardless of extenuating cosmic, geological, or other environmental circumstances (Chapters 9, 12; Table 13.10; Morrison et al. 1994; Toon et al. 1994).

Table 13.10. Impact effects as a function of energy yielded and bolide/crater diameter (modified from Morrison et al. 1994).

Yield [Mt]	Diameter of Bolide	Diameter of Crater	Environmental Consequences
<10		No crater	Upper atmosphere detonation of stones and comets; only irons penetrate to surface of Earth
10^1–10^2	75 m	1.5 km	Irons make craters (like Meteor Crater); Stones produce airbursts (like Tunguska); land impacts destroy area the size of a large city (Paris)
10^2–10^3	160 m	3 km	Irons and stones produce groundbursts; comets produce airbursts; land impacts destroy area the size of a large urban area (New York)
10^3–10^4	350 m	6 km	Impacts on land produce craters; ocean tsunamis become significant; land impacts destroy area the size of a small state (Delaware)
10^4–10^5	700 m	12 km	Tsunamis reach oceanic scales, exceed damage from land impacts; land impacts destroy area the size of a moderate state (Virginia)
10^5–10^6	1.7 km	30 km	Land impacts raise enough dust to affect climate, freeze crops; oceanic impacts generate hemispheric-scale tsunamis; global destruction of ozone; land impacts destroy area the size of a large state (California) or small country (Japan)
10^6–10^7	3 km	60 km	Both land and ocean impacts raise dust, change climate; impact ejecta are global, triggering widespread wildfires; land impacts destroy area the size of a large nation (Mexico)
~10^7	3–5 km	85 km	Chesapeake Bay and Popigai impacts
10^7–10^8	7 km	125 km	Prolonged climatic effects, global conflagration, probable mass extinction; direct destruction approaches continental scale (United States)
10^8–10^9	16 km	250 km	Catastrophic mass extinction (Chicxulub impact and K-T extinctions)
>10^9	?	?	Threatens survival of all advanced life forms

13.2.1 Hypothetical Short-Term Effects

Potential short-term global environmental effects of the Chesapeake Bay impact are hypothetical, based primarily on models and predicted atmospheric disruptions extrapolated from the results of nuclear explosions (Tables 13.10, 13.11). The scaling calculations for an 85-km-diameter crater predict that the Chesapeake Bay bolide must have been 3–5 km in diameter (Table 13.10). An impactor of this size would produce an explosion equivalent to ~10^7 Mt of TNT (Morrison et al. 1994). Several authors (Adushkin and Nemchinov 1994; Rampino and Haggerty 1994;

Table 13.11. Estimated environmental damage from Chesapeake Bay-sized bolide impact on deep continental shelf (modified from Toon et al. 1994).

Disruptive Agent	Disruptive Mechanism(s)	Duration of Disruption	Geographic Scale of Disruption
Dust loading	Cooling; cessation of photosynthesis; loss of vision	Years	Global
Fires	Burning; soot cooling; pyrotoxins; acid rain	Months	Global
NO_x generation	Ozone loss; acid rain; cooling	Months to years	[1]Regional to global
Shock wave	Mechanical pressure; acoustic fluidization	Seconds to minutes	[1]Regional to global
Tsunamis	Drowning	Hours to days	[1]Regional
Heavy metals	Poisoning	Years	Global
Water and CO_2 injections	Warming	Decades	Global

[1]regional means an area of 10^6 km^2

Toon et al. 1994) have concluded that a bolide impact of this magnitude would disperse ejecta, water vapor, and submicrometer dust on a global scale. The resultant atmospheric opacity would appreciably cool the atmosphere and Earth's surface for months to years (Tinus and Roddy 1990; Bailey et al. 1994; Toon et al. 1994), and would limit photosynthesis for several months (Gerstl and Zardecki 1982; Grieve and Shoemaker 1994), though Pope (2002) has argued against the submicrometer-dust scenario. Such severe atmospheric deterioration could be exacerbated by an immediate, shock-induced, heat pulse (Rampino and Haggerty 1994), short-term enhanced greenhouse warming (Emiliani et al. 1982; Toon et al. 1994), global wildfires (Melosh et al. 1990), acid rain (Toon et al. 1994), and dense photochemical fog (Wolbach et al. 1988). A substantial increase in atmospheric CO_2 derived from carbonate target rocks should, in turn, have created decades of greenhouse warming after the cooling effects of atmospheric dust had dissipated (O'Keefe and Ahrens 1989; Sigurdsson et al. 1992; Covey et al. 1994). So far, though, little compelling evidence of such short-term atmospheric perturbations has been derived from studies of late Eocene impacts. However, Vonhof et al. (2000) and Spezzaferri et al. (2002) have noted evidence among dinoflagellates and benthic foraminifera, respectively, for a possible short-term cooling associated with deposition of late Eocene ejecta at Massignano, Italy, and at ODP Site 689B in the Southern Ocean.

13.2.2 Possible Long-Term Effects

The weakness or lack of an immediate or short-term atmospheric response to the late Eocene impacts does not, however, preclude a longer-term response. There is ample evidence that the Chesapeake Bay, Popigai, and the Toms Canyon impacts took place during the late stages of a long-term, step-wise climatic cooling event. This cooling event is evidenced by the buildup of Antarctic ice sheets, which culminated in a sharp temperature decline, accompanied by mass extinction, in the early Oligocene (Keller et al. 1987; Marty et al. 1988; Miller et al. 1991; Wise et al. 1991; Prothero 1994; Miller 1992; Prothero and Berggren 1992; Clymer et al. 1996). Poag (1997b) and Poag et al. (2003) pointed out that marine isotopic signatures and changes in the fossil record of late Eocene terrestrial and marine organisms seem to suggest that the long-term temperature decline was interrupted by a global pulse of atmospheric warmth, which began at approximately the time of the Chesapeake Bay, Popigai, and Toms Canyon impacts (regardless of whether they were simultaneous or sequential; Fig. 13.13). Diverse evidence for this late Eocene warming includes a 6–8 ^{o}C temperature increase deduced from leaf-margin analysis of North American land plants (Wolfe 1978), a 0.5 $^{o}/_{oo}$ decrease in δ^{18}O measured in Southern Ocean cores (Miller 1992), and migration of low-latitude nannofloras into high latitudes (Haq and Lohmann 1976). Pearson et al. (2001) and Kobashi et al. (2001) have recently strengthened the case for warm late Eocene oceanic surface waters on the basis of δ^{18}O analyses of planktonic foraminifera and shallow-water molluscs, respectively. Poag (1997b) proposed that impact-generated greenhouse warming interrupted the progressive, long-term Eocene cooling, and, ironically thereby, may have postponed a pending late Eocene mass extinction until the early Oligocene. In fact, the particularly large temperature differential between a late Eocene greenhouse and an early Oligocene ice-house may have triggered the mass extinction (Fig. 13.13).

The work of Farley et al. (1998), who recorded the relative abundance of extraterrestrial helium isotopes (^{3}He) in late Eocene sediments at Massignano (Fig. 13.14A), supports Poag's (1997b) supposition of a late Eocene heat pulse. Farley et al. found that the concentration of ^{3}He in the Massignano section increased dramatically in the late Eocene, reached a peak (~5.5 times the baseline value) coincident with the impacts at Chesapeake Bay, Popigai, and Toms Canyon, and gradually declined to near baseline values 1–2 myr later in the early Oligocene. Farley et al. (1998) interpreted this distinctively high ^{3}He concentration to represent accelerated deposition of interplanetary dust particles (by which the ^{3}He was carried) when a comet shower bombarded Earth in the late Eocene. If this interpretation is correct, additional late Eocene impact craters may be expected to be found.

If successive impacting continued for 1–2 myr following the Chesapeake Bay, Popigai, and, Toms Canyon collisions, a resultant production of long-term atmospheric warming may have extended into the early Oligocene. When the comet shower abated, global temperatures declined along an unusually steep gradient, accelerated by ice-sheet buildup on Antarctica (Miller 1992).

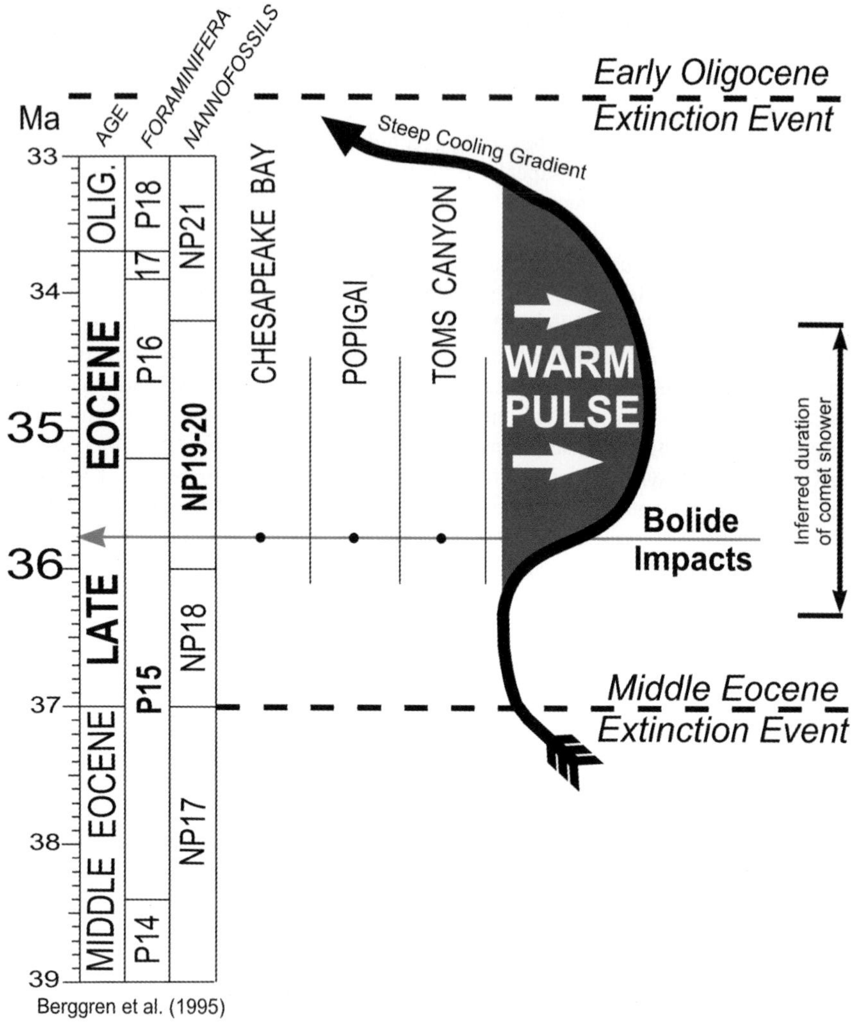

Fig. 13.13. Stratigraphic chart showing proposed relations between late Eocene pulse of warm climate, bolide impacts at Chesapeake Bay, Popigai, and Toms Canyon, and inferred comet shower (modified from Poag 1997b).

13.2.3 Implications of $\delta^{18}O$ Data

Poag (1997b) used Miller's (1992) isotopic data (benthic foraminifera) from Southern Ocean ODP Sites 689B and 703, along with other evidence of warm late Eocene paleoclimates (Haq and Lohmann 1976; Wolfe 1978, 1992; Wolfe and Poore 1982; Aubry 1992; Keller et al. 1992; Prothero 1994; Bestland et al. 1996), to support the speculation that the three known late Eocene bolide impacts produced overriding greenhouse effects and triggered a long-term (~2 myr) pulse of late Eocene climatic warmth. Miller (1992) used a few widely spaced samples to infer an interval of relatively negative $\delta^{18}O$ (warm bottom water) between the ejecta layer in ODP core 689B and the Eocene-Oligocene boundary at that site (Fig. 13.14B). Zachos et al. (2001) used additional $\delta^{18}O$ data to indicate a similar late Eocene warm pulse.

A broad, late Eocene warm pulse also can be inferred from the $\delta^{18}O$ records at numerous other localities in the Atlantic, Pacific, and Indian Oceans (e.g., Miller et al. 1985, 1987; Keigwin and Corliss 1986; Shackleton 1986; Steckler et al. 1999; Zachos et al. 2001; Fig. 13.15), which attests to its near-global effects.

The $\delta^{18}O$ record in the Chickahominy Formation comes from a thicker, more rapidly deposited upper Eocene sedimentary section than any other impactite site studied (63.4 m thick at Kiptopeke, 52.4 m at NASA Langley, vs 45 m at DSDP Site 612, 25 m at Bath Cliff, 14 m at ODP Site 689B, and 12 m at Massignano). The expanded record at Kiptopeke and NASA Langley shows a three-fold subdivision of the inferred late Eocene warm pulse (Figs. 13.14A,B 13.15). The oldest warm subdivision, W-1, is expressed by negative $\delta^{18}O$ values in Chron 16n.2n and the lower part of Chron 16r.1r. Subpulse W-1 correlates with the Chesapeake Bay, Toms Canyon, and Popigai impacts, but has not been identified at Massignano or Bath Cliff (insufficient data; Fig. 13.14A).

The next youngest Chickahominy warm subpulse, W-2, coincides with Chron 16n.1n and the lower two-thirds of Chron 15r (Figs. 13.14A,B 13.15). Subpulse W-2 correlates with the lower part of the lower warm interval at Massignano, but has not been identified at Bath Cliff (insufficient data; Fig. 13.14A).

The youngest Chickahominy warm subpulse, W-3, occupies the middle to upper part of Chron 13r, and correlates with the upper warm interval at Massignano and the single warm interval identified at Bath Cliff (Figs. 13.14A, 13.15).

Relatively positive $\delta^{18}O$ values in the Chickahominy record during Chron 15n and early Chron 13r indicate cooler water, and correlate with a similar interval at Bath Cliff. This interval, however, is difficult to recognize at Massignano (Figs. 13.14A, 13.15).

Fig. 13.14. (Next two pages) *A* stable isotope records in cores and outcrops containing late Eocene impact deposits (Chesapeake Bay; Kiptopeke corehole), DSDP Site 612 (deep-sea core), Massignano, Italy (outcrop), and Bath Cliff, Barbados (outcrop). Shaded intervals of $\delta^{18}O$ curve show three negative (warm) excursions (W-1, W-2, W-3); dotted pattern of $\delta^{13}C$ marks two negative excursions. *B* stable isotope records in cores and outcrops containing late Eocene impact deposits at Chesapeake Bay (Kiptopeke corehole), in Antarctic (deep-sea cores), and at ODP Site 689B (deep-sea core). Symbols as in Fig. 13.14A. Modified from Poag et al. (2003).

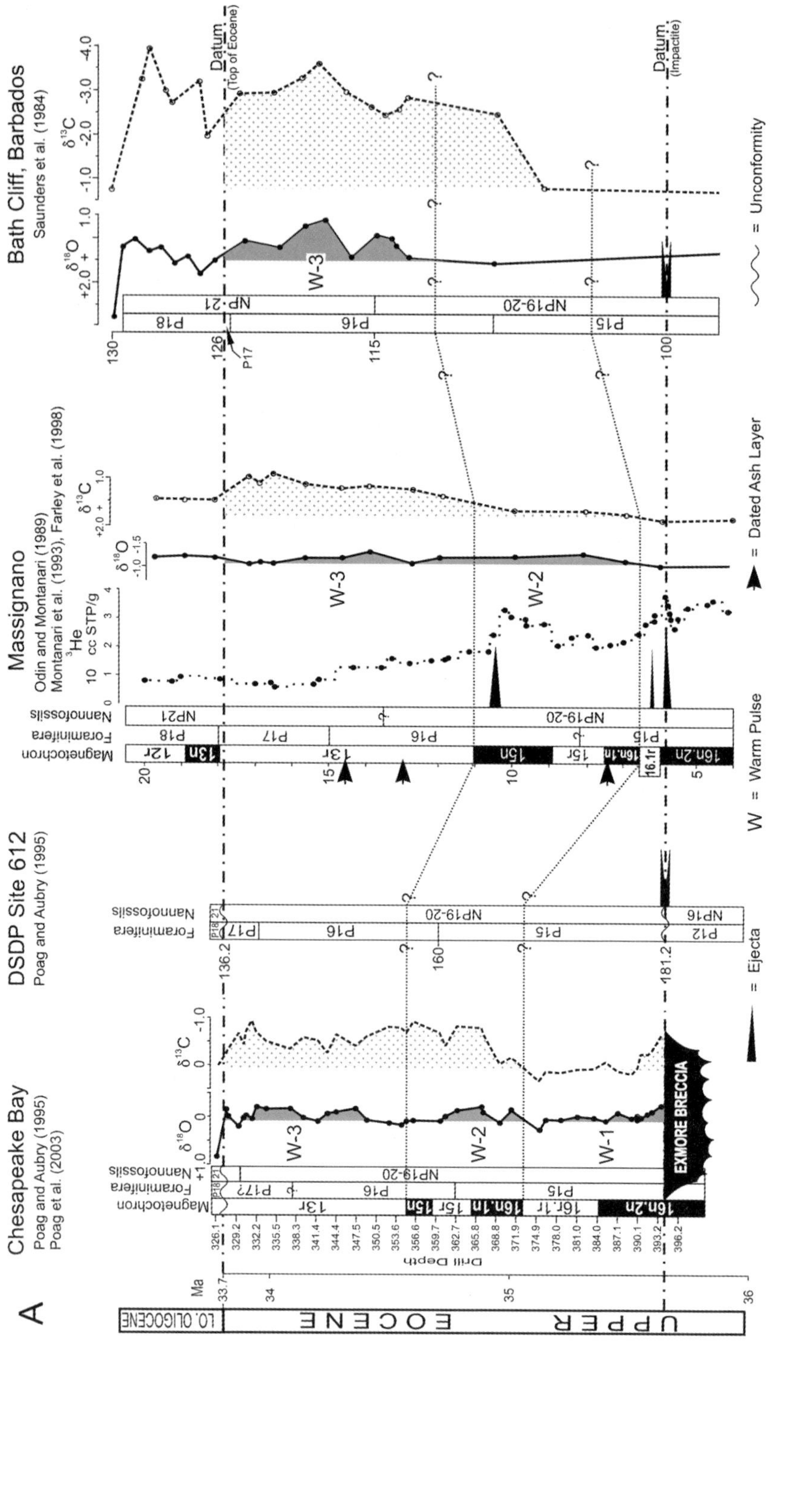

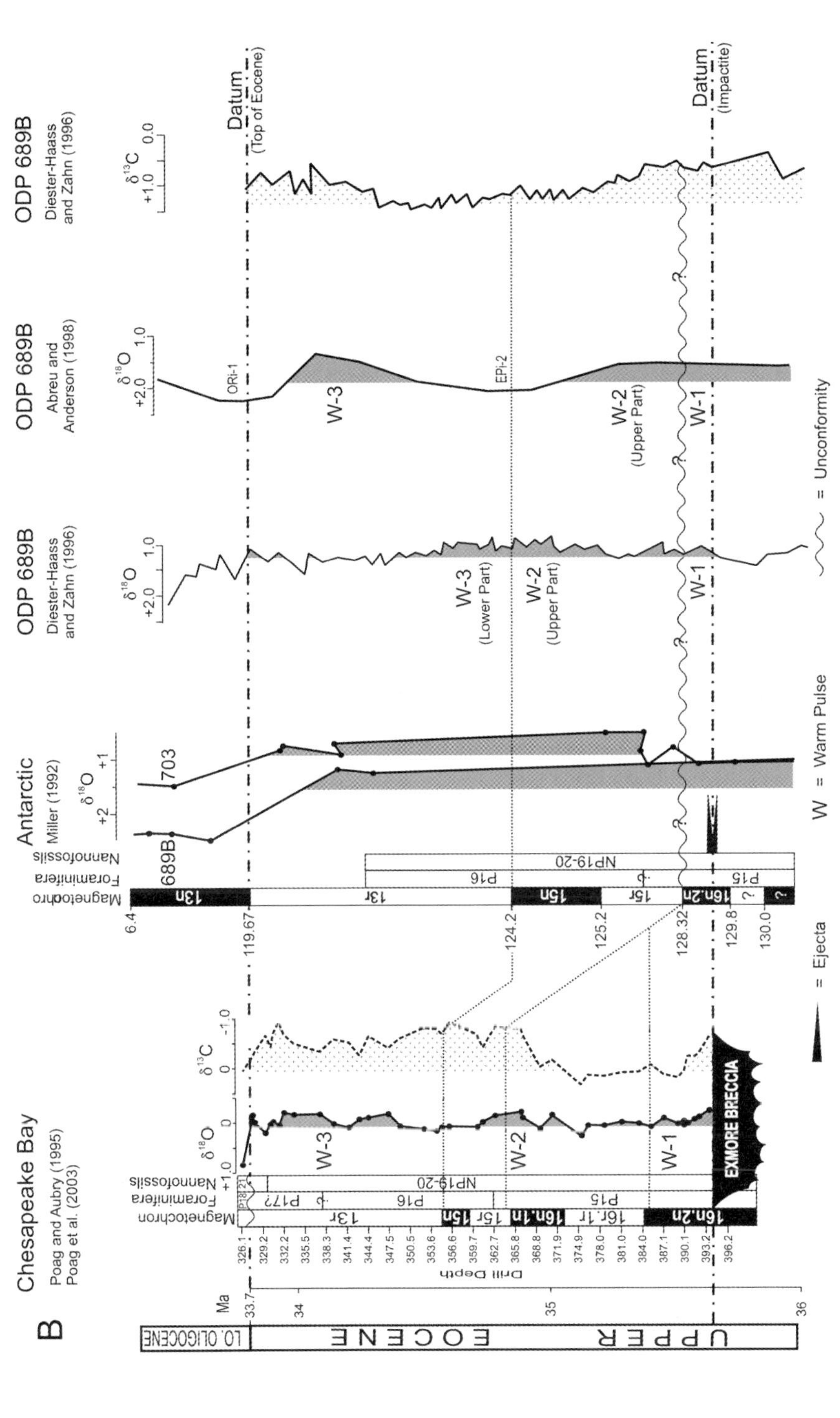
B
Chesapeake Bay
Poag and Aubry (1995)
Poag et al. (2003)
Antarctic
Miller (1992)
689B
703
ODP 689B
Diester-Haass and Zahn (1996)
ODP 689B
Abreu and Anderson (1998)
ODP 689B
Diester-Haass and Zahn (1996)
Datum (Top of Eocene)
Datum (Impactite)
W-1
W-2
W-3
W-3 (Lower Part)
W-2 (Upper Part)
EXMORE BRECCIA
Magnetochron
Foraminifera
Nannofossils
Drill Depth
Ma
LO. OLIGOCENE
UPPER EOCENE
= Ejecta
W = Warm Pulse
= Unconformity

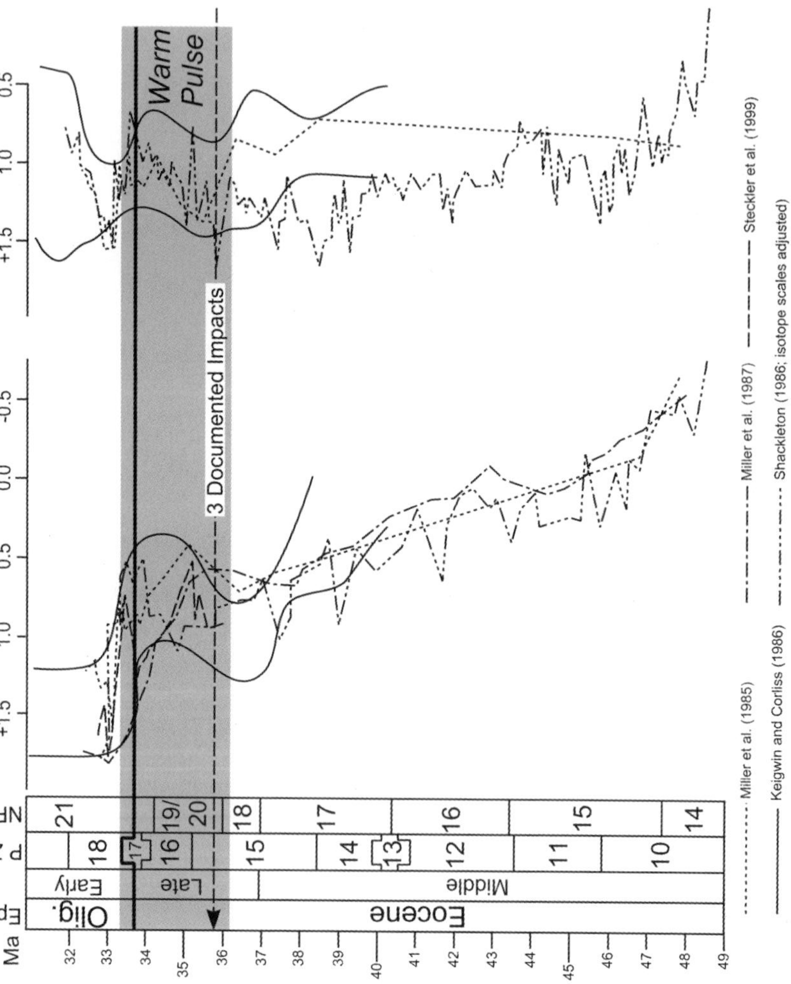
Ma
Epoch
P Zone
NP Zone
Olig.
Eocene
Early
Late
Middle
δ18O
δ13C
Warm Pulse
3 Documented Impacts
Miller et al. (1985)
Keigwin and Corliss (1986)
Miller et al. (1987)
Shackleton (1986; isotope scales adjusted)
Steckler et al. (1999)

Upon further consideration of the published record, we conclude that the presence of at least two to three late Eocene warm pulses also can be inferred from Antarctic cores (Fig. 13.14B). We found that the isotopic record at Site 689B is more complicated than indicated by Miller's (1992) coarse sampling. For example, in a subsequent isotopic study (benthic foraminifera) of the same core, Diester-Haass and Zahn (1996) sampled at 20-cm intervals, and produced a much different $\delta^{18}O$ profile from that of Miller (1992). Warm subpulses W-1 and W-2 appear to be represented (Fig. 13.14B), and perhaps the lower part of W-3. There is no clear separation of W-2 and the lower part of W-3, however, and the upper part of W-3 is only faintly represented.

Abreu and Anderson (1998) composited and smoothed the late Eocene $\delta^{18}O$ record in 19 Southern Ocean sites (including Site 689B) and presented a third paleoenvironmental interpretation. They recognized two relatively negative intervals (ice retreat; warmer bottom water) separated by a distinct positive excursion (ice advance; colder bottom water) during late Chron 15n and early Chron 13, which they designated as event EPi-2 (Fig. 13.14B). The $\delta^{18}O$ correlations between the Kiptopeke core, Massignano, and 689B suggest that the magnetostratigraphy at 689B is correct in the interval interpreted to represent Chrons 15r to 13r. If one assumes that the ejecta layer represents the upper part of Chron 16n.2n, however, then Chrons 16n.1n and 16r.1r appear not to be represented at Site 689B (Fig. 13.14B). We infer, then, that the stratigraphically lower warm subpulse of Abreu and Anderson (1998) represents, in reality, two warm subpulses separated by an unconformity. Thus, three warm pulses appear to be represented in the Antarctic region, although not all can be distinguished at every site studied.

In summary, there is increasing evidence that the late Eocene warm pulse proposed by Poag (1997b) is composed of at least three subpulses. A plot of the two $\delta^{18}O$ records from Chesapeake Bay and the ^{3}He record from Massignano on a time scale (Fig. 13.16) indicates an especially strong correlation between oxygen and ^{3}He abundance for subpulse W-1. The ^{3}He abundance decreases in the lower half of W-2, but reaches a second major peak in the upper half of W-2. The correlation is weakest for subpulse W-3. The ^{3}He abundance begins to decline in the lower half of W-3, and maintains near-baseline values for the rest of the Eocene. We have modified Poag's (1997b) pulse diagram to show the three subpulses of greenhouse warming (Fig. 13.16), which produced a particularly steep cooling gradient in the early Oligocene, as ice sheets built up rapidly on Antarctica (Miller, 1992). It is this steep gradient, which Poag (1997b) suggested might have triggered threshold events that initiated the early Oligocene mass extinction. McGhee (2001) termed this proposed extinction mechanism, the "lag-time multiple impacts hypothesis." The significance of this late Eocene warm pulse (a succession of at least three subpulses) has been overshadowed in all previous studies by the spectacular cooling event that followed in the early Oligocene.

In contrast to the isotopic studies, Vonhof et al. (2000) speculated (on the basis of an unusual assemblage of dinoflagellates) that a short interval of global cooling

Fig. 13.15. (Opposite page) Distribution of $\delta^{18}O$ and $\delta^{13}C$ isotope data for global ocean sites (deep-sea cores; from Poag et al. 2003).

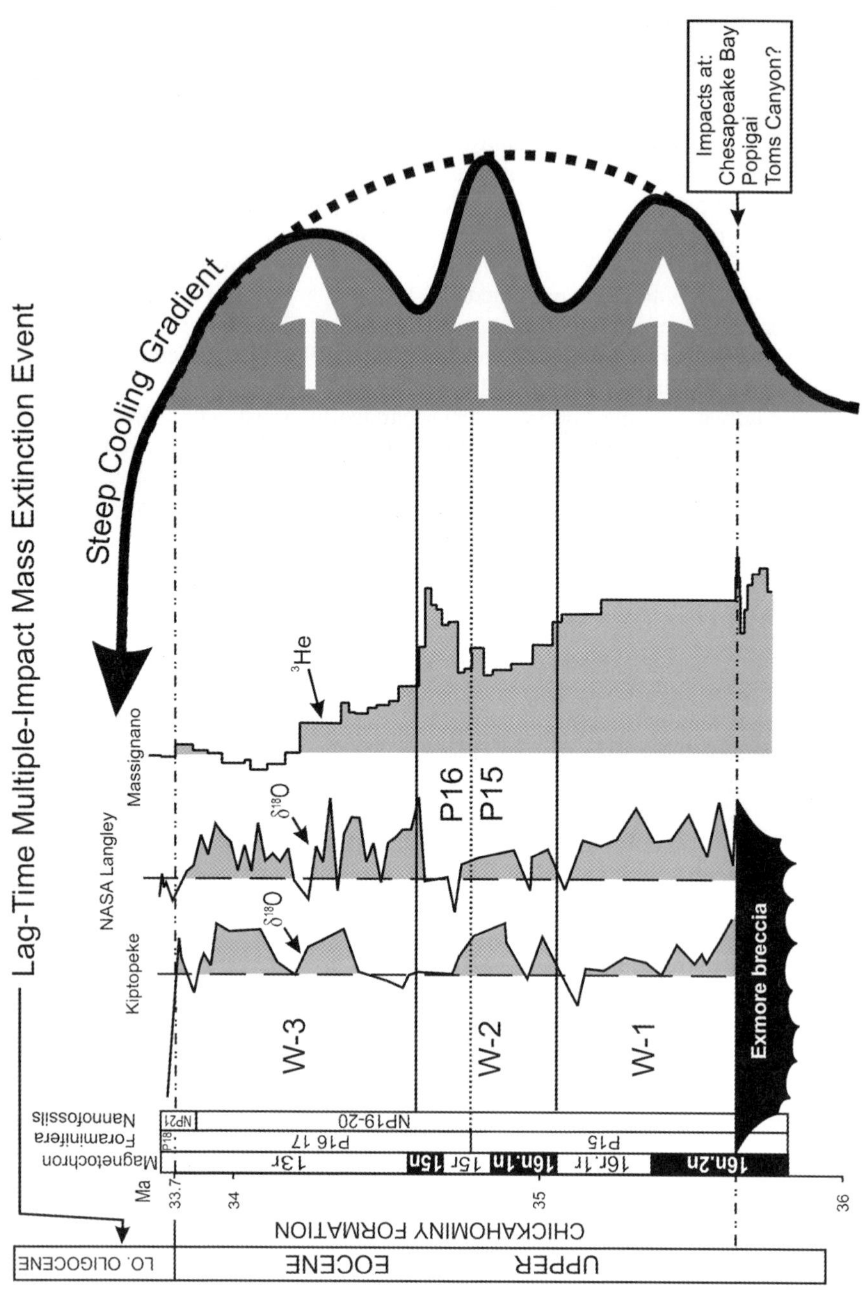
Lag-Time Multiple-Impact Mass Extinction Event
Steep Cooling Gradient
Impacts at:
Chesapeake Bay
Popigai
Toms Canyon?
Kiptopeke
NASA Langley
Massignano
$\delta^{18}O$
$\delta^{18}O$
^3He
P16
P15
W-3
W-2
W-1
Exmore breccia
Nannofossils
Foraminifera
Magnetochron
NP21
NP19-20
P18
P16 17
P15
13r
15n
15r
16n.1n
16r.1r
16n.2n
Ma
33.7
34
35
36
CHICKAHOMINY FORMATION
LO. OLIGOCENE
UPPER EOCENE

might have been associated with deposition of the impact ejecta at Massignano and at Site 689B (though there remains a strong probability that the late Eocene stratigraphic record at Site 689B is incomplete). Most recently, Spezzaferri et al. (2002) studied foraminiferal trends in a short section that includes the ejecta layer at Massignano (a 4-m section representing ~526 kyr). These authors found foraminiferal evidence of an ~20-kyr cool-water interval immediately following deposition of the late Eocene ejecta at that site. Above that 4-m interval, however, the foraminiferal assemblages, as well as the dinoflagellate assemblages, indicate a pulse of warm water that lasted 27–40 kyr, before cooler waters returned.

We speculate that this warm pulse may be correlative with W-1 at Kiptopeke. Unfortunately, Spezzaferri et al. (2002) did not provide microfossil paleoenvironmental interpretations for the remaining 15 m of upper Eocene section at Massignano.

13.2.4 Implications of $\delta^{13}C$ Data

The late Eocene $\delta^{13}C$ record has not received as much attention as the $\delta^{18}O$ record because of the latter's importance in defining the early Oligocene ice-advance signal and initial development of the oceanic cryosphere. The Chickahominy $\delta^{13}C$ curve contains easily distinguishable subdivisions, however, and holds promise for identifying fundamental late Eocene environmental changes and for improving correlations with other upper Eocene sections and related impactites (Figs. 13.14A,B, 13.15).

The most obvious characteristic of the late Eocene $\delta^{13}C$ record is the persistence of distinctly negative values in the upper two-thirds of the section at Chesapeake Bay, Massignano and Bath Cliff (encompassing Chron 16n.1n through most of Chron 13r; Fig. 13.14A). A coincident negative $\delta^{13}C$ excursion characterizes the late Eocene record at many additional deep-sea sites (often the same sites showing the negative $\delta^{18}O$ excursion; Figs. 13.14A,B). Zachos et al. (2001) noted this excursion, for example, in their synthesis of the Cenozoic marine stable isotope record, but made no attempt to explain its origin.

A second, older, less prominent, negative $\delta^{13}C$ excursion is present at Chesapeake Bay (Kiptopeke and NASA Langley coreholes; upper part of Chron 16n.2n), and can be seen in some (e.g., Shackleton 1986), but not all the other sites discussed herein (Figs. 13.14A,B).

Several notable negative excursions of $\delta^{13}C$ have been documented in other parts of the geologic record, also coincident, or nearly so, with negative $\delta^{18}O$ excursions. Perhaps the most dramatic is the Late Paleocene Thermal Maximum (Norris and Röhl 1999). The origins of such rapid $\delta^{13}C$ excursions have been attributed mainly to either a catastrophic release of biogenic methane from global dissociation of submarine gas hydrates (Bains et al. 1999; Norris and Röhl 1999)

Figure 13.16. (Opposite page) Chronostratigraphic chart showing correlations of $\delta^{18}O$ at Chesapeake Bay and ^3He at Massignano with proposed subpulses of greenhouse warmth for the late Eocene. At right is revision of Poag's (1997b) single-pulse diagram to show the three subpulses (white arrows).

or a massive injection of atmospheric CO_2 by sustained, intense, volcanic degassing (Zachos et al. 1993; Thomas and Shackleton 1996). The late Eocene negative excursion of $\delta^{13}C$, however, was long lasting (1.7 myr), and has been interpreted in paleoceanographic contexts to indicate an incursion of colder, more vigorous bottom waters (Miller et al. 1985) and increased paleoproductivity (Diester-Haass and Zahn 1996).

Theoretically, atmospheric disturbances due to dust and particulates from a single projectile the size of the Chesapeake Bay impactor (3–5 km) would last, at most, for only a few decades (Table 13.14; Toon et al. 1994). However, individual late Eocene impacts could be expected to also change atmospheric gas composition (e.g., the global carbon budget) by injecting massive volumes of CO_2 into the atmosphere from vaporized seawater (e.g., Chesapeake Bay and Toms Canyon) and carbonate-rich target rocks (e.g., Popigai – Masaitis et al. 1975; and Toms Canyon – Poag and Poppe 1998). A succession of surface impacts (accompanied, presumably, by many more atmospheric bursts) during a 2-myr-long comet shower, on the other hand, could plausibly have sustained atmospheric perturbations for most of the late Eocene.

14 Residual Effects of Chesapeake Bay Impact

The Chesapeake Bay impact crater is unusual in that more people live above or adjacent to it than at most other large impact craters. The only exception is the Vredefort multiring basin of South Africa, where some eight million people inhabit the deeply eroded crater region. The seven coastal cities of Virginia Beach, Norfolk, Portsmouth, Chesapeake, Suffolk, Hampton, and Newport News incorporate a population of two million people – far more than any other comparable area of Virginia – clustered in an arcuate corridor around the crater rim (Fig. 14.1). Among this urban sprawl is one of the U.S. East Coast's premier seaports (Norfolk), as well as numerous universities, hospitals, historical tourist attractions, and a particularly dense concentration of military bases, airfields, research installations, and weapons depots. Therefore, it is societally relevant to document and evaluate any lingering effects of that ancient impact upon the existing geology, geohydrology, geomorphology, and hydrography of southeastern Virginia.

14.1 Hypersaline Groundwater

Excavation of the Chesapeake Bay crater eliminated the antecedent freshwater aquifer system over an area twice the size of the state of Rhode Island (~6,400 km^2; Poag 1997a). In place of the previous system of vertically alternating aquifers and confining units, deposition of the Exmore breccia within the crater created a huge complex reservoir whose sediment volume (~4,300 km^3) is great enough to cover the states of Virginia and Maryland with a layer 30 m thick. Groundwater tests in the main body of the Exmore breccia and in the surrounding breccia apron have revealed that the breccia contains interstitial brine, with existent chlorinities as high as 25,700 mg/L inside the inner basin (at the Kiptopeke core site; Poag 1997a; Powars and Bruce 1999; Fig. 14.2). The presence of shallow brine aquifers around the lower part of Chesapeake Bay was known nearly 100 years ago (Sanford 1911, 1913), but early investigators did not know its detailed distribution and origin. The high-salinity problem was particularly highlighted by the studies of Cederstrom (1943, 1945a,b), who synthesized groundwater data for the Middle Neck and York-James Peninsulas. Additional analyses carried out during the 1980s and 1990s showed that isochlors for shallow aquifers near and inside the crater curve concentrically around the western rim of the buried crater, and that the chlorinity gradient steepens markedly at the crater rim (Larson 1981; Meisler 1981, 1989; Trapp 1992; Focazio et al. 1993; Poag 1997a; Powars and Bruce 1999; Fig. 14.2).

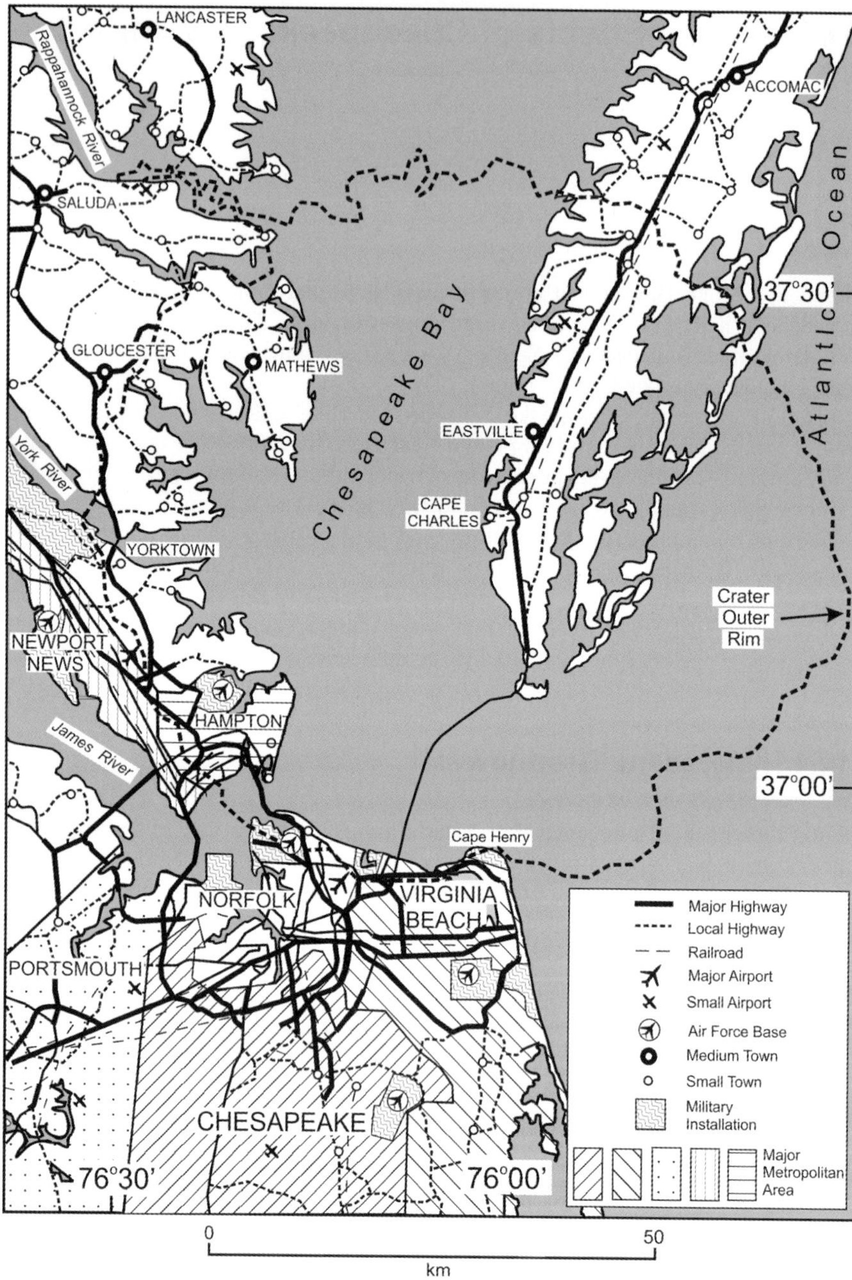

Fig. 14.1. Distribution of municipalities and other cultural features around Chesapeake Bay impact crater, which are still affected by presence of underlying crater.

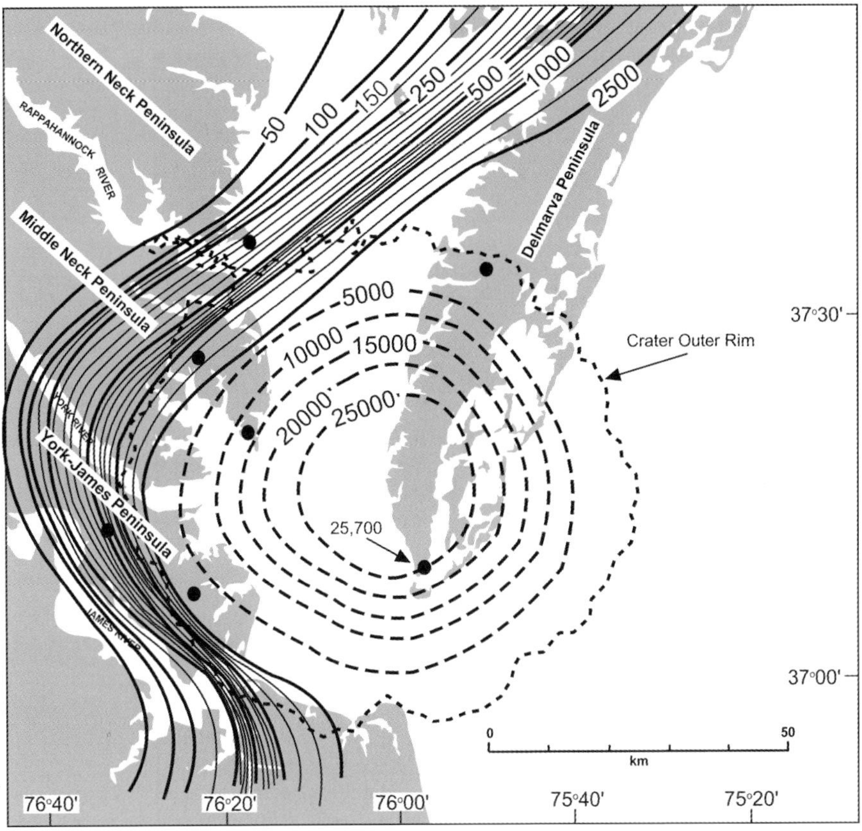

Fig. 14.2. Isopleths (lines of equal value) of groundwater chlorinity values for upper part of Exmore breccia documented for southeastern Virginia, showing relationship to geometry of Chesapeake Bay impact crater. Solid contours represent extrapolation of measured values; dashed isopleths represent interpolation of values between 2,500 mg/L contour and 25,700 mg/L value measured at Kiptopeke (irregular dashed line represents position of crater outer rim; solid dots represent locations of coreholes that sampled Exmore breccia *in situ.* Solid isopleths from Focazio et al. (1993); contour interval variable.

Possible explanations for the presence of the brine include the following: (1) short-distance migration of seawater from the adjacent Chesapeake Bay or Atlantic Ocean (Larson 1981; Leahy and Martin 1993); (2) long-distance migration of brines from Jurassic salt beds in the Baltimore Canyon trough, the largest postrift sedimentary basin of the U.S. Atlantic margin (Larson 1981; Meisler 1989); (3) incomplete flushing of ancient seawater (Cederstrom 1946; Powars and Bruce 1992; McFarland 2002); (4) concentration of dissolved salts through reverse chemical osmosis and membrane filtration, enhanced by rapid sedimentation (Larson 1981; Meisler 1989); and (5) presence of evaporite beds in the impact target rocks (Poag 1997a). Only the latter hypothesis explains why the brine is localized

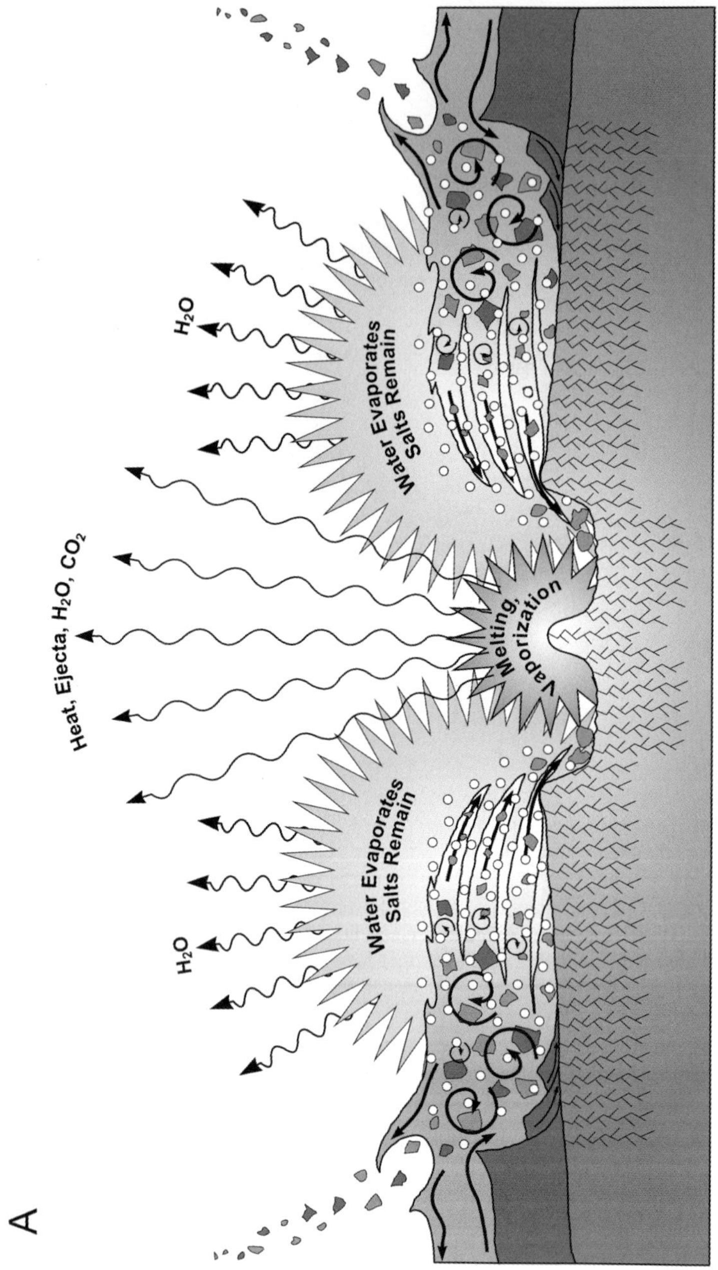
A
Heat, Ejecta, H_2O, CO_2
H_2O
Water Evaporates
Salts Remain
Melting,
Vaporization
H_2O
Water Evaporates
Salts Remain

B

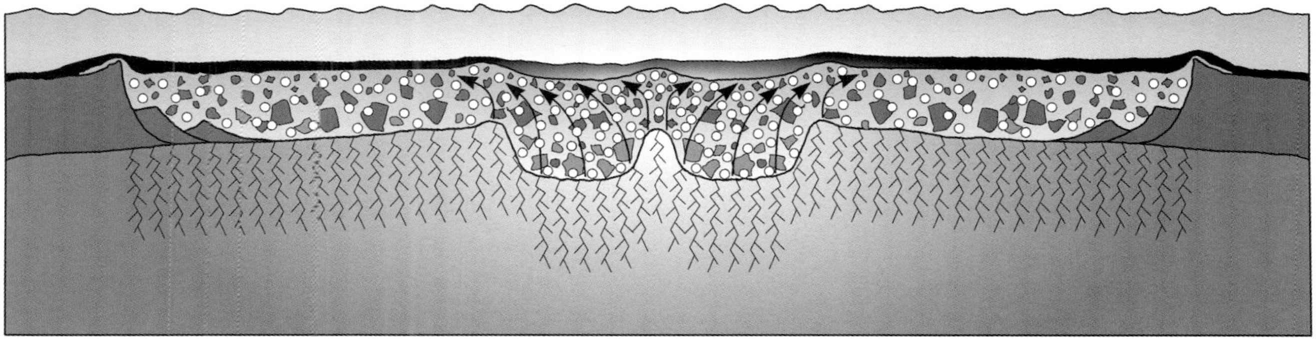

Fig. 14.3. Two-step cartoon of conceptual model for brine formation during Chesapeake Bay impact event. *A* (opposite page), marine waters around periphery of impact (outside vaporization and melting zone) evaporate, leaving behind salts to be incorporated within Exmore breccia during surgeback process. *B*, superheated basement within peak ring remains hot for thousands of years, continues to heat interstitial waters within the breccia, and thereby causes further concentration of salts (upper surface of breccia now sealed by clay-rich sediments of Chickahominy Formation).

in southeastern Virginia within the Exmore breccia, but this hypothesis remains unsupported by other evidence.

We support an hypothesis that depends upon flash evaporation of a huge volume of seawater at the time of impact followed by thousands of years in which the breccia was heated by the underlying shock-heated basement rocks (Sanford 2002; Figs. 14.3A,B). At the center of the bolide explosion, the seawater would have been completely vaporized, but as impact temperatures declined radially away from the center, a broad zone of seawater boil-off (evaporation) would surround the vaporization zone. Salts left behind would elevate salinity in the water-saturated sediments of the Exmore breccia. Preliminary hydrochemical studies indicate that the Exmore brine is depleted in ^{2}H and ^{18}O, a condition consistent with a late Eocene origin of the brine (McFarland 2002; Sanford 2002). If one assumes a porosity of ~20% in the breccia, and that it is a single saturated reservoir with a volume of 4,300 km^3, then it would require evaporation of ~430 km^3 of seawater to bring the entire breccia reservoir to ~1.5 times the salinity of normal seawater. The Exmore interstitial fluids were subsequently sealed in by postimpact resumption of clay deposition as the Chickahominy Formation began to accumulate. Calculations using Darcy's law (Sanford 2002) indicate flow rates within the breccia of <1 mm/yr, which would be insufficient to move the brine out of the crater within the 36 myr since the impact took place. Sanford (2002) also applied Fick's law to estimate that at typical molecular diffusion rates, >100 myr would be required for the brine solutes to escape from the Exmore breccia into shallower aquifers.

Initial studies of interstitial fluids in the Exmore breccia from the NASA Langley, North, and Bayside coreholes (McFarland and Bruce 2002; McFarland 2002) indicate a general increase in salinity toward the center of the crater, as previous investigations had predicted (Poag 1977a). It is quite clear, at this point, however, that the Exmore breccia is not a single fully saturated reservoir (McFarland 2002). The downhole geophysical logs and preliminary analyses of pore water squeezed from cores at NASA Langley, North, and Bayside, reveal that salinity values vary too randomly to be stratigraphically correlated from one corehole to the next.

Regardless of how it formed, the presence of the shallow Exmore brine reservoir has two significant implications for the citizens living around southern Chesapeake Bay (Fig. 14.4): (1) The shallow stratigraphic position of the top of the breccia and the existence of hundreds of faults within its overlying seal (the Chickahominy Formation), create a contamination hazard for overlying freshwater aquifers; and (2) The shallow depth to the top of the brine reservoir and its great thickness inside the crater limit the availability of freshwater aquifers to communities overlying the crater. It is especially limiting on the Delmarva Peninsula, which extends over the thickest and saltiest parts of the brine system. Here, post-impact sedimentary beds may be the only potential sources of fresh groundwater. If we assume, as current investigations indicate, that the brine occupies the entire lateral extent of the Exmore breccia, including the breccia apron outside the crater, then the brine covers an area the size of the state of Connecticut (~8,000 km^2). Our maps of the breccia distribution offer a means of predicting the lateral extent and potential thickness of the brine reservoirs in areas not yet drilled. This know-

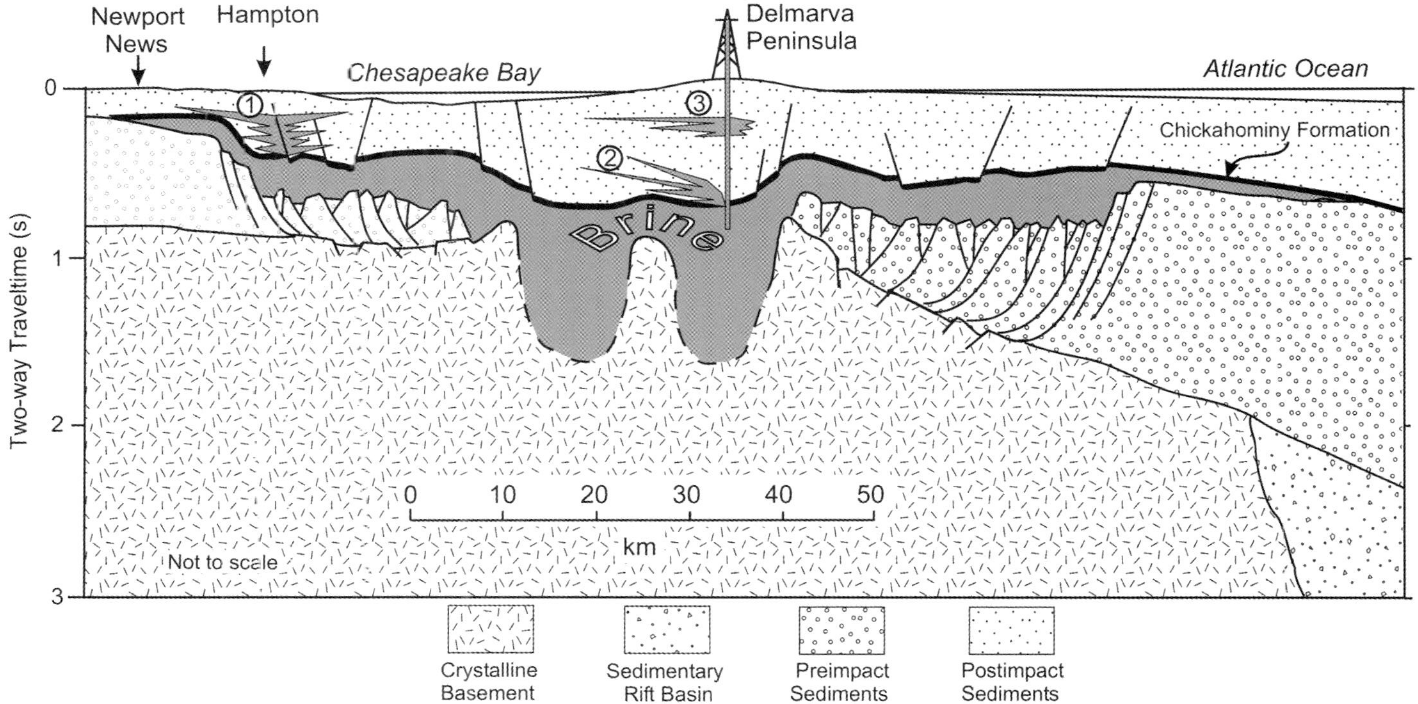

Fig. 14.4. Diagrammatic cross section of brine reservoir, showing three possible implications for groundwater resources: (1) Possible leakage of brine upward along compaction faults, where they breach the Chickahominy seal; (2) Possible upward leakage of brine around drill pipes if drill inadvertently penetrates breccia; (3) Possible leakage of brine from faulty joints or flaws in drill pipe or production pipe.

ledge, along with future geochemical and hydrogeological studies, will help guide prudent management of fresh groundwater resources in this region.

14.2 Near-Surface Compaction Faults

The presence of an extensive system of impact-related, near-surface, compaction faults also must be considered a potential geohazard for the region around lower Chesapeake Bay (Poag 1997a; Fig. 7.11). Most of the postimpact faults appear to have long histories of dip-slip movement, showing the typical upward decrease in throw attributable to growth faults (Fig. 7.12A,B,C). Most of these faults begin at the base of the Chickahominy Formation and extend upward into the Miocene and Pliocene sections, and some are traceable into the base of the Pleistocene units. High-resolution seismic reflection profiles published by Colman and Mixon (1988) show several faults that can be traced to within 15 m of the bay floor (Figs. 7.13A,B). At shallower depths, the Quaternary sediments are generally too soft and water-saturated to preserve the upper extent of fault planes. Thus, we are not sure that any of the faults actually cut the bay floor or surrounding land surfaces.

Nevertheless, the compaction faults have created a zone of structural weakness over the crater, which is more susceptible to earthquake displacement than areas outside the crater. Though historical earthquakes in southeastern Virginia have been rare and relatively mild (Sibol et al. 1996, 1997), Johnson et al. (1998) pointed out that surface projections of the epicenters of all four significant historical earthquakes in this region (1884, 1899, 1918, 1995) were near or inside the trace of the rim of the crater (Fig. 14.5).

14.3 Surface Expression of Crater

Though the Chesapeake Bay crater is now buried beneath 300–500 m of postimpact sedimentary formations, its presence is expressed on the modern land surface in the distribution of outcropping sedimentary units and in the configuration of certain topographic features (Poag 1997a). In terms of geologic expression, the geologic map of Virginia (Mixon et al. 1989) shows clearly that the sedimentary units inside the perimeter of the crater are of late Pleistocene and Holocene age, whereas most of those outside the crater perimeter are of middle Pleistocene and older ages (Fig. 14.6). Furthermore, on the Middle Neck and York-James Peninsulas, the contact separating the lower and middle Pleistocene units (Shirley, Windsor, and Chuckatuck Formations) from the upper Pleistocene units (Sedgefield, Lynnhaven, and Poquoson Members of the Tabb Formation) arches (convex to the west) around the western perimeter of the crater within approximately one kilometer of where the crater outer rim projects to the surface (Fig. 14.6). Outside the crater rim, in contrast, the contact between these older and younger units trends almost due north-south.

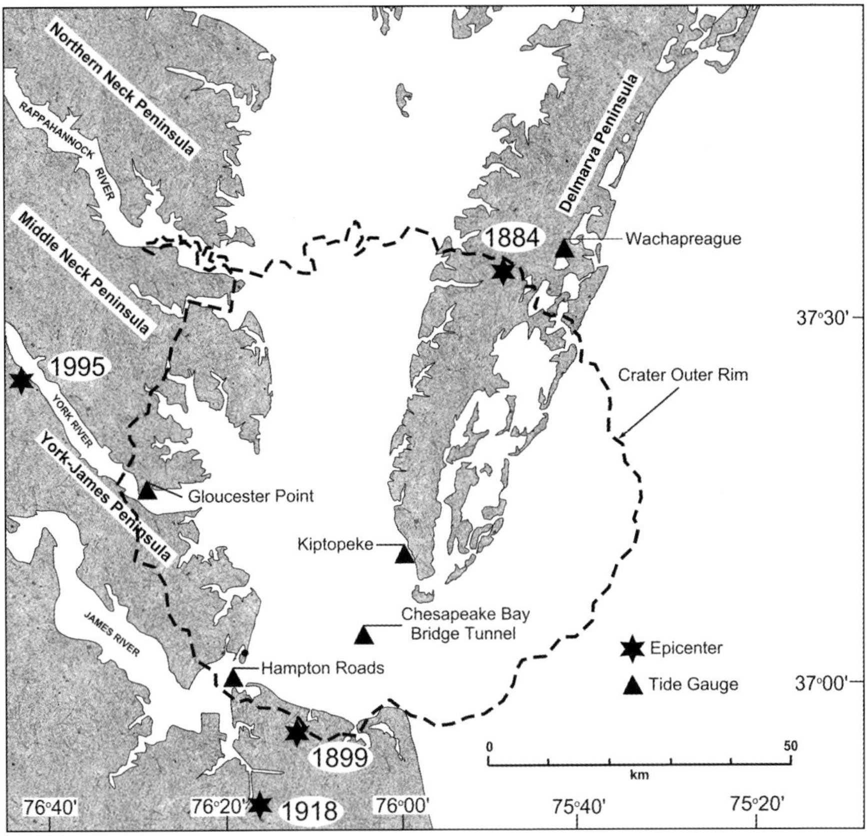

Fig. 14.5. Distribution of epicenters of historical earthquakes recorded in southeastern Virginia (year given), and location of selected tide gauges, showing geographic positions relative to Chesapeake Bay impact crater. Epicenter data from Johnson et al. (1998).

Similarly, on the Delmarva Peninsula, the contact separating middle Pleistocene sediments (Accomac Member of the Omar Formation) from upper Pleistocene sediments (Occohannock and Butlars Bluff Members of the Nassawadox Formation) marks the surface projection of the crater outer rim (Fig. 14.6). In addition, along the southern shore of Chesapeake Bay between Norfolk and Cape Henry, the surface projection of the outer rim is marked approximately by the contact between the Sedgefield Member of the Tabb Formation (upper Pleistocene) and Holocene shoreline sands.

The outer rim of the crater also is expressed in the coastal topography. Peebles (1984) showed that two Pleistocene shoreline erosional features on the Middle Neck and York-James Peninsulas, known as the Suffolk scarp (Harpersville scarp, in part, of Johnson 1976) and the Big Bethel scarp (Johnson 1972), mark the contact between older and younger Pleistocene units on the western side of the bay (Figs. 14.6, 14.7). Thus the positions of these scarps (with topographic relief as

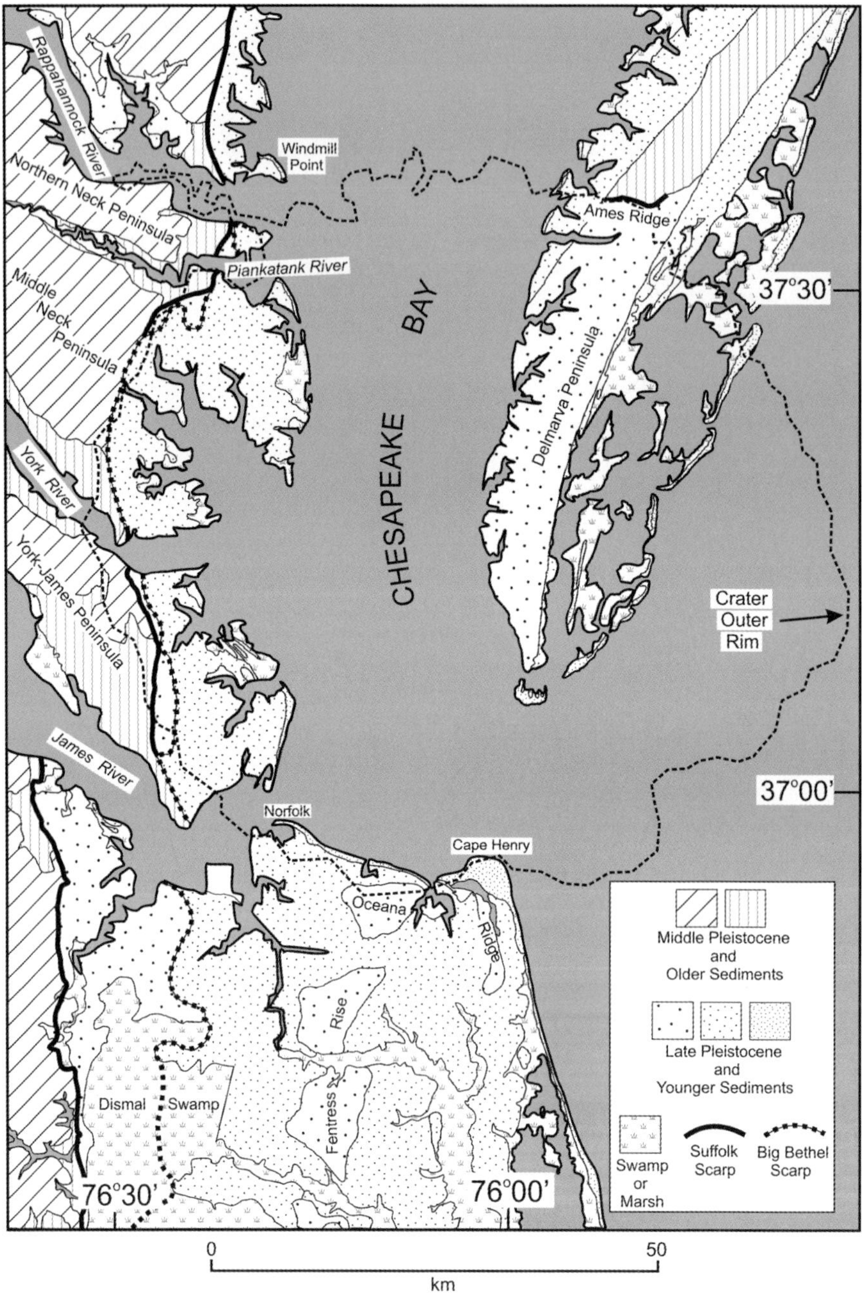

Fig. 14.6. Geologic map (modified from Mixon et al. 1989), showing relationship of surface trace of crater rim to: (1) distribution of formations and their boundary contacts, and; (2) location of prominent topographic features (scarps and Quaternary shorelines).

great as 22 m) also approximate the western boundary of the crater. The buried southern rim of the crater is marked at the surface by the arcuate southern shore of Chesapeake Bay and by the parallel slope of the Diamond Springs scarp (Powars and Bruce 1999).

On the Delmarva Peninsula, a landform comparable to the Suffolk scarp is known as the Ames Ridge; relief there is ~5 m (Fig. 14.6). Though these scarps have been interpreted to be erosional features (Peebles, 1984), we infer that long-term differential subsidence across the crater's outer rim played a major role in controlling the maximum westward extent of the two successive Pleistocene transgressions that produced these erosional shoreline features. Further evidence of this westward topographic limitation is the fact that on the Middle Neck and York-James Peninsulas, the spatial positions of the Suffolk and Big Bethel scarps are nearly coincident. That is, the two successive transgressions that eroded them reached the same westward limit in this region. In contrast, south of the James River, the Suffolk shoreline was able to transgress 22 km farther west than the Big Bethel shoreline (Fig. 14.6).

Between Norfolk and Cape Henry, the location of an additional landform approximates the southern edge of the crater rim. The northern boundary of the Oceana Ridge essentially mimics the curve of the crater's outer rim in this area (Fig. 14.6).

Fig. 14.7. Photograph of Suffolk scarp, a modern topographic expression of the outer rim of Chesapeake Bay impact crater. Topographic relief at this location is ~14 m (arrow). Location is on west side of Virginia State Highway 14, between towns of Gloucester and James Store, ~0.5 km north of Mt. Zion Church. See CD-ROM for color version of this figure.

14.4 Altered River Courses

The lower courses of the four largest modern rivers that cross the surface trace of the crater's outer rim bear signs that long-term differential subsidence also has determined the locations of their channels (Poag 1997a). The James, York, Piankatank, and Rappahannock Rivers all make sharp bends approximately at the outer rim of the crater (Figs. 14.6, 14.8). The upstream courses of the James and York head directly southeastward to the Atlantic, but in each case, the river turns abruptly to the northeast near the crater rim (an acute angle in the case of the James), and its channel heads toward the center of the crater. In the case of the Piankatank, its channel makes a right-angle turn to the northeast and parallels the outer rim for about 5 km before turning 90 degrees back to the southeast toward the crater's center.

The channel of the Rappahannock River makes its eastward turn toward the crater farther upstream, approximately 20 km from its mouth at Windmill Point. The location of this bend approximates the position of the northwesternmost extension of the crater rim, which we have termed the Rappahannock Canyon.

The lower course of the modern Susquehanna River also apparently has been altered by differential subsidence over the buried crater. The Susquehanna makes a sharp turn to the east where it exits Chesapeake Bay over the southern part of the crater (Fig. 14.8). This bend to the east is a reaction to the eastward tilt of the crater rim. Furthermore, subsurface studies by Colman and Mixon (1988) show that three successive older (buried) channels of the Pleistocene Susquehanna River also responded to differential subsidence in a similar fashion (Fig. 14.8). Each makes a distinct course change toward the lower eastern side of the crater upon entering the boundaries of the crater (Fig. 14.8).

Though differential subsidence across the outer rim may be the crater's primary influence on the river courses, it may not be the only influence thereon. The remarkable similarity between the position and geometry of the radial basement faults and the lower courses of the James, York, and Rappahannock Rivers, introduces the possibility that the faults also exert partial control on the courses of these rivers (Figs. 9.1, CD-ROM.5).

14.5 Relative Change of Sea Level

Long-term tide (sea-level) gauges in the Chesapeake Bay region indicate that during the last 70 years, the rate of relative sea-level rise in this area has been among the highest values (4 mm/yr) in the continental United States (Nerem et al. 1998; Table 14.1). This rate is twice the global average of 1.8 mm/yr (Douglas 1991; Douglas et al. 2001). Several authors have attributed about half of this high value to structural relaxation and rebound of a large regional periglacial bulge created by Wisconsinan ice sheets during the last glacial maximum in eastern North America (Davis and Mitrovica 1996; Nerem et al. 1998). In addition, as most of these gauges are near population centers, some of the relative rise may be the result of

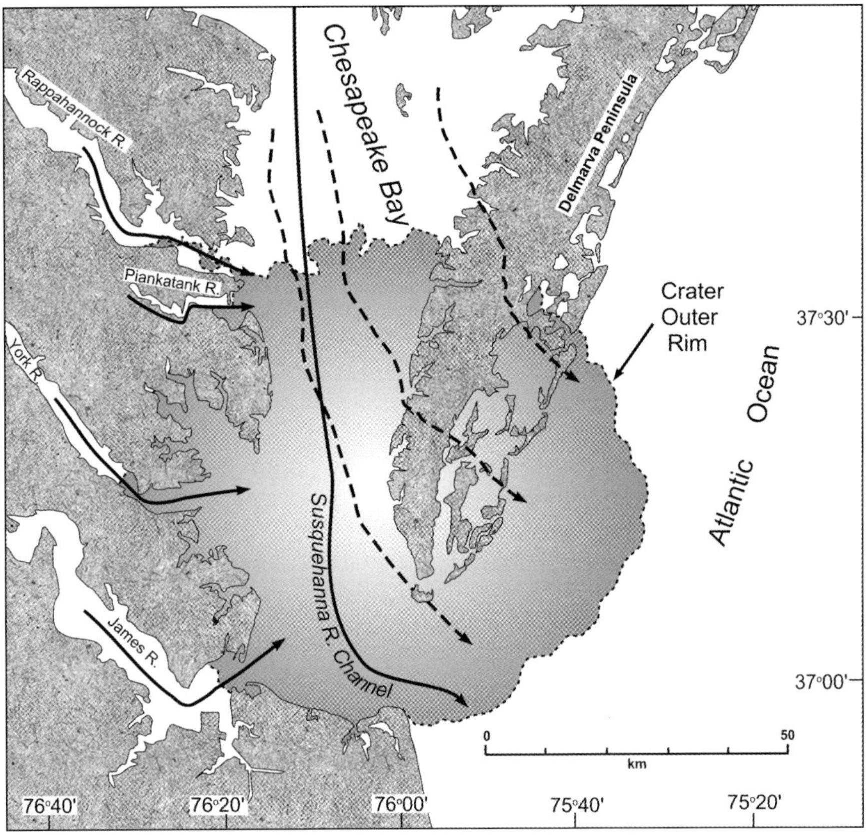

Fig. 14.8. Map showing distinctive bends in lower courses of modern (solid lines) and Pleistocene (dashed lines) river channels. Bends inferred to have been caused by differential subsidence of Exmore breccia within Chesapeake Bay crater.

land subsidence due to groundwater extraction (Gornitz and Seeber 1990; Holdal and Morrison 1974). Nevertheless, when Nerem et al. (1998) corrected for periglacial rebound, they found that sites inside and near the Chesapeake Bay crater still averaged 4 mm/yr relative sea-level rise (Table 14.1). Rates at all four sites inside the crater are higher than the global average, and at the Gloucester Point station, the value is as high as 6.7 mm/yr. We infer that differential subsidence of the Exmore breccia has contributed an additional component to these high relative sea-level rise values, by differentially lowering the ground surface and bay floor over the crater.

The rate of only 4 mm/yr in relative sea-level rise seems trivial at face value, but the gentle slope of the coastal plain surface around Chesapeake Bay makes this rate a major threat in the form of wetland loss. Slope values of 1:1000 are applicable to much of the bay margin, which would result in a 1-m horizontal loss of wetlands (transgression of the shoreline) for every 1-mm rise in relative sea level.

Thus it is important for planning future land use in the Chesapeake Bay region to understand fully all the causative components of relative sea-level rise and their interrelationships in this region.

Table 14.1. Trends of relative sea-level rise measured at tide gauges inside and near buried outer rim of Chesapeake Bay impact crater, 1930–1993. Data from Nerem et al. (1998).

Location	Latitude [degrees N]	Longitude [degrees W]	Relative sea-level trend [mm/yr]	Periglacial rebound trend [mm/yr]	Corrected sea-level trend [mm/yr]
Chesapeake Bay Bridge tunnel (inside crater)	37.000	76.003	7.5 ± 1.1	0.9	2.5
Gloucester Point, Va. (inside crater)	37.247	76.500	6.2 ± 2.2	0.8	6.7
Hampton Roads, Va. (inside crater)	36.927	76.006	4.1 ± 0.2	0.8	3.3
Kiptopeke, Va. (inside crater)	37.167	75.988	3.2 ± 0.3	1.1	2.1
Wachapreague, Va. (near crater)	37.607	75.687	6.7 ± 1.4	1.1	5.6
Average			5.54	0.9	4.0

15 Summary and Conclusions

We have presented a comprehensive geological and geophysical synthesis of the Chesapeake Bay crater, the largest known bolide impact structure in the United States. The structure, morphology, stratigraphy, and age of the crater and the nature and depositional history of the crater-fill rocks are documented by >2,000 km of seismic reflection profiles and >2,000 m of continuously cored and logged borehole sections (Chapter 1).

The Chesapeake Bay bolide struck the ~300-m-deep continental shelf of eastern North America ~35.78 Ma at a site now covered by the lower part of Chesapeake Bay, the low-lying peninsulas of southeastern Virginia, and the shallow marine waters of the inner Atlantic Continental Shelf. The impactor struck a three-layered target (Chapter 2). The upper layer comprised a column of seawater ~300 m deep; the middle layer encompassed 600–1000 m of poorly consolidated, water-saturated, sedimentary rocks (Early Cretaceous to late Eocene strata); the basal layer was a crystalline basement composed of metasedimentary and metaigneous rocks (Proterozoic to Paleozoic in age).

The bolide impact created a crater 85 km wide and 1.3–2.0 km deep (Chapter 4). Today the crater features a steep sedimentary outer-rim escarpment (300–1200 m high), a relatively flat, crystalline-floored annular trough (15–28 km wide), a crystalline peak ring (35–45 km wide; 40–300 m high), a deep, crystalline-floored inner basin (10–18 km wide; 1.3–2 km deep), and an irregular crystalline central peak (12 km wide; 200–600 m high), all attributes typical of other large complex craters found on Earth and its planetary neighbors. The Chesapeake Bay crater is filled with an orderly succession of inferred and documented synimpact deposits (Chapters 6, 11, 12). Filling the lower part of the inner basin is an inferred layer of fallback breccia, dekameters thick, presumably dominated by meter-to-dekameter-sized clasts of crystalline basement rocks. Such fallback breccia is known from the deep inner basins of other complex craters, but the inner basin at Chesapeake Bay has not yet been cored. One of the Chesapeake Bay coreholes, however, the Bayside corehole, contains ~20 m of matrix-supported breccia above the basement surface, whose abundant crystalline and sandstone lithoclasts appear to represent fragments of rocks from deep within the inner basin, and thus may constitute a modest section of fallback breccia.

The basal synimpact deposit in the annular trough at Chesapeake Bay is an ~300-m-thick layer of hectometer-to-kilometer-sized, displaced, sedimentary megablocks (slumpback lithofacies; Chapter 6). These megablocks are derived from the shock-generated collapse and basal fluidization of poorly consolidated, mainly Lower Cretaceous sediments that sloughed off the crater's outer rim.

Seismic reflection profiles indicate also that kilometer-sized megablocks of crystalline basement have slumped from the walls of the inner basin.

The next highest crater-fill deposit, 100–200 m thick, is surgeback breccia, a sediment-dominated, subaqueous deposit, which covers the entire crater, burying both the fallback and megablock deposits, as well as the peak ring and central peak. Surgeback breccia was formed by hydraulic erosion and gravity-driven collapse of the sedimentary crater rim and the tops of the displaced megablocks. An enormous hydraulic head developed as the 300-m-thick oceanic water column plunged back into the crater cavity.

Above the surgeback deposits is a sediment-dominated, matrix-supported, upward-fining, washback breccia, dekameters thick. The matrix is characteristically a greenish gray to nearly black, glauconite/quartz sand, containing stratigraphically mixed microfossils. This washback breccia not only covers the entire crater, but also extends as a breccia apron a few kilometers outside the crater rim. The washback breccia is a tsunamiite, created by runup and washback processes as impact-generated tsunami wave trains eroded and redistributed shock-weakened sediments from the inner continental shelf and coastal plain.

Both the surgeback and washback breccias contain granitoid clasts derived from the crystalline basement, which have been variably shock metamorphosed from <5 to > 45 (~60) GPa (Chapter 6). The geochemistry of these two breccia deposits indicates derivation from a sedimentary, upper crustal, post-Archean source, similar to the source inferred for the North American tektite strewn field (Chapter 6).

The antepenultimate synimpact crater-fill deposit is a clayey silt unit, a few meters thick, which displays evidence of multidirectional sediment flow during deposition. This is a flowin unit, attributable to hypercanes that moved across the continental shelf and triggered successions of small debris flows from the crater rim.

The final synimpact crater-fill deposit is a thin (1–5 cm thick), clayey silt, which contains evidence of impact-derived microspherules (Chapter 6). The 1-mm cavities that originally contained the microspherules are preserved in distinctive pyrite lattices, from which glass-derivative clay may have been inadvertently washed away during routine sample preparation. We infer that this microspherule layer is a fallout product of the condensing impact vapor plume.

Outside the primary crater, seismic profiles reveal 23 small structures that appear to be secondary craters (3–6-km diameters), because they display characteristic downfaulted sedimentary rims, raised lips, and chaotic crater-fill reflections (Chapter 5). Though no recent coreholes have penetrated any of the secondary craters, there is evidence from older boreholes that at least one of the possible secondaries contains crater-fill deposits lithologically equivalent to the Exmore breccia.

Perhaps the most dramatic aspect of the impact process is the enormous speed with which it took place. Computer simulations of the impact indicate that the 85 x 1.3 km excavation (4,300 km^3) was created and refilled within a geological blink-of-the-eye (a few minutes to hours; Chapter 12).

The age of the Chesapeake Bay impact structure has been determined indirectly by biochronological and magnetochronological studies of sediments (the Chickahominy Formation) directly overlying the crater-fill (Chapter 7). Microfossil biochronology indicates that the Chesapeake Bay impact took place during a 0.8-myr interval in which the top of planktonic foraminiferal biochron P15 (upper boundary at 35.2 Ma) overlaps the base of calcareous nannofossil biochron NP19-20 (lower boundary at 36.0 Ma; Chapter 8). A similar crater age (35.2 ±0.3 to 35.5 ±0.3 Ma) has been derived from radiometric analyses ($^{40}Ar/^{39}Ar$) of distal ejecta from the North American tektite strewn field (DSDP Site 612 and Bath Cliff, Barbados), currently thought to be a product of the Chesapeake Bay impact. Extrapolation of a magnetochronologically-derived sediment-accumulation rate from the lower part of the Chickahominy Formation at the Kiptopeke site refines the impact age to 35.78 Ma. This age for the Chesapeake Bay impact is statistically indistinguishable from the 35.7 ±0.4 Ma radiometric age of the Popigai crater in Northern Siberia and the 35.7 ±0.4 age of the distal ejecta that crops out near Massignano, Italy. The stratigraphic separation of microkrystite ejecta (derived from Popigai) from microtektite ejecta (derived from Chesapeake Bay) in deep-sea cores (Atlantic Ocean and Caribbean Sea), however, indicates that the Chesapeake Bay impact is younger than that of Popigai by 10–20 kyr.

The Chesapeake Bay crater and its sedimentary fill are buried now by 300–500 m of postimpact (late Eocene to Holocene) siliciclastic, mainly marine, sediments (Chapters 2, 7, 13). The initial postimpact deposit is a 20-cm-thick, laminated silt layer, which contains no indigenous biota, and represents the first ~0–3 kyr of lifeless marine deposition following the bolide impact (Chapter 7). Thereafter, normal marine deposition resumed and formed the Chickahominy Formation, a sandy-to-silty, massive-to-laminated, glauconitic, micaceous, highly microfossiliferous marine clay, of relatively deep-water origin (~300 m paleodepth). The Chickahominy represents the final 2.1 myr of Eocene sediment accumulation over the crater. Three distinct episodes of sedimentation (distinguished by different rates of accumulation) can be documented within the Chickahominy clay (Chapter 13). These three depositional intervals correspond roughly to three cycles of low-to-high species richness among the benthic foraminiferal community. Culmination of the first cycle represents full recovery of the benthic foraminiferal community ~36 kyr following the bolide impact. Superimposed on these three cycles of species richness are five biotic subzones defined by characteristic associations of benthic foraminiferal species. As a whole, the Chickahominy benthic foraminifera record a succession of paleoenvironments characterized by oxygen deficiency and an abundant supply of organic detritus at the seafloor and in shallow interstitial waters. Phytodetrital feeders were prominent members of this benthic community, especially in the upper part of the formation.

Though no immediate global loss of marine or terrestrial species comparable to that of the K-T mass extinctions arose from the Chesapeake Bay impact, there is evidence that long-term climatic changes may have resulted from it. The climatic perturbations, in turn, may have triggered a major extinction event in the early Oligocene, ~2 myr after the Chesapeake Bay impact (Chapter 13). Stable isotope records ($\delta^{18}O$ and $\delta^{13}C$) derived from the tests of the benthic foraminifer *Cibicidoides pippeni* indicate that postimpact climate at the impact site was punctuated by at least three warm pulses. The final pulse was accompanied by a notable negative excursion in $\delta^{13}C$ values. The $\delta^{18}O$ results are best understood in the context of a late Eocene comet shower, which produced unusually high concentrations of extraterrestrial ^{3}He at the late Eocene outcrop near Massignano, Italy, which contains 35.7-myr-old impact ejecta. We infer that a succession of impacts during the comet shower (including those at Chesapeake Bay and Popigai) produced the climatic warming indicated by the $\delta^{18}O$ record.

Though buried for the last ~36 myr, the Chesapeake Bay crater and its related deposits still have important consequences for the citizens of southeastern Virginia (Chapter 14). The Exmore breccia subsided differentially as it compacted under a load of postimpact sediments, and this subsidence, in turn, produced a vast network of near-surface faults. The pervasive fault systems have destabilized the bayfloor, seafloor, and low-lying wetlands above and near the crater, contributing to rapid rates of relative sea-level rise that characterize the Chesapeake Bay region.

The most important modern consequence of the ancient impact may be the presence of high-salinity groundwater (derived from flash-evaporation of huge volumes of seawater during the bolide impact) at shallow depths within the Exmore breccia. This brine limits the quality and availability of potable shallow groundwater for more than two million citizens in the rapidly growing urban corridor surrounding lower Chesapeake Bay (Chapter 14).

Comparison of the Chesapeake Bay crater and its associated deposits with other complex craters of comparable submarine origin reveals some significant similarities (Chapters 10,11,12). On the other hand, each known crater has distinct characteristics that set it apart from all the rest. Our analyses lead us to emphasize the following principal points: (1) The succession of marine modification processes associated with surgeback, washback, and flowin depositional regimes appear to be unique to submarine impacts on shallow continental margins. These processes are responsible for the unusually thick body of sediment-clast breccia that fills the Chesapeake Bay crater and several other submarine craters. Surgeback processes may also operate in deeper, open-ocean settings, but washback and flowin processes require a nearby, easily erodable, land surface or shallow continental shelf. (2) The density differential between crystalline basement rocks and overlying sedimentary rocks is important in constraining both the excavation and modification processes of submarine crater development. This differential appears to account for the great structural and morphological disparity in between the Chesapeake Bay and Mjølnir craters, for example. (3) This density differential depends in large part on the degree of water saturation and lithification of the sedimentary target rocks. In the case of Chesapeake Bay, the sedimentary target rocks are primarily loosely consolidated quartz sands and silts, most of which today are (and presumably were in the late Eocene) important freshwater or saline aquifers. Their weak consolidation must have facilitated acoustic fluidization of the basal target sediments by the impact shock wave, thereby promoting widespread sliding and slumping of megablocks along a basement décollement, without producing pervasive brittle deformation features, such as faults, which ordinarily are expected in décollement zones. This displacement of megablocks significantly widened the crater. (4) Though there is scattered evidence of an upturned lip on the outer rim of the Chesapeake Bay crater, the lip is insubstantial compared to the lips of well-preserved subaerial craters on Earth and other planetary bodies. This appears to be, in part, due to intense modification of the outer rim by surgeback and washback processes. The lack of a well-defined outer-rim lip appears to be common to all known submarine impact craters.

The principal structural, morphological, depositional, and paleoenvironmental aspects of the Chesapeake Bay impact crater are now thoroughly documented by borehole, seismic-reflection, and gravimetric data. Acquisition and analysis of new cores and geophysical surveys continue at Chesapeake Bay, however, and, undoubtedly, will help to refine and revise some of the interpretations we have presented. Several critical questions remain to be answered, especially regarding the central features of the structure: (1) What is the nature (composition, shock history) of the crystalline basement that comprises the peak ring, central peak, and floor of the inner basin? (2) Does fallout breccia dominated by large crystalline clasts occupy the floor of the inner basin? (3) Are large melt bodies or melt sheets associated with this structure? (4) What is the radiometric age of the crater? (5) Is there a breach in the southeastern margin of the peak ring, as suggested by the pattern of gravity anomalies? (6) What is the configuration of the basement surface in the eastern sector of the crater? and (7) Are displaced sedimentary megablocks, which are common to the western sector, also present in the eastern sector?

Questions 1–4 can best be answered by obtaining cores from the central features of the crater. The cores can be obtained from a series of deep coreholes (700–2,000 m deep) drilled on the Delmarva Peninsula near the town of Cape Charles, Virginia. Questions 5–7 require additional deep seismic reflection surveys across the southern part of the Delmarva Peninsula and the inner continental shelf east of Delmarva. The search for these answers will provide stimulating challenges for a new generation of planetary geologists. The answers themselves will contribute significantly to understanding the essential role of bolide impacts in the history of our solar system and their implications for its living species.

Appendix

Data Collection, Processing, Analysis, and Interpretation

Seismic Reflection Surveys

We include here only the most recent three marine seismic reflection surveys, which provided digital data, and upon which we relied most heavily for detailed interpretation of the structure and morphology of the Chesapeake Bay impact crater.

Seaward Explorer

The *Seaward Explorer* collected 875 km of seismic reflection data over the Chesapeake Bay impact crater during the interval of April 21–30, 1996. The primary imaging system for the crater survey consisted of single-channel reflection profiling using a GI gun (Generator-Injector airgun). This system provided reflections from the basement surface (at about 1.0 s two-way travel time) and overlying sediments. Components for this system were: (1) GI gun (Seismic Systems Inc.) configured in the harmonic mode with generator and injector chambers both set to 45 in^3. The depth of the gun beneath the surface was fixed at 2.5 m. Firing pressure was maintained at 2000 psi; (2) Teledyne 2-channel streamer reconfigured for one channel. The single active section, 50 m long, was attached to a weighted dead section (25 m long), which, in turn, was attached to a feathered lead-in cable. The front of the dead section was approximately even with the gun. The center of the active trace was approximately 50 m behind the gun.

Differential GPS navigation was recorded every 10 s. An Ashtech GPS XII receiver accepted the GPS information and a Megapule Accufix D100 receiver provided the differential information. We used a USGS PC-based logger, which also provided real-time line and steering information for the ship's laboratory and a remote display for the ship's bridge.

We acquired digital seismic data for the GI-gun profiles within Chesapeake Bay using USGS Masscomp hardware and software. Time of each shot was logged from a True Time GPS Model XL-DC clock. A Digital Delay Generator (Model 7010) provided the master trigger to fire the gun. Recording parameters were: firing rate 13 s; filters 20–500 Hz; sample rate 0.5 ms; record length 2s.

The *Seaward Explorer* data were processed by Myung Lee (USGS, Denver, Colorado), in the following steps: (1) Trace DC removal; (2) AGC 200 ms; (3) Direct arrival removal; (4) 2-nd zero-crossing deconvolution filter with length of operator 240 ms; (5) Wavelet processing; (6) F-X deconvolution; (7) Plot. Filtered by a 12–18–250–300 Hz band-pass filter.

R/V Maurice Ewing

The *R/V Maurice Ewing* collected 220 km of 2-channel seismic reflection data over the crater on October 15–16, 1998. The seismic source for this crater survey consisted of a tuned array of six airguns with chamber capacities of 80, 120, 145, 200, 305, and 500 in^3, at 2000 psi. The airgun array was towed 35–45 m behind the ship at a depth of 6 $\pm$2 m. The hydrophone streamer was 263 m long, with two 50-m active sections, and was towed 100 m behind the ship at a depth of 5 m. Shooting interval was 12 seconds.

Continuous marine gravity data were collected simultaneously over the same tracklines used for the seismic survey. The recording instrument was a Bell gravity meter, with a logging interval of 1 s. The shipboard gravity meter was tied to a reference base station by using a portable Lacoste Model G#70 gravity meter. The base station was a brass plate located at dockside in Norfolk, Virginia, where the ship was docked (at the NOAA Atlantic Marine Center, 439 West York Street, Dock 3, 150 ft south of the SW corner of Warehouse Bollard 2; lat 36° 51.193 N, long 76° 17.896 W). The drift in the Bell gravity meter was determined from the difference between the precruise station tie and the postcruise station tie.

Texaco Data

The 310 km of seismic data kindly provided to the USGS by Texaco, Inc., were collected in 1986 by Teledyne Exploration Co., using a DFS IV recording instrument and an array of 6 air guns (984 in^3 at 2000 psi) as the energy source. The data were filtered at 8–128 Hz, and recorded on 96 channels and processed digitally to produce 48-fold, migrated CDP profiles. Record length was 6 s, sample rate was 2 ms, and spread array was 125–1312 m. Shot point interval was 25 m, with a group interval of 12 m.

Paleomagnetics

We took 32 samples for paleomagnetic study from the 66-m section of the Chickahominy Formation cored continuously at a site near Kiptopeke, Virginia (Poag et al. 1994; Poag and Aubry 1995; Poag 1997a). We sampled the Kiptopeke core at approximately 2-m intervals, representing approximately every 25–38 kyr, with the following results:

Sample Number	Sample Depth [m]	Normal Polarity	Reversed Polarity
1	330.1	yes	
2	332.0	?	?
3	333.8		yes
4	335.7		yes
5	338.6		yes
6	339.9		yes
7	344.5		yes
8	346.3		yes
9	348.1		yes
10	351.7	yes	
11	353.2		yes
12	354.6		yes
13	356.4	yes	
14	358.2	yes	
15	360.1		yes
16	362.7		yes
17	364.3		yes
18	366.8	yes	
19	368.2	yes	
20	370.0		yes
21	371.9	yes	
22	373.7		yes
23	375.4		yes
24	378.3	yes	
25	379.8		yes
26	381.7		yes
27	383.6		yes
28	385.3	yes	
29	387.2	yes	
30	388.6	yes	
31	390.5	yes	
32	392.1		yes

The core was oriented only with respect to top and bottom, so the geomagnetic polarity of each sample is based on magnetic inclination. The natural remanent magnetization (NRM) of each sample was measured with a superconducting magnetometer housed in a magnetically sheltered room (Ed Mankinen, USGS, Menlo Park, California). The NRM was very weak, with a geometric mean intensity of 0.055 mA2/kg for 31 samples. Progressive alternating-field (AF) demagnetization experiments on a few selected samples indicated that this method would be ineffective for recovering the primary remanence within this core. Remaining samples, therefore, were subjected to progressive thermal demagnetization in a partial vacuum (50 μm) to remove any gases emitted during the experiments and to inhibit chemical alteration. Ovens used for the heating experiments were enclosed in a Rubens coil array, which maintained an ambient field of approximately 40 nT. All samples began to chemically alter after heating to temperatures higher than 350°C. Isothermal remanent magnetization (IRM) curves were generated for representative samples by applying direct magnetic fields varying from 0 to 700 nT.

Normalized magnetic intensities were determined for each sample during thermal demagnetization, and geometric means of the entire sample population were calculated at several temperature steps. The range of 1 standard deviation about each mean is quite large, because of an increased amount of experimental noise due to the very weak magnetic intensity, and because of varying thermal properties of the individual samples. An initial rise in intensity often occurred at low temperatures due to the removal of an anti-directed secondary component of magnetization. Some samples generally decreased in intensity until a sharp rise occurred at about 200°C and formed a peak between 200°C and 260°C, which is characteristic of Fe_9S_{10} (Schwarz 1973). Finally, chemical alteration of the samples occurred at different times during the heating experiments.

Despite these complicating factors, the general pattern of behavior for the sample population as a whole is readily apparent. There is an almost monotonic decrease in intensity, characteristic of Fe_7S_8 (Schwarz 1973), with maximum unblocking temperatures reached in the experiments occurring between 300°and 350°C. These unblocking temperatures are characteristic of pyrrhotite, which has a Curie temperature of about 325°C. The temperature-step curve also provides a hint of a small intensity rise at 200°C, indicating that both Fe_7S_8 and Fe_9S_{10} must be present in a significant number of samples. The very strong intensity increase beginning above 350°C also is characteristic as the sulfides begin to oxidize and form magnetite. The oxidation seen here is an actual breakdown of the remanence carriers, because, although pyrite also oxides to form magnetite, it does so at a higher temperature (450°C; Nguyen Tkhi Kim Tkhoa and Pecherskiy 1984) than was reached in our experiments.

We conclude that pyrrhotite is the most likely carrier of remanence in the Chickahominy Formation where sampled in the Kiptopeke corehole. Resistance of the samples to AF demagnetization further indicates that this pyrrhotite must be fine-grained (Clark 1984; Rochette et al. 1990). Because pyrrhotite is capable of carrying a strong remanent magnetization comparable to that of magnetite, the weak intensities encountered in the Chickahominy samples must mean that the concentration of pyrrhotite is very low. Perhaps conditions existing within the crater basin were more favorable for the formation of pyrite rather than pyrrhotite. Alternatively, perhaps the original iron sulfide was greigite, which has long since altered to generally less magnetic materials, including pyrite and marcasite, along with small amounts of pyrrhotite and(or) magnetite, depending on redox conditions (Krs et al. 1992). A significant lag time in the acquisition of remanent magnetization is not likely, regardless of whether pyrrhotite formed as one of the early authigenic minerals, or resulted from breakdown of greigite, because both usually occur in the early stages of diagenesis.

Because of the low signal-to-noise ratio in these samples, it often was difficult to ascertain whether or not magnetization had reached a stable end point, or to allow us to rely entirely upon principal component analysis (Kirschvink 1980). Directional trends during the experiments, however, generally were unambiguous and allowed us to determine the geomagnetic polarity with some degree of confidence. Samples in which the magnetic direction was inclined below the horizontal are considered to be of normal polarity; those with inclinations above the horizontal are interpreted to be of reversed polarity. Because of alteration occurring at relatively low temperatures, it is possible that all secondary components of magnetization were not removed. Interpretation of one-sample reversals, therefore, should be avoided, although some of these may represent actual geomagnetic field behavior (e.g., the "tiny wiggles" or "cryptochrons" of Cande and Kent 1992a,b). Misinterpretation of a single sample's magnetic polarity also may result from incorrect orientation or physical disturbance of the core segment. The polarity zonation we use emphasizes the predominant polarity of any given interval within the Kiptopeke core.

Stable Isotopes

We performed stable-isotope analyses for oxygen and carbon on 40 samples taken from the same cored interval sampled for the foraminiferal study at Kiptopeke. The raw data are given below. We used monospecific samples [~3–20 individuals of *Cibicidoides pippeni* f. speciosus (Cushman and Cederstrom) 1949] in the >63 μm grain-size fraction. Mass spectrometry was achieved using a Finnigan MAT 252 instrument with an on-line automated carbonate reaction Kiel device (Dick Norris, Woods Hole Oceanographic Institution). Analytical precision based on repeated analysis of standards (NBS-19, Carrara Marble, and B-1 marine carbonate) was better than 0.03 ‰ for $\delta^{13}C$ and 0.08 ‰ for $\delta^{18}O$ relative to the Peedee belemnite (PDB) standard. Poag and Norris (in press) used identical processing on 66 samples of the Chickahominy Formation from the NASA Langley corehole.

Depth [m]	$\delta^{13}C$	$\delta^{18}O$	Depth [m]	$\delta^{13}C$	$\delta^{18}O$	Depth [m]	$\delta^{13}C$	$\delta^{18}O$
326.3	+0.061	+0.808	347.4	-0.43	-0.192	384.7	-0.06	+0.056
327.7	-0.27	-0.152	349.1	-0.65	+0.046	387.2	+0.128	-0.130
328.0	-0.34	-0.013	352.6	-0.85	+0.106	389.2	+0.21	-0.011
329.6	-0.71	+0.201	354.4	-0.83	+0.148	390.2	+0.04	-0.070
330.4	-0.44	+0.027	355.2	-0.71	+0.071	376.6	+0.15	+0.022
330.7	-0.75	-0.037	365.3	-0.98	+0.048	378.6	+0.147	+0.029
331.6	-0.97	+0.036	360.3	-0.72	+0.063	381.0	+0.104	-0.040
332.3	-0.75	-0.218	361.2	-0.46	-0.040	383.5	+0.058	-0.019
333.8	-0.53	-0.165	363.0	-0.84	-0.166	390.7	-0.22	-0.020
337.5	-0.39	-0.182	366.7	-0.85	-0.243	391.7	-0.25	-0.111
339.6	-0.60	+0.006	366.9	-0.68	-0.127	392.2	-0.28	-0.146
341.7	-0.56	+0.080	369.4	-0.04	+0.086	393.7	-0.68	-0.275
343.2	-0.27	-0.089	371.2	-0.16	-0.173			
344.5	-0.70	-0.120	375.3	-0.33	+0.252			

Foraminiferal Assemblages

We extracted foraminiferal assemblages from ~20-cm³ core sections using standard micropaleontological procedures. Core samples were boiled in a solution of water and sodium hexametaphosphate to disperse the clays, and then wet-sieved over a 63-μm screen to separate both planktonic and benthic foraminiferal tests.

Oven-dried (70°C) specimens were examined and identified to species level, using optical and scanning-electron microscopy.

Taxonomic nomenclature employed by us is based on comparison with published literature and the senior author's personal collections of US Gulf and Atlantic Coastal Plain foraminiferal assemblages. Species-level taxonomy is primarily from Cushman (1935) and Cushman and Cederstrom (1949). Generic-level taxonomy is mainly from Loeblich and Tappan (1988). We have used quotation marks (see Chapter 13) to indicate informal species names that have no valid taxonomic status.

Isopach and Structure Maps

We constructed isopach (equal thickness) and structure maps from a 2,000-km network of intersecting seismic reflection profiles, after converting 2-way traveltime scales to depth scales, as described in Chapter 3.

Gravity Modeling

For comparison with the observed Chesapeake Bay gravity anomalies, P. Molzer (USGS 1998) translated models (blocks of constant-density rocks) into gravity responses with the software program GM-SYS (Northwest Geophysical Associates, Inc. 1999). We illustrate the gravity model with three cross sections through the inner basin, which also pass through, or near, as many data points as possible. Each section is ~90 km long and contains ~500 gridded gravity points. For modeling, Molzer constructed profiles with simple geometries along each line. Each model's upper, topographic surface was extracted from a Defense Mapping Agency 5-second horizontal resolution grid. Molzer picked flat basement surfaces for each line by connecting basement depth picks at the end of each line, outside the crater's disturbed zone. The depth values come from the structure map of the basement surface defined by scattered well penetrations and numerous seismic reflection profiles (Figs. 4.1, CD-ROM.1,2). Because no direct measurements of rock density were available for this area, Molzer chose generic density values used in gravity investigations of other impact craters (Plescia 1996; Pohl et al. 1977; crystalline basement rock = 2.67 g/cm^3; layered siliciclastic sedimentary rock = 2.3 g/cm^3; crater fill = 2.56 g/cm^3). Because the regional gravity anomalies are quite heterogeneous, consistent with probable heterogeneity in the basement composition (Lefort and Max 1991), Molzer made no attempt to model the gravity beyond the seismically-defined outer periphery of the peak ring. Cross sections of the starting model conform to observations from nearby seismic reflection profiles, with the exception of disturbances in the basement due to the impact (Fig. 4.36). Seismic profiles through the impact structure show disturbance and excavation of the basement surface inside the peak ring, characterized by loss of the basement surface reflector. The reflection data indicate uplift of the basement surface to form the peak ring. Outside the crater, the basement rock is overlain by an undisturbed sedimentary section, which sags into the crater. Continuous reflectors in the postimpact sedimentary section continue relatively undisturbed across the crater, with no indications of significant lateral facies change that could complicate the shallow density structure. Modeling thus focused on variations in basement density and shape of the basement surface, assuming a laterally homogeneous overlying sedimentary section.

References

Abreu VS, Anderson JB (1998) Glacial eustasy during the Cenozoic: Sequence stratigraphic implications. American Association of Petroleum Geologists Bulletin 82:1385–1400

Adushkin VV, Nemchinov IV (1994) Consequences of impacts of cosmic bodies on the surface of the Earth. In: Gehrels T (ed) Hazards due to comets and asteroids. University of Arizona Press, Tucson, pp 721–778

Ahrens TJ, O'Keefe JD (1987) Impact on the Earth, ocean and atmosphere. International Journal of Impact Engineering 5:13–32

Albin EF, Norman, MD, Roden M (2000) Major and trace element compositions of georgiaites: Clues to the source of North American tektites. Meteoritics and Planetary Science 335:795–806

Alexopolous JS, McKinnon WB (1992) Multiringed impact craters on Venus: An overview from Arecibo and Venera images and initial Magellan data. Icarus 100:347–363

Alexopolous JS, McKinnon WB (1994) Large impact craters and basins on Venus, with implications for ring mechanics on the terrestrial planets. In: Dressler BO, Grieve RAF, Sharpton VL (eds) Large meteorite impacts and planetary evolution. Geological Society of America Special Paper 293:29–50

Anderson RR, Witzke BJ (1996) Sedimentary-clast breccias of the Manson impact structure. In: Koeberl C, Anderson RR (eds) The Manson impact structure, Iowa: Anatomy of an impact crater. Geological Society of America Special Paper 302:115–144

Anderson RR, Witzke BJ, Roddy DJ (1996) The drilling of the 1991–1992 Geological Survey Bureau and U.S. Geological Survey Manson impact structure research cores. In: Koeberl C, Anderson RR (eds) The Manson impact structure, Iowa: Anatomy of an impact crater. Geological Society of America Special Paper 302:45–88

Artemieva NA, Shuvalov VV (2002) Shock metamorphism on the ocean floor (numerical simulations). In: Gersonde R, Deutsch A, Ivanov BA, Kyte FT (eds) Oceanic impacts: Mechanisms and environmental perturbations. Deep-Sea Research Part II 49:959–968

Asquith GB (1982) Basic well log analysis for geologists. American Association of Petroleum Geologists, Tulsa, 216 pp

Aubry M-P (1992) Late Paleogene calcareous nannoplankton evolution: A tale of climatic deterioration. In: Prothero DR, Berggren WA (eds) Eocene-Oligocene climatic and biotic evolution. Princeton University Press, Princeton, pp 272–309

Aubry M-P (1993) Data report: Eocene to upper Miocene calcareous nannofossil stratigraphy. In: Mountain GS, Miller KG, Blum P, Poag CW, Twichell DC (eds) Proceedings of the Ocean Drilling Program, scientific results. Ocean Drilling Program, College Station, Texas, 150:435–437

Avermann M (1992) Die Genese der allochthonen, polymikten Breccien der Onaping-Formation, Sudbury-Struktur, Ontario, Kanada. Ph.D. thesis, University of Münster, Germany, 175 pp

Avermann M (1994) Origin of the polymict, allochthonous breccias of the Onaping Formation, Sudbury structure, Ontario, Canada. In: Dressler BO, Grieve RAF, Sharpton VL (eds) Large meteorite impacts and planetary evolution. Geological Society of America Special Paper 293:265–274

Avermann M (1999) The Green Member of the Onaping Formation, the collapsed fireball layer of the Sudbury impact structure, Ontario, Canada. In: Dressler BO, Sharpton VL Large meteorite impacts and planetary evolution II. Geological Society of America Special Paper 339:323–330

Bailey ME, Clube SVM, Gahn G, Napier WM, Valsecchi GB (1994) Hazards due to giant comets: Climate and short-term catastrophism. In: Gehrels T (ed) Hazards due to comets and asteroids. University of Arizona Press, Tucson, pp 479–533

Bains S, Corfield RM, Norris RD (1999) Mechanisms of climate warming at the end of the Paleocene. Science 285:724–727

Baker VR, Milton DJ (1974) Erosion by catastrophic floods on Mars and Earth. Icarus 23:27–41

Bandy OL (1949) Eocene and Oligocene foraminifera from Little Stave Creek, Clarke County, Alabama. Bulletins of American Paleontology 32:1–210

Barker PF, Kennett JP O'Connell S, Berkowitz S, Bryant WR, Burckle LH, Egeberg PK, Fütterer DK, Gersonde RE, Golovchenko X, Hamilton N, Lawver L, Lazarus DB, Lonsdale M, Mohr B, Nagao T, Pereira CPG, Pudsey CJ, Robert CM, Schaudl E, Speiss V, Stott LD, Thomas E, Thompson KFM, Wise SW Jr (1990) Proceedings of the Ocean Drilling Program, scientific results. Ocean Drilling Program, College Station, Texas, 113:1–785

Bayer KC Milici RC (1987) Geology and petroleum potential of Mesozoic and Cenozoic rocks, offshore Virginia. Virginia Division of Mineral Resources Publication 73 (D):1–111

Beatty JK, Petersen CC, Chaikin A (1999) The new solar system. Cambridge University Press, Cambridge, 421 pp

Beckman JP (1954) Die Foraminferen der Oceanic Formation (Eocaen-Oligocaen) von Barbados, Kl. Antillen. Eclogae Geologicae Helvetiae 46:301–412

Bennett RR, Collins GG (1952) Brightseat Formation, a new name for sediments of Paleocene age in Maryland. Journal of the Washington Academy of Sciences 42:114–116

Berggren WA, Kent DV, Swisher CC III, Aubry M-P (1995) A revised Cenozoic geochronology and chronostratigraphy. In: Berggren WA, Kent DV, Aubry M-P, Hardenbol J (eds) Geochronology, time scales, and global stratigraphic correlation: SEPM Special Publication 54:129–212

Bernhard JM, Sen Gupta BK (1999) Foraminifera of oxygen-depleted environments. In: Sen Gupta BK (ed) Modern foraminifera. Kluwer Academic Publishers, Dordrecht, pp 201–216

Bestland EA, Retallack GJ, Rice AE, Mindszenty A (1996) Late Eocene detrital laterites in central Oregon: Mass balance, geochemistry, depositional setting, and landscape evaluation. Geological Society of America Bulletin 108:285–302

Bice DM, Montanari A (1988) Magnetic stratigraphy of the Massignano section across the Eocene-Oligocene boundary. In: Premoli Silva I, Coccioni R, Montanari A (eds) The Eocene-Oligocene boundary in the Marche-Umbria Basin (Italy). International Subcommission on Paleontology and Stratigraphy Special Publication II, pp 111–117

Bick KF, Coch NK (1969) Geology of the Williamsburg, Hog Island, and Bacons Castle quadrangles, Virginia. Virginia Division of Mineral Resources Report of Investigations pp 1–28

Blackwelder BW (1981) Stratigraphy of upper Pliocene and lower Pleistocene marine and estuarine deposits of northeastern North Carolina and southeastern Virginia. US Geological Survey Bulletin 1502-B:1–16

Bohor BF (1990) Shocked quartz and more; Impact signatures in Cretaceous/Tertiary boundary clays. In: Sharpton VL, Ward PD (eds) Global catastrophes in Earth history: An interdisciplinary conference on impacts, volcanism, and mass mortality. Geological Society of America Special Paper 247:335–342

Bottomley R, Grieve R, York D, Masaitis V (1997) The age of the Popigai impact event and its relation to events of the Eocene/Oligocene boundary. Nature 388:365–368

Brady HB (1879) Notes on some of the Reticularian Rhizopoda of the "Challenger" Expedition. Part 1. On new or little known arenaceous types. Quarterly Journal of Microscopical Science, new series, 19:20–63

Brinkhuis H, Coccioni R (1995) Is there a relation between dinocysts changes and iridium after all? Preliminary results from the Massignano section. In: Montanari A, Coccioni R (eds) The effects of impacts on the atmosphere and biosphere in regard to short- and long-term changes [abs.]. Abstracts and Field Trips, Ancona, Italy, p 214

Brinkhuis H, Zacharaiasse WJ (1988) Dinoflagellate cysts, sea level changes and planktonic foraminifers across the Cretaceous-Tertiary boundary at El Haria, Northwest Tunisia. Marine Micropaleontology 13:153–199

Brittan J, Morgan J, Warner M, Marin, L (1999) Near-surface seismic expression of the Chicxulub impact crater. In: Dressler BO, Sharpton VL (eds) Large meteorite impacts and planetary evolution II. Geological Society of America Special Paper 339:269–279

Brockmeyer P (1990) Petrographie, Geochemie, und Isotopenuntersuchungen an der Onaping-Formation im Nordteil der Sudbury-Struktur (Ontario, Kanada) und ein Modell zur Genese der Struktur. Ph.D. thesis, University of Münster, Germany, 228 pp

Brown PM, Miller JA, Swain FM (1972) Structural and stratigraphic framework and spatial distribution of permeability of the Atlantic Coastal Plain, North Carolina to New York. US Geological Survey Professional Paper 796:1–79

Bunch TE, Becker L, Des Marais D, Tharpe A, Schultz PH, Wolbach W, Glavin DP, Brinton KL, Bada JL (1999) Carbonaceous matter in the rocks of the Sudbury Basin, Ontario, Canada. In: Dressler BO, Sharpton VL (eds) Large meteorite impacts and planetary evolution II. Geological Society of America Special Paper 339:331–343

Buzas MA, Culver SJ, Jorissen FJ (1993) A statistical evaluation of the microhabitats of living (stained) infaunal benthic foraminifera. Marine Micropaleontology 20:311–320

Cande SC, Kent DV (1992a) A new geomagnetic polarity time scale for the Late Cretaceous and Cenozoic. Journal of Geophysical Research 97:13917–13951

Cande SC, Kent DV (1992b) Ultrahigh resolution marine magnetic anomaly profiles: A record of continuous paleointensity variations? Journal of Geophysical Research 97:15075–15083

Caralp MH (1989) Abundance of *Bulimina exilis* and *Melonis barleeanum*: Relationship to the quality of marine organic matter. Geo-Marine Letters 9:37–43

Cederstrom DJ (1943) Chloride in ground water in the Coastal Plain of Virginia. Virginia Geological Survey Bulletin 58:1–384

Cederstrom DJ (1945a) Selected well logs in the Virginia Coastal Plain north of James River. Virginia Geological Survey Circular 3:1–81

Cederstrom DJ (1945b) Geology and ground-water resources of the coastal plain in southeastern Virginia. Virginia Geological Survey Bulletin 63:1–384

Cederstrom DJ (1945c) Structural geology of southeastern Virginia. American Association of Petroleum Geologists Bulletin 29:71–95

Cederstrom DJ (1946) Chemical character of ground water in the coastal plain of Virginia. Virginia Geological Survey Bulletin 4:13–322

Cederstrom DJ (1957) Geology and ground-water resources of the York-James Peninsula, Virginia. US Geological Survey Water-Supply Paper 1361:1–237

Charletta AC (1980) Eocene benthic foraminiferal paleoecology and paleobathymetry of the New Jersey continental margin. Ph.D. thesis, Rutgers University, 84 pp

Christeson GL, Buffler RT, Nakamura Y (1999) Upper crustal structure of the Chicxulub impact crater from wide-angle ocean bottom seismograph data. In: Dressler BO, Sharpton VL (eds) Large meteorite impacts and planetary evolution II. Geological Society of America Special Paper 339:291–298

Clark DA (1984) Hysteresis properties of sized dispersed monoclinic pyrrhotite grains. Geophysical Research Letters 11:173–176

Clark WB (1896) The Eocene deposits of the middle Atlantic slope in Delaware, Maryland, and Virginia. US Geological Survey Bulletin 141:1–167

Clark WB, Martin GC (1901) The Eocene deposits of Maryland. Maryland Geological Survey, Eocene Volume, pp 1–92, 122–204

Clark WB, Miller BL (1906) A brief summary of the geology of the Virginia Coastal Plain. Virginia Geological Survey Bulletin 2:11–24

Clark WB, Miller BL (1912) Physiography and geology of the Coastal Plain province of Virginia. Virginia Geological Survey Bulletin 4:1–274

Clement SC, Bice KL (1982) Andesine anorthosite in the eastern Piedmont of Virginia [abs.]. Geological Society of America Abstracts with Programs 14(7):10

Clymer AK, Bice DM, Montanari A (1996) Shocked quartz from the late Eocene: Impact evidence from Massignano, Italy. Geology 24:483–486

Coccioni R, Galeotti S (1994) K-T boundary extinction: Geologically instantaneous or gradual event? Evidence from deep-sea benthic foraminifera. Geology 22:779–782

Coccioni R, Monaco P, Monechi S, Parisi G (1988) Biostratigraphy of the Eocene-Oligocene boundary at Massignano (Ancona, Italy). In: Premoli Silva I, Coccioni R, Montanari A (eds) The Eocene-Oligocene Boundary in the Marche-Umbria Basin (Italy). International Subcommittee on Paleontology and Stratigraphy Special Publication II, pp 59–80

Coch NK (1968) Geology of the Benns Church, Smithfield, Windsor, and Chuckatuck quadrangles, Virginia. Virginia Division of Mineral Resources Report of Investigations, pp 1–39

Cohen BA, Swindle TD, Kring DA (2000) Support for the lunar cataclysm hypothesis from lunar meteorite impact melt ages. Science 290:1754–1756

Collins GS, Melosh HJ, Morgan JV, Warner MR (2002) Hydrocode simulations of Chicxulub crater collapse and peak-ring formation. Icarus 157:24–33

Colman SM, Hobbs CH III (1987) Quaternary geology of the southern Virginia part of the Chesapeake Bay. US Geological Survey Miscellaneous Field Investigations Map MF-1948-A

Colman SM, Mixon RB (1988) The record of major Quaternary sea-level changes in a large coastal estuary, Chesapeake Bay, Eastern United States. Palaeogeography Palaeoclimatology Palaeoecology 68:99–116

Conrad TA (1832) Fossil shells of the Tertiary formations of North America. Judah Dobson, Philadelphia, 20 pp

Corliss BH (1985) Microhabitats of benthic foraminifera within deep-sea sediments. Nature 314:435–438

Corliss BH (1991) Morphology and microhabitat preferences of benthic foraminifera from northwest Atlantic Ocean: Marine Micropaleontology 17:195–236

Corliss BH, Fois E (1990) Morphotype analysis of deep-sea benthic foraminifera from the northwest Gulf of Mexico. PALAIOS 5:589–605

Corliss BH, Silva K (1993) Rapid growth of deep sea benthic foraminifera. Geology 21:991–994

Costain JK (1979) Geothermal exploration methods and results: Atlantic Coastal Plain. In: Geothermal resources council, Davis, California: A symposium on geothermal energy and its direct uses in the eastern United States. Special Report 5:13–22

Costain JK, Glover L III (1976–1982) Evaluation and targeting of geothermal energy resources in the southeastern United States – Final report. Department of Geological Sciences Virginia Polytechnical Institute and State University, Blacksburg

Costain JK, Lambiase JJ, Bielanski E, Coruh C, Dysart PS (1981) Seismic evidence for the extension of the Baltimore Canyon "lower acoustical zone" sediments beneath the Northern Atlantic Coastal Plain [abs.]. Geological Society of America Abstracts with Programs 13(7):431

Covey C, Thompson SL, Weissman PR, MacCracken MC (1994) Global climatic effects of atmospheric dust from an asteroid or comet impact on Earth. Global and Planetary Change 9:262–272

Culver SJ, Buzas MA (1980) Distribution of Recent benthic foraminifera off the North American Atlantic coast. Smithsonian Contributions to Marine Science 6:1–512

Culver SJ, Buzas MA (1981) Distribution of Recent benthic foraminifera in the Gulf of Mexico. Smithsonian Contributions to Marine Science 8:1–898

Culver SJ, Buzas MA (1982) Distribution of Recent benthic foraminifera in the Caribbean Region. Smithsonian Contributions to Marine Science 14:1–382

Cushman JA (1925a) New foraminifera from the upper Eocene of Mexico. Contributions from the Cushman Laboratory of Foraminiferal Research 1:4–8

Cushman JA (1925b) Eocene foraminifera from the Cocoa Sand of Alabama. Contributions from the Cushman Laboratory for Foraminiferal Research 1:65–70

Cushman JA (1926) Some fossil Bolivinas from Mexico. Contributions from the Cushman Laboratory for Foraminiferal Research 1:81–84

Cushman JA (1933) New foraminifera from the Upper Jackson Eocene of the southeastern coastal plain of the United States. Contributions from the Cushman Laboratory for Foraminiferal Research 9:1–21

Cushman JA (1935) Upper Eocene foraminifera of the southeastern United States. US Geological Survey Professional Paper 181:1–88

Cushman JA, Applin ER (1926) Texas Jackson foraminifera. American Association of Petroleum Geologists Bulletin 10:154–189

Cushman JA, Bermúdez PT (1948) Some Paleocene foraminifera from the Madruga Formation of Cuba. Contributions from the Cushman Laboratory for Foraminiferal Research 24:68–75

Cushman JA, Cederstrom DJ (1949) An upper Eocene foraminiferal fauna from deep wells in York County, Virginia. Virginia Geological Survey Bulletin 67:1–58

Cushman JA, Garrett JB (1938) Three new rotaliform foraminifera from the lower Oligocene and upper Eocene of Alabama. Contributions from the Cushman Laboratory for Foraminiferal Research 14:62–66

Cushman JA, Todd R (1946) A foraminiferal fauna from the Byram Marl at its type locality. Contributions from the Cushman Laboratory for Foraminiferal Research 22:76–102

Dailey DH (1983) Late Cretaceous and Paleocene benthic foraminifers from Deep Sea Drilling Project Site 516, Rio Grande Rise, western South Atlantic Ocean. In: Barker PJ, Johnson DA, Carlson RL, Cepek P, Coulbourn WT, Gamboa LA, Hamilton N, Melo U de, Pujol C, Shor AN, Suzyumov AE, Tjalsma LRC, Walton WH, Initial reports of the Deep Sea Drilling Project. US Government Printing Office, Washington, DC, 72:757–782

Darby DA (1984) Origin and deposition of transgressive coarse sediments marking unconformities in the Tertiary section exposed along the Pamunkey River, Virginia. In: Ward LW, Krafft K (eds) Stratigraphy and paleontology of the outcropping Tertiary beds in the Pamunkey River region, central Virginia Coastal Plain – Guidebook for Atlantic Coastal Plain Geological Association 1984 Field Trip. Atlantic Coastal Plain Geological Association, pp 93–110

Darton NH (1891) Mesozoic and Cenozoic formations of eastern Virginia and Maryland. Geological Society of America Bulletin 2:431–450

Davis JL, Mitrovica JX (1996) Glacial isostatic adjustment and the anomalous tide gauge record of eastern North America. Nature 379:331–333

Deutsch A, Grieve RAF, Avermann L, Bischoff L, Brockmeyer P, Buhl D, Lakomy R, Müller-Mohr V, Ostermann M, Stöffler D (1995) The Sudbury structure (Ontario, Canada): A tectonically deformed multi-ring basin. Geologische Rundschau 84:697–709

Diester-Haass L, Zahn R (1996) Eocene-Oligocene transition in the Southern Ocean: History of water mass circulation and biological productivity. Geology 24:163–166

Dodd JR, Stanton RJ Jr (1981) Paleoecology, concepts and applications. John Wiley and Sons, New York, 559 pp

Douglas BC (1991) Global sea level rise. Journal of Geophysical Research 96:6981–6992

Douglas BC, Kearney MS, Leatherman SP (2001) Sea level rise. Academic Press, London, 272 pp

Douglas RG, Heitman HL (1979) Slope and basin benthic foraminifera of the California borderland. In: Doyle LJ, Pilkey OH (eds) Geology of continental slopes. SEPM Special Publication 27:231–246

Dressler BO, Reimold WU (2001) Terrestrial impact melt rocks and glasses. Earth-Science Reviews 56:205–284

Dypvik H, Gudlaugsson ST, Tsikalas F, Attrep M Jr, Ferrell RE Jr, Krinsley DH, Mørk A, Faleide JI, Nagy J (1996) Mjølnir structure: An impact crater in the Barents Sea. Geology 24:779–782

Dysart PS, Coruh C, Costain JK (1983) Seismic response of major regional unconformities in Atlantic Coastal Plain sediments at Smith Point, Virginia. Geological Society of America Bulletin 94:305–311

Emanuel KA, Speer K, Rotunno R, Srivistava R, Molina M (1995) Hypercanes: A possible link in global extinction scenarios. Journal of Geophysical Research 100:13,755–13,765

Emiliani C, Kraus EB, Shoemaker EM (1982) Sudden death at the end of the Mesozoic. Earth and Planetary Science Letters 55:317–334

Engelhardt Wv (1990) Distribution, petrography, and shock metamorphism of the ejecta of the Ries crater in Germany – A review. Tectonophysics 171:259–273

Engelhardt Wv (1997) Suevite breccia of the Ries crater, Germany: Petrography, chemistry, and shock metamorphism of crystalline rock clasts. Meteoritics and Planetary Science 32:545–554

Engelhardt Wv, Bertsch W (1969) Shock induced planar deformation features in quartz from the Ries crater, Germany. Contributions to Mineralogy and Petrology 20:203–234

Engelhardt Wv, Arndt J, Fecker B, Pankau HG (1995) Suevite breccia from the Ries crater, Germany: Origin, cooling history, and devitrification of impact glasses. Meteoritics 30:279–293

Ewing M, Crary AP, Rutherford HM (1937) Geophysical investigations in the emerged and submerged Atlantic Coastal Plain, Part I, Methods and results. Geological Society of America Bulletin 48:753–802

Farley KA, Montanari A, Shoemaker EM, Shoemaker CS (1998) Geochemical evidence for a comet shower in the late Eocene. Science 280:1250–1253

Farre JA, Ryan WBF (1985) 3-D view of erosional scars on U.S. mid-Atlantic continental margin. American Association of Petroleum Geologists Bulletin 69:923–932

Farrar SS (1984) The Goochland granulite terrane: Remobilized Grenville basement in the eastern Virginia Piedmont. Geological Society of America Special Paper 194:215–227

Focazio MJ, Speiran GK, Rowan ME (1993) Quality of ground water in the coastal plain physiographic province of Virginia. US Geological Survey Water Resources Investigations Report 92-4175:1–20

French BM (1998) Traces of catastrophe: A handbook of shock-metamorphic effects in terrestrial meteorite impact structures: LPI Contribution 954, Lunar and Planetary Institute, Houston, 120 pp

Füchtbauer H, Brelie Gvd, Dehm R, Förstner U, Gall H, Höfling R Hoefs J, Hollerbach A, Hufnagel H, Jankowski B, Jung W, Malz H, Mertes H, Rothe P, Salger M, Wehner H, Wolf M (1977) Tertiary lake sediments of the Ries research borehole Nördlingen 1973 – A summary. Geologica Bavarica 75:13–19

Gault DE, Sonett CP (1982) Laboratory simulation of pelagic asteroid impact: Atmospheric injection, benthic topography, and the surface wave radiation field. In: Silver LT, Schultz PH (eds) Geological implications of impacts of large asteroids and comets on the Earth. Geological Society of America Special Paper 190:69–92

Gault DE, Quaide WL, Oberbeck VR (1968) Impact cratering mechanics and structures. In: French BM, Short NM (eds) Shock metamorphism of natural materials. Mono Book Co., Baltimore, pp 87–99

Gersonde R, Hodell DA, Blum P, Andersson C, Austin WEN, Billups K, Channell JET, Charles CD, Diekmann B, Filippelli GM, Flores J-A, Hewitt AT, Howard WR, Ikehara M, Janecek TR, Kanfoush SL, Kemp AES, King SL, Kleiven HF, Kuhn G, Marino M, Ninnemann US, O'Connell S, Ortiz JD, Stoner JS, Sugiyama K, Warnke D, Zielinski U (1999) Proceedings of the Ocean Drilling Program, initial reports. Ocean Drilling Program, College Station, Texas, 177:1–67

Gerstl SAW, Zardecki A (1982) Reduction of photosynthetically active radiation under extreme atmospheric aerosol loads. In: Silver LT, Schultz PH (eds) Geological implications of impacts of large asteroids and comets on the Earth. Geological Society of America Special Paper 190:201–210

Gibson TG (1967) Stratigraphy and paleoenvironment of the phosphatic Miocene strata of North Carolina. Geological Society of America Bulletin 78:631–650

Gibson TG (1971) Miocene of the Middle Atlantic Coastal Plain. In: Gernant RE, Gibson TG, Whitmore FC Jr (eds) Environmental history of Maryland Miocene, Field Trip 5 of Geological Society of America Annual Meeting. Maryland Geological Survey Guidebook 3:1–15

Gibson TG (1983) Key foraminifera from upper Oligocene to lower Pleistocene strata of the central Atlantic Coastal Plain. In: Ray CE (ed) Geology and paleontology of the Lee Creek Mine, North Carolina. Smithsonian Contributions to Paleobiology 53:355–453

Gibson TG (1989) Planktonic benthonic foraminiferal ratios: Modern patterns and Tertiary applicability. Marine Micropaleontology 15:29–52

Glaser JD (1971) Geology of mineral resources of southern Maryland. Maryland Geological Survey Report of Investigations 15:1–84

Glass BP (1989) North American tektite debris and impact ejecta from DSDP Site 612. Meteoritics 24:209–218

Glass BP, Burns CA (1987) Late Eocene crystal-bearing spherules: Two layers or one? Meteoritics 22:265–279

Glass BP, Koeberl C (1999a) Ocean Drilling Project Hole 689B spherules and upper Eocene microtektite and clinopyroxene-bearing spherule strewn fields. Meteoritics and Planetary Science 34:197–208

Glass BP, Koeberl C (1999b) North American tektites in the Indian Ocean? [abs.]. Meteoritics and Planetary Science 34(Supplement):A43–A44

Glass BP, Wu J (1993) Coesite and shocked quartz discovered in the Australasian and North American microtektite layers. Geology 21:435–438

Glass BP, Burns CA, Crosbie JR, DuBois DL (1985) Late Eocene North American microtektites and clinopyroxene-bearing spherules. Proceedings of the 16th Lunar and Planetary Science Conference. Journal of Geophysical Research 90:D175–D196

Glass BP, Hall CM, York D (1986) $^{40}Ar/^{39}Ar$ laser probe dating of North American tektite fragments from Barbados and the age of the Eocene-Oligocene boundary. Chemical Geology 59:181–186

Glass BP, Koeberl C, Blum JD, Senftle F, Izett GA, Evans BJ, Thorpe AN, Povenmire H, Strange RL (1995) A Muong Nong-type Georgia tektite. Geochimica et Cosmochimica Acta 59:4071–4082

Glass BP, Koeberl C, Blum JD, McHugh CMG (1998) Upper Eocene tektite ejecta layer on the continental slope off New Jersey. Meteoritics and Planetary Science 33:229–241

Gleason RJ (1979) Study of the pre-Cretaceous basement below the Atlantic Coastal Plain. In: Costain JK, Glover L III (principal investigators) Evaluation and targeting of geothermal energy resources in the southeastern United States. Progress Report to US Department of Energy, October 1, 1978–March 30, 1979. Contract No. DE-AC05-78ET27001, pp A60–A83

Gleason RJ (1982) Basement rocks beneath Atlantic Coastal Plain: Isle of Wright, Virginia. In: Costain JK, Glover L III (principal investigators) Evaluation and targeting of geothermal energy resources in the southeastern United States. Final Report to US Department of Energy, May 1, 1976–June 30, 1982, Contract DE-AC05-78ET27001, pp 71–77

Glover L III, Mose DG, Poland FB, Bobyarchick AR, Bourland WC (1978) Grenville basement in the eastern Piedmont of Virginia: Implications for orogenic models [abs.]. Geological Society of America Abstracts with Programs 10(7):1–69

Gohn GS (1988) Late Mesozoic and early Cenozoic geology of the Atlantic Coastal Plain: North Carolina to Florida. In: Sheridan RE, Grow JA (eds) The geology of North America, the Atlantic continental margin, US. Geological Society of America, Boulder, Colorado, I-2:107–130

Gohn GS (ed) (in press) Geological and geohydrological studies of the USGS-NASA Langley corehole, Hampton, Virginia. US Geological Survey Professional Paper

Gooday AJ (1986) Meiofaunal foraminiferans from the bathyal Porcupine Seabight (northeast Atlantic): Size, structure, standing crop, taxonomic composition, species diversity, and vertical distribution in the sediment. Deep-Sea Research 33:1345–1373

Gooday AJ (1993) Deep-sea benthic foraminiferal species which exploit phytodetritus: Characteristic features and controls on distribution. Marine Micropaleontology 22:187–205

Gooday AJ (1994) The biology of deep-sea foraminifera: A review of some advances and their applications to paleoceanography. PALAIOS 9:14–31

Gooday AJ (1996) Epifaunal and shallow infaunal foraminiferal communities at three abyssal NE Atlantic sites subject to differing phytodetritus input regimes. Deep-Sea Research 43:1395–1421

Gornitz V, Seeber L (1990) Vertical crustal movements along the east coast, North America, from historic and late Holocene sea level data. Tectonophysics 178:127–150

Govindaraju K (1993) Automated rock analysis for forty elements. In: Hall GEM (ed) Geoanalysis 90. Geological Survey of Canada Report 451, pp 18–19.

Greeley R (1994) Planetary landscapes. Chapman and Hall, New York, 286 pp

Grieve RAF (1987) Terrestrial impact structures. Annual Review of Earth and Planetary Science 15:245–270

Grieve RAF (1991) Terrestrial impact: The record in the rocks. Meteoritics 26:175–194

Grieve RAF (1994) An impact model of the Sudbury structure, Ontario. In: Lightfoot PG, Naldrett AJ (eds) Proceedings of the Sudbury-Noril'sk symposium. Ontario Geological Survey Special Volume 5:119–132

Grieve RAF, Robertson PB (1979) The terrestrial cratering record, 1, Current status of observations. Icarus 38:212–229

Grieve RAF, Shoemaker EM (1994) The record of past impacts on Earth. In: Gehrels T (ed) Hazards due to comets and asteroids. University of Arizona Press, Tucson, pp 417–462

Grieve RAF, Therriault A (2000) Vredefort, Sudbury, Chicxulub: Three of a kind? Annual Reviews of Earth and Planetary Science 28:305–338

Grieve RAF, Dence MR, Robertson PB (1977) Cratering processes: As interpreted from the occurrence of impact melts. In: Roddy DJ, Pepin RO, Merrill RB (eds) Impact and explosion cratering. Pergamon Press, New York, pp 791–814

Grieve RAF, Rupert J, Smith J, Therriault A (1995) The record of terrestrial impact cratering. GSA Today 5:189–196

Grieve RAF, Langenhorst F, Stöffler D (1996) Shock metamorphism of quartz in nature and experiment; II, Significance in geoscience. Meteoritics 31:6–35

Grow JA, Klitgord KD, Schlee JS (1988) Structure and evolution of the Baltimore Canyon trough. In: Sheridan RE, Grow JA (eds) The geology of North America, Atlantic continental margin, US. Geological Society of America, Boulder, Colorado, I-2:269–290

Gudlaugsson ST (1993) Large impact crater in the Barents Sea. Geology 21:291–294

Hajnal Z, Scott D, Robertson PB (1988) Reflection study of the Haughton impact crater. Journal of Geophysical Research 93:11,930–11,942

Hale WS, Grieve RAF (1982) Volumetric analysis of complex lunar craters: Implications for basin ring formation. Proceedings of the 17th Lunar and Planetary Science Conference, Journal of Geophysical Research 87:A65–A76

Hale WS, Head JW (1979) Central peaks in lunar craters: Morphology and morphometry. Proceedings of the 10th Lunar and Planetary Science Conference, Pergamon Press, New York, pp 2623–2633

Hale WS, Head JW (1980) Central peaks in Mercurian craters: Comparisons to the moon. Proceedings of the 11th Lunar and Planetary Science Conference, Pergamon Press, New York, pp 2191–2205

Hansen HJ (1978) Upper Cretaceous (Senonian) and Paleocene (Danian) pinchouts on the south flank of the Salisbury Embayment, Maryland, and their relationship to antecedent basement structures. Maryland Geological Survey Report of Investigations 29:1–36

Hansen HJ, Wilson JM (1984) Summary of hydrogeologic data from a deep (2,687 ft) well at Lexington Park, St. Marys County, Maryland. Maryland Geological Survey Open File Report 84-02-1:1–61

Haq BU, Lohmann GP (1976) Cenozoic calcareous nannoplankton biogeography of the Atlantic Ocean. Marine Micropaleontology 1:119–194

Hart MB (ed) (1996) Recovery from mass extinction events. Geological Society Special Publication 102:1–392

Hawarth RT, Daniels DL, Williams H, Zietz I (1980) Bouguer gravity map of the Appalachian orogen. Map 3, Memorial University of Newfoundland, St. Johns, scale 1:1,000,000

Hayek L, Buzas MA (1997), Surveying natural populations: Columbia University Press, New York, 563 pp

Higgins GH, Butkovich TR (1967) Effects of water content, yield, medium, and depth of burst on crater radii. Lawrence Radiation Laboratory Report UCRL5023

Hildebrand AR (1997) Contrasting Chicxulub crater structural models: What can seismic velocities differentiate? Journal of Conference Proceedings 1:37–46

Hildebrand AR, Penfield GT, Kring DA, Pilkington M, Camargo AZ, Jacobsen SB, Boynton WV (1991) A possible Cretaceous-Tertiary boundary impact crater on the Yucatan peninsula, Mexico. Geology 19:867–871

Hildebrand A R, Pilkington M, Ortiz-Aleman C, Chavez RE, Urrutia-Fucugauchi J, Connors M, Graniel-Castro E, Camara-Zi A, Halpenny JF, Niehaus D (1998) Mapping Chicxulub crater structure with gravity and seismic reflection data. In: Grady MM, Hutchison R, McCall GJH, Rothery DA (eds) Meteorites: Flux with time and impact effects. Geological Society Special Publication 140:155–176

Hills JG, Goda MP (1999) Damage from comet-asteroid impacts with earth. Physica D 133:189–198

Hills JG, Nemchinov IV, Popov SP, Teterev AV (1994) Tsunami generated by small asteroid impacts. In: Gehrels T (ed) Hazards due to comets and asteroids. University of Arizona Press, Tucson, pp 779–789

Holdahl SR, Morrison NL (1974) Regional investigations of vertical crustal movements in the U.S., using precise relevelings and mareograph data. Tectonophysics 23:373–390

Horton JW Jr, Drake AA Jr, Rankin DV, Dallmeyer RD (1991) Preliminary tectonostratigraphic terrane map of the central and southern Appalachians. US Geological Survey Miscellaneous Investigations Series Map I-2163

Horton JW Jr, Aleinikoff JN, Izett GA, Naeser CW, Naeser ND (2001) Crystalline rocks from the first corehole to basement in the Chesapeake Bay impact structure, Hampton, VA [abs.]. Geological Society of America Abstracts with Programs 33(7):A448

Horton JW Jr, Kunk MJ, Naeser CW, Naeser ND, Aleinkoff JN, Izett GA (2002) Petrography, geochronology, and significance of crystalline basement rocks and impact-derived clasts in the Chesapeake Bay impact structure, southeastern Virginia [abs.]. Geological Society of America Abstracts with Programs 34 (6):466

Hörz F, Banholzer GS (1980) Deep-seated materials in the continuous deposits of the Ries crater, Germany. In: Papike JJ, Merrill RB (eds) Proceedings of the Conference on Lunar Highlands Crust. Pergamon Press, New York, pp 211–231

Hörz F, Ostertag R, Rainey DA (1983) Bunte Breccia of the Ries: Continuous deposits of large impact craters. Review of Geophysics and Space Physics 21:1667–1725

Huber H, Koeberl C, King DT Jr, Petruny LW, Montanari A (2002) Effects of bioturbation through the late Eocene impactoclastic layer near Massignano, Italy. In: Buffetaut E, Koeberl C (eds) Geological and biological effects of impact events. Springer, Berlin, pp 197–216

Huffman AR, Reimold WU (1996) Experimental constraints on shock-induced microstructures in naturally deformed silicates. Tectonophysics 256:165–217

Ivanov BA, Deutsch A (1999) Sudbury impact event: cratering mechanics and thermal history. In: Dressler BO, Sharpton VL (eds) Large meteorite impacts and planetary evolution II. Geological Society of America Special Paper 339:389–397

Izett GA, Cobban WA, Obradovich JD, Kunk MJ (1993) The Manson impact structure: $^{40}Ar/^{39}Ar$ age and its distal impact ejecta in the Pierre Shale in southeastern South Dakota. Science 262:729–732

Jackson J (1997) Glossary of geology. American Geological Institute, Alexandria, Virginia, 769 pp

Jansa LF (1993) Cometary impacts into ocean: Their recognition and the threshold constraint for biological extinctions. Palaeogeography Palaeoclimatology Palaeoecology 104:271–286

Jansa LF, Pe-Piper G (1987) Identification of an underwater extraterrestrial impact crater. Nature 327:612–614

Jansa LF, Pe-Piper G, Robertson BP, Friedenreich O (1989) Montagnais: A submarine impact structure on the Scotian shelf, eastern Canada. Geological Society of America Bulletin 101:450–463

Jansa LF, Aubry M-P, Gradstein FM (1990) Comets and extinctions: Cause and effect? In: Sharpton VL, Ward PD (eds) Global catastrophes in earth history. Geological Society of America Special Paper 247:223–232

Jarosewich E, Clarke RS, Barrows JN (1987) The Allende meteorite reference sample. Smithsonian Contributions to the Earth Sciences 27:49

Johnson GH (1972) Geology of the Yorktown, Poquoson West, and Poquoson East quadrangles Virginia. Virginia Division of Mineral Resources Report of Investigations 30:1–57

Johnson GH (1976) Geology of the Mulberry Island, Newport News North, and Hampton quadrangles Virginia. Virginia Division of Mineral Resources Report of Investigations 41:1–72

Johnson GH, Goodwin BK (1969) Guidebook to the geology of the York-James Peninsula and south bank of the James River. Atlantic Coastal Plain Geological Society 10th Annual Field Conference, Williamsburg, Virginia, College of William and Mary (Department of Geology Guidebook 1), pp 1–33

Johnson GH, Kruse SE, Vaughn AW, Lucey JK, Hobbs CH III, Powars DS (1998) Postimpact deformation associated with the late Eocene Chesapeake Bay impact structure in southeastern Virginia. Geology 26:507–510

Johnson SS (1977) Gravity map of Virginia, simple Bouguer anomaly. Virginia Division of Mineral Resources, scale 1:500,000

Jones MH (1990) Middle Eocene foraminifera from the Piney Point Formation of the Virginia and Maryland coastal plain. MS thesis, Old Dominion University, 244 pp

Jorissen FJ (1999) Benthic foraminiferal microhabitats below the sediment-water interface In: Sen Gupta BK (ed) Modern foraminifera. Kluwer Academic Publishers, Dordrecht, pp 161–179

Jorissen FJ, Barmawidjaja DM, Puskaric S, van der Zwaan GJ (1992) Vertical distribution of benthic foraminifera in the northern Adriatic Sea: The relation with high organic flux. Marine Micropaleontology 19:131–146

Jorissen FJ, De Stigter HC, Widmark JGV (1995) A conceptual model explaining benthic foraminiferal microhabitats: Marine Micropaleontology 26:3–15

Jorissen FJ, Wittling I, Peypouquet JP (1998) Live benthic foraminiferal faunas off Cape Blanc, NW Africa: Community structure and microhabitats. Deep-Sea Research 45:2157–2188

Kaiho K (1992) A low extinction rate of intermediate water benthic foraminifera at the Cretaceous/Tertiary boundary. Marine Micropaleontology 18:229–259

Kaminski MA, Boersma A, Tyszka J, Holhourn AEL (1995) Response of deep-water agglutinated foraminifera to dysoxic conditions in the California borderland basins. In: Kaminski MA, Geroch S, Gasinski MA (eds) Proceedings of the fourth international workshop on agglutinated foraminifera. Grzybowski Foundation Special Publication 3:131–140

Keigwin LD, Corliss BH (1986) Stable isotopes in late middle Eocene to Oligocene foraminifera. Geological Society of America Bulletin 97:335–345

Keller G (1986) Stepwise mass extinctions and impact events: Late Eocene to early Oligocene. Marine Micropaleontology 10:267–293

Keller G (1988) Biotic turnover in benthic foraminifera across the Cretaceous/Tertiary boundary at El Kef, Tunisia. Palaeogeography Palaeoclimatology Palaeoecology 66:153–177

Keller G (1992) Paleoecologic response of Tethyan benthic foraminifera to the Cretaceous/Tertiary boundary transition. In: Takayanagi Y, Saito T (eds) Studies in benthic foraminifera, Benthos '90, Sendai, Japan, 1990. Tokai University Press, pp 77–91

Keller G, D'Hondt SL, Orth CJ, Gilmore JS, Oliver PQ, Shoemaker EM, Molina E (1987) Late Eocene impact microspherules: Stratigraphy, age, and geochemistry. Meteoritics 22:25–60

Keller G, MacLeod N, Barrera E (1992) Eocene-Oligocene faunal turnover in planktic foraminifera, and Antarctic glaciation. In: Prothero DR, Berggren WA (eds) Eocene-Oligocene climatic and biotic evolution. Princeton University Press, Princeton, pp 218–243

Kendall CG St C, Levine PA, Erlich R (1995) The enigma of third-order sea level cycles: A cosmic connection? In: Haq BU (ed) Sequence stratigraphy and depositional response to eustatic, tectonic, and climatic forcing. Kluwer Academic Publishers, Dordrecht, pp 367–376

Kidwell SM (1984) Outcrop features and origin of basin margin unconformities in the lower Chesapeake Group (Miocene), Atlantic Coastal Plain. In: Schlee JS (ed) Interregional unconformities and hydrocarbon accumulation. American Association of Petroleum Geologists Memoir 36:37–58

Kieffer SW, Simonds CH (1980) The role of volatiles and lithology in the impact cratering process. Review of Geophysics and Space Physics 18:143–181

Kirschvink JL (1980) The least-squares line and plane and the analysis of paleomagnetic data. Geophysical Journal of the Royal Astronomical Society 62:699–718

Kiselev NP, Korotuschenko Yu G (1986) The Bigach astrobleme eastern Kazkhstan [in Russian]. Meteoritika 45:119–121

Klitgord, KD, Schneider CM (1994) Geophysical database of the East Coast of the United States Northern Atlantic Margin: Velocity analyses. US Geological Survey Open-File Report 94-19:1–74

Kobashi T, Grossman EL, Yancey TE, Dockery DT III (2001) Reevaluation of conflicting Eocene tropical temperature estimates: Molluskan oxygen isotope evidence for warm low latitudes. Geology 29:983–986

Koeberl C (1986) Geochemistry of tektites and impact glasses. Annual Review of Earth and Planetary Science 14:323–350

Koeberl C (1989) New estimates of area and mass for the North American tektite strewn field. Proceedings of the 19th Lunar and Planetary Science Conference, Houston, Cambridge University Press, pp 745–751

Koeberl C (1993) Instrumental neutron activation analysis of geochemical and cosmochemical samples: A fast and reliable method for small sample analysis. Journal of Radioanalytical and Nuclear Chemistry 168:47–60

Koeberl C (1994) Tektite origin by hypervelocity asteroidal or cometary impact: Target rocks, source craters, and mechanisms. In: Dressler BO, Grieve RAF, Sharpton VL (eds) Large meteorite impacts and planetary evolution. Geological Society of America Special Paper 293:133–151

Koeberl C (1998) Identification of meteoritic components in impactites. In: Grady MM, Hutchison R, McCall GJH, Rothery DA (eds) Meteorites: Flux with time and impact effects. Geological Society Special Publication 140:133–154

Koeberl C, Anderson RR (eds) (1996a) The Manson impact structure, Iowa: Anatomy of an impact crater: Geological Society of America Special Paper 302:1–468

Koeberl C, Anderson RR (1996b) Manson and company: Impact craters in the United States. In: Koeberl C, Anderson RR (eds) The Manson impact structure, Iowa: Anatomy of an impact crater. Geological Society of America Special Paper 302:1–29

Koeberl C, Glass BP (1988) Chemical composition of North American microtektites and tektite fragments from Barbados and DSDP Site 612 on the continental slope off New Jersey. Earth and Planetary Science Letters 87:286–292

Koeberl C, Poag CW, Reimold WU, Brandt D (1996) Impact origin of the Chesapeake Bay structure, and source of the North American tektites. Science 271:1263–1266

Koeberl C, Kruger FJ, Poag CW (2001) Geochemistry of surficial sediments near the Chesapeake Bay impact structure and the search for source rocks of the North American tektites [abs.]. Lunar and Planetary Science 32, abs. no. 1333 (CD-ROM)

Krinov EL (1966) Giant meteorites. Pergamon Press, New York, 397 pp

Kring DA (2000) Impact events and their effect on the origin, evolution, and distribution of life. GSA Today 10(8):1–7

Krs M, Novák F, Krsová M, Pruner P, Kouklíková L, Jansa J (1992) Magnetic properties and metastability of greigite-smythite mineralization in brown-coal basins of the Krušné hory Piedmont, Bohemia. Physics of the Earth and Planetary Interiors 70:273–287

Kuhnt W, Kaminski MA (1993) Changes in the community structure of deep water agglutinated foraminifers across the Cretaceous/Tertiary boundary in the Basque basin (Northern Spain). Revue Española Micropaleontologia 25:57–92

Lanci L, Lowrie W, Montanari A (1996) Magnetostratigraphy of the Eocene/Oligocene boundary in a short drill-core. Earth and Planetary Science Letters 143:37–48

Langenhorst F (1996) Characteristics of shocked quartz in the late Eocene impact ejecta from Massignano (Ancona, Italy): Clues to shock conditions and source crater. Geology 24:487–490

Larson JD (1981) Distribution of saltwater in the coastal plain aquifers of Virginia. US Geological Survey Open-File Report 81-1013:1–25

Lash G (1989) Documentation and significance of progressive microfabric changes in middle Ordovician trench mudstones. Geological Society of America Bulletin 101:1268–1279

Leahy PP, Martin M (1993) Geohydrology and simulation of ground-water flow in the northern Atlantic Coastal Plain aquifer system. US Geological Survey Professional Paper 1404-K:1–81

Lefort JP, Max M (1991) Is there an Archean crust beneath Chesapeake Bay? Tectonics 10:213–226

LeVan DC, Pharr RF (1963) A magnetic survey of the coastal plain in Virginia. Virginia Division of Mineral Resources Report of Investigations 4:1–17

Lindström M, Sturkell EFF, Törnberg R, Ormö J (1996) The marine impact crater at Lockne, central Sweden. GFF 118:193–206

Linke P, Altenbach A, Graf G, Heeger T (1995) Response of deep-sea benthic foraminifera to a simulated sedimentation event. Journal of Foraminiferal Research 25:75–82

Liu S, Glass BP (2002) The distribution of unmelted impact ejecta associated with the upper Eocene clinopyroxene-bearing (cpx) spherule layer [abs.]. Geological Society of America Abstracts with Programs 34(6):402

Liu S, Kyte FT, Glass BP, Gersonde R (2000) Upper Eocene spherules at Ocean Drilling Site 1090B [abs.]. Meteoritics and Planetary Science 35 (Supplement):A98

Liu S, Glass BP, Montanari A (2001) Detailed stratigraphy across the impact layer at Massignano, Italy [abs.]. Geological Society of America Abstracts with Programs 33(7):A310

Loeblich AR Jr, Tappan H (1988) Foraminiferal genera and their classification. Van Nostrand Reinhold, New York, 970 pp

Loubere P, Fariduddin M (1999) Benthic foraminifera and the flux of organic carbon to the seabed. In: Sen Gupta BK (ed) Modern foraminifera. Kluwer Academic Publishers, Dordrecht, pp 181–199

Lueck KLO, Snyder SW (1997) Lateral variations among populations of stained benthic foraminifera in surface sediments of the North Carolina continental shelf (U.S.A.). Journal of Foraminiferal Research 27:20–41

Lutze GF, Thiel H (1989) Epibenthic foraminifera from elevated microhabitats: *Cibicidoides wuellerstorfi* and *Planulina ariminensis*. Journal of Foraminiferal Research 19:153–158

Lyons PL, O'Hara NW (1982) Gravity anomaly map of the United States. Society of Exploration Geophysics, scale 1:2,500,000, 2 sheets

Mackensen A, Douglas RG (1989) Down-core distribution of live and dead deep-water benthic foraminifera in box cores from the Weddell Sea and the California continental borderland. Deep-Sea Research 36:879–900

MacLeod N (1990) Effects of late Eocene impacts on planktonic foraminifera. In: Sharpton VL, Ward PD (eds) Global catastrophes in Earth history. Geological Society of America Special Paper 247:595–606

MacLeod N, Keller G, Kitchell JA (1990) Progenesis in late Eocene populations of *Subbotina linaperta* (Foraminifera) from the western Atlantic. Marine Micropaleontology 16:219–240

Maher JC (1965) Correlations of subsurface Mesozoic and Cenozoic rocks along the Atlantic Coast. American Association of Petroleum Geologists, Cross Section Publication 3:1–18

Maher JC (1971) Geologic framework and petroleum potential of the Atlantic Coastal Plain and Continental Shelf. US Geological Survey Professional Paper 659:1–98

Maltman AJ (1994a) Prelithification deformation. In: Hancock PL (ed) Continental deformation. Pergamon Press, Oxford, pp 143–158

Maltman AJ (ed) (1994b) The geological deformation of sediments. Chapman and Hall, London, 362 pp

Mansfield WC (1928) New fossil mollusks from the Miocene of Virginia and North Carolina, with a brief outline of the divisions of the Chesapeake Group. US National Museum Proceedings 74:1–11

Manspeizer W (1988) Triassic-Jurassic rifting and opening of the Atlantic: An overview. In: Manspeizer W (ed) Triassic-Jurassic rifting. Elsevier, New York, pp 41–80

Mark K (1987) Meteorite craters. University of Arizona Press, Tucson, 302 pp

Marty R, Dunbar R, Martin JB, Baker P (1988) Late Eocene diatomite from the Peruvian coastal desert, coastal upwelling in the eastern Pacific, and Pacific circulation before the terminal Eocene event. Geology 16:818–822

Masaitis VL (1994) Impactites from Popigai crater. In: Dressler BO, Grieve RAF, Sharpton VL (eds) Large meteorite impacts and planetary evolution. Geological Society of America Special Paper 293:153–162

Masaitis VL (1999) Impact structures of northeastern Eurasia: The territories of Russia and adjacent countries. Meteoritics and Planetary Science 34:691–712

Masaitis VL (2002) The middle Devonian Kaluga impact crater (Russia): new interpretation of marine setting. Deep-Sea Research Part II 49:1157–1169

Masaitis VL, Mikhailov MV, Selivanovskaya TV (1975) Popigai meteorite crater [in Russian]. Nauka Press, Moscow, 124 pp

Masaitis VL, Naumov MV, Mashchak MS (1999) Anatomy of the Popigai impact crater, Russia. In: Dressler BO, Sharpton VL (eds) Large meteorite impacts and planetary evolution II. Geological Society of America Special Paper 339:1–17

Maurrasse F, Glass BP (1976) Radiolarian stratigraphy and North American microtektites in Caribbean RC9-58: Implications concerning late Eocene radiolarian chronology and the age of the Eocene-Oligocene boundary. Transactions of the Seventh Caribbean Geological Conference, Guadalupe, 1974, pp 205–212

McFarland EM (2002) Hydrochemical evidence for the origin of elevated ground-water salinity in the Chesapeake Bay impact structure, southeastern Virginia [abs.]. Geological Society of America Abstracts with Programs 34(6):466

McFarland EM, Bruce TS (2002) Distribution, origin, and relations to flow of salty ground water along the western margin of the Chesapeake Bay impact structure in eastern Virginia [abs.]. American Geophysical Union, Spring Meeting Abstract T21A-08, p 8

McGhee GR Jr (2001) The "multiple impacts hypothesis" for mass extinction: a comparison of the Late Devonian and the late Eocene. Palaeogeography Palaeoclimatology Palaeoecology 176:47–58

McGlaun JM, Thompson SL (1990) CTH: A three-dimensional shock-wave physics code. International Journal of Impact Engineering 10:351–360

McHugh CM, Ryan WBF, Schreiber C (1993) The role of diagenesis in exfoliation of submarine canyons. American Association of Petroleum Geologists Bulletin 77:145–172

McHugh CMG, Snyder SW, Deconinck J-F, Saito Y, Katz ME, Aubry M-P (1996) Upper Eocene tektites of the New Jersey continental margin, Site 904. In: Mountain GS, Miller KG, Blum P, Poag CW, Twichell DC (eds) Proceedings of the Ocean Drilling Program, scientific results. Ocean Drilling Program, College Station, Texas, 150:241–265

McKinnon WB (1982) Impact into the Earth's ocean floor: Preliminary experiments, a planetary model, and possibilities for detection. In: Silver LT, Schultz PH (eds) Geological implications of impacts of large asteroids and comets on the Earth. Geological Society of America Special Paper 190:129–142

McKinnon WB, Goetz P (1981) Impact into the earth's ocean floor during the last billion years: Preliminary experiments, theoretical models, and possibilities for geological detection. Conference on large body impacts and terrestrial evolution: Geological, climatological, and biological implications. Lunar and Planetary Institute, Houston, pp 1–34

Meisler H (1981) Preliminary delineation of salty ground water in the northern Atlantic Coastal Plain. US Geological Survey Open-File Report 81-71:1–39

Meisler H (1989) The occurrence and geochemistry of salty ground water in the Northern Atlantic Coastal Plain. US Geological Survey Professional Paper 1404-D:1–51

Melosh HJ (1981) Atmospheric breakup of terrestrial impactors. In: Schultz PH, Merrill RB (eds) Multiring basins. Pergamon Press, New York, pp 29–35

Melosh HJ (1982) A schematic model of crater modification by gravity. Journal of Geophysical Research 87:371–380

Melosh HJ (1989) Impact cratering – A geologic process. Oxford University Press, New York, 245 pp

Melosh HJ, Ivanov BA (1999) Impact crater collapse: Annual Reviews of Earth and Planetary Science 27:385–415

Melosh HJ, Schneider NM, Zahnle KJ, Latham D (1990) Ignition of global wildfires at the Cretaceous/Tertiary boundary. Nature 343:251–254

Meng AA III, Harsh JF (1988) Hydrogeologic framework of the Virginia Coastal Plain. US Geological Survey Professional Paper 1404-C:1–82

Milici RC, Costain JK, Coruh C, Pappano PA (1995) Structural section across the Atlantic Coastal Plain of Virginia and southeasternmost Maryland. Virginia Department of Mines, Minerals, and Energy Publication 140:2 plates

Miller KG (1992) Middle Eocene to Oligocene stable isotopes, climate, and deep-water history: The terminal Eocene event? In: Prothero DR, Berggren WA (eds) Eocene-Oligocene climatic and biotic evolution. Princeton University Press, Princeton, pp 160–177

Miller KG, Curry WB, Ostermann DR (1985) Late Paleogene (Eocene to Oligocene) benthic foraminiferal oceanography of the Goban Spur region, Deep-Sea Drilling Project Leg 80. In: Graciansky PC de, Poag CW, Cunningham R Jr, Loubere P, Masson DG,

Mazzullo JM, Montadert L, Müller C, Otsuka K, Reynolds L, Sigal J, Snyder S, Townsend HA, Vaos SP, Waples D, Initial reports of the Deep Sea Drilling Project. Washington, DC, US Government Printing Office 80:505–531

Miller KG, Fairbanks RG, Mountain GS (1987) Tertiary oxygen isotope synthesis, sea level history, and continental margin erosion. Paleoceanography 2:1–19

Miller KG, Berggren WA, Zhang J, Palmer-Julson AA (1991) Biostratigraphy and isotope stratigraphy of upper Eocene microtektites at Site 612; How many impacts? PALAIOS 6:17–38

Miller KG, Sugarman PJ, Browning JV, Olsson RK, Pekar SF, Reilly TJ, Cramer BS, Aubry M-P, Lawrence RP, Curran J, Stewart M, Metzger J.M., Uptegrove J, Buckry D, Burckle LH, Wright JD, Feigenson MD, Brenner GJ, Dalton RF (1998) Bass River site. In: Miller KG, Sugarman PJ, Browning JV, Aubry M-P, Brenner GJ, Bukry D, Burckle LH, Cramer BS, Curran J, Dalton RF, Feigenson MD, Lawrence RP, Metzger JM, Olsson RK, Pekar SF, Reilly TJ, Stewart M, Uptegrove J, Wright JD, Proceedings of the Ocean Drilling Program, scientific results. Ocean Drilling Program, College Station, Texas 174AX:5–43

Mixon RB (1985) Stratigraphic and geomorphic framework of uppermost Cenozoic deposits in the southern Delmarva Peninsula, Virginia and Maryland. US Geological Survey Professional Paper 1067-G:1–53

Mixon RB, Berquist CR Jr, Newell WL, Johnson GH, Powars DS, Schindler JS, Rader EK (1989) Geologic map and generalized cross sections of the coastal plain and adjacent parts of the Piedmont, Virginia. US Geological Survey Miscellaneous Investigations Series Map I-2033, 2 sheets, scale 1:250,000

Montanari A, Asaro F, Michel HV, Kennett JP (1993) Iridium anomalies of late Eocene age at Massignano (Italy) and ODP Site 689B, Maud Rise, Antarctic. PALAIOS 8:420–437

Moore JC, Roeske S, Lundberg N (1986) Scaly fabrics from Deep Sea Drilling Project cores from forearcs. Geological Society of America Memoir 166:55–73

Morgan J, Warner M (1999a) Chicxulub: The third dimension of a multi-ring impact basin. Geology 27:407–410

Morgan J, Warner M (1999b) Morphology of the Chicxulub impact: Peak-ring crater or multi-ring basin? In: Dressler BO, Sharpton VL (eds) Large meteorite impacts and planetary evolution II. Geological Society of America Special Paper 339:281–290

Morgan J, Warner M, Chicxulub Working Group (1997) Size and morphology of the Chicxulub impact crater. Nature 390:472–476

Morgan J, Warner M, Grieve R (2002) Geophysical constraints on the size and structure of the Chicxulub impact crater. In Koeberl C, McLeod KG (eds) Catastrophic events and mass extinctions: Impacts and beyond. Geological Society of America Special Paper 356:39–46

Morigi C, Jorissen FJ, Gervais A, Guichard S, Borsetti AM (2001) Benthic foraminiferal faunas in surface sediments off NW Africa: Relationship with organic flux to the ocean floor. Journal of Foraminiferal Research 31:350–368

Morrison D, Chapman CR, Slovic P (1994) The impact hazard. In: Gehrels T (ed) Hazards due to comets and asteroids. University of Arizona Press, Tucson, pp 59–91

Morrison RH, Oberbeck VR (1978) A composition and thickness model for lunar impact crater and basin deposits. Proceedings of the 9th Lunar and Planetary Conference, Journal of Geophysical Research 83:3763–3785

Movshovich EV, Milyavsky AE (1990) Morphology and inner structure of Kamensk and Gusev astroblemes. In: Masaitis VL (ed) Impact craters of the Mesozoic-Cenozoic Boundary [in Russian]. Nauka Press, Leningrad, pp 110–146

Murali AV, Zolensky ME, Blanchard DP (1987) Tektite-like bodies at Lonar crater, India: Implications for the origin of tektites. Proceedings of the17th Lunar and Planetary Science Conference, Journal of Geophysical Research 92:E729–E735

Murosky MW, Snyder SW (1992) Epifaunal and infaunal distribution of living (stained) foraminifera from the mid-Atlantic continental shelf [abs.]. Geological Society of America Abstracts with Programs 24(7):75

Murray GE (1961) Geology of the Atlantic and Gulf Coastal Province of North America. New York, Harper and Brothers, 692 pp

Nemtchinov IV, Loseva TV, Teterev AV (1996) Impacts into oceans and seas. Earth, Moon, and Planets 72:405–418

Nerem RS, van Dam TM, Schenewerk MS (1998) Chesapeake Bay subsidence monitored as wetlands loss continues. EOS Transactions of the American Geophysical Union 79(149):156–157

Neukum G, Oberst J, Hoffmann H, Wagner R, Ivanov BA (2001) Geologic evolution and cratering of Mercury. Planetary and Space Science 49:1507–1521

Newsom HE, Graup G, Iseri DA, Geissman JW, Keil K (1990) The formation of the Ries crater, West Germany: Evidence of atmospheric interactions during large cratering event. In: Sharpton VL, Ward PD (eds) Global catastrophes in earth history. Geological Society of America Special Paper 247:195–206

Nguyen Tkhi Kim Tkhoa, Pecherskiy DM (1984) Experimental study of remanent crystallization magnetization of magnetite formed during the oxidation of pyrite. Izvestia Academia Nauk SSSR, Fizika Zemli 20:362–373.

Nogan DS (1964) Foraminifera, stratigraphy, and paleoecology of the Aquia Formation of Maryland and Virginia. Cushman Foundation for Foraminiferal Research Special Publication 7:1–50

Nomura R (1991) Paleoceanography of upper Maastrichtian to Eocene benthic foraminiferal assemblages at Sites 752, 753, and 754, Eastern Indian Ocean. In: Weissel J, Pierce J, Taylor J, Dehn J, Driscoll N, Farrell, J, Fortanier E, Frey F, Ganson PD, Gee JS, Gibson IL, Janacek T, Klootwijk C, Lawrence JR, Littke R, Newman JS, Nomura R, Owen RM, Pospichal JJ, Rea DK, Resiwati P, Saunders AD, Smit J, Smith GM, Tamaki K, Weis D, Wilkinson C, Proceedings of the Ocean Drilling Program, scientific results. Ocean Drilling Program, College Station, Texas 121:3–29

North American Stratigraphic Commission (1983) North American stratigraphic code. American Association of Petroleum Geologists Bulletin 67:841–875

Norris RD, Röhl U (1999) Carbon cycling and chronology of climate warming during the Paleocene/Eocene transition. Nature 401:775–778

Oaks RQ Jr, Coch NK (1973) Post-Miocene stratigraphy and morphology, southeastern Virginia. Virginia Division of Mineral Resources Publication 82:1–116

Obradovich J, Snee LW, Izett GA (1989) Is there more than one glassy impact layer in the late Eocene? [abs.]. Geological Society of America Abstracts with Programs 21(7):134

Oberbeck VR (1975) The role of ballistic erosion and sedimentation in lunar stratigraphy. Review of Geophysics and Space Physics 13:337–362

Oberbeck VR, Marshall JR, Aggarwal H (1993) Impacts, tillites, and the breakup of Gondwanaland. Journal of Geology 101:1–19

Odin GS, Montanari A (1989) Age radiométrique et stratotype de la limite Éocène Oligocène. CR Academie du Sciences Paris 309:1039–1045

O'Keefe JD, Ahrens TJ (1989) Impact production of CO_2 by K-T extinction bolide and the resultant heating of the Earth. Nature 338:247–249

O'Keefe JD, Ahrens TJ (1997) Complex craters: Relationship of inner rings to the impactor and transient crater evolution [abs.]. Lunar and Planetary Science 28:1043

O'Keefe JD, Ahrens TJ (1999) Complex craters: Relationship of stratigraphy and rings to the impact conditions. Journal of Geophysical Research 104:27,091–27,104

Olsson RK (1960) Foraminifera of Cretaceous and earliest Tertiary age in the New Jersey Coastal Plain. Journal of Paleontology 34:1–58

Olsson RK, Miller KG, Browning JV, Habib D, Sugarman PJ (1997) Ejecta layer at the Cretaceous-Tertiary boundary, Bass River, New Jersey (Ocean Drilling Program, Leg 174AX). Geology 25:759–762

Orbigny AD d' (1846) Foraminifères fossiles du bassin Tertiare de Vienne (Autriche). Gide et Comp, Paris, 312 pp

Ormö J (1998) Impact cratering at sea. Meddelanden från Stockholms Universitets Institution för Geologi Och Geokemi 300, Ph.D. thesis, University of Stockholm, 206 pp

Ormö J, Lindström M (2000) When a cosmic impact strikes the seabed. Geological Magazine 137:67–80

Ormö J, Miyamoto H (2002) Computer modeling of the water resurge at a marine impact: The Lockne crater, Sweden. Deep-Sea Research Part II 49:983–994

Ormö J, Shuvalov VV, Lindström M (2002) Numerical modeling for target water depth estimation of marine-target impact craters. Journal of Geophysical Research 107(E12), 5120, doi:10.1029/2002JE001865

Otton EG (1955) Ground-water resources of the southern Maryland Coastal Plain. Maryland Department of Geology, Minerals, and Water Resources Bulletin 15:1–347

Owens JP, Gohn GS (1985) Depositional history of the Cretaceous Series in the U.S. Atlantic Coastal Plain; Stratigraphy, paleoenvironments, and tectonic controls of sedimentation. In: Poag CW (ed) Geologic evolution of the United States Atlantic Margin. Van Nostrand Reinhold, New York, pp 25–86

Pearson PN, Ditchfield PW, Singano J, Harcourt-Brown KG, Nicholas CJ, Olsson RK, Shackleton NJ, Hall MA (2001) Warm tropical sea surface temperatures in the Late Cretaceous and Eocene epochs. Nature 413:481–487

Peebles PC (1984) Late Cenozoic landforms, stratigraphy, and history of sea level oscillations of southeastern Virginia and northeastern North Carolina. Ph.D. thesis, College of William and Mary, 148 pp

Peucker-Ehrenbrink B, Schmitz B (eds) (2001) Accretion of extraterrestrial matter throughout Earth's history. Plenum Publishers Inc, New York, 492 pp

Pfannkuche O (1993) Benthic response to the sedimentation of particulate organic matter at the BIOTRANS station, 47°N, 20°W. Deep-Sea Research 40:135–139

Phleger FB, Soutar A (1973) Production of benthic foraminifera in three east Pacific oxygen minima. Micropaleontology 19:110–115

Pierrard O, Robin E, Rocchia R, Montanari A (1998) Extraterrestrial Ni-rich spinel in late Eocene sediments from Massignano. Geology 26:307–310

Pierrard O, Robin E, Rocchia R, Lefevre I, Smit J, Vonhof H (1999) Late Eocene Ni-rich spinel from LL44-GPC3 (Central Pacific), ODP Site 689B (Maud Rise, Antarctic), DSDP Sites 94 (Gulf of Mexico) and 612 (U.S. East Coast) [abs.]. Lunar and Planetary Science 30, abs. no. 1674 (CD-ROM)

Pike RJ (1988) Geomorphology of impact craters on Mercury. In: Vilas F, Chapman CR, Matthews MS (eds) Mercury. University of Arizona Press, Tucson, pp 165–273

Pilkington M, Grieve RAF (1992) The geophysical signature of terrestrial impact craters. Review of Geophysics 30:161–181

Pilkington M, Hildebrand AR, Ortiz-Aleman C (1994) Gravity and magnetic field modeling and structure of the Chicxulub crater, Mexico. Journal of Geophysical Research 99:13,147–13,162

Pilkington M, Jansa LF, Grieve RAF (1995) Geophysical studies of the Montagnais impact crater, Canada. Meteoritics 30:446–450

Plescia JB (1996) Gravity investigation of the Manson impact structure, Iowa. In: Koeberl C, Anderson RR (eds) The Manson impact structure: Anatomy of an impact crater. Geological Society of America Special Paper 302:89–104

Plummer HJ (1927) Foraminifera of the Midway Formation in Texas. Texas University Bulletin 2644:1–205

Poag CW (1981) Ecologic atlas of benthic foraminifera of the Gulf of Mexico. Marine Science International, Woods Hole, Massachusetts, 174 pp

Poag CW (1985) Depositional history and stratigraphic reference section for central Baltimore Canyon Trough. In: Poag CW (ed) Geologic evolution of the United States Atlantic Margin. Van Nostrand Reinhold, New York, pp 217–263

Poag CW (1993) U.S. middle Atlantic continental rise: Provenance, dispersal, and deposition of Jurassic to Quaternary sediments. In: Poag CW, Graciansky PC de (eds) Geologic evolution of Atlantic Continental Rises. Van Nostrand Reinhold, New York, pp 100–156

Poag CW (1996) Structural outer rim of the Chesapeake Bay impact crater: Seismic and bore hole evidence. Meteoritics and Planetary Science 31:218–226

Poag CW (1997a) The Chesapeake Bay bolide impact: A convulsive event in Atlantic Coastal Plain evolution. In: Segall MP, Colquhoun DJ, Siron D (eds) Evolution of the Atlantic Coastal Plain – Sedimentology, stratigraphy, and hydrogeology. Sedimentary Geology 108:45–95

Poag CW (1997b) Roadblocks on the kill curve: Testing the Raup hypothesis. PALAIOS 12: 582–590

Poag CW (1999a) Secondary craters from the Chesapeake Bay impact [abs.]. Lunar and Planetary Science 30, abs. no. 1047 (CD-ROM)

Poag CW (1999b) Depositional regimes during emplacement of the Exmore breccia (Chesapeake Bay impact crater): Clues from downhole geophysical logs [abs.]. Geological Society of America Abstracts with Programs 31(7):A-64

Poag CW (1999c) Chesapeake Invader. Princeton University Press, Princeton, 183 pp

Poag CW (2000) Nature and distribution of deposits derived from the Chesapeake Bay submarine impact [abs.]. Geological Society of America Abstracts with Programs 32(7):A163

Poag CW (2002a) Tsunamis and hypercanes: Late-stage impact depositional systems at Chesapeake Bay crater [abs.]. American Geophysical Union, Spring Meeting Abstract T21A-06, p 6

Poag CW (2002b) Synimpact-postimpact transition inside Chesapeake Bay crater. Geology 30:995–998

Poag CW, Aubry M-P (1995) Upper Eocene impactites of the U.S. East Coast: Depositional origins, biostratigraphic framework, and correlation. PALAIOS 10:16–43

Poag CW, Chesapeake Coring Team (2001) Drilling to basement inside the Chesapeake Bay crater [abs.]. Lunar and Planetary Science 32, abs. no. 1203 (CD ROM)

Poag CW, Commeau JA (1995) Paleocene to middle Miocene planktonic foraminifera of the southwestern Salisbury Embayment, Virginia and Maryland: Biostratigraphy, allostratigraphy, and sequence stratigraphy. Journal of Foraminiferal Research 25:134–155

Poag CW, Foster DS (2000) Chesapeake Bay impact crater: New seismic evidence of a central peak [abs.]. Lunar and Planetary Science 31, abs. no. 1358 (CD-ROM)

Poag CW, Norris RD (in press) Early postimpact deposition and paleoenvironments near the southwestern margin of Chesapeake Bay crater: The Chickahominy Formation in the USGS-NASA Langley corehole. In: Gohn GS (ed) Geological and geohydrological studies of the USGS-NASA Langley corehole, Hampton, Virginia. US Geological Survey Professional Paper

Poag CW, Poppe LJ (1998) The Toms Canyon structure, New Jersey Outer Continental Shelf: A possible late Eocene impact crater. Marine Geology 145:23–60

Poag CW, Sevon WD (1989) A record of Appalachian denudation in postrift Mesozoic and Cenozoic sedimentary deposits of the U.S. Middle Atlantic continental margin. Geomorphology 2:119–157

Poag CW, Ward LW (1993) Allostratigraphy of the U.S. Middle Atlantic continental margin – Characteristics, distribution, and depositional history of principal unconformity-bounded Upper Cretaceous and Cenozoic sedimentary units. US Geological Survey Professional Paper 1542:1–81

Poag CW, Watts AB, Cousin M, Goldberg D, Hart MB, Miller KG, Mountain GS, Nakamura Y, Palmer A, Schiffelbein PA, Schreiber BC, Tarafa M, Thein JE, Valentine PC, Wilkens RH (1987) Initial reports of the Deep Sea Drilling Project. US Government Printing Office, Washington, DC 95:1–817

Poag CW, Powars DS, Poppe LJ, Mixon RB, Edwards LE, Folger DW, Bruce S (1992) Deep Sea Drilling Project Site 612 bolide event: New evidence of a late Eocene impact-wave deposit and a possible impact site. Geology 20:771–774

Poag CW, Powars DS, Poppe LJ, Mixon RB (1994) Meteoroid mayhem in Ole Virginny: Source of the North American tektite strewn field. Geology 22:691–694

Poag CW, Hutchinson D R, Colman SM, Lee MW (1999) Seismic expression of the Chesapeake Bay impact crater: Structural and morphologic refinements based on new seismic data. In: Dressler BO, Sharpton VL (eds) Impact cratering and planetary evolution II. Geological Society of America Special Paper 339:149–164

Poag CW, Gohn GS, Powars DS (2001) From shocked basement to fallout spherules: The coring record at the Chesapeake Bay crater [abs.]. Geological Society of America Abstracts with Programs 33(7):433

Poag CW, Plescia JB, Molzer PC (2002) Ancient impact structures on modern continental shelves: The Chesapeake Bay, Montagnais, and Toms Canyon craters, Atlantic margin of North America. Deep-Sea Research Part II 49:1081–1102

Poag CW, Mankinen E, Norris RD (2003) Late Eocene impacts: Geologic record, correlation, and paleoenvironmental consequences. In: Prothero DR, Yvany LC, Nesbitt E (eds) From greenhouse to icehouse: The marine Eocene-Oligocene transition. Columbia University Press, New York, pp 495–510

Pohl J, Stöffler D, Gall H, Ernstson K (1977) The Ries impact crater. In: Roddy DJ, Pepin RO, Merrill RB (eds) Impact and explosion cratering. Pergamon Press, New York, pp 343–404

Pope KO (2002) Impact dust not the cause of the Cretaceous-Tertiary mass extinction. Geology 30:99–102

Posamentier HW, Vail PR (1988) Eustatic controls on clastic deposition II: Sequence and systems tract models. In: Wilgus CK, Hastings BS, Kendall CGStC, Posamentier HW, Ross CA, Van Wagoner JC (eds) Sea-level changes: An integrated approach. SEPM Special Publication 42:125–154

Pospichal JJ (1994) Calcareous nannofossils and the K/T boundary, El Kef: No evidence for stepwise, gradual, or sequential extinctions. Geology 22:99–102

Powars DS (2000) The effects of the Chesapeake Bay impact crater on the geologic framework and the correlation of hydrogeologic units of southeastern Virginia, south of the James River. US Geological Survey Professional Paper 1622:1–53

Powars DS, Bruce TS (1999) The effects of the Chesapeake Bay impact crater on the geologic framework and the correlation of hydrogeologic units of the lower York-James Peninsula, Virginia. US Geological Survey Professional Paper 1612:1–82

Powars DS, Mixon RB, Edwards LE, Poag C W, Bruce S (1990) Cross section of Cretaceous and Cenozoic strata, Norfolk Arch to Salisbury Basin, outer coastal plain of Virginia [abs.]. Geological Society of America Abstracts with Programs 22(7):57

Powars DS, Poag CW, Bruce S (1991) Uppermost Mesozoic and Cenozoic stratigraphic framework of the central and outer coastal plain of Virginia [abs.]. Geological Society of America Abstracts with Programs 23(7):117

Powars DS, Mixon RB, Bruce S (1992) Uppermost Mesozoic and Cenozoic geologic cross section, outer coastal plain of Virginia. In: Gohn GS (ed) Proceedings of the 1988 US Geological Survey workshop on the geology and geohydrology of the Atlantic Coastal Plain. US Geological Survey Circular 1059:85–101

Powars DS, Bruce TS, Bybell LM, Cronin TM, Edwards LE, Frederiksen, NO, Gohn GS, Horton JW Jr, Izett GA, Johnson GH, Levine JS, McFarland ER, Poag CW, Quick J E, Schindler JS, Self-Trail JM, Smith MJ, Stamm RG, Weems RE (2001a) Preliminary geologic summary for the USGS-NASA Langley corehole, Hampton, Virginia. US Geological Survey Open-File Report 01-87-B:1–19

Powars DS, Gohn GS, Edwards LE, Bruce TS, Catchings RD, Emry SR, Johnson GH, Levine JS, Poag CW, Pierce HA (2001b) Structure and composition of the southwest margin of the buried Chesapeake Bay impact structure, Virginia [abs.]. Geological Society of America Abstracts with Programs 33(6):463

Prothero DR (1994) The Eocene-Oligocene transition: Paradise lost. Columbia University Press, New York, 291 pp

Prothero DR, Berggren WA (eds) (1992) Eocene-Oligocene climatic and biotic evolution. Princeton University Press, Princeton, pp 1–28

Prothero DR, Emry RJ (1996) The terrestrial Eocene-Oligocene transition in North America. Cambridge University Press, Cambridge, 688 pp

Quaide WL, Oberbeck VR (1968) Thickness determinations of the lunar surface layer from lunar impact craters. Journal of Geophysical Research 73:5247–5270

Rampino MR, Haggerty BM (1994) Extraterrestrial impacts and mass extinctions of life. In: Gehrels T (ed) Hazards due to comets and asteroids. University of Arizona Press, Tucson, pp 827–858

Rathburn AE, Corliss BH (1994) The ecology of living (stained) deep-sea benthic foraminifera from the Sulu Sea. Paleoceanography 9:87–150

Raup DM, (1991a) A kill curve for Phanerozoic marine species. Paleobiology 17:37–48

Raup DM (1991b) Extinction – bad genes or bad luck? WW Norton, New York, 210 pp

Reimold WU (1988) Shock experiments with preheated Witwatersrand quartzite and the Vredefort microdeformation controversy [abs.]. Lunar and Planetary Science 19, Lunar and Planetary Institute 970–971

Reimold WU (1995) Pseudotachylite-generation by friction melting and shock brecciation? – A review and discussion. Earth-Science Reviews 39:247–264

Reimold WU (1998) Exogenic and endogenic breccias: A discussion of major problematics. Earth-Science Reviews 43:25–47

Reimold WU, Koeberl C, Bishop J (1994) Roter Kamm impact crater, Namibia: Geochemistry of basement rocks and breccias. Geochemica et Cosmochemica Acta 58:2689–2710

Reinhardt J, Newell WL, Mixon RB (1980) Tertiary lithostratigraphy of the core. In: Geology of the Oak Grove core. Virginia Division of Mineral Resources Publication 20:1–13

Richards HG (1945) Subsurface stratigraphy of Atlantic Coastal Plain between New Jersey and Georgia. American Association of Petroleum Geologists Bulletin 29:885–995

Richards HG (1967) Stratigraphy of the Atlantic Coastal Plain between Long Island and Georgia: Review. American Association of Petroleum Geologists Bulletin 51:2400–2429

Richards HG, Straley WH III (1953) Geophysical and stratigraphic investigations of the Atlantic Coastal Plain. Georgia Geological Survey Bulletin 60:101–115

Robb JM, Kirby JR, Hampson JC Jr, Gibson PR, Hecker B (1983) Furrowed outcrops of Eocene chalk on the lower continental slope offshore New Jersey. Geology 11:182–186

Robbins EI, Perry WJ Jr, Doyle JA (1975) Palynological and stratigraphic investigations of four deep wells in the Salisbury Embayment of the Atlantic Coastal Plain. US Geological Survey Open-File Report 75-307:1–120

Rochette P, Fillion G, Mattéi J-L, Dekkers MJ (1990) Magnetic transition at 30–40 Kelvin in pyrrhotite: Insight into a widespread occurrence of this mineral in rocks. Earth and Planetary Science Letters 98:319–328

Roddy DJ (1977) Large-scale impact and explosion craters: Comparison of morphological and structural analogs. In: Roddy DJ, Pepin RO, Merrill RB (eds) Impact and explosion cratering. Pergamon Press, New York, pp 185–246

Roddy DJ, Schuster SH, Rosenblatt M, Grant LB, Hassig PJ, Kreyenhagen KN (1987) Computer simulations of large asteroid impacts into oceanic and continental sites: Preliminary results on atmospheric, cratering, and ejecta dynamics. International Journal of Impact Engineering 5:525–541

Rollinson H (1993) Using geochemical data: Evaluation, presentation, interpretation. Longman, New York, 352 pp

Russell GS, Russell C (1980) Rb-Sr isotopic dating of granitic core from Portsmouth, Virginia. In: Costain JK, Glover L III (principal investigators), Evaluation and targeting of geothermal energy resources in the southeastern United States. Progress report to US Department of Energy, October 1, 1979–June 30, 1980, Contract No. DE-AC05-78ET27001:A64–A68

Ryder G (1990) Lunar samples, lunar accretion and early bombardment of the Moon. Eos Transactions of the American Geophysical Union 71:322–323

Ryder G, Fastovsky D, Gartner S (eds) (1996) The Cretaceous-Tertiary event and other catastrophes in Earth history. Geological Society of America Special Paper 307:1–569

Ryder G, Koeberl C, Mojzsis SJ (2000) Heavy bombardment on the Earth at ~3.85 Ga: The search for petrographic and geochemical evidence. In: Canup RM, Righter K (eds) Origin of the Earth and Moon. University of Arizona Press, pp 475–492

Sabet MA (1973) Gravity and magnetic investigation, Eastern Shore area, Virginia. Geological Society of America Bulletin 84:2119–2126

Sanfilippo A, Riedel WR, Glass BP, Kyte FT (1985) Late Eocene microtektites and radiolarian extinctions on Barbados. Nature 314:613–615

Sanford S (1911) Saline artesian waters of the Atlantic Coastal Plain. In: Fuller ML (ed) Underground-water Papers, 1910. US Geological Survey Water-Supply Paper 258:75–86

Sanford S (1913) The underground water resources of the Coastal Plain Province of Virginia. Virginia Geological Survey Bulletin 5:1–361

Sanford WE (2002) Possible hydrothermal activity following the Chesapeake Bay bolide impact [abs.]. Geological Society of America Abstracts with Programs 34(6):466

Saunders JB, Bernoulli D, Müller-Merz E, Oberhänsli H, Perch-Nielsen K, Riedel WR, Sanfilippo A, Torrini R Jr (1984) Stratigraphy of the late middle Eocene to early Oligocene in the Bath Cliff section, Barbados, West Indies. Micropaleontology 30:390–425

Schmitz B, Keller G, Stenvall O (1992) Stable isotope and foraminiferal changes across the Cretaceous-Tertiary boundary at Stevns Klint, Denmark: Arguments for long-term oceanic instability before and after bolide-impact event. Palaeogeography Palaeoclimatology Palaeoecology 96:233–260

Schoenberg R, Kamber BS, Collerson KD, Moorbath S (2002) Tungsten isotope evidence from ~3.8-Gyr metamorphosed sediments for early meteorite bombardment of the Earth. Nature 418:403–405

Schultz PH, Anderson RR (1996) Asymmetry of the Manson impact structure: Evidence for impact angle and direction. In: Koeberl C, Anderson RR (eds) The Manson impact structure, Iowa: Anatomy of an impact crater. Geological Society of America Special Paper 302:397–417

Schwarz EJ (1973) Magnetic characteristics of massive sulfide ore bodies near Sudbury, Canada. Canadian Journal of Earth Science 10:1735–1743

See TH, Mittlefehldt DW, Hörz F (1989) Analysis of aeroballistically dispersed glass samples from Wabar Crater, Saudi Arabia [abs.]. Lunar and Planetary Science 20:980

Sen Gupta BK (ed) (1999) Modern foraminifera. Kluwer Academic Publishers, Dordrecht, 371 pp

Sen Gupta BK, Turner RE, Rabalais NN (1996) Seasonal oxygen depletion in continental-shelf waters of Louisiana: Historical record of benthic foraminifers. Geology 24:227–230

Shackleton NJ (1986) Paleogene stable isotope events. Palaeogeography Palaeoclimatology Palaeoecology 57:91–102

Sharpton VL, Burke K, Camargo-Z A, Hall SA, Lee S, Marin LE, Suarez-R G, Quezada-M JM, Spudis PD, Urrutia-F J (1993) Chicxulub multi-ring impact basin: Size and other characteristics derived from gravity analysis. Science 261:1564–1567

Sharpton VL, Marin LE, Carney JL, Lee S, Ryder G, Schuraytz BC, Sikora P, Spudis PD (1996) Model of the Chicxulub impact basin. In: Ryder G, Fastovsky D, Gartner S (eds) The Cretaceous-Tertiary event and other catastrophes in earth history. Geological Society of America Special Paper 307:55–74

Shattuck GB (1902) The Miocene Formation of Maryland. Science 15:906

Shattuck GB (1904) Geological and paleontological relations with a review of earlier investigations. Maryland Geological Survey Miocene Volume, pp 33–94

Shoemaker EM (1962) Interpretation of lunar craters. In: Kopal Z (ed) Physics and astronomy of the Moon. Academic Press, New York, pp 283–330

Shoemaker EM (1963) Impact mechanics at Meteor Crater, Arizona. In: Middlehurst BM, Kuiper GP (eds) The Moon, meteorites, and comets. University of Chicago Press 4:301–336

Shuvalov VV (1999) Multi-dimensional hydrodynamic code SOVA for interfacial flows: Application to thermal layer effect. Shock Waves 9:381–390

Shuvalov V, Dypvik H, Tsikalas P (2002) Numerical simulations of the Mjølnir marine impact crater. Journal of Geophysical Research 107(E7), 5047, doi:10.1029/2001JE001698

Sibol MS, Snoke JA, Mathena EC (1996) Southeastern United States Seismic Network Bulletin. Blacksburg, Virginia, Seismological Observatory. Virginia Polytechnical Institute and State University 30:1–62

Sibol MS, Bollinger GA, Chapman MC (1997) Catalog of southeastern United States seismicity compiled from various sources: Blacksburg, Virginia, Seismological Observatory. Virginia Polytechnical Institute and State University (electronic format)

Sigurdsson H, D'Hondt SL, Carey S (1992) The impact of the Cetaceous/Tertiary bolide on evaporite terrane and generation of major sulfuric acid aerosol. Earth and Planetary Science Letters 109:543–559

Silver LT (1982) Introduction. In: Silver LT, Schultz PH (eds) Geological implications of impacts of large asteroids and comets on the earth. Geological Society of America Special Paper 190:xii–xix

Simon S (1987) Stratigraphie, Petrographie, und Entstehungsbedingungen von Grobklastika in der autochtonen ordovizischen Schichtenfolge Jämtlands (Schweden). Sveriges Geologiska Undersökning C 815:1–156

Smelror M, Kelly SRA, Dypvik H, Mørk A, Nagy J, Tsikalas F (2001) Mjølnir (Barents Sea) meteorite impact ejecta offers a Volgian-Ryazanian boundary marker. Newsletters on Stratigraphy 38:129–140

Smit J, Dressler B, Buffler R, Moran Zeteno D, Sharpton VL, Stöffler D, Urrutia J, Morgan J (2002) Yaxcopoil-1 drillhole in the Chicxulub crater [abs.]. Geological Society of America Abstracts with Programs 34(6):317

Snyder DB, Hobbs RW (1999a) Deep seismic reflection profiles across the Chicxulub crater. In: Dressler BO, Sharpton VL (eds) Large meteorite impacts and planetary evolution II. Geological Society of America Special Paper 339:269–279

Snyder DB, Hobbs RW (1999b) The BIRPS Atlas II: A second decade of deep seismic reflection profiling. The Geological Society, London, 3 CD-ROMs

Snyder SW, Miller KG, Saperson E (1993) Paleogene and Neogene planktonic foraminiferal biostratigraphy of the New Jersey continental slope: Sites 902, 903, and 904 (Leg 150). In: Mountain GS, Miller KG, Blum P, Poag CW, Twichell DC (eds), Proceedings of the Ocean Drilling Program, Scientific Results. Ocean Drilling Program, College Station, Texas 150:3–15

Speijer RP, van der Zwaan GJ (1996) Extinction and survivorship of southern Tethyan benthic foraminifera across the Cretaceous/Paleogene boundary. In: Hart MB (ed) Biotic recovery from mass extinction events. Geological Society Special Publication 102:343–372

Speiss V (1990) Cenozoic magnetostratigraphy of Leg 113 drill sites, Maud Rise, Weddell Sea, Antarctica. In: Barker PF, Kennett JP O'Connell S, Berkowitz S, Bryant WR, Burckle LH, Egeberg PK, Fütterer DK, Gersonde RE, Golovchenko X, Hamilton N, Lawver L, Lazarus DB, Lonsdale M, Mohr B, Nagao T, Pereira CPG, Pudsey CJ, Robert CM, Schaudl E, Speiss V, Stott LD, Thomas E, Thompson KFM, Wise SW Jr, Proceedings of the Ocean Drilling Program, scientific results. Ocean Drilling Program, College Station, Texas, 113:261–318

Spezzaferri S, Basso D, Coccioni R (2002) Late Eocene planktonic foraminiferal response to an extraterrestrial impact at Massignano GSSP (northeastern Appennines, Italy). Journal of Foraminiferal Research 32:188–199

Spudis PD (1993) The Geology of multi-ring impact basins. Cambridge University Press, Cambridge, 263 pp

Steckler MS, Watts AB, Thorne JA (1988) Subsidence and basin modeling at the U.S. Atlantic passive margin. In: Sheridan RE, Grow JA (eds) The Geology of North America, The Atlantic Continental Margin, US. Geological Society of America, Boulder, Colorado, I-2:399–415

Steckler MS, Mountain GS, Miller KG, Christie-Blick N (1999) Reconstruction of Tertiary progradation and clinoform development on the New Jersey passive margin by 2-D backstripping. Marine Geology 154:399–420

Steiner MB, Shoemaker EM (1996) A hypothesized Manson impact tsunami: Paleomagnetic and stratigraphic evidence in the Crow Creek Member, Pierre Shale. In: Koeberl C, Anderson RR (eds) The Manson impact structure, Iowa: Anatomy of an impact crater. Geological Society of America Special Paper 302:419–432

Stöffler D (1977) Research drilling Nördlingen 1973: Polymict breccias, crater basement, and cratering model of the Ries impact structure. Geologica Bavarica 75:443–458

Stöffler D, Grieve RAF (1994) Classification and nomenclature of impact metamorphic rocks: A proposal to the IUGS Subcommission on the Systematics of Metamorphic Rocks. In: Montanari A, Smit J (eds) Post-Oestersund Newsletter, European Science Foundation Scientific Network on Impact Cratering and Evolution of Planet Earth, Strasbourg, pp 1–15

Stöffler D, Deutsch A, Avermann M, Bischoff L, Brockmeyer P, Buhl D, Lakomy R, Müller-Mohr V (1994) The formation of the Sudbury Structure, Canada: Towards a unified impact model. In: Dressler BO, Grieve RAF, Sharpton VL (ed) Large meteorite impacts and planetary evolution. Geological Society of America Special Paper 293:303–318

Stott LD, Kennett JP (1990) Antarctic Paleogene planktonic foraminifer biostratigraphy: ODP Leg 113, Sites 689 and 690. In: Barker PF, Kennett JP O'Connell S, Berkowitz S, Bryant WR, Burckle LH, Egeberg PK, Fütterer DK, Gersonde RE, Golovchenko X, Hamilton N, Lawver L, Lazarus DB, Lonsdale M, Mohr B, Nagao T, Pereira CPG, Pudsey CJ, Robert CM, Schaudl E, Speiss V, Stott LD, Thomas E, Thompson KFM, Wise SW Jr, Proceedings of the Ocean Drilling Program, scientific results. Ocean Drilling Program, College Station, Texas, 113:549–569

Sturkell EFF (1998) Resurge morphology of the marine Lockne impact crater, Jämtland, central Sweden. Geological Magazine 135:121–127

Sturkell EFF, Ormö J (1997) Impact-related clastic injections in the marine Ordovician Lockne impact structure, central Sweden. Sedimentology 44:793–804

Sturkell E, Ormö J, Nõlvak J, Wallin Å (2000) Distant ejecta from the Lockne marine-target impact crater, Sweden. Meteoritics and Planetary Science 35:929–936

Suuroja K, Suuroja S, All T, Floden T (2002) Kärdla (Hiiumaa Island, Estonia) – the buried and well-preserved Ordovician marine impact structure. Deep-Sea Research Part II 49:1121–1144

Taylor PT, Zietz I, Dennis LS (1968) Geologic implications of aeromagnetic data for the eastern continental margin of the United States. Geophysics 33:755–780

Taylor SR, McLennan S (1985) The continental crust: Its composition and evolution. Blackwell, Oxford, 312 pp

Teifke H (1973) Geologic studies, coastal plain of Virginia. Virginia Division of Mineral Research Bulletin 83:1–101

Thein JE (1985) A tektite layer in upper Eocene sediments of the New Jersey continental slope (Site 612, Leg 95). In: Poag CW, Watts AB, Cousin M, Goldberg D, Hart MB, Miller KG, Mountain GS, Nakamura Y, Palmer A, Schiffelbein PA, Schreiber BC, Tarafa M, Thein JE, Valentine PC, Wilkens RH Initial reports of the Deep Sea Drilling Project. US Government Printing Office, Washington, D.C., 95:565–579

Thomas E (1990) Late Cretaceous-early Eocene mass extinctions in the deep sea. In: Sharpton VL, Ward PD (eds) Global catastrophes in earth history: An interdisciplinary conference on impacts, volcanism, and mass mortality. Geological Society of America Special Paper 247:481–496

Thomas E, Shackleton NJ (1996) The Paleocene-Eocene benthic foraminiferal extinction and stable isotope anomalies. In: Knox RW, Corfield RM, Dunay RE (eds) Correlation of the early Paleogene in Northwest Europe. Geological Society Special Publication 101:401–441

Thomas PJ, Chyba CF, McKay CP (eds) (1997) Comets and the origin and evolution of life. Springer-Verlag, New York, 296 pp

Thomas WA, Chowns TM, Daniels DL, Neathery TL, Glover L III, Gleason RJ (1989) The subsurface Appalachians beneath the Atlantic and Gulf Coastal Plains. In: Hatcher RD Jr, Thomas WA, Viele GW (eds) Geology of North America: The Appalachian-Ouachita Orogen in the United States. Geological Society of America, Boulder, Colorado, F-2:445–458

Tinus RW, Roddy DJ (1990) Effects of global atmospheric perturbations on forest ecosystems in the North Temperate Zone: Predictions of seasonal depressed-temperature kill mechanisms, biomass production, and wildfire soot emissions. In: Sharpton VL, Ward PD (eds) Global catastrophes in earth history. Geological Society of America Special Paper 247:77–86

Tjalsma LA, Lohmann GP (1983) Paleocene-Eocene bathyal and abyssal benthic foraminifera from the Atlantic Ocean. Micropaleontology Special Publication 4:1–90

Toon OB, Zahnle K, Turco RP, Covey C (1994) Environmental perturbations caused by asteroid impacts. In: Gehrels T (ed) Hazards due to comets and asteroids. University of Arizona Press, Tucson, pp 791–826

Trapp H Jr (1992) Hydrogeologic framework of the northern Atlantic Coastal Plain in parts of North Carolina, Virginia, Maryland, Delaware, New Jersey, and New York. US Geological Survey Professional Paper 1404-G:1–59

Tsikalas F (1996) A geophysical study of Mjølnir: A proposed impact structure in the Barents Sea. Ph.D. thesis, University of Oslo, 93 pp

Tsikalas F, Gudlaugsson ST, Faleide JI (1998a) Collapse, infilling, and postimpact deformation at the Mjølnir impact structure, Barents Sea. Geological Society of America Bulletin 110:537–552

Tsikalas F, Gudlaugsson ST, Eldholm O, Faleide JI (1998b) Integrated geophysical analysis supporting the impact origin of the Mjølnir structure, Barents Sea. Tectonophysics 289:257–280

Tsikalas F, Gudlaugsson ST, Faleide JI, Eldholm O (1999) Mjølnir structure, Barents Sea: A marine impact crater laboratory. In: Dressler BO, Sharpton VL (eds) Impact cratering and planetary evolution II. Geological Society of America Special Paper 339:193–204

Tsikalas F, Faleide JI, Eldholm O, Dypvik H (2002) Seismic correlation of the Mjølnir marine impact crater to shallow boreholes. In: Plado J, Pesonen LJ (eds) Impacts in Precambrian shields. Springer-Verlag, Berlin, pp 307–321

Vail PR, Mitchum RM Jr, Todd RG, Widmier JM, Thompson S III, Sangree JB, Bubb JN, Hatlelid WG (1977) Seismic stratigraphy and global changes of sea level. In: Payton CE (ed) Seismic stratigraphy: Applications to hydrocarbon exploration. American Association of Petroleum Geologists Memoir 26:49–212

Van Morkhoven FPCM, Berggren WA, Edwards AS (1986) Cenozoic cosmopolitan deep-water benthic foraminifera. Bulletin Centre Rechearches Exploration-Production Elf-Aquitaine Memoir 11:1–421

von Dalwigk I, Ormö J (2001) Formation of resurge gullies at impacts at sea: The Lockne crater, Sweden. Meteoritics and Planetary Science 36:359–369

von der Borch CC, Sclater JG, Gartner S Jr, Hekinian R, Johnson DA, McGowran B, Pimm AC, Thompson RW, Veevers JJ, Waterman LS (1974) Initial reports of the Deep Sea Drilling Project. US Government Printing Office, Washington, DC, 22:1–890

Vonhof HB, Smit J (1999) Late Eocene microkrystites and microtektites at Maud Rise (Ocean Drilling Project Hole 689B; Southern Ocean) suggest a global extension of the approximately 35.5 Ma Pacific impact ejecta strewn field. Meteoritics and Planetary Science 34:747–756

Vonhof HB, Smit J, Brinkhuis H, Montanari A, Nederbragt AJ (2000) Global cooling accelerated by early late Eocene impacts? Geology 28:687–690

Ward LW (1992) Molluscan biostratigraphy of the Miocene, Middle Atlantic Coastal Plain of North America. Virginia Museum of Natural History Memoir 2:1–159

Ward LW (1984) Stratigraphy of outcropping Tertiary beds along the Pamunkey River, Central Virginia Coastal Plain. In: Ward LW, Krafft K (eds) Stratigraphy and paleontology of the outcropping Tertiary beds in the Pamunkey River Region, Central Virginia Coastal Plain: Guidebook for Atlantic Coastal Plain Geological Association 1984 Field Trip. Atlantic Coastal Plain Geological Association, pp 11–78

Ward LW, Blackwelder BW (1976) The Claremont, Cobham, and Yorktown Formations. In: Stratigraphy of the Chesapeake Group of Maryland and Virginia. Geological Society of America Guidebook, Field Trip 7b:21–31

Ward LW, Blackwelder BW (1980) Stratigraphic revision of upper Miocene and lower Pliocene beds of the Chesapeake Group, Middle Atlantic Coastal Plain. US Geological Survey Bulletin 1482-D:1–61

Ward LW, Krafft K (eds) (1984) Stratigraphy and paleontology of the outcropping Tertiary beds in the Pamunkey River region, central Virginia Coastal Plain: Guidebook for Atlantic Coastal Plain Geological Association 1984 Field Trip. Atlantic Coastal Plain Geological Association, 280 pp

Ward LW, Strickland GL (1985) Outline of Tertiary stratigraphy and depositional history of the U.S. Atlantic Coastal Plain. In: Poag CW (ed) Geologic evolution of the United States Atlantic Margin. Van Nostrand Reinhold, New York, pp 87–123

Watts AB, Steckler MS (1979) Subsidence and eustasy at the continental margin of eastern North America. American Geophysical Union Maurice Ewing Series 3:218–234

Wessel P, Smith WHF (1991) Free software helps map and display data. Eos Transactions of the American Geophysical Union 72:441, 445–446

Whitehead J, Papanastassiou DA, Spray JG, Grieve RAF, Wasserburg GJ (2000) Late Eocene impact ejecta: geochemical and isotopic connections with the Popigai impact structure. Earth and Planetary Science Letters 181:473–487

Whitehead J, Grieve RAF, Spray J (2002) Mineralogy and petrology of melt rocks from the Popigai impact structure, Siberia. Meteoritics and Planetary Science 37:623–648

Widmark JGV, Malmgren BA (1992) Benthic foraminiferal changes across the Cretaceous/Tertiary boundary in the deep sea: DSDP Sites 525, 527, and 465. Journal of Foraminiferal Research 22:81–113

Wilgus CK, Hastings BS, Kendall CG St C, Posamentier HW, Ross CA, Van Wagoner JC (1988) Sea-level changes: An integrated approach. SEPM Special Publication 42:1–407

Wise SW Jr, Breza JR, Harwood DM, Wei W (1991) Paleogene glacial history of Antarctica. In: Mueller DW, McKenzie JA, Weissert H (eds) Controversies in modern geology. Academic Press, London, pp 133–171

Witzke BJ, Anderson RR (1996) Sedimentary-clast breccias of the Manson impact structure. In: Koeberl C, Anderson RR (eds) The Manson impact structure, Iowa: Anatomy of an impact crater. Geological Society of America Special Paper 302:115–144

Wolbach WS, Lewis RS, Anders E (1988) Cretaceous extinctions: Evidence for wildfires and search for meteoritic material. Science 230:167–170

Wolfe JA (1978) A paleobotanical interpretation of Tertiary climates in the northern hemisphere. American Scientist 66:694–703

Wolfe JA (1992) Climatic, floristic, and vegetational changes near the Eocene/Oligocene boundary in North America. In: Prothero DR, Berggren WA (eds) Eocene-Oligocene climatic and biotic evolution. Princeton University Press, Princeton, pp 421–436

Wolfe JA (1994) Tertiary climatic changes at middle latitudes of western North America. Palaeogeography Palaeoclimatology Palaeoecology 108:195–205

Wolfe JA, Poore RZ (1982) Tertiary marine and nonmarine climatic trends. In: Berger W, Crowell JC (eds) Climate in earth history. National Academy of Science, Washington, D.C., pp 154–158

Wood CA (1980) Martian double ring basins: New observations. Proceedings of the 11th Lunar and Planetary Science Conference. Pergamon Press, New York, pp 2221–2241

Wood CA, Tam W (1993) Morphologic classes of impact basins on Venus. [abs.] Lunar and Planetary Science 24:1535–1536

Woollard GP, Bonini WE, Meyer RP (1957) A seismic refraction study of the Atlantic Coastal Plain and continental shelf between Virginia and Florida. Wisconsin University Department Geology Report, Contract N7 ONR-28512:1–128

Wright JE, Sinha AK, Glover L III (1975) Age of zircons from the Petersburg Granite, Virginia; with comments on belts of plutons in the Piedmont. American Journal of Science 275:848–856

Wünnemann K, Lange MA (2002) Numerical modeling of impact-induced modifications of the deep-sea floor. Deep-Sea Research Part II 49:969–982

Zachos JC, Lohmann KC, Walker JCG, Wise SW (1993) Abrupt climate change and transient climates during the Paleogene: A marine perspective. Journal of Geology 101:191–213

Zachos J, Pagani M, Sloan L, Thomas E, Billups K (2001) Trends, rhythms, and aberrations in global climate 65 Ma to present. Science 292:686–693

Index

*In order to conserve space, we have used CBC as an abbreviation for "Chesapeake Bay impact crater" throughout the index.

*In order to conserve space, we have used CBC as an abbreviation for "Chesapeake Bay impact crater" throughout the index.

Printing: Mercedes-Druck, Berlin
Binding: Stein+Lehmann, Berlin